Operator Theory
Advances and Applications
Vol. 90

Editor
I. Gohberg

Singular Integral Operators and Related Topics

Joint German-Israeli Workshop, Tel Aviv, March 1–10, 1995

Edited by

A. Böttcher
I. Gohberg

Birkhäuser Verlag
Basel · Boston · Berlin

Volume Editorial office:

Raymond and Beverly Sackler Faculty of Exact Sciences
School of Mathematical Sciences
Tel Aviv University
69978 Tel Aviv
Israel

1991 Mathematics Subject Classification 47-06, 45-06

A CIP catalogue record for this book is available from the Library of Congress, Washington D.C., USA

Deutsche Bibliothek Cataloging-in-Publication Data
Singular integral operators and related topics : joint German
Israeli workshop, Tel Aviv, March 1–10, 1995 / ed. by A.
Böttcher ; I. Gohberg. – Basel ; Boston ; Berlin : Birkhäuser,
1996
  (Operator theory ; Vol. 90)
  ISBN-13:978-3-0348-9881-2   e-ISBN-13:978-3-0348-9040-3
  DOI: 10.1007/978-3-0348-9040-3

NE: Böttcher, Albrecht [Hrsg.]; GT

© 1996 Birkhäuser Verlag, P.O. Box 133, CH-4010 Basel, Switzerland
Softcover reprint of the hardcover 1st edition 1996

Printed on acid-free paper produced from chlorine-free pulp. TCF ∞
Cover design: Heinz Hiltbrunner, Basel

ISBN-13:978-3-0348-9881-2

9 8 7 6 5 4 3 2 1

Table of Contents

A. Böttcher, Yu.I. Karlovich
**Toeplitz and singular integral operators on general Carleson
Jordan curves** 119

M.R. Capobianco, P. Junghanns, U. Luther, G. Mastroianni
**Weighted uniform convergence of the quadrature method for
Cauchy singular integral equations** 153

T. Ehrhardt, S. Roch, B. Silbermann
**Symbol calculus for singular integrals with operator-valued
PQC-coefficients** 182

T. Ehrhardt, S. Roch, B. Silbermann
**Finite section method for singular integrals with
operator-valued PQC-coefficients** 204

EDITORIAL INTRODUCTION

This volume contains the proceedings of the Joint German-Israeli Workshop on Linear One-Dimensional Singular Integral Equations which was held at the Tel Aviv University, March 1 to 10, 1995.

The main topics of the Workshop were symbol calculus, index formulas, projection and quadrature methods for Toeplitz and singular integral operators with different symbols, algebras generated by such operators and algebras generated by indempotents. The other topics discussed were inverse scattering problems for differential operators, distribution of zeros for orthogonal functions, factorization of matrix functions and calculation of norms.

This workshop was held concurrently with the ninth in the series of Toeplitz Lectures held biennially at the Tel Aviv University. This time the lecturers were Professors J.A. Ball (Blacksburg), B. Silbermann (Chemnitz), and M.E. Vishik (Moscow).

Participating in the workshop was a group of leading experts from Germany and Israel as well as invited guests from the Netherlands, Ukraine and the United States.

The Lectures and the Workshop were generously supported by

> Deutsche Forschungsgemeinschaft
> President of the Tel Aviv University.

With gratitude we also acknowledge the financial support of the following organizations and authorities:

> Alfried Krupp Foundation,
> Israel Academy of Sciences and Humanities,
> Edmund Landau Center for Research in Mathematical Analysis,
> Rector of Tel Aviv University,
> Raymond and Beverly Sacker Faculty of Exact Sciences,
> School of Mathematical Sciences of Tel Aviv University.

Special thanks to Lily and Nathan Silver for their generous support of these activities throughout the years.

A. Böttcher

I. Gohberg

Operator Theory
Advances and Applications, Vol. 90
© 1996 Birkhäuser Verlag Basel/Switzerland

INVERSE SCATTERING PROBLEM FOR DIFFERENTIAL OPERATORS WITH RATIONAL SCATTERING MATRIX FUNCTIONS

D. ALPAY and I. GOHBERG

In this paper we study the inverse scattering problem for linear canonical differential equations of a special type in the case when the scattering function is a rational matrix–valued function. The main result here is the form of the potential in terms of a realisation of the scattering function. The difference between this publication and the previous one [3] consists in the fact that here is used the Marchenko method instead of Kreĭn's method. For rational scattering functions we also prove the equivalence of the two methods.

1   Introduction

In the present paper we pursue our study of inverse problems for differential operators with rational scattering function [3], [2], [4]. Our starting point is an operator of the form

$$(1.1) \qquad (Df)(t) = -iJ\frac{df}{dt} - \begin{pmatrix} 0 & k(t) \\ k(t)^* & 0 \end{pmatrix} f(t), \quad t \geq 0.$$

In this expression, the function k is $\mathbb{C}^{n \times n}$–valued and with entries in $L_1(0, \infty)$, the function f is $\mathbb{C}^{2n \times p}$–valued, and

$$J = \begin{pmatrix} I_n & 0 \\ 0 & -I_n \end{pmatrix}.$$

The function

$$V(t) = \begin{pmatrix} 0 & k(t) \\ k(t)^* & 0 \end{pmatrix}$$

is called the potential of the differential equation (1.1). The scattering function and the spectral function are two functions which play an important role in the study of the operator D: they are defined in terms of the $\mathbb{C}^{2n \times n}$–valued function $X(t, \lambda)$ solution to the eigenvalue problem

$$-iJ\frac{d}{dt}X(t, \lambda) - \begin{pmatrix} 0 & k(t) \\ k(t)^* & 0 \end{pmatrix} X(t, \lambda) = \lambda X(t, \lambda), \quad \lambda \in \mathbb{R},$$

subject to the boundary conditions

$$(I_n \quad - I_n)X(0, \lambda) = 0$$

and

$$(I_n \quad 0)X(t, \lambda) = e^{-i\lambda t}I_n + o(1), \quad t \to +\infty.$$

For every $\lambda \in \mathbb{R}$, such a solution $X(t, \lambda)$ exists and is unique. It has the following supplementary property that

$$(0 \quad I_n)X(t, \lambda) = e^{i\lambda t}S(\lambda) + o(1) \quad (t \to +\infty)$$

where S is a $\mathbb{C}^{n \times n}$-valued function.

The function S is called the scattering matrix function of the operator D. To give the properties of S we first recall that the Wiener algebra $\mathcal{W}^{n \times n}$ consists of the functions of the form

(1.2) $$f(\lambda) = \alpha + \int_{-\infty}^{+\infty} e^{i\lambda u}x(u)du$$

where $\alpha \in \mathbb{C}^{n \times n}$ and $x \in L_1^{n \times n}(\mathbb{R})$. For $f \in \mathcal{W}^{n \times n}$, the limits $\lim_{\lambda \to \pm\infty} f(\lambda)$ exist and are equal to α. We will use the notation $\alpha = f(\infty)$. The subalgebras $\mathcal{W}_+^{n \times n}$ and $\mathcal{W}_-^{n \times n}$ consist of the elements of the form (1.2) with the support of f in $\mathbb{R}_+$ and $\mathbb{R}_-$ respectively.

The scattering function has the following properties:

(a) It takes unitary values.

(b) It belongs to $\mathcal{W}^{n \times n}$, with $S(\infty) = I_n$.

(c) It admits a Wiener–Hopf factorization

$$S(\lambda) = S_-(\lambda)S_+(\lambda)$$

where S_- and its inverse are in $\mathcal{W}_-^{n \times n}$ and S_+ and its inverse are in $\mathcal{W}_+^{n \times n}$.

Not every function satisfying these three properties is the scattering function of a differential operator of the form (1.1). A sufficient condition for S to be the scattering function of a differential operator of the form (1.1) is that $S(\lambda) - I_n$ is the Fourier transform of a function which has entries in $L_1(\mathbb{R}) \cap L_2(\mathbb{R})$; see [14]. This condition holds in particular when S is rational. The inverse scattering problem consists in recovering the function k (and hence the potential) from the scattering function S. The spectral function associated to the operator D with scattering function S is defined by

(1.3) $$W(\lambda) = S_-(\lambda)^{-1}S_-(\lambda)^{-*},$$

and the inverse spectral problem consists in computing the function k from the spectral function. In the paper [3] we computed k in terms of a minimal realization of the spectral function, and proved the following:

Theorem 1.1 *Let W be a $\mathbb{C}^{n\times n}$-valued rational function analytic at infinity and on the real line and assume that $W(\infty) = I_n$ and $W(\lambda) > 0$ for all $\lambda \in \mathbb{R}$. Then W is the spectral function of a differential operator of the form (1.1). Let $W(\lambda) = I_n + C(\lambda I_m - A)^{-1}B$ be a minimal realization of W. The function k is given by the formula*

$$(1.4) \qquad k(t) = 2C(Pe^{-2itA^{\times}}|_{\operatorname{Im} P})^{-1}PB.$$

In this expression, $A^{\times} = A - BC$ and P is the Riesz projection corresponding to the eigenvalues of A in the open upper half-plane.

This formula was proved using Kreĭn's approach to the inverse spectral problem [11], [10]. Formula (1.4) is in particular suitable for a study in a neighborhood of $t = 0$. In [4] we show that the matrices A, B, C (and hence the function W) can be recovered from the values $k^{(j)}(0)$, $j = 0, \ldots, 2m - 1$, where $m = \deg W$.

In the present work we consider the inverse scattering problem (i.e. start from S rather than W) and use Marchenko's approach to inverse scattering [14], [7, section 9, p. 211]. We obtain another expression for $k(t)$, in terms of a minimal realization of S, and prove:

Theorem 1.2 *Let S be a $\mathbb{C}^{n\times n}$-valued rational function analytic at infinity with $S(\infty) = I_n$. Assume that S takes unitary values on the real line and admits a Wiener–Hopf factorization. Then S is the scattering function of a differential operator of the form (1.1). Let $S(\lambda) = I_n + C(\lambda I_m - A)^{-1}B$ be a minimal realization of S. Then the potential is equal to*

$$V(t) = \begin{pmatrix} 0 & k(t) \\ k(t)^* & 0 \end{pmatrix}$$

where

$$(1.5) \qquad k(t) = -2CP\left(\left(Pe^{-2itA}P - PX_2P^*e^{-2itA^*}P^*X_1P\right)|_{\operatorname{Im} P}\right)^{-1}PB.$$

In this expression, P denotes the Riesz projection corresponding to the eigenvalues of A in the open upper half plane $\mathbb{C}_+$, and X_1 and X_2 are such that

$$(1.6) \qquad i((P^*X_1P)(AP) - (AP)^*(P^*X_1P)) = -P^*C^*CP$$

and

$$(1.7) \qquad i((AP)(PX_2P^*) - (PX_2P^*)(AP)^*) = -PBB^*P^*.$$

Furthermore, the asymptotic equality holds

$$(1.8) \qquad k(t) = -2CPe^{-2itA}PB(I + o(e^{-(\alpha+\epsilon)t})) \quad t \to +\infty$$

where ε is any strictly positive number and

$$\alpha = 4\inf\{\operatorname{Im}\lambda;\quad \lambda \in \sigma(A) \cap \mathbb{C}_+\}.$$

The asymptotic (1.8) is unique, in a sense to be made precise: see Theorem 7.1. This allows us to reconstruct the potential from its value and the values of a number of its derivatives at a given large enough point. Of course, both expressions (1.4) and (1.5) agree. We check this as follows: we start from the factor S_- of a right spectral factorization $S = S_-S_+$ of the scattering function (all matrix functions in $W^{n\times n}$ are here normalized to take value I_n at infinity) and take $S_-(\lambda) = I_n + c(\lambda I_p - a)^{-1}b$ a minimal realization of S_-. Then:

Theorem 1.3 *Let S be a scattering rational function and let $S_-(\lambda) = I_n + c(\lambda I_p - a)^{-1}b$ be a minimal realization of its left factor S_- in the Wiener Hopf factorization $S = S_- S_+$. Then*

(1.9) $$k(t) = -2ce^{ita}(I + \Omega(Y - e^{-2ita^*}Ye^{2ita}))^{-1}(b + i\Omega c^*).$$

In this expression, Ω and Y are the solutions of the Lyapunov equations

(1.10) $$i(\Omega a^{\times *} - a^\times \Omega) = bb^*$$

and

(1.11) $$i(Ya - a^*Y) = -c^*c.$$

Formula (1.9) was proved in [3] using (1.4). We prove it here using (1.5), showing thus directly, for the rational case, the equivalence between Kreĭn's and Marchenko's approaches.

The outline of the paper is as follows. The paper consists of eight sections. This introduction is the first. In the second we review a few facts on the theory of realization and factorization of rational matrix valued functions. In Section 3, we study minimal realizations of scattering functions. Section 4 is devoted to Marchenko's equation in the rational case. The formula (1.5) is proved in Section 5; In Section 6 we prove the equivalence of formulas (1.4) and (1.5). Finally, Section 7 gives a result on uniqueness of asymptotic expansions and in Section 8 we explain how to reconstruct the potential from the values of a finite number of derivatives at a given point.

2   Rational matrix valued functions.

Let R be a $\mathbb{C}^{n \times n}$-valued rational function analytic at infinity. It can be written as

(2.1) $$R(\lambda) = D + C(\lambda I_m - A)^{-1}B$$

where $m \in \mathbb{N}^*$, and $(A, B, C, D) \in \mathbb{C}^{m \times m} \times \mathbb{C}^{m \times n} \times \mathbb{C}^{n \times m} \times \mathbb{C}^{n \times n}$. Expression (2.1) is called a realization of R; a realization for which m is as small as possible is called minimal. A realization is minimal if it is both observable: $\cap_{\ell=0}^\infty \text{Ker } CA^\ell = \{0\}$ and controllable: $\cup_{\ell=0}^\infty \text{Ran } A^\ell B = \mathbb{C}^m$. See [6]. In this section we will present without proofs the main theorems on factorization of rational matrix–valued functions which are used in this paper.

A factorization $R = R_- R_+$ of R into two $\mathbb{C}^{n \times n}$-valued functions analytic at infinity is called a (right) canonical (Wiener-Hopf) factorization if R_- and its inverse are analytic in the closed lower half plane and R_+ and its inverse are analytic in the closed upper half plane. Wiener–Hopf factorizations need not exist, and we refer to [6] for a complete discussion. More generally if Γ is a contour which consists of finitely many non intersecting closed rectifiable curves, $R = R_- R_+$ is called right spectral if R_- and its inverse are analytic in the closure of the interior F_- of Γ and R_+ and its inverse are analytic in the closure of the exterior F_+ of Γ. The factorizations $R = L_+ L_-$ will be called left spectral if L_+ and its inverse are analytic in F_+ and L_- and its inverse are analytic in F_-.

Theorem 2.1 [6, Section 4.5]. *Let R be a $\mathbb{C}^{n \times n}$-valued rational function analytic at infinity with $R(\infty) = I_n$ and let $R(\lambda) = I_n + C(\lambda I_m - A)^{-1}B$ be a minimal realization of R. Assume that A has no real eigenvalues. Then R admits a right canonical Wiener–Hopf factorization relative to the real line if and only if the following two conditions are fulfilled:*

(i) $A^\times = A - BC$ has no real eigenvalues.

(ii) $\mathbb{C}^m = M \oplus M^\times$

where M (resp. $M^\times$) is the space spanned by the eigenvectors and generalized eigenvectors corresponding to the eigenvalues of A (resp. $A^\times$) in the upper (resp. lower) half plane. Furthermore, in that case, R admits a canonical factorization $R(\lambda) = R_-(\lambda)R_+(\lambda)$ with $R_-(\lambda) = I_n + C(\lambda I_m - A)^{-1}(I - \pi)B$ and $R_+(\lambda) = I_n + C\pi(\lambda I_m - A)^{-1}B$. Then

$$
\begin{aligned}
R_-(\lambda)^{-1} &= I_n - C(I - \pi)(\lambda I_m - A^\times)^{-1}B \\
R_+(\lambda)^{-1} &= I_n - C(\lambda I_m - A^\times)^{-1}\pi B
\end{aligned}
$$

where π is the projection of $\mathbb{C}^m$ along M onto $M^\times$.

See [9, p. 2], for this theorem and further discussion. The book [9] also contains a paper of Ball and Ran where the following problem is considered: how to compute a right spectral factorization when a left spectral factorization is given. The result, which we will need in the sequel, is:

Theorem 2.2 [5, *Theorem 2.1 p. 13*]. *Let* Γ *consists of finitely many non intersecting closed rectifiable Jordan curves. Suppose that the rational* $\mathbb{C}^{n\times n}$*-valued function* W *admits a left canonical factorization* $W(\lambda) = Y_+(\lambda)Y_-(\lambda)$ *where*

$$
Y_+(\lambda) = I_n + C_+(\lambda I_{m_+} - A_+)^{-1}B_+,
$$

and

$$
Y_-(\lambda) = I_n + C_-(\lambda I_{m_-} - A_-)^{-1}B_-.
$$

We assume that both A_- *and* $A_-^\times = A_- - B_-C_-$ *have their spectra in the interior* F_- *of* Γ *and* A_+ *and* $A_+^\times = A_+ - B_+C_+$ *have their spectra in the exterior* F_+ *of* Γ. *Let* P *and* Q *denote the unique solutions of the Lyapunov equations*

$$
A_-^\times P - PA_+^\times = B_-C_-
$$

$$
A_+Q - QA_- = -B_+C_+.
$$

Then, W *admits a right canonical factorization if and only if* $I - QP$ *is invertible. When this is the case the factors* W_- *and* W_+ *for the right factorization* $W(\lambda) = W_-(\lambda)W_+(\lambda)$ *are given by*

$$
W_-(\lambda) = I_n + (C_+Q + C_-)(\lambda I_{m_-} - A_-)^{-1}(I_{m_-} - PQ)^{-1}(-PB_+ + B_-)
$$

and

$$
W_+(\lambda) = I_n + (C_+ + C_-P)(I_{m_+} - QP)^{-1}(\lambda I_{m_+} - A_+)^{-1}(B_+ - QB_-).
$$

Not every matrix–valued function S analytic at infinity with $S(\infty) = I_n$ and which takes unitary values on the real line is the scattering function of a differential operator of the form (1.1). From formula (1.3) it is seen that S should have the same number of poles in $\mathbb{C}_+$ and $\mathbb{C}_-$, but even this is not sufficient for S to admit a Wiener-Hopf factorization. To see this, consider the function

$$
S(\lambda) = \begin{pmatrix} \frac{\lambda-\omega}{\lambda-\bar{\omega}} & 0 \\ 0 & \frac{\lambda-\bar{\omega}}{\lambda-\omega} \end{pmatrix}
$$

where w is a nonreal number. A minimal realization of S is given by $S(\lambda) = I_2 + C(\lambda I_2 - A)^{-1}B$ where

$$A = \begin{pmatrix} \bar{w} & 0 \\ 0 & w \end{pmatrix}, \quad B = \begin{pmatrix} -1 & 0 \\ 0 & 1 \end{pmatrix} \quad \text{and} \quad C = (w - \bar{w})I_2.$$

Thus

$$A^\times = A - BC = \begin{pmatrix} w & 0 \\ 0 & \bar{w} \end{pmatrix}.$$

Let $w \in \mathbb{C}_+$. Then, in the notation of Theorem 2.1,

$$M = M^\times = \text{span}\left\{ \begin{pmatrix} 0 \\ 1 \end{pmatrix} \right\}$$

and hence S does not admit a right canonical factorization.

Finally we recall the following result of [1], where minimal realizations of rational matrix-valued functions unitary on the real line are studied.

Theorem 2.3 *Let S be a matrix valued rational function analytic at infinity with $S(\infty) = I_n$ and let $S(\lambda) = I_n + C(\lambda I_m - A)^{-1}B$ be a minimal realization of S. Then S takes unitary values on the real line if and only if there exists an invertible solution H to the system of equations*

(2.2) $$i(AH - HA^*) = -BB^*$$
(2.3) $$CH = iB^*.$$

The matrix H is then uniquely determined from the given realization of S. It is hermitian, and is called the associated hermitian matrix (to the given realization).

3   Minimal realizations of rational scattering functions

Theorem 3.1 *Let S be the scattering function of a differential operator (1.1) and let $S = S_- S_+$ be its right Wiener-Hopf factorization. Let $S_-(\lambda) = I_n + c(\lambda I_p - a)^{-1}b$ be a minimal realization of the spectral factor S_-. Then*

(1)

(3.1) $$S_+(\lambda) = I_n - (ic\Omega - b^*)(\lambda - a^{\times *})^{-1}(I_p + Y\Omega)^{-1}(c^* + iYb)$$

is a minimal realization of the spectral factor S_+. In this expression, Ω and Y are the solutions of the Lyapunov equations (1.10) and (1.11).

(2) *A minimal realization of S is given by $S(\lambda) = I_n + C(\lambda I_{2p} - A)^{-1}B$, where*

(3.2) $$A = \begin{pmatrix} a & -b(ic\Omega - b^*) \\ 0 & a^{\times *} \end{pmatrix}, \quad B = \begin{pmatrix} b \\ (I_p + Y\Omega)^{-1}(c^* + iYb) \end{pmatrix}$$

(3.3) $$C = (c \quad ic\Omega - b^*).$$

The associated hermitian matrix is equal to

(3.4) $$H = \begin{pmatrix} -\Omega & iI_p \\ -iI_p & -Y(I_p + \Omega Y)^{-1} \end{pmatrix}$$

(3) *A rational matrix valued function is the scattering function of a differential operator of the form (1.1) if and only if it has a minimal realization similar to a realization of the form (3.2)–(3.3).*

Let us first make a few remarks. If S is a scattering matrix function, with right Wiener–Hopf factorization $S = S_- S_+$, the factor S_- uniquely determines S_+ since $S_-(\lambda)^{-1} S_-(\lambda)^{-*} = S_+(\lambda) S_+(\lambda)^*$, and $S_+(\infty) = S_-(\infty) = I_n$. Starting from a minimal realization of S_- we can therefore obtain a minimal realization of S_+ using Theorem 2.1. This is indeed the way we first obtained (3.1). Once (3.1) is available, we can prove Theorem 3.1 by applying directly Theorem 2.3 to the realization (3.2)–(3.3). and H given by (3.4). This is the way we now prove Theorem 3.1.

Proof of Theorem 3.1: It is easily seen that the triple (A, B, C) defined by (3.2)–(3.3) is minimal since the triple (a, b, c) is minimal. Since $\sigma(a)$ and $\sigma(a^\times)$ are included in $\mathbb{C}_+$, the matrices Ω and Y are strictly positive and thus $I_p + \Omega Y$ is invertible. The matrix H is hermitian and invertible, as is seen from

$$H = \begin{pmatrix} -\Omega & iI_p \\ -iI_p & -Y(I_p + \Omega Y)^{-1} \end{pmatrix} = \begin{pmatrix} I_p & 0 \\ +i\Omega^{-1} & I_p \end{pmatrix} \begin{pmatrix} -\Omega & 0 \\ 0 & (I_p + \Omega Y \Omega)^{-1} \end{pmatrix} \begin{pmatrix} I_p & -i\Omega^{-1} \\ 0 & I_p \end{pmatrix}.$$

We now show that (2.2) and (2.3) hold for the present choice of A, B, C and H. We begin with (2.3), i.e. $CH = iB^*$, which can be rewritten as

$$\begin{pmatrix} c & ic\Omega - b^* \end{pmatrix} \begin{pmatrix} -\Omega & iI_p \\ -iI_p & -Y(I_p + \Omega Y)^{-1} \end{pmatrix} = i(b^*(c - ib^*Y)(I_p + \Omega Y)^{-1})$$

i.e.
(3.5) $$\qquad\qquad -c\Omega + (ic\Omega - b^*)(-iI) = ib^*$$

and
(3.6) $$\qquad ic - (ic\Omega - b^*)Y(I + \Omega Y)^{-1} = i(c - ib^*Y)(I + \Omega Y)^{-1}$$

which hold true. We now turn to (2.2). From

$$AH = \begin{pmatrix} -a^\times\Omega + ibb^* & ia - b(ic\Omega - b^*)Y(I + \Omega Y)^{-1} \\ -ia^{\times *} & -a^{\times *}Y(I + \Omega Y)^{-1} \end{pmatrix}$$

we see that to check (2.2) is equivalent to checking the following three equalities:

$$i(\Omega a^{\times *} - a^\times \Omega + 2ibb^*) = -bb^*$$
$$i(ia - b(ic\Omega - b^*)Y(I + \Omega Y)^{-1}) = -b(c - ib^*Y)Y(I + \Omega Y)^{-1}i \times$$
$$\times \left\{(I + Y\Omega)^{-1}Ya^\times - a^{\times *}Y(I + \Omega Y)^{-1}\right\}$$
$$i\left\{(I + Y\Omega)^{-1}Ya^\times - a^{\times *}Y(I + \Omega Y)^{-1}\right\} = -(I + Y\Omega)^{-1}(c^* + iYb)(c - ib^*Y)(I + \Omega Y)^{-1}.$$

These in turn are readily checked, taking into account that Ω and Y are solutions of (1.10) and (1.11). By Theorem 2.3 we thus see that S takes unitary values on the real line. The last claim follows from the fact that two minimal realizations are similar. ∎

In the next lemma, we present some computations which will be needed in the sequel. The proof is straightforward and will be omitted.

Lemma 3.2 *Let S be a scattering matrix function with realization (3.2)–(3.3). The Riesz projections P associated to the eigenvalues of A in the open upper half plane is equal to*

$$P = \begin{pmatrix} I_p & i\Omega \\ 0 & 0 \end{pmatrix}$$

(where Ω is the solution to the Lyapunov equation (1.10)).
Furthermore,

$$AP = \begin{pmatrix} a & ia\Omega \\ 0 & 0 \end{pmatrix}$$

$$PB = \begin{pmatrix} (I_p + \Omega Y)^{-1}(b + i\Omega c^*) \\ 0 \end{pmatrix}$$

and

$$CP = \begin{pmatrix} c & ic\Omega \end{pmatrix}.$$

4   Marchenko's equation in the rational case.

In this section we study Marchenko's equation in the rational case. Let us first briefly review Marchenko's approach to inverse scattering. Here we follow the paper [14]. For more information we refer to [7, Section 9] and [13], [12].
Thus let

$$S(\lambda) = I_n - \int_{-\infty}^{\infty} \sigma(u)e^{-i\lambda u} du \quad \lambda \in \mathbb{R}$$

be a scattering matrix function, where $\sigma \in L_1^{n \times n}(\mathbb{R}) \cap L_2^{n \times n}(\mathbb{R})$. Set

$$\xi(u) = \begin{pmatrix} 0 & \upsilon(u)^* \\ \sigma(u) & 0 \end{pmatrix}.$$

Marchenko's approach consists in solving the equation

(4.1) $$M(t,s) - \xi(t+s) - \int_t^{\infty} M(t,u)\xi(u+s) du = 0$$

for $0 \le t \le s < \infty$ (see [14, equation (1.10)]).
The potential is then given by $V(t) = \begin{pmatrix} 0 & k(t) \\ k(t)^* & 0 \end{pmatrix}$ where

$$k(t) = -2im_{21}(t,t)$$

(where $M = (m_{ij})$ is the decomposition of M into four $\mathbb{C}^{n \times n}$-valued blocks).
When S is rational, with minimal realization

(4.2) $$S(\lambda) = I_n + C(\lambda I_m - A)^{-1}B$$

we have

(4.3) $$\sigma(u) = \begin{cases} iCe^{iuA}(I_m - P)B & u < 0 \\ -iCe^{iuA}PB & u > 0 \end{cases}$$

where P is the Riesz projection corresponding to the eigenvalues of A in $\mathbb{C}_+$. In particular, $\sigma \in L_1^{n \times n}(\mathbb{R}) \cap L_2^{n \times n}(\mathbb{R})$. The function ξ can be written as

$$\xi(u) = \begin{pmatrix} 0 & \sigma(u)^* \\ \sigma(u) & 0 \end{pmatrix} = \begin{pmatrix} 0 & iB^*P^*e^{-iuA^*}P^*C^* \\ -iCe^{iuA}PB & 0 \end{pmatrix} = Fe^{uT}G$$

where

(4.4)
$$F = i\begin{pmatrix} 0 & B^*P^* \\ -CP & 0 \end{pmatrix},$$

(4.5)
$$T = \begin{pmatrix} iAP & 0 \\ 0 & -iA^*P^* \end{pmatrix},$$

and

(4.6)
$$G = \begin{pmatrix} PB & 0 \\ 0 & P^*C^* \end{pmatrix}.$$

Note that the spectrum of the restriction of T to $\operatorname{Im} P \oplus \operatorname{Im} P^*$ is in the open left half plane Π_-. We will give a formula for the solution $M(t,s)$ of Marchenko's equation when ξ is of the form $\xi(u) = Fe^{uT}G$, with $\sigma(T) \subset \Pi_-$. We begin with a lemma.

Lemma 4.1 Let $(T_1, T_2, M) \in \mathbb{C}^{m \times m} \times \mathbb{C}^{m \times m} \times \mathbb{C}^{m \times m}$ and assume that both T_1 and T_2 have spectrum included in the open left half plane Π_-. Let Z be the (unique) solution of

(4.7)
$$- M = T_1 Z + Z T_2.$$

Then, for every $t \geq 0$

(4.8)
$$\int_t^\infty e^{uT_1} M e^{uT_2} \, du = e^{tT_1} Z e^{tT_2}.$$

Proof: We first remark that the Lyapunov equation (4.7) has indeed a unique solution, since $\sigma(T_1) \cup \sigma(T_2) \subset \mathbb{C}_-$. This last condition also insures that the integral in the left side of (4.8) converges for every finite t. To prove (4.8), we differentiate both sides of it and obtain

$$-e^{tT_1} M e^{tT_2} = e^{tT_1}(T_1 Z + Z T_2)e^{tT_2}.$$

In view of (4.7), it follows that $\int_t^\infty e^{uT_1} M e^{uT_2} \, du - e^{tT_1} Z e^{tT_2}$ is a constant independent of t. Setting $t \to +\infty$, we see that this constant is equal to 0, which proves (4.8). ∎

The solution to Marchenko's equation in the rational case is given in the next theorem.

Theorem 4.2 Let ξ be of the form $\xi(u) = Fe^{uT}G$ where $(F, T, G) \in \mathbb{C}^{n \times m} \times \mathbb{C}^{m \times m} \times \mathbb{C}^{m \times n}$, with $\sigma(T) \subset \Pi_-$, and consider the equation

(4.9)
$$M(t,s) - \xi(t+s) - \int_t^\infty M(t,u)\xi(u+s)\,du = 0,$$

with $0 \leq t \leq s < \infty$.
Let Z be the solution of
(4.10)
$$- GF = TZ + ZT.$$

Then, equation (4.9) has a unique solution on $[t, \infty)$ *if and only if*

(4.11) $I - e^{tT} Z e^{tT}$

is invertible (this holds for t large enough since $\sigma(T) \subset \Pi_-$*). When this condition is in force,* $M(t, s)$ *is given by the formula*

(4.12) $M(t, s) = F e^{tT} (I - e^{tT} Z e^{tT})^{-1} e^{sT} G.$

Proof: For $t \geq 0$, let α_t denote the operator from $\mathbb{C}^{n \times m}$ into $L_2^{n \times m}[t, \infty)$ defined by

$$(\alpha_t M)(s) = M e^{sT} G \quad s \geq t,$$

and let β_t denote the operator from $L_2^{n \times m}[t, \infty)$ into $\mathbb{C}^{n \times m}$ defined by

$$\beta_t f = \int_t^\infty f(u) F e^{uT} du.$$

Both α_t and β_t are bounded, since $\sigma(T) \subset \Pi_-$. Furthermore, for $f \in L_2^{n \times m}[t, \infty)$

$$.(\alpha_t \beta_t)(f)(s) = \int_t^\infty f(u) F e^{(u+s)T} G du \quad s \geq t.$$

Thus, with $\xi_t(s) = \xi(t+s)$, $s \geq t$, and $f(s) = M(t, s)$, $s \geq t$, equation (4.9) becomes

$$(I - \alpha_t \beta_t) f = \xi_t.$$

The operator $(I - \alpha_t \beta_t)$ is invertible if and only if the operator $(I - \beta_t \alpha_t)$ is invertible. For $M \in \mathbb{C}^{n \times m}$ we have

$$(\beta_t \alpha_t)(M) = M \Gamma_t$$

where

(4.13) $\Gamma_t = \int_t^\infty e^{uT} G F e^{uT} du.$

By Lemma 4.1, $\Gamma_t = e^{tT} Z e^{tT}$, where Z is the solution of (4.10).Thus, $(I - \alpha_t \beta_t)$ is invertible if and only if $\det(I - e^{tT} Z e^{tT}) \neq 0$. This condition always holds for t large enough, since $\sigma(T) \subset \Pi_-$. When this condition holds, and using the formula

$$(I - \alpha_t \beta_t)^{-1} = I + \alpha_t (I - \beta_t \alpha_t)^{-1} \beta_t,$$

we obtain

$$f = (I - \alpha_t \beta_t)^{-1} \xi_t$$

$$= \xi_t + \alpha_t (I - \beta_t \alpha_t)^{-1} \beta_t \xi_t.$$

To compute f, we first compute $\beta_t \xi_t$.

$$\beta_t \xi_t = \int_t^\infty \xi_t(u) F e^{uT} du = \int_t^\infty \xi(u+t) F e^{uT} du = F e^{uT} \Gamma_t$$

where Γ_t is defined by (4.13). Hence, since $(\beta_t \alpha_t) M = M \Gamma_t$,

$$f = \xi_t + \alpha_t (F e^{tT} \Gamma_t (I - \Gamma_t)^{-1})$$

i.e.

$$f(s) = \xi_t + F e^{tT} \Gamma_t (I - \Gamma_t)^{-1} e^{sT} G \qquad\qquad s \geq t$$

$$= F e^{(s+t)T} G + F e^{tT} \Gamma_t (I - \Gamma_t)^{-1} e^{sT} G$$

$$= F e^{tT} (I - \Gamma_t)^{-1} e^{sT} G.$$

5 Formula for the potential

We now prove formula (1.5) for the potential of the canonical differential equation (1.1) in terms of a minimal realization of the scattering matrix function. To that purpose we compute the function $M(t, s)$ solution of the integral equation (4.1) using formula (4.12) for the realization (4.4)–(4.6). We first compute the solution Z of the Lyapunov equation (4.10).

Lemma 5.1 *Let F, T and G be given by (4.4)–(4.6). The equation*

$$-GF = TZ + ZT$$

has a unique solution Z such that

$$\begin{pmatrix} P & 0 \\ 0 & P^* \end{pmatrix} Z \begin{pmatrix} P^* & 0 \\ 0 & P \end{pmatrix} = Z.$$

It is given by

$$Z = \begin{pmatrix} 0 & iX_2 \\ 0 & -iX_1 \end{pmatrix}$$

where X_2 is the unique solution of the Lyapunov equation (1.7) which satisfies $X_2 = PX_2P^*$ and where X_1 is the unique solution of the Lyapunov equation (1.6) which satisfies $X_1 = P^*X_1P$.

Proof: Let $Z = \begin{pmatrix} Z_1 & Z_2 \\ Z_3 & Z_4 \end{pmatrix}$ be the decomposition of Z into four $\mathbb{C}^{n \times n}$-valued blocks. With the present choice of F, T, G, equation (4.10) becomes

$$\begin{pmatrix} APZ_1 + Z_1AP & APZ_2 - Z_2(AP)^* \\ Z_3AP - (AP)^*Z_3 & Z_4(AP)^* - Z_4AP \end{pmatrix} = \begin{pmatrix} 0 & -iPBB^*P^* \\ P^*C^*CP & 0 \end{pmatrix}.$$

Since $\sigma(AP) \subset \mathbb{C}_+$, we obtain that PZ_4P^* and PZ_1P^* are equal to 0. Setting $PZ_2P^* = iX_2$ and $P^*Z_3P = -iX_1$ we find that X_1 and X_2 are the solutions of the equations (1.6) and (1.7). ∎

The next step toward the computation of $M(t, s)$ is to compute $(I - \Gamma_t)^{-1}$ where $\Gamma_t = e^{tT} Z e^{tT}$. We have

$$(I - \Gamma_t) = I - \begin{pmatrix} e^{itAP} & 0 \\ 0 & e^{-itA^*P^*} \end{pmatrix} \begin{pmatrix} 0 & iX_2 \\ -iX_1 & 0 \end{pmatrix} \begin{pmatrix} e^{itAP} & 0 \\ 0 & e^{-itA^*P^*} \end{pmatrix}$$

$$= \begin{pmatrix} I & -ie^{itAP}X_2e^{-itA^*P^*} \\ ie^{-itA^*P^*}X_1e^{itAP} & I \end{pmatrix}$$

To compute $(I - \Gamma_t)^{-1}$ we use the formula

$$\begin{pmatrix} I & \alpha \\ \beta & I \end{pmatrix}^{-1} = \begin{pmatrix} (I - \alpha\beta)^{-1} & -\alpha(I - \beta\alpha)^{-1} \\ -\beta(I - \alpha\beta)^{-1} & (I - \beta\alpha)^{-1} \end{pmatrix}$$

with

(5.1)
$$\alpha = -ie^{itAP}X_2 e^{-itA^*P^*}$$
(5.2)
$$\beta = ie^{-itA^*P^*}X_1 e^{itAP}.$$

We postpone to the next section the proof that $I - \alpha\beta$ is invertible for every choice of $t \in \mathbb{R}$. We have:

$$
\begin{aligned}
M(t,s) &= Fe^{tT}(I - e^{tT}Ze^{tT})^{-1}e^{sT}G \\
&= i\begin{pmatrix} 0 & B^*P^* \\ -CP & 0 \end{pmatrix}\begin{pmatrix} e^{itAP} & 0 \\ 0 & e^{-itA^*P^*} \end{pmatrix} \times \\
&\quad \times \begin{pmatrix} (1-\alpha\beta)^{-1} & -\alpha(1-\beta\alpha)^{-1} \\ -\beta(1-\alpha\beta)^{-1} & (1-\beta\alpha)^{-1} \end{pmatrix}\begin{pmatrix} e^{itAP} & 0 \\ 0 & e^{-itA^*P^*} \end{pmatrix}\begin{pmatrix} PB & 0 \\ 0 & P^*C^* \end{pmatrix} \\
&= i\begin{pmatrix} 0 & B^*P^*e^{-itA^*P^*} \\ -CPe^{itAP} & 0 \end{pmatrix}\begin{pmatrix} (1-\alpha\beta)^{-1} & -\alpha(1-\beta\alpha)^{-1} \\ -\beta(1-\alpha\beta)^{-1} & (1-\beta\alpha)^{-1} \end{pmatrix} \times \\
&\quad \times \begin{pmatrix} e^{isAP}PB & 0 \\ 0 & e^{-isA^*P^*}P^*C^* \end{pmatrix} \\
&= i\begin{pmatrix} -B^*P^*e^{-itA^*P^*}\beta(1-\alpha\beta)^{-1} & B^*P^*e^{-itA^*P^*}(1-\beta\alpha)^{-1} \\ -CPe^{itAP}(1-\alpha\beta)^{-1} & CPe^{itA}\alpha(1-\beta\alpha)^{-1} \end{pmatrix} \times \\
&\quad \times \begin{pmatrix} e^{isAP}PB & 0 \\ 0 & e^{-isA^*P^*}P^*C^* \end{pmatrix} \\
&= i\begin{pmatrix} -B^*P^*e^{-itA^*P^*}\beta(1-\alpha\beta)^{-1}e^{isAP}PB & B^*P^*e^{-itA^*P^*}(1-\beta\alpha)^{-1}e^{-isA^*P^*}P^*C^* \\ -CPe^{itAP}(1-\alpha\beta)^{-1}e^{isAP}PB & CPe^{itA}\alpha(1-\beta\alpha)^{-1}e^{-isA^*P^*}P^*C^* \end{pmatrix}.
\end{aligned}
$$

So
$$m_{21}(t,s) = -iCPe^{itAP}(1-\alpha\beta)^{-1}e^{isAP}PB$$

$$= -iCPe^{itAP}(1 - e^{itAP}X_2 e^{-2itA^*P^*}X_1 e^{itAP})^{-1}e^{isAP}PB,$$

and hence the formula (1.5) for $k(t) = -2im_{21}(t,t)$.

6 Equivalence with Kreĭn's approach

To show that the approach of Marchenko is equivalent to Kreĭn's approach we compute $k(t)$ via formula (1.5) for the special realization (3.2)–(3.3) of S.

Lemma 6.1 *Let A, B, C be given by (3.2)–(3.3). The solutions X_1 and X_2 of the Lyapunov equations (1.6) and (1.7) are given by*

(6.1)
$$P^*X_1P = \begin{pmatrix} Y & iY\Omega \\ -i\Omega Y & \Omega Y\Omega \end{pmatrix}$$

and

$$PX_2P^* = (I_p + \Omega Y)^{-1}\Omega$$

respectively (where Y and Ω are the solutions of (1.10) and (1.11) respectively).

Proof: Write $X_1 = \begin{pmatrix} X_{11} & X_{12} \\ X_{21}^* & X_{22} \end{pmatrix}$. Then

$$P^* X_1 P = \begin{pmatrix} X_{11} & i X_{11} \Omega \\ -i \Omega X_{11} & \Omega X_{11} \Omega \end{pmatrix}.$$

From (1.6) we have

$$i(X_{11} a - a^* X_{11}) = -c^* c$$

and so X_{11} is equal to the solution Y of (1.10) and hence (6.1) holds. Next, $P X_2 P^*$ is of the form $P X_2 P^* = \begin{pmatrix} Z & 0 \\ 0 & 0 \end{pmatrix}$.

Equation (1.7) leads to

(6.2) $$i(aZ - Za^*) = -(b + i\Omega^* c)(b + i\Omega c)^* (I_p + Y\Omega)^{-1}.$$

This equation has a unique solution since $\sigma(a) \cap \sigma(a^*) = \{0\}$. To check that $Z = \Omega(I_p + Y\Omega)^{-1}$ we see that (6.2) is equivalent to checking that

$$i(a\Omega(I_p + Y\Omega)^{-1} - (I_p + \Omega Y)^{-1} \Omega a^*) = -(I_p + \Omega Y)^{-1}(b + i\Omega^* c)(b + i\Omega^* c)^* (I_p + Y\Omega)^{-1}$$

i.e.

$$i\{(I_p + \Omega Y)\Omega - \Omega a^*(I_p + Y\Omega)\} = -bb^* - \Omega c^* c\Omega - i\Omega c^* b^* + ibc\Omega,$$

that is,

$$i(a\Omega - \Omega a^* + \Omega(Ya - a^*Y)\Omega) = i\{(a - bc)\Omega - \Omega(a - bc)^* + \Omega(Ya - a^*Y)\Omega\} = -bb^* - \Omega c^* c$$

which hold since Ω and Y solution of (1.10) and (1.11). ∎

Note that, with PA given by

$$P e^{-2itA} P^* = \begin{pmatrix} e^{-2ita} & i e^{-2ita} a\Omega \\ 0 & I_p \end{pmatrix}$$

and

$$P e^{-2itA^*} P^* = \begin{pmatrix} e^{-2ita^*} & 0 \\ -i\Omega a e^{-2ita^*} & I_p \end{pmatrix}.$$

Thus

$$(e^{-2itA} - P X_2 P^* e^{-2itA^*} P^* X_1 P) = \begin{pmatrix} e^{-2ita} - (I_p + \Omega Y)^{-1} \Omega e^{-2ita^*} & * \\ 0 & * \end{pmatrix}$$

where $*$ denotes nonrelevant entries and, taking into account the formulas for PB and CP

$$k(t) = -2CP(e^{-2itA} - P X_2 P^* e^{-2itA^*} P^* X_1 P)^{-1} PB$$

$$= -2c(e^{-2ita} - (I_p + \Omega Y)^{-1} \Omega e^{-2ita^*} Y)^{-1} (I_p + \Omega Y)^{-1} (b + i\Omega c^*)$$

which is equal to (1.9). We note that

$$\Delta(t) = e^{-2ita} + (I_p + \Omega Y)^{-1} \Omega e^{-2ita^*} Y$$

is invertible for every $t \in \mathbb{R}$. Indeed,

$$\Delta(t) = (I_p + \Omega Y)^{-1}((I_p + \Omega Y) - \Omega e^{-2ita^*} Y e^{2ita}) e^{-2ita}$$

$$= (I_p + \Omega Y)^{-1}(I_p + \Omega(Y - e^{-2ita^*} Y e^{2ita})) e^{-2ita}$$

As shown in [3],

$$Y - e^{-2ita^*} Y e^{2ita} = \int_0^t e^{-iua^*} c^* c e^{iua} \, du \geq 0$$

and so $\Delta(t)$ is invertible. ∎

To conclude this section we show that the matrix $I_p - \alpha\beta$ defined in Section 5 via (5.1) and (5.2) is indeed invertible for every choice of $t \in \mathbb{R}$. This property is independent of the given minimal realization of S. The one considered in this section leads to the fact that $I_p - \alpha\beta$ is invertible if and only if $e^{2ita} + (I_p + \Omega Y)^{-1} \Omega e^{-2ita^*} Y$ is invertible. But this is indeed the case, as just shown above.

7 · A result on asymptotic expansions.

We now study the uniqueness of the asymptotic (1.8). First note that the function

$$-2B^* e^{-2itA^*} P^* C^*$$

is equal to a finite sum of terms of the form $p(t)e^{\gamma t}$ where p is a $\mathbb{C}^{n \times n}$-valued polynomial and where γ is in the open right half plane.

Theorem 7.1 *Let $f(t)$ be a $\mathbb{C}^{n \times n}$-valued function, $t \geq 0$, and suppose that f admits the following two asymptotic expansions*

$$(7.1) \qquad\qquad f(t) = \left(\sum_{i \in I} M_i(t) e^{-\alpha_i t}\right)(1 + o(e^{-\alpha t}))$$

$$(7.2) \qquad\qquad\quad = \left(\sum_{i \in I'} M_i'(t) e^{-\alpha_i' t}\right)(1 + o(e^{-\alpha' t}))$$

as $t \to \infty$; in these expressions, M_i and M_i' are $\mathbb{C}^{n \times n}$-valued polynomials and the $\alpha_i, \alpha_i', \alpha, \alpha'$ are in the open right half plane, with

$$0 < \operatorname{Re} \alpha_i < \alpha, \quad \alpha > 0$$

$$0 < \operatorname{Re} \alpha_i' < \alpha', \quad \alpha' > 0.$$

Assume that $\alpha < \alpha'$ and let $I' = I_1' \cup I_2'$ where $I_1' = \{j \in I'; \operatorname{Re} \alpha_j' < \alpha\}$ and $I_2' = I' \setminus I_1'$. Then

$$\sum_{j \in I} M_i(t) e^{-\alpha_j t} = \sum_{j \in I_1'} M_j(t) e^{-\alpha_j t}.$$

Proof: From (7.1) and (7.2) we obtain

$$(\sum_{j\in I} M_j(t)e^{-\alpha_j t} - \sum_{j\in I_1'} M_j(t)e^{-\alpha_j t})(1 + o(e^{-\alpha t})) = (\sum_{j\in I_1'} M_j(t)e^{-\alpha_j t})(o(e^{-\alpha' t}) - o(e^{-\alpha t}))$$

(7.3)
$$+ (\sum_{j\in I_1'} M_j(t)e^{-\alpha_j t})(1 + o(e^{-\alpha t})).$$

Chose $\varepsilon > 0$ such that

$$\mathrm{Re}\,\alpha_j' < \alpha - \varepsilon \quad j \in I_1',$$
$$\mathrm{Re}\,\alpha_j < \alpha - \varepsilon \quad j \in I,$$

and multiply both sides of (7.3) by $e^{t(\alpha-\varepsilon)}$. The right side of (7.3) now becomes the sum of the two expressions

(7.4)
$$e^{t(\alpha-\varepsilon)}(\sum_{j\in I_1'} M_j(t)e^{-\alpha_j t})(o(e^{-\alpha' t}) - o(e^{-\alpha t}))$$

and

(7.5)
$$e^{t(\alpha-\varepsilon)}(\sum_{j\in I_2'} M_j(t)e^{-\alpha_j t})(1 + o(e^{-\alpha t})).$$

Since $\alpha < \alpha'$, $\lim e^{t(\alpha-\varepsilon)}o(e^{-\alpha' t}) = 0$, and hence expression (7.4) goes to 0 as $t \to \infty$. Since, $\alpha \le \mathrm{Re}\,\alpha_j'$ for $j \in I_2'$ we have

$$\lim_{t\to+\infty} e^{t(\alpha-\varepsilon)} M_j'(t)e^{-\alpha_j' t} = 0,$$

for such j, and so expression (7.5) also tends to zero as $t \to +\infty$.
Hence, from (7.3) we obtain

(7.6)
$$\lim_{t\to+\infty} e^{t(\alpha-\varepsilon)}(\sum_{j\in I} M_j(t)e^{-\alpha_j t} - \sum_{j\in I_1'} M_j(t)e^{-\alpha_j t}) = 0.$$

It follows that $\Delta(t) = \sum_{j\in I} M_j(t)e^{-\alpha_j t} - \sum_{j\in I_1'} M_j(t)e^{-\alpha_j t} = 0$. Indeed, assume that the above difference is nonzero. It can be rewritten as

$$\Delta(t) = \sum_{j\in J} N_j(t)e^{-((u_j+iv_j)t)}$$

where the $N_j(t)$ are matrix polynomials, $u_j, v_j \in \mathbb{R}$ and $u_j > 0$. Let

$$\eta = \sup_{j\in J}(\alpha - \varepsilon - u_j).$$

Since $\mathrm{Re}\,\alpha_j' < \alpha - \varepsilon$, $g \in I_1'$ and $\mathrm{Re}\,\alpha_j < \alpha - \varepsilon$, $j \in I$, we have $\eta > 0$. Hence (7.6) implies that

$$\lim_{t\to\infty} e^{-\eta t} e^{t(\alpha-\varepsilon)} \Delta(t) = 0.$$

Thus

$$\lim_{t\to+\infty} \sum_{j\in K} N_j(t)e^{-iv_j t} = 0$$

where the sum runs on the j for which $\eta = \alpha - \varepsilon - n_j$.
Rewrite $N_j(t) = \sum N_{jn}t^n$. Then

$$\sum N_j(t)e^{-i\nu_j t} = \sum_0^M t^n \varphi_n(t)$$

where the last sum is finite and the φ_n are finite sums of terms of the form $Ae^{-i\nu t}$, $A \in \mathbb{C}^{n\times n}$, $\nu \in \mathbb{R}$. Since $\lim_{t\to+\infty} \sum N_j(t)e^{-i\nu_j t} = 0$ we also have $\lim_{t\to+\infty} t^{-M} \sum N_j(t)e^{-i\nu_j t} = 0$ and in particular $\lim_{t\to+\infty} \varphi_M(t) = 0$. But φ_M is an almost periodic function. Its limit will be zero if and only if it is identically equal to zero, and the theorem follows.

8   Reconstructing the potential.

By the analysis in the previous section, the function

(8.1) $k_{app}(t) = -2CPe^{-2itA}PB$

is a good approximation of the function $k(t)$ (defining the potential $V(t) = \begin{pmatrix} 0 & k(t) \\ k(t)^* & 0 \end{pmatrix}$)
as $t \to \infty$. In this section we show how we can reconstruct $k_{app}(t)$ (and hence, to the extent of the asymptotic (1.8), the function $k(t)$) from the values of $k_{app}(t)$ and of the values of a number of its derivatives at a point t_0 large enough. The function k_{app} is of the form

$$k_{app}(t) = Fe^{tT}G$$

for a minimal triple $(F, G, T) \in \mathbb{C}^{n\times p} \times \mathbb{C}^{p\times p} \times \mathbb{C}^{p\times n}$, with $p = \dim \text{ Ran } P$.
Thus

$$k_{app}^{(\ell)}(t) = FT^\ell e^{tT}G$$

and, in particular, at $t = t_0$

$$k_{app}^{(\ell)}(t_0) = FT^\ell G_0,$$

where we have set $G_0 = e^{t_0 T}G$.
The problem of reconstructing F, T and G_0 from a finite number of derivatives $k_{app}^{(\ell)}(t_0)$ is an instance of the partial realization problem; if we start from the $L + 1$ first values $k_{app}^{(\ell)}(t_0)$, $\ell = 0, \ldots, L$, one can always find a minimal triple (F_1, T_1, G_1) such that

(8.2) $k_{app}^{(\ell)}(t_0) = F_1 T_1^\ell G_1.$

In general, such a triple is not uniquely determined, even up to similarity. In [4], using the analysis of [8], we proved:

Proposition 8.1 *Let $M_0, \ldots, M_{2m-1}$ be $\mathbb{C}^{n\times n}$ matrices and assume that*

(8.3) $M_j = FT^\ell G$

for some minimal triple with $T \in \mathbb{C}^{m\times m}$. Then, the representation (8.3) is unique up to similarity.

As a direct corollary of this proposition we have:

Theorem 8.2 *Let S be the scattering function of a differential operator (1.1) with $k(t) \in L_1^{n \times n}(0, \infty)$. Assume S rational and analytic at infinity with $S(\infty) = I_n$ and that the McMillan degree of S is $m = 2p$. Then the approximate potential $k_{app}(t)$ is uniquely determined from*

$$k_{app}^{(\ell)}(t_0) \quad \ell = 0, \ldots, 2p - 1.$$

It is reconstructed from these matrices in the following way:

Step 1: Compute F_1, T_1, G_1, minimal triple such that

$$(8.4) \qquad k_{app}^{(\ell)}(t_0) = F_1 T_1^\ell G_1 \quad \ell = 0, \ldots, 2p - 1.$$

Step 2: The function k_{app} is given by:

$$(8.5) \qquad k_{app}(t) = F_1 e^{tT_1} e^{-t_0 T_1} G_1.$$

Proof: From Proposition 8.1 one knows that the $2p$ first values $k_{app}^{(\ell)}(t_0)$ are enough to find a minimal triple such that (8.4) holds. Writing

$$
\begin{aligned}
k_{app}(t) &= F e^{tT} G \\
&= F e^{(t-t_0)T} e^{t_0 T} G
\end{aligned}
$$

and taking into account the uniqueness of the triple satisfying (8.2), we obtain (8.5). ∎

References

[1] D. Alpay and I. Gohberg. *Unitary rational matrix functions*, volume 33 of *Operator Theory: Advances and Applications*, pages 175–222. Birkhäuser Verlag, Basel, 1988.

[2] D. Alpay and I. Gohberg. Inverse spectral problems for difference operators with rational scattering matrix function. *Integral Equations Operator Theory*, 20:125–170, 1994.

[3] D. Alpay and I. Gohberg. Inverse spectral problem for differential operators with rational scattering matrix functions. *Journal of Differential Equations*, 118:1–19, 1995.

[4] D. Alpay, I. Gohberg, and M. Kaashoek. Potential associated to rational weights. Preprint.

[5] J. Ball and A. Ran. *Left versus right canonical Wiener-Hopf factorization*, volume 21 of *Operator Theory: Advances and Applications*, pages 9–38. Birkhäuser–Verlag, Basel, 1986.

[6] H. Bart, I. Gohberg, and M. Kaashoek. *Minimal factorization of matrix and operator functions*, volume 1 of *Operator Theory: Advances and Applications*. Birkhäuser Verlag, Basel, 1979.

[7] H. Dym and A. Iacob. *Positive definite extensions, canonical equations and inverse problem*, volume 12 of *Operator Theory: Advances and Applications*, pages 141–240. Birkhäuser Verlag, Basel, 1984.

[8] I. Gohberg, M. Kaashoek, and L. Lerer. On minimality in the partial realization problem. *Systems and Control Letters*, 9:97–104, 1987.

[9] I. Gohberg and M.A Kaashoek, editors. *Constructive methods of Wiener–Hopf factorizations*, volume 21 of *Operator theory: Advances and Applications*. Birkhäuser Verlag, Basel, 1986.

[10] M.G. Kreĭn. On the determination of a potential of a particle from its *s*–function. *Dokl. Akad. Nauk. SSSR*, 105:637–640, 1955.

[11] M.G. Kreĭn. *Topics in differential and integral equations and operator theory*, volume 7 of *Operator theory: Advances and Applications*. Birkhäuser Verlag, 1983.

[12] M.G. Kreĭn and F.E. Melik-Adamyan. On the theory of *S*-matrices of canonical equations with summable potentials. *Dokl. Akad. Nauk. SSSR*, 16:150–159, 1968.

[13] F.E. Melik-Adamyan. Canonical differential operators in Hilbert space. *Izvestya Akademii Nauk. Armyanskoi SSR Matematica*, 12:10–31, 1977.

[14] F.E. Melik-Adamyan. On a class of canonical differential operators. *Izvestya Akademii Nauk. Armyanskoi SSR Matematica*, 24:570–592, 1989. English translation in: Soviet Journal of Contemporary Mathematics, vol. 24, pages 48–69 (1989).

Daniel Alpay
Department of Mathematics
Ben–Gurion University of the Negev
POB 653. 84105 Beer-Sheva
Israel

Israel Gohberg
School of Mathematical Sciences
The Raymond and Beverly Sackler Faculty
of Exact Sciences
Tel–Aviv University
Tel–Aviv, Ramat–Aviv 69989, Israel

MSC: 34L25, 81U40, 47A56

Operator Theory
Advances and Applications, Vol. 90
© 1996 Birkhäuser Verlag Basel/Switzerland

BANACH ALGEBRAS GENERATED BY N IDEMPOTENTS AND APPLICATIONS

A. Böttcher, I. Gohberg, Yu. Karlovich, N. Krupnik,
S. Roch[1], B. Silbermann, I. Spitkovsky

It is well known that for Banach algebras generated by two idempotents and the identity all irreducible representations are of order not greater than two. These representations have been described completely and have found important applications to symbol theory. It is also well known that without additional restrictions on the idempotents these results do not admit a natural generalization to algebras generated by more than two idempotents and the identity. In this paper we describe all irreducible representations of Banach algebras generated by N idempotents which satisfy some additional relations. These representations are of order not greater than N and allow us to construct a symbol theory with applications to singular integral operators.

Preface and acknowledgement

The study of the structure of Banach algebras generated by two idempotents and the identity has a long history of more than 30 years and has found interesting applications to Banach algebras of singular integral operators on simple contours.

Further advances in the theory of Banach algebras of singular integral operators on non-simple contours required developing a structure theory for Banach algebras generated by N idempotents which satisfy certain relations. The authors of this paper, working in different groups, developed several approaches to this problem.

At the request of the other authors, Steffen Roch, a member of one of the groups, unified these approaches and styles, closed the gaps, and brought the paper to the form in which it is presented. All authors express their sincere gratitude to Steffen Roch for the outstanding task he performed.

1 Introduction

In the last 15 - 20 years, notable advance in understanding the structure of Banach algebras generated by singular integral operators has been made. Many new insights are essentially based on two observations which are characteristic for a large variety of concrete algebras.

---

[1]supported by a DFG Heisenberg grant

The first one is that the Calkin image of operator algebras often contains a non-trivial center, which offers the opportunity of applying local techniques such as Allan's local principle (see below). This principle associates with each of these algebras a whole family of smaller, so-called *local*, algebras which are labeled by the points of a compact space, namely the maximal ideals of the center. Now the second observation enters the scene: in many cases, these local algebras are generated by two (concrete) idempotent cosets, and so they are subject to so-called two projections theorems (see [26], [34], [36], [38], [46], [48] for the C^*-case, [46] for the W^*-case, and [18], [21], [22], [40], and [49], for the general Banach algebra case). Two projection theorems describe *abstract* algebras generated by two idempotents either completely (the C^*-case) or yield at least necessary and sufficient invertibility criteria for the elements of the algebra (the Banach algebra case) by associating with each element of the algebra a certain 2×2 or 1×1 matrix function. The correspondence between the elements of the algebra and the matrix function is either an isometric isomorphism (C^*-case) or a spectrum-preserving homomorphism (Banach algebra case).

Since Douglas' pioneering paper [14], the idea of combining local principles with two projections theorems has been successfully employed, e.g., for algebras generated by one-dimensional singular integral operators with piecewise continuous coefficients, for algebras of Wiener-Hopf and multiplication operators, for algebras of Toeplitz and Hankel operators with piecewise continuous or piecewise quasicontinuous generating functions, for algebras of Fourier integral operators, and for algebras of operators with Carleman shifts (see [2], [3] - [7], [8](Chapter 4), [9], [15], [32], [36], [37], [40], [41], [44]). In all these situations, effective symbol calculi for Fredholmness are available.

Moreover, during the last few years it has become clear that the same approach also applies to certain algebras of approximating sequences for operator equations, the symbol now telling us something about the stability of the sequence. For this topic see [25], [39], [42], [43] and the monograph [24].

In the present paper we consider Banach algebras which are generated by more than two idempotents. Algebras of this type appear as local algebras of concrete operator or sequence algebras. We recall that, in general, there is *no* matrix-valued symbol calculus even for algebras generated by only three idempotents. However, under certain additional conditions, we establish an N projections theorem which yields exactly the two projections theorem (without additional conditions) in case $N = 2$. We also illustrate the application of our N projections theorem to the construction of a symbol calculus for algebras generated by singular integral operators with piecewice continuous coefficients.

For a first discussion of the N projections problem (but without deriving effective invertibility criteria) see [47].

The paper is organized as follows. In Section 2, we remind the reader of some known results on algebras generated by two idempotents and on so-called local principles. Section 3 is devoted to algebras generated by three idempotents. We there point out that such algebras do not possess a matrix symbol in general, but that a matrix symbol exists under certain additional hypotheses. Section 4 contains the main theorem (Theorem 9 in Subsection 4.4) and its proof. In Section 5, we illustrate how the main theorem may be applied to singular integral operators on composed curves. In Section 6, we record several special cases, modifications, and extensions of the main theorem.

2   Algebras generated by two idempotents

The following theorem is one of the main results of [40], with a completion by [21].

Theorem 1 *Let $\mathcal{A}$ be a Banach algebra with identity e, and let p and q be idempotents in $\mathcal{A}$ (i.e. $p^2 = p$ and $q^2 = q$). Let further $\mathcal{B}$ stand for the smallest closed subalgebra of $\mathcal{A}$ which contains p, q and e. Then*
(a) for each
$$x \in \sigma_{\mathcal{B}}(pqp + (e - p)(e - q)(e - p)) \setminus \{0, 1\},$$
the mapping
$$F_x : \{e, p, q\} \rightarrow \mathbb{C}^{2 \times 2},$$
given by
$$F_x(e) = \begin{pmatrix} 1 & 0 \\ 0 & 1 \end{pmatrix}, \quad F_x(p) = \begin{pmatrix} 1 & 0 \\ 0 & 0 \end{pmatrix}, \quad F_x(q) = \begin{pmatrix} x & \sqrt{x(1-x)} \\ \sqrt{x(1-x)} & 1-x \end{pmatrix},$$
where $\sqrt{x(1-x)}$ denotes any number with $\left(\sqrt{x(1-x)}\right)^2 = x(1-x)$ and $\sigma_{\mathcal{B}}(a)$ refers to the spectrum of a in $\mathcal{B}$, extends to a continuous algebra homomorphism from $\mathcal{B}$ into $\mathbb{C}^{2 \times 2}$;
(b) for each
$$m \in \sigma_{\mathcal{B}}(p + 2q) \cap \{0, 1, 2, 3\},$$
the mapping
$$G_m : \{e, p, q\} \rightarrow \mathbb{C}^{1 \times 1},$$
given by
$$G_0(e) = 1, \ G_0(p) = G_0(q) = 0, \quad G_1(e) = G_1(p) = 1, \ G_1(q) = 0,$$
$$G_2(e) = G_2(q) = 1, \ G_2(p) = 0, \quad G_3(e) = G_3(p) = G_3(q) = 1$$
extends to a continuous algebra homomorphism from $\mathcal{B}$ into $\mathbb{C}^{1 \times 1}$;
(c) an element $a \in \mathcal{B}$ is invertible in $\mathcal{B}$ if and only if the matrices $F_x(a)$ are invertible for all $x \in \sigma_{\mathcal{B}}(pqp + (e - p)(e - q)(e - p)) \setminus \{0, 1\}$, and the numbers $G_m(a)$ are non-zero for all $m \in \sigma_{\mathcal{B}}(p + 2q) \cap \{0, 1, 2, 3\}$.
(d) an element $a \in \mathcal{B}$ is invertible in $\mathcal{A}$ if and only if the matrices $F_x(a)$ are invertible for all $x \in \sigma_{\mathcal{A}}(p\dot{q}p + (e - p)(e - q)(e - p)) \setminus \{0, 1\}$, and the numbers $G_m(a)$ are non-zero for all $m \in \sigma_{\mathcal{A}}(p + 2q) \cap \{0, 1, 2, 3\}$.

For a proof see [40], [18], [21], [22] and compare also [49].

The known proofs of the two projections theorem make use of at least one of the following basic properties of the abstract two projections algebra $\mathcal{B} = \text{alg}(e, p, q)$.

(a) The algebra $\mathcal{B}$ possesses a non-trivial center. In particular, the element $pqp + (e - p)(e - q)(e - p)$ commutes with each other element of $\mathcal{B}$ (recall that the center of an algebra consists of all elements which commute with each other element of the algebra).

(b) The algebra $\mathcal{B}$ is an algebra with a polynomial identity. More precisely, it satisfies the standard polynomial F_4 where

$$F_{2n}(a_1, \ldots, a_{2n}) = \sum_{\sigma \in S_{2n}} (\text{sign } \sigma) \, a_{\sigma(1)} \cdots a_{\sigma(2n)}$$

and S_{2n} refers to the group of all permutations of the set $\{1, 2, \ldots, 2n\}$, which means that

$$F_4(b_1, b_2, b_3, b_4) = 0 \quad \text{for all} \quad b_1, \ldots, b_4 \in \mathcal{B}.$$

The first property renders the algebra $\mathcal{B}$ accessible to the local principle by Allan and Douglas (see [1] and [13]), which reads as follows.

Theorem 2 *Let $\mathcal{A}$ be a Banach algebra with identity e, and let $\mathcal{C}$ be a subalgebra of the center of $\mathcal{A}$ which contains e. For each maximal ideal x of the (commutative) Banach algebra $\mathcal{C}$, let I_x denote the smallest closed two-sided ideal of $\mathcal{A}$ which contains x. Then an element a of $\mathcal{A}$ is invertible if and only if the cosets $a + I_x$ are invertible in the quotient algebra $\mathcal{A}/I_x$ for all maximal ideals of $\mathcal{C}$.*

(In case $I_x = \mathcal{A}$, the coset $a + I_x$ is invertible *by definition* for all a.)

Property (b) shows that the two projections algebra is also subject to another local principle, which is due to one of the authors (see [29]):

Theorem 3 *Let $\mathcal{A}$ be a Banach algebra with identity which satisfies the standard polynomial F_{2n}. Then*

(a) for each two-sided maximal ideal M of $\mathcal{A}$, the quotient algebra $\mathcal{A}/M$ is isomorphic to the matrix algebra $\mathbb{C}^{l \times l}$ with a certain $l = l(M)$ less than or equal to n;

(b) an element $a \in \mathcal{A}$ is invertible if and only if the matrices $f_M(a)$ are invertible for all two-sided maximal ideals M where $f_M = \varphi_M \pi_M$, π_M is the canonical homomorphism from $\mathcal{A}$ onto $\mathcal{A}/M$, and φ_M is the isomorphism from $\mathcal{A}/M$ onto $\mathbb{C}^{l \times l}$ given by (a).

Let us remark that this theorem remains true if $\mathcal{A}$ only satisfies a certain power F_{2n}^m of F_{2n} (see [17]).

3 Algebras generated by three idempotents

Let $\mathcal{A}$ be a Banach algebra with identity. We say that $\mathcal{A}$ possesses a *matrix symbol of order n* if there is a family $(f_t)_{t \in T}$ of continuous algebra homomorphisms f_t from $\mathcal{A}$ into the algebra $\mathbb{C}^{l(t) \times l(t)}$ with $l(t) \leq n$ such that an element $a \in \mathcal{A}$ is invertible in $\mathcal{A}$ if and only if the matrices $f_t(a)$ are invertible for all $t \in T$. By Theorem 3, each F_{2n}^m-algebra has a matrix symbol of order n and, in particular, each algebra generated by two idempotents has a matrix symbol of order 2.

The following result is taken from [30]. It shows that the (abstract) algebra generated by three idempotents cannot possess a matrix symbol of a certain fixed order.

Theorem 4 *If $n \geq 3$ then the algebra $\mathbb{C}^{n \times n}$ is generated by three idempotents.*

Moreover, one has the following characterization of algebras generated by three idempotents. Recall that a Banach algebra is called *separable* if it possesses a countable dense subset.

Theorem 5 *(a) Every Banach algebra generated by three idempotents is separable.*

(b) Every separable Banach algebra is isomorphic to a subalgebra of an algebra generated by three idempotents.

Proof. The first assertion is evident. For the second one, we first prove that every separable Banach algebra is isomorphic to a subalgebra of a finitely generated Banach algebra.

Let $\mathcal{A}$ be a separable Banach algebra with a dense subset $\{a_1, a_2, \ldots\}$ and suppose without loss of generality that $a_n \neq 0$ for all n. For $n = 1, 2, \ldots$ and $k = 1, 2, \ldots, 2^n$ set $c_{2^n-2+k} := a_k / \|a_k\|$. Let further $l^2(\mathcal{A})$ stand for the Banach space of all sequences $(x_n)_{n=1}^{\infty}$ of elements of $\mathcal{A}$ such that

$$\|(x_n)\|^2 := \sum_{n=1}^{\infty} \|x_n\|^2 < \infty,$$

and write $L(l^2(\mathcal{A}))$ for the Banach algebra of all bounded linear operators on $l^2(\mathcal{A})$. On $l^2(\mathcal{A})$ we consider the following operators:

$$A : (x_n) \mapsto (y_n), \quad y_n = c_n x_n,$$

$$V_1 : (x_n) \mapsto (y_n), \quad y_n = \begin{cases} 0 & \text{if} \quad n = 1 \\ x_{n-1} & \text{if} \quad n > 1, \end{cases}$$

$$V_{-1} : (x_n) \mapsto (y_n), \quad y_n = x_{n+1},$$

$$W_1 : (x_n) \mapsto (y_n), \quad y_n = \begin{cases} x_k & \text{if} \quad n = 2^k - 1 \\ 0 & \text{if} \quad n \neq 2^k - 1, \end{cases}$$

$$W_{-1} : (x_n) \mapsto (y_n), \quad y_n = x_{2^n-1}.$$

Obviously, $A, V, V_{-1}, W, W_{-1} \in L(l^2(\mathcal{A}))$, and so it makes sense to consider the smallest closed subalgebra $\mathcal{B}$ of $L(l^2(\mathcal{A}))$ which contains the operators A, V, V_{-1}, W, W_{-1} and the identity operator I. The algebra $\mathcal{B}$ is finitely generated, and we claim that $\mathcal{A}$ is isomorphic to a subalgebra of $\mathcal{B}$. Since

$$W_{-1} A W_1 : (x_n) \mapsto (y_n), \quad y_n = c_1 x_n,$$

$$W_{-1} V_{-1} A V_1 W_1 : (x_n) \mapsto (y_n), \quad y_n = c_2 x_n,$$

we conclude that the diagonal matrix $\operatorname{diag}(c_k, c_k, \ldots)$ lies in $\mathcal{B}$ for $k = 1, 2$. In order to arrive at this conclusion for $k > 2$, set

$$r_k = 2^{\{\log_2 k\}} - 3 + k \quad \text{and} \quad s_k = 2^{\{\log_2 k\}} - 2$$

where $\{z\}$ refers to the smallest integer which is greater than or equal to z. Then

$$V_{-1}^{s_k} W_{-1} V_1^{s_k} \cdot V_{-1}^{r_k} A V_1^{r_k} \cdot V_{-1}^{s_k} W_1 V_1^{s_k} \; : \; (x_n) \mapsto (y_n), \quad y_n = c_k x_n$$

for all $k > 2$. Hence, $\operatorname{diag}(c_k, c_k, \ldots)$ is in $\mathcal{B}$ for all k and consequently, $\operatorname{diag}(a_k, a_k, \ldots)$ belongs to $\mathcal{B}$ for all k. Now it is easy to check that the mapping

$$T : \mathcal{A} \to \mathcal{B}, \; a \mapsto \operatorname{diag}(a, a, \ldots)$$

is the desired isomorphism from $\mathcal{A}$ onto a subalgebra of $\mathcal{B}$.

To finish the proof it remains to remark that, for *each* finitely generated Banach algebra $\mathcal{B}$, the algebra $\mathcal{B}^{r \times r}$ of all $r \times r$ matrices with entries in $\mathcal{B}$ is generated by three idempotents if only r is large enough (see [31]). Thus, each finitely generated Banach algebra is isomorphic to a subalgebra of a Banach algebra generated by three idempotents (the isomorphism simply being given by

$$\mathcal{B} \to \mathcal{B}^{r \times r}, \; b \mapsto \operatorname{diag}(b, b, \ldots, b)),$$

and this result in combination with what has already been proved gives our claim. ∎

Theorem 5 indicates that the variety of all Banach algebras generated by three idempotents is extremely large and that these algebras can show a rather involved structure. This observation suggests the study of Banach algebras generated by three (or more) idempotents with additional relations between their generators. For example, let $L^2(J)$ denote the Hilbert space of the squared integrable functions on some (finite or infinite) interval J. On $L^2(\mathbb{R})$, we introduce the operator $S_{\mathbb{R}}$ of singular integration,

$$(S_{\mathbb{R}} f)(t) = \frac{1}{\pi i} \int_{-\infty}^{\infty} \frac{f(s)}{s - t} \, ds, \quad t \in \mathbb{R},$$

and the operators $\chi_{\mathbb{R}^+} I$ and $\chi_{[0,1]} I$ of multiplication by the characteristic functions of the intervals $\mathbb{R}^+$ and $[0, 1]$, respectively. Let $\mathcal{A}$ denote the smallest closed subalgebra of $L(L^2(\mathbb{R}))$ which contains the operators $S_{\mathbb{R}}$, $\chi_{\mathbb{R}^+} I$ and $\chi_{[0,1]} I$. Since $S_{\mathbb{R}}^2 = I$ and $S_{\mathbb{R}}^* = S_{\mathbb{R}}$ (see [20]), we conclude that $P_{\mathbb{R}} := (I + S_{\mathbb{R}})/2$ is a projection and hence, the algebra $\mathcal{A}$ is generated by three projections and the identity operator. Let further $\mathcal{B}$ refer to the smallest closed subalgebra of $\mathcal{A}$ which contains all operators $\chi_{[0,1]} (\chi_{\mathbb{R}^+} S_{\mathbb{R}} \chi_{\mathbb{R}^+})^k \chi_{[0,1]} I$ with $k = 0, 1, \ldots$. Clearly, one can think of $\mathcal{B}$ as a subalgebra of $L(L^2([0, 1]))$.

Theorem 6 *The algebra $\mathcal{B}$ (which is a subalgebra of an algebra generated by three idempotents) contains all compact operators on $L^2([0, 1])$.*

For a proof see, e.g., Theorem 8.7 in [16].

Taking into account that the ideal of all compact operators on $L^2([0, 1])$ contains a copy of $\mathbb{C}^{l \times l}$ for all l or having recourse to Corollary 22.1 in [29]), we arrive at the conclusion that the algebra $\mathcal{B}$ (and hence the algebra $\mathcal{A}$) cannot possess a matrix symbol of any fixed order.

Thus, even if the three idempotents are projections, and even if two of them commute, a matrix symbol need not exist. This highlights that the additional conditions we look for in order to guarantee the existence of matrix symbols have to be rather strong.

Here are two (positive) examples of algebras generated by three idempotents which *possess* a matrix symbol. Observe the strong relations between the generating elements.

Theorem 7 *Let $\mathcal{A}$ be a Banach algebra with identity e, and let p, q and j be elements in $\mathcal{A}$ such that*

$$p^2 = p, \quad q^2 = q, \quad j^2 = e \quad and \quad jpj = e - p, \quad jqj = e - q.$$

Then the smallest closed subalgebra $\mathcal{B}$ of $\mathcal{A}$ which contains e, p, q and j is F_4, and it possesses a matrix symbol of order 2.

For a proof (and also for the explicit derivation of the matrix symbol under an additional condition) see [40]. Let us emphasize that the algebra $\mathcal{B}$ in Theorem 7 is indeed generated be three idempotents since p and q are idempotent and $(e + j)/2$ is idempotent, too.

Theorem 8 *Let $\mathcal{A}$ be a Banach algebra with identity e, and let p, q and j be elements in $\mathcal{A}$ such that*

$$p^2 = p, \quad q^2 = q, \quad j^2 = e \quad and \quad jpj = p, \quad jqj = e - q.$$

Then the smallest closed subalgebra $\mathcal{B}$ of $\mathcal{A}$ which contains e, p, q and j possesses a matrix symbol of order 4.

For a proof, and for an explicit matrix symbol, see [32].

4 An N projections theorem

4.1 Choice of the additional conditions

We are going to describe a class of Banach algebras which are generated by a large number of idempotents and possess a matrix symbol. Our choice of the additional conditions between the generating elements of the algebras is motivated by the situation considered in Section 5 (and, in a sense, by the approach of the papers [19] and [23]).

Let $\mathcal{A}$ be a Banach algebra with identity element I, and let $\{p_1, \ldots, p_{2N}\}$ be a *partition of unity into projections*, i.e. suppose $p_i \neq 0$ for all i,

$$p_i \cdot p_j = \delta_{ij} p_i \quad \text{for all } i, j, \tag{1}$$

where δ_{ij} is the Kronecker delta, and

$$\sum_{i=1}^{2N} p_i = I. \tag{2}$$

Let further $P \in \mathcal{A}$, put $Q = I - P$, and suppose that

$$P(p_{2i-1} + p_{2i})P = (p_{2i-1} + p_{2i})P \tag{3}$$

and

$$Q(p_{2i} + p_{2i+1})Q = (p_{2i} + p_{2i+1})Q \qquad (4)$$

for all $i = 1, \ldots, N$, where $p_{2N+1} := p_1$. In what follows we use the convention $p_k := p_r$ with $r \in \{1, \ldots, 2N\}$ whenever $k - r$ is divisible by $2N$. It is clear that then (3) and (4) hold for all integers i.

The algebra $\mathcal{B}$ we are interested in is the smallest closed subalgebra of $\mathcal{A}$ which contains the set $\{p_i\}_{i=1}^{2N}$ as well as the element P. Observe that $\mathcal{B}$ contains the identity I (due to (2)) and that P and Q are complementary idempotents. Indeed, adding the identities (3) for $i = 1, \ldots, N$ yields

$$P \cdot \sum_{i=1}^{2N} p_i \cdot P = \sum_{i=1}^{2N} p_i \cdot P,$$

that is, $P^2 = P$, whence $Q^2 = Q$. Thus, $\mathcal{B}$ is actually an algebra generated by $2N + 1$ idempotents (or by $2N$ idempotents and the identity).

We will show that the algebra $\mathcal{B}$ possesses a matrix symbol of order $2N$.

4.2   Algebraic structure of $\mathcal{B}$

We start with examining the smallest (not necessarily closed) subalgebra $\mathcal{B}^0$ which contains the partition of unity into projections $\{p_i\}_{i=1}^{2N}$ and the idempotent P. Set

$$X := \sum_{i=1}^{N} (p_{2i-1} P p_{2i-1} + p_{2i} Q p_{2i}).$$

Proposition 1 *The element X is in the center of $\mathcal{B}^0$.*

Proof. Evidently, X commutes with each of the idempotents p_i. It remains to show that $PX = XP$. Let us first prove that

$$X = \sum_{i=1}^{N} \left((p_{2i} + p_{2i+1}) Q p_{2i} Q + (p_{2i-1} + p_{2i}) P p_{2i-1} P \right). \qquad (5)$$

Since the p_j form a partition of unity into projections, it is sufficient to prove that

$$p_j X = p_j \sum_{i=1}^{N} \left((p_{2i} + p_{2i+1}) Q p_{2i} Q + (p_{2i-1} + p_{2i}) P p_{2i-1} P \right)$$

for $j = 1, \ldots, 2N$ or, equivalently, that

$$p_{2i} Q p_{2i} = p_{2i} Q p_{2i} Q + p_{2i} P p_{2i-1} P \qquad (6)$$

and

$$p_{2i-1} P p_{2i-1} = p_{2i-1} Q p_{2i-2} Q + p_{2i-1} P p_{2i-1} P \qquad (7)$$

for all $i = 1, \ldots, N$. For (6) observe that

$$p_{2i} Q p_{2i} Q + p_{2i} P p_{2i-1} P = p_{2i} Q - p_{2i} P p_{2i} Q + p_{2i} P p_{2i-1} P$$

$$= p_{2i}Q - p_{2i}Pp_{2i} + p_{2i}Pp_{2i}P + p_{2i}Pp_{2i-1}P$$

$$= p_{2i}Q - p_{2i}Pp_{2i} + p_{2i}P(p_{2i-1} + p_{2i})P$$

$$= p_{2i}Q - p_{2i}Pp_{2i} + p_{2i}P$$

$$= p_{2i} - p_{2i}Pp_{2i} = p_{2i}(P + Q)p_{2i} - p_{2i}Pp_{2i} = p_{2i}Qp_{2i},$$

and (7) follows analogously. Thus (5) holds. Further, axioms (3) and (4) say that

$$Q(p_{2i-1} + p_{2i})P = P(p_{2i} + p_{2i+1})Q = 0 \tag{8}$$

for $i = 1, \ldots, N$, and the axioms (3), (4) together with the identities (5), (8) yield

$$PX = P \cdot \sum_{i=1}^{N}((p_{2i} + p_{2i+1})Qp_{2i}Q + (p_{2i-1} + p_{2i})Pp_{2i-1}P) = \sum_{i=1}^{N}(p_{2i-1} + p_{2i})Pp_{2i-1}P$$

and

$$XP = \sum_{i=1}^{N}((p_{2i} + p_{2i+1})Qp_{2i}Q + (p_{2i-1} + p_{2i})Pp_{2i-1}P) \cdot P = \sum_{i=1}^{N}(p_{2i-1} + p_{2i})Pp_{2i-1}P$$

and, hence, $PX = XP$. ∎

Proposition 2 *Considered as module over its center, the algebra $\mathcal{B}^0$ is generated by the $(2N)^2$ elements $(p_i)_{i=1}^{2N}$ and $(p_iPp_j)_{i,j=1}^{2N}$ with $i \neq j$. To be more precise, given $A \in \mathcal{B}^0$, there are polynomials R_{ij} in X such that*

$$A = \sum_{i=1}^{2N} R_{ii}(X)p_i + \sum_{\substack{i,j=1 \\ i \neq j}}^{2N} R_{ij}(X)p_iPp_j. \tag{9}$$

Proof. Let $\mathcal{B}^1$ denote the set of all elements in $\mathcal{B}^0$ which can be written as in (9). First we show that the generating elements of $\mathcal{B}^0$ belong to $\mathcal{B}^1$. This is evident for the idempotents p_i. Since further

$$p_iPp_i = p_iPp_i \cdot p_i = \begin{cases} X \cdot p_i & \text{if } i \text{ is odd} \\ (I - X) \cdot p_i & \text{if } i \text{ is even,} \end{cases} \tag{10}$$

the assertion for P can be seen as follows:

$$P = \sum_{i,j=1}^{2N} p_iPp_j = \sum_{i=1}^{N} p_{2i}Pp_{2i} + \sum_{i=1}^{N} p_{2i-1}Pp_{2i-1} + \sum_{\substack{i,j=1 \\ i \neq j}}^{2N} p_iPp_j$$

$$= \sum_{i=1}^{N}(I - X)p_{2i} + \sum_{i=1}^{N} Xp_{2i-1} + \sum_{\substack{i,j=1 \\ i \neq j}}^{2N} p_iPp_j.$$

In the second step we are going to show that $\mathcal{B}^1$ is actually an algebra. Since the generating elements of $\mathcal{B}^0$ belong to $\mathcal{B}^1$ this automatically yields that $\mathcal{B}^0 = \mathcal{B}^1$.

The set $\mathcal{B}^1$ is evidently closed under addition. In order to get its closedness under multiplication we have to show that the product of each two of the elements $(p_i)_{i=1}^{2N}$ and $(p_i P p_j)_{i,j=1}^{2N}$ with $i \neq j$ is in $\mathcal{B}^1$ again. This is obvious if one of these elements is p_i, and so we have only to deal with the products $p_i P p_j \cdot p_k P p_l$ with $i \neq j$ and $k \neq l$. This product is 0 (which is in $\mathcal{B}^1$) if $j \neq k$ and equal to $p_i P p_j P p_l$ in case $j = k$. If j is even (say, $j = 2n$) then

$$p_i P p_{2n} P p_l = p_i P(p_{2n-1} + p_{2n}) P p_l - p_i P p_{2n-1} P p_l$$

$$= p_i(p_{2n-1} + p_{2n}) P p_l - p_i P p_{2n-1} P p_l \tag{11}$$

by axiom (3), whereas in case j is odd ($j = 2n - 1$),

$$p_i P p_{2n-1} P p_l = p_i P(p_{2n-2} + p_{2n-1}) P p_l - p_i P p_{2n-2} P p_l$$

$$= p_i P(p_{2n-2} + p_{2n-1}) p_l - p_i P p_{2n-2} P p_l \tag{12}$$

by (8). The first items in (11) and (12) are in $\mathcal{B}^1$. Indeed, they are either 0 or equal to $p_i P p_l$ (in dependence on j). If $i \neq l$ then $p_i P p_l \in \mathcal{B}^1$ by definition, whereas the inclusion $p_i P p_i \in \mathcal{B}^1$ follows from (10).

Thus, identities (11) and (12) reduce the question whether $p_i P p_j P p_l \in \mathcal{B}^1$ to the problem whether $p_i P p_{j-1} P p_l \in \mathcal{B}^1$. Repeated application of this argument finally yields an element of the form $p_i P p_i P p_l$. This element is in $\mathcal{B}^1$ since

$$p_i P p_i P p_l = p_i P p_i \cdot p_i P p_l = \begin{cases} X \cdot p_i P p_l & \text{if } i \text{ is odd} \\ (I - X) \cdot p_i P p_l & \text{if } i \text{ is even} \end{cases}$$

and by (10). ∎

Let us have a closer look at the products $p_i P p_j \cdot p_j P p_l$ in case $i \neq j$ and $j \neq l$.

Proposition 3 (a) If $l > j > i$ or $j > i > l$ or $i > l > j$ then

$$p_i P p_j P p_l = (-1)^{j-1}(X - I) p_i P p_l.$$

(b) If $l > i > j$ or $j > l > i$ or $i > j > l$ then

$$p_i P p_j P p_l = (-1)^{j-1} X \, p_i P p_l.$$

(c) If $i = l$ and $i \neq j$ then

$$p_i P p_j P p_i = (-1)^{j-i} X(X - I) \, p_i.$$

Proof. Let $j \neq i, l$. Then

$$p_i P p_j P p_l = p_i P(p_{j-i} + p_j) P p_l - p_i P p_{j-1} P p_l. \tag{13}$$

If, moreover, $j - i \neq i, l$, then we conclude from (3) and (8) that $p_i P(p_{j-i} + p_j) P p_l = 0$ and, hence,

$$p_i P p_j P p_l = -p_i P p_{j-1} P p_l. \tag{14}$$

Suppose now the conditions of assertion (a) to be satisfied. Then there is a smallest positive integer k such that (all computations modulo $2N$) $j \neq i, l, j-1 \neq i, l, \ldots, j-(k-1) \neq i, l$ but $j - k = i$. Consequently, repeated application of (14) gives

$$p_i P p_j P p_l = (-1)^{k-1} p_i P p_{j-(k-1)} P p_l$$

whence by virtue of (13),

$$p_i P p_j P p_l = (-1)^{k-1} (p_i P(p_{j-k} + p_{j-(k-1)}) P p_l - p_i P p_{j-k} P p_l)$$

$$= (-1)^{k-1} (p_i P(p_i + p_{i+1}) P p_l - p_i P p_i P p_l).$$

Observe that our assumptions imply that $l \neq i$ and $l \neq i+1$ (otherwise $j - (k-1)$ would be equal to l). Thus

$$p_i P p_j P p_l = \begin{cases} (-1)^{k-1} (p_i(p_i + p_{i+1}) P p_l - p_i P p_i P p_l) & \text{if } i \text{ is odd} \\ (-1)^{k-1} (p_i P(p_i + p_{i+1}) p_l - p_i P p_i P p_l) & \text{if } i \text{ is even} \end{cases}$$

$$= \begin{cases} (-1)^{k-1} (p_i P p_l - p_i P p_i P p_l) & \text{if } i \text{ is odd} \\ (-1)^{k-1} (-p_i P p_i P p_l) & \text{if } i \text{ is even} \end{cases}$$

$$= \begin{cases} (-1)^{k-1} (I - X) p_i P p_l & \text{if } i \text{ is odd} \\ (-1)^{k-1} (-1)(I - X) p_i P p_l & \text{if } i \text{ is even} \end{cases}$$

(again take into account (10)). Replacing k by $j - i$ yields assertion (a). The proof for (b) and (c) is analogous. ∎

4.3   Localization, and identification of the local algebras

The element X belongs to the center of the algebra $\mathcal{B}^0$ (Proposition 1) and thus to the center of $\mathcal{B}$ itself. Hence, the smallest closed subalgebra $\mathcal{C}$ of $\mathcal{B}$ which contains the identity element I and the element X is in the center of $\mathcal{B}$, and this offers the possibility of localizing $\mathcal{B}$ over $\mathcal{C}$ by the local principle of Allan and Douglas (Theorem 2). It is well known that the maximal ideal space of the singly (by X) generated Banach algebra $\mathcal{C}$ is homeomorphic to the spectrum $\sigma_C(X)$ of its generator (see [12], 15.3.6) and that under this homeomorphism the point $x \in \sigma_C(X)$ corresponds to the smallest closed ideal of $\mathcal{C}$ which contains $X - xI$. In accordance with Theorem 2, we introduce ideals I_x in $\mathcal{B}$ for all $x \in \sigma_C(X)$.

Proposition 4 (a) If $x \in \sigma_B(X)$ $(\subseteq \sigma_C(X))$ then $I_x \neq \mathcal{B}$.
(b) If $x \in \sigma_C(X) \setminus \sigma_B(X)$ then $I_x = \mathcal{B}$.

For a proof see [18]. Thus, by Theorem 2, an element $b \in \mathcal{B}$ is invertible if and only if the cosets $b + I_x$ are invertible for all $x \in \sigma_B(X)$.
    For $x \in \sigma_B(X)$, let $\mathcal{B}_x := \mathcal{B}/I_x$ denote the local algebra associated with x and let $\Phi_x :$ $\mathcal{B} \to \mathcal{B}_x$ be the canonical homomorphism. Let us remark once more that, by Proposition 4, each algebra $\mathcal{B}_x$ contains at least two different elements (the zero and the identity). Our next goal is the explicit description of the local algebras $\mathcal{B}_x$.

Proposition 5 *If $x \in \sigma_B(X) \setminus \{0, 1\}$, then $\mathcal{B}_x$ is isomorphic to $\mathbb{C}^{2N \times 2N}$.*

Proof. Consider the image $\Phi_x(\mathcal{B}^0)$ of the algebra $\mathcal{B}^0$ in $\mathcal{B}_x$. Since each element of $\mathcal{B}^0$ can be written in the form (9) and since

$$\Phi_x(X) = x\Phi_x(I)$$

by definition, it follows that

$$\Phi_x(R(X)) = R(x)\Phi_x(I)$$

for each polynomial R. Consequently, we conclude that each element of $\Phi_x(\mathcal{B}^0)$ is a complex linear combination of the elements

$$\Phi_x(p_i) \quad (i = 1, \ldots, 2N) \quad \text{and} \quad \Phi_x(p_i P p_j) \quad (i, j = 1, \ldots, 2N, i \neq j). \tag{15}$$

Conversely, each linear combination of the elements (15) is in $\Phi_x(\mathcal{B}^0)$. Thus, $\Phi_x(\mathcal{B}^0)$ is a finite dimensional linear space (of dimension $\leq (2N)^2$). In particular, $\Phi_x(\mathcal{B}^0)$ is closed in $\mathcal{B}_x$. On the other hand, $\mathcal{B}^0$ is dense in $\mathcal{B}$ and, hence, $\Phi_x(\mathcal{B}^0)$ is dense in $\Phi_x(\mathcal{B}) = \mathcal{B}_x$. Thus, $\mathcal{B}_x = \Phi_x(\mathcal{B}^0)$, and $\mathcal{B}_x$ is a linear space of dimension $\leq (2N)^2$.

We claim that the dimension of $\mathcal{B}_x$ is exactly $(2N)^2$ and that the elements (15) form a basis of this space. Given $i, j = 1, \ldots, 2N$, define $a_{ij} \in \mathcal{B}_x$ by

$$a_{ij} = \begin{cases} (-1)^{i-1}(x-1)^{-1}\Phi_x(p_i P p_j) & \text{if } i < j \\ (-1)^{i-1}x^{-1}\Phi_x(p_i P p_j) & \text{if } i > j \\ \Phi_x(p_i) & \text{if } i = j. \end{cases}$$

(This definition is correct since $x \neq 0$ and $x \neq 1$.) Proposition 3 implies that

$$a_{ij}a_{kl} = \delta_{jk} \cdot a_{il} \quad \text{for all} \quad 1 \leq i, j, k, l \leq 2N. \tag{16}$$

We check (16), for example, in case $j = k$ and $j > i > l$:

$$a_{ij}a_{jl} = (-1)^{i-1}(x-1)^{-1}\Phi_x(p_i P p_j) \cdot (-1)^{j-1}x^{-1}\Phi_x(p_j P p_l)$$

$$= (-1)^{i-1}(-1)^{j-1}x^{-1}(x-1)^{-1}\Phi_x(p_i P p_j P p_l)$$

$$= (-1)^{i-1}(-1)^{j-1}x^{-1}(x-1)^{-1}\Phi_x((-1)^{j-1}(X-I)p_i P p_l)$$

$$= (-1)^{i-1}x^{-1}\Phi_x(p_i P p_l) = a_{il}.$$

The other cases can be disposed of analogously.

Now suppose the elements a_{ij} are linearly dependent. Then there are numbers c_{ij} with

$$\sum_{i,j=1}^{2N} c_{ij}a_{ij} = 0 \tag{17}$$

but $c_{i_0 j_0} \neq 0$ for certain $i_0 j_0$. Multiplying (17) from the left by $a_{k i_0}$ and from the right by $a_{j_0 k}$ yields that $c_{i_0 j_0} a_{k i_0} a_{i_0 j_0} a_{j_0 k} = c_{i_0 j_0} a_{kk} = 0$ and hence, $a_{kk} = 0$ for all $k = 1, \ldots, 2N$. Consequently,

$$\Phi_x(I) = \Phi_x(\sum_{k=1}^{2N} p_k) = \sum_{k=1}^{2N} a_{kk} = \Phi_x(0)$$

which contradicts Proposition 4(a) (see also the remark following this proposition).

Thus, the elements $(a_{ij})_{i,j=1}^{2N}$ are linearly independent. It follows that so are also the elements (15), and therefore both sets of elements form a basis of $\mathcal{B}_x$. Finally, it is immediate from (16) that the mapping

$$\Psi_x : (a_{ij})_{i,j=1}^{2N} \to \mathbb{C}^{2N \times 2N}, \quad a_{ij} \mapsto E_{ij},$$

where E_{ij} refers to the $2N \times 2N$ matrix whose i, j entry is 1 and all other entries of which are zero, extends to an algebra isomorphism from $\mathcal{B}_x$ onto $\mathbb{C}^{2N \times 2N}$. ∎

Here are the images of the generating elements of the algebra $\mathcal{B}$ under the homomorphism $F_x := \Psi_x \circ \Phi_x : \mathcal{B} \to \mathbb{C}^{2N \times 2N}$.

Corollary 1 *Let $x \in \sigma_\mathcal{B}(X) \setminus \{0, 1\}$. Then*

$$F_x(p_i) = \operatorname{diag}(0, \ldots, 0, 1, 0, \ldots, 0), \tag{18}$$

the 1 standing at the ith place, and

$$F_x(P) = \operatorname{diag}(1, -1, 1, -1, \ldots, 1, -1) \times$$

$$\times \begin{pmatrix} x & x-1 & x-1 & x-1 & \cdots & x-1 & x-1 \\ x & x-1 & x-1 & x-1 & \cdots & x-1 & x-1 \\ x & x & x & x-1 & \cdots & x-1 & x-1 \\ x & x & x & x-1 & \cdots & x-1 & x-1 \\ \vdots & \vdots & \vdots & \vdots & \ddots & \vdots & \vdots \\ x & x & x & x & \cdots & x & x-1 \\ x & x & x & x & \cdots & x & x-1 \end{pmatrix}. \tag{19}$$

Proof. To verify (18) recall that $\Phi_x(p_i) = a_{ii}$, and to get (19) observe that

$$F_x(P) = F_x\left(\sum_{i,j=1}^{2N} p_i P p_j\right)$$

$$= (\Psi_x \circ \Phi_x)\left(\sum_{\substack{i,j=1 \\ i<j}}^{2N} p_i P p_j\right) + (\Psi_x \circ \Phi_x)\left(\sum_{\substack{i,j=1 \\ i>j}}^{2N} p_i P p_j\right) + (\Psi_x \circ \Phi_x)\left(\sum_{i=1}^{2N} p_i P p_i\right)$$

$$= \Psi_x\left(\sum_{\substack{i,j=1 \\ i<j}}^{2N} (-1)^{i-1}(x-1)a_{ij}\right) + \Psi_x\left(\sum_{\substack{i,j=1 \\ i>j}}^{2N} (-1)^{i-1} x a_{ij}\right) + (\Psi_x \circ \Phi_x)\left(\sum_{i=1}^{2N} p_i P p_i\right)$$

and take into account (10). ∎

Our next subject is the local algebras $\mathcal{B}_x$ associated with the points in $\sigma_\mathcal{B}(X) \cap \{0, 1\}$. These algebras will not be identified completely; we will only show that all irreducible representations are one-dimensional and will compute them.

Proposition 6 *If $x \in \sigma_\mathcal{B}(X) \cap \{0, 1\}$ then $\mathcal{B}_x$ is an F_2^{N+1}-algebra.*

Proof. Instead of working with the polynomial $F_2^{N+1}(a, b) = (ab - ba)^{N+1}$ in two variables, which is non-linear, let us consider the polynomial

$$F_2^{\langle N+1 \rangle}(a_1, b_1, \ldots, a_{N+1}, b_{N+1}) := \prod_{k=1}^{N+1} (a_k b_k - b_k a_k)$$

in $2(N+1)$ variables, which is linear in each variable. Notice that if $\mathcal{B}_x$ is an $F_2^{\langle N+1 \rangle}$-algebra, then it is also an F_2^{N+1}-algebra.

Since $\mathcal{B}_x$ is a linear space (recall the proof of Proposition 5) and since $F_2^{\langle N+1 \rangle}$ is multi-linear, it remains to prove that

$$\prod_{k=1}^{N+1} (a_k b_k - b_k a_k) = 0$$

for all choices of cosets a_k, b_k $(k = 1, \ldots, N+1)$ among the (possible) basis elements of the algebra $\mathcal{B}_x$:

$$\Phi_x(p_i) \quad (i = 1, \ldots, 2N) \quad \text{and} \quad \Phi_x(p_i P p_j) \quad (i, j = 1, \ldots, 2N, i \neq j).$$

Proposition 3 entails that each commutant $a_k b_k - b_k a_k$ can be written as $c_k \Phi_x(p_{i_k} P p_{j_k})$ where $i_k, j_k \in \{1, \ldots, N+1\}$ and $c_k \in \mathbb{C}$ can be zero. Hence,

$$\prod_{k=1}^{N+1} (a_k b_k - b_k a_k) = c \Phi_x \Big(\prod_{k=1}^{N+1} p_{i_k} P p_{j_k} \Big).$$

Since the partition of unity into projections (p_i) consists of $2N$ elements, there are two of the elements p_{i_k} and p_{j_k} with $k = 1, \ldots, N+1$ which coincide. Thus, $\prod_{k=1}^{N+1} p_{i_k} P p_{j_k}$ contains at least one subproduct of the form $p_i P p_{l_1} P p_{l_2} \ldots P p_{l_r} P p_i$ with $r \geq 1$, and invoking Proposition 3 once more, one easily gets $\Phi_x(p_i P p_{l_1} P p_{l_2} \ldots P p_{l_r} P p_i) = 0$. Thus,

$$\Phi_x \Big(\prod_{k=1}^{N+1} p_{i_k} P p_{j_k} \Big) = 0 \quad \text{for} \quad x \in \sigma_{\mathcal{B}}(X) \cap \{0, 1\}. \quad \blacksquare$$

By Theorem 3 (extended version), the algebras $\mathcal{B}_x$ possess matrix symbols of order 1, i.e. scalar-valued symbols. Since each algebra homomorphism $\Psi : \mathcal{B}_x \to \mathbb{C}$ gives rise to an algebra homomorphism $\Psi \circ \Phi_x : \mathcal{B} \to \mathbb{C}$, we proceed with determining the one-dimensional representations of the algebra $\mathcal{B}$.

Clearly, each homomorphism $G : \mathcal{B} \to \mathbb{C}$ maps idempotents to idempotents. Thus, if $p \in \mathcal{B}$ is idempotent, then $G(p)$ is either 0 or 1. Moreover, since $G(I) = 1$ for each non-zero homomorphism G, we conclude that, given a partition of unity into projections $(p_i)_{i=1}^{2N}$, there is an i_0 such that $G(p_{i_0}) = 1$ and $G(p_i) = 0$ for all $i \neq i_0$. Hence, the restriction of a non-zero homomorphism $G : \mathcal{B} \to \mathbb{C}$ to the set $\{P, p_1, p_2, \ldots, p_{2N}\}$ coincides with one of the following mappings G_n with $n \in \{1, 2, \ldots, 4N\}$:

$$G_{4m}(p_i) = \begin{cases} 1 & \text{if } i = 2m \\ 0 & \text{if } i \neq 2m \end{cases}, \qquad G_{4m}(P) = 0,$$

$$G_{4m-1}(p_i) = \begin{cases} 1 & \text{if } i = 2m \\ 0 & \text{if } i \neq 2m \end{cases}, \qquad G_{4m-1}(P) = 1,$$

$$(20)$$

$$G_{4m-2}(p_i) = \begin{cases} 1 & \text{if } i = 2m-1 \\ 0 & \text{if } i \neq 2m-1 \end{cases}, \quad G_{4m-2}(P) = 1,$$

$$G_{4m-3}(p_i) = \begin{cases} 1 & \text{if } i = 2m-1 \\ 0 & \text{if } i \neq 2m-1 \end{cases}, \quad G_{4m-3}(P) = 0,$$

where $m = 1, \ldots, N$. Set

$$Y := \sum_{i=1}^{N}(p_{2i-1}P + p_{2i}Q) + \sum_{i=1}^{2N}(2i-1)p_i.$$

Proposition 7 *If $m \in \sigma_B(Y) \cap \{1, 2, \ldots, 4N\}$ then the mapping $G_m : \{P, p_1, p_2, \ldots, p_{2N}\} \to \mathbb{C}$, given by (20), extends to an algebra homomorphism from B onto $\mathbb{C}$.*

Proof. First of all notice that if G_m extends to an algebra homomorphism, then

$$G_m(Y) = m. \tag{21}$$

We claim that, for $m \in \sigma_B(Y) \cap \{1, 2, \ldots, 4N\}$ and $x \in \sigma_B(X) \setminus \{0, 1\}$,

$$m \notin \sigma_{B_x}(\Phi_x(Y)). \tag{22}$$

What we have to prove is, by Corollary 1, that the $2N \times 2N$ matrices

$$(\Psi_x \circ \Phi_x)(Y) - \mathrm{diag}\,(m, m, \ldots, m) =$$

$$= \begin{pmatrix} x & x-1 & x-1 & x-1 & \cdots & x-1 & x-1 \\ x & x & x-1 & x-1 & \cdots & x-1 & x-1 \\ x & x & x & x-1 & \cdots & x-1 & x-1 \\ x & x & x & x & \cdots & x-1 & x-1 \\ \vdots & \vdots & \vdots & \vdots & \ddots & \vdots & \vdots \\ x & x & x & x & \cdots & x & x-1 \\ x & x & x & x & \cdots & x & x \end{pmatrix} +$$

$$+ \mathrm{diag}\,(1-m, 3-m, \ldots, 4N-1-m)$$

are invertible. For this goal we compute the determinant of the slightly more general $M \times M$ matrix

$$\begin{pmatrix} x+\lambda_1 & x-1 & x-1 & x-1 & \cdots & x-1 & x-1 \\ x & x+\lambda_2 & x-1 & x-1 & \cdots & x-1 & x-1 \\ x & x & x+\lambda_3 & x-1 & \cdots & x-1 & x-1 \\ x & x & x & x+\lambda_4 & \cdots & x-1 & x-1 \\ \vdots & \vdots & \vdots & \vdots & \ddots & \vdots & \vdots \\ x & x & x & x & \cdots & x+\lambda_{M-1} & x-1 \\ x & x & x & x & \cdots & x & x+\lambda_M \end{pmatrix} \tag{23}$$

where $\lambda_1, \ldots, \lambda_M, x \in \mathbb{C}$. Consider x as being variable and denote the determinant of the matrix (23) by $D(x)$. Subtracting in (23) the first row from all other rows, and then the last column from all other columns, one gets a matrix the $1, N$ entry of which is $x - 1$ while all other entries are independent of x. Thus, $D(x)$ is a polynomial of first degree in x and, since $D(0) = \prod_{i=1}^{M} \lambda_i$ and $D(1) = \prod_{i=1}^{M}(1 + \lambda_i)$, one has

$$D(x) = x \prod_{i=1}^{M}(1 + \lambda_i) + (1 - x) \prod_{i=1}^{M} \lambda_i. \tag{24}$$

Now let $m \in \{1, \ldots 4N\}$, $M = 2N$, and $\lambda_i = 2i - 1 - m$ for $i = 1, \ldots, 2N$. If m is odd, then one of the numbers λ_i is equal to zero, but $\prod_{i=1}^{M}(1 + \lambda_i) \neq 0$. If m is even, then one of the numbers $1 + \lambda_i$ is zero, but $\prod_{i=1}^{M} \lambda_i \neq 0$. Hence, in any case,

$$x \prod_{i=1}^{M}(1 + \lambda_i) + (1 - x) \prod_{i=1}^{M} \lambda_i \neq 0$$

whenever $x \notin \{0, 1\}$. This proves our claim (22).

Now the assertion can be obtained as follows. Let $m \in \sigma_\mathcal{B}(Y) \cap \{1, 2, \ldots, 4N\}$. Then, by the local principle,

$$m \in \cup_{x \in \sigma_\mathcal{B}(X)} \sigma_{\mathcal{B}_x}(\Phi_x(Y))$$

whereas, by (22),

$$m \notin \cup_{x \in \sigma_\mathcal{B}(X) \backslash \{0,1\}} \sigma_{\mathcal{B}_x}(\Phi_x(Y)).$$

Hence,

$$m \in \cup_{x \in \sigma_\mathcal{B}(X) \cap \{0,1\}} \sigma_{\mathcal{B}_x}(\Phi_x(Y)).$$

But the algebras $\mathcal{B}_x$ with $x \in \sigma_\mathcal{B}(X) \cap \{0, 1\}$ possess a scalar-valued symbol (Proposition 6 and Theorem 3). Thus, if $m \in \sigma_{\mathcal{B}_{x_0}}(\Phi_{x_0}(Y))$ with a certain $x_0 \in \sigma_\mathcal{B}(X) \cap \{0, 1\}$ then there is an algebra homomorphism G' from $\mathcal{B}_{x_0}$ onto $\mathbb{C}$ with $G'(\Phi_{x_0}(Y)) = m$. Then $G := G' \circ \Phi_{x_0}$ is an algebra homomorphism from $\mathcal{B}$ onto $\mathbb{C}$ with $G(Y) = m$. The restriction of G to the set $\{P, p_1, \ldots, p_{2N}\}$ coincides with one of the mappings G_n introduced in (20) and, by (21), this restriction is just G_m. In other words, G_m extends to an (evidently continuous) algebra homomorphism from $\mathcal{B}$ onto $\mathbb{C}$. ∎

For $m \in \sigma_\mathcal{B}(Y) \cap \{1, 2, \ldots, 4N\}$, let us denote the extension of G_m by G_m again. One easily checks that

$$G_m(X) = \begin{cases} 0 & \text{if } m \text{ is odd} \\ 1 & \text{if } m \text{ is even.} \end{cases}$$

Thus, if $0 \in \sigma_\mathcal{B}(X)$ and m is odd, then the local ideal I_0 lies in the kernel of G_m and consequently, for each $A \in \mathcal{B}$ the number $G_m(A)$ depends on the coset $\Phi_0(A)$ only. So the quotient mapping

$$G'_m : \mathcal{B}_0 \to \mathbb{C}, \quad \Phi_0(A) \mapsto G_m(A)$$

is correctly defined, and it is an algebra homomorphism from $\mathcal{B}_0$ onto $\mathbb{C}$. Analogously, if $0 \in \sigma_{\mathcal{B}}(X)$ and m is even, then

$$G'_m : \mathcal{B}_1 \to \mathbb{C}, \quad \Phi_1(A) \mapsto G_m(A)$$

is a correctly defined and non-trivial algebra homomorphism.

Proposition 8 (a) If $0 \in \sigma_{\mathcal{B}}(X)$ then the set $\{G_m\}$, consisting of all mappings G_m with $m \in \sigma_{\mathcal{B}}(Y) \cap \{1, 2, \ldots, 4N\}$ and m odd, forms a scalar-valued symbol for $\mathcal{B}_0$.
(b) If $1 \in \sigma_{\mathcal{B}}(X)$ then the set $\{G_m\}$, consisting of all mappings G_m with $m \in \sigma_{\mathcal{B}}(Y) \cap \{1, 2, \ldots, 4N\}$ and m even, forms a scalar-valued symbol for $\mathcal{B}_1$.

Proof. The mappings G'_m with m odd (even) are the *only* non-trivial algebra homomorphisms from $\mathcal{B}_0$ (resp. $\mathcal{B}_1$) into $\mathbb{C}$. But since the algebras $\mathcal{B}_0$ (resp. $\mathcal{B}_1$) *possess* a scalar-valued symbol by Theorem 3 and Proposition 6, we conclude that for all $A \in \mathcal{B}$ the coset $\Phi_0(A)$ (resp. $\Phi_1(A)$) is invertible whenever all $G'_m(\Phi_0(A)) = G_m(A)$ with m odd (resp. even) are invertible. ∎

4.4　The N projections theorem

Now we are in a position to state our main result.

Theorem 9 *Let $\mathcal{A}$ be a Banach algebra with identity I. Let $p_1, p_2, \ldots, p_{2N}$ and P be nonzero elements of $\mathcal{A}$ satisfying*

$$p_i \cdot p_j = \delta_{ij} p_i \quad \text{for all } i, j \quad \text{and} \quad p_1 + p_2 + \ldots + p_{2N} = I,$$

where δ_{ij} is the Kronecker delta, and

$$P(p_{2i-1} + p_{2i})P = (p_{2i-1} + p_{2i})P \quad \text{and} \quad Q(p_{2i} + p_{2i+1})Q = (p_{2i} + p_{2i+1})Q$$

for all $i = 1, \ldots, N$, where $Q := I - P$ and $p_{2N+1} := p_1$. Let further $\mathcal{B}$ stand for the smallest closed subalgebra of $\mathcal{A}$ containing the elements P and $p_1, \ldots, p_{2N}$. Then the following assertions hold.
(a) *If $x \in \sigma_{\mathcal{B}}(X) \setminus \{0, 1\}$ where $X := \sum_{i=1}^{N}(p_{2i-1}Pp_{2i-1} + p_{2i}Qp_{2i})$, then the mapping $F_x : \{P, p_1, \ldots, p_{2N}\} \to \mathbb{C}^{2N \times 2N}$ given by*

$$F_x(p_i) = \operatorname{diag}(0, \ldots, 0, 1, 0, \ldots, 0),$$

with the 1 standing at the ith place, and

$$F_x(P) = \operatorname{diag}(1, -1, 1, -1, \ldots, 1, -1) \times$$

$$\times \begin{pmatrix} x & x-1 & x-1 & x-1 & \cdots & x-1 & x-1 \\ x & x-1 & x-1 & x-1 & \cdots & x-1 & x-1 \\ x & x & x & x-1 & \cdots & x-1 & x-1 \\ x & x & x & x-1 & \cdots & x-1 & x-1 \\ \vdots & \vdots & \vdots & \vdots & \ddots & \vdots & \vdots \\ x & x & x & x & \cdots & x & x-1 \\ x & x & x & x & \cdots & x & x-1 \end{pmatrix}$$

extends to a continuous algebra homomorphism from $\mathcal{B}$ onto $\mathbb{C}^{2N \times 2N}$.

(b) If $m \in \sigma_B(Y) \cap \{1, \ldots, 4N\}$ where $Y := \sum_{i=1}^{N} (p_{2i-1}P + p_{2i}Q) + \sum_{i=1}^{2N} (2i-1)p_i$, then the mapping $G_m : \{P, p_1, \ldots, p_{2N}\} \to \mathbb{C}$ defined by

$$G_{4m}(p_i) = \begin{cases} 1 & if \;\; i = 2m \\ 0 & if \;\; i \neq 2m \end{cases}, \qquad G_{4m}(P) = 0,$$

$$G_{4m-1}(p_i) = \begin{cases} 1 & if \;\; i = 2m \\ 0 & if \;\; i \neq 2m \end{cases}, \qquad G_{4m-1}(P) = 1,$$

$$G_{4m-2}(p_i) = \begin{cases} 1 & if \;\; i = 2m - 1 \\ 0 & if \;\; i \neq 2m - 1 \end{cases}, \qquad G_{4m-2}(P) = 1,$$

$$G_{4m-3}(p_i) = \begin{cases} 1 & if \;\; i = 2m - 1 \\ 0 & if \;\; i \neq 2m - 1 \end{cases}, \qquad G_{4m-3}(P) = 0$$

where $m = 1, \ldots, N$, extends to a continuous algebra homomorphism from $\mathcal{B}$ onto $\mathbb{C}$.

(c) An element $B \in \mathcal{B}$ is invertible in $\mathcal{B}$ if and only if the matrices $F_x(B)$ are invertible for all $x \in \sigma_B(X) \setminus \{0, 1\}$ and the numbers $G_m(B)$ are non-zero for all $m \in \sigma_B(Y) \cap \{1, \ldots, 4N\}$.

(d) An element $B \in \mathcal{B}$ is invertible in $\mathcal{A}$ if and only if the matrices $F_x(B)$ are invertible for all $x \in \sigma_A(X) \setminus \{0, 1\}$ and the numbers $G_m(B)$ are non-zero for all $m \in \sigma_A(Y) \cap \{1, \ldots, 4N\}$.

Proof. The proof of assertions (a), (b) and (c) is immediate from the local principle in combination with the description of the local algebras given in the preceding subsection. Concerning the continuity of the mappings F_x and G_m we refer to a general result by Johnson (see, e.g., [28], Chapter 6, Theorem 2.65) stating that an algebra homomorphism from a Banach algebra onto a semi-simple Banach algebra is always continuous.

For a proof of assertion (d) recall that the algebra $\mathcal{B}^0$ is a $(2N)^2$ dimensional module over its center. Thus, Corollary 1.2 in [22] tells us that there is a set $\{\nu_t\}$, $t \in T$, of representations of $\mathcal{B}$ such that $\operatorname{Im} \nu_t = \mathbb{C}^{l \times l}$ with $l = l(t) \leq 2N$ and such that an element B of $\mathcal{B}$ is invertible in $\mathcal{A}$ if and only if $\det \nu_t(B) \neq 0$ for all $t \in T$. The very same arguments as in the proof of assertion (c) entail that each of these representations is of the form F_x (with an $x \in \mathbb{C} \setminus \{0, 1\}$) as defined in Corollary 1 or G_m (with an $m \in \{1, 2, \ldots, 4N\}$) as defined after Proposition 6. Hence, there exist two sets $\xi = \xi(\mathcal{A}, \mathcal{B}) \subset \mathbb{C} \setminus \{0, 1\}$ and $\mu = \mu(\mathcal{A}, \mathcal{B}) \subseteq \{1, 2, \ldots, 4N\}$ such that

$$\sigma_A(B) = \cup_{x \in \xi} \sigma(F_x(B)) \cup \{G_m(B) : m \in \mu\} \tag{25}$$

for all $B \in \mathcal{B}$. We claim that $\xi = \sigma_A(X) \setminus \{0, 1\}$ and $\mu = \sigma_A(Y) \cap \{1, \ldots, 4N\}$.

Since $G_m(X) \in \{0, 1\}$ and $\xi \cap \{0, 1\} = \emptyset$, one has

$$\sigma_A(X) \setminus \{0, 1\} = \cup_{x \in \xi} \sigma(F_x(X)) \cup \{G_m(X) : m \in \mu\} \setminus \{0, 1\} = \cup_{x \in \xi} \{x\} = \xi. \tag{26}$$

For the second claim note that, for any $\lambda \in \mathbb{C}$, the matrix $F_x(Y - \lambda I)$ coincides with the matrix (21) with the λ_i in (21) replaced by $2i - 1 - \lambda$. It follows from the explicit form (24) of the determinant of this matrix that every eigenvalue λ of $F_x(Y)$ solves the equation

$$x \prod_{i=1}^{2N}(2i - \lambda) + (1 - x) \prod_{i=1}^{2N}(2i - 1 - \lambda) = 0.$$

But, if $x \notin \{0, 1\}$, then $\sigma(F_x(Y)) \cap \{1, 2, \ldots, 2N\} = \emptyset$. Thus,

$$\sigma_{\mathcal{A}}(Y) \cap \{1, \ldots, 4N\} = \{G_m(Y) : m \in \mu\} \cap \{1, \ldots, 4N\} = \{m\}_{m \in \mu} = \mu. \tag{27}$$

Now assertion (d) follows immediately from (25), (26) and (27). ∎

Observe that assertion (d) is evident in case the algebra $\mathcal{B}$ is inverse closed in $\mathcal{A}$. However, this is not always satisfied as the following example indicates.

Example. Let $\mathbf{T}$ be the unit circle $\{t \in \mathbb{C} : |t| = 1\}$ and consider the algebra $\mathcal{A}$ of all continuous 2×2 matrix functions on $\mathbf{T}$. If t denotes the identical mapping of $\mathbf{T}$ then

$$P = \begin{pmatrix} t & 1-t \\ t & 1-t \end{pmatrix}, \quad p_1 = \begin{pmatrix} 1 & 0 \\ 0 & 0 \end{pmatrix}, \quad \text{and} \quad p_2 = \begin{pmatrix} 0 & 0 \\ 0 & 1 \end{pmatrix}$$

are elements of $\mathcal{A}$ which satisfy the assumptions of Theorem 9 (with $N = 1$). The element

$$X = p_1 P p_1 + p_2 (I - P) p_2 = \begin{pmatrix} t & 0 \\ 0 & t \end{pmatrix}$$

is invertible in $\mathcal{A}$ but not invertible in $\mathcal{B}$ since the latter algebra consists of matrix functions holomorphic in the unit disk only. ∎

In this connection, let us emphasize an evident consequence of assertions (c) and (d) of the previous theorem.

Corollary 2 *If $\sigma_{\mathcal{B}}(X) = \sigma_{\mathcal{A}}(X)$ and $\sigma_{\mathcal{B}}(Y) = \sigma_{\mathcal{A}}(Y)$, then the algebra $\mathcal{B}$ is inverse closed in $\mathcal{A}$.*

The following additional assertions are often useful.

Proposition 9 *(a) If $0 \notin \sigma_{\mathcal{B}}(X)$ and $1 \notin \sigma_{\mathcal{B}}(X)$ then $\sigma_{\mathcal{B}}(Y) \cap \{1, \ldots, 4N\} = \emptyset$.*
(b) If $0 \in \sigma_{\mathcal{B}}(X)$ and $1 \in \sigma_{\mathcal{B}}(X)$, and if both points are not isolated in $\sigma_{\mathcal{B}}(X)$, then the family (F_x) with $x \in \sigma_{\mathcal{B}}(X)$ is a matrix symbol for $\mathcal{B}$.
(c) If $0 \notin \sigma_{\mathcal{A}}(X)$ and $1 \notin \sigma_{\mathcal{A}}(X)$ then $\sigma_{\mathcal{A}}(Y) \cap \{1, \ldots, 4N\} = \emptyset$.
(d) If $0 \in \sigma_{\mathcal{A}}(X)$ and $1 \in \sigma_{\mathcal{A}}(X)$, and if both points are not isolated in $\sigma_{\mathcal{A}}(X)$, then the family (F_x) with $x \in \sigma_{\mathcal{A}}(X)$ is a matrix symbol for the invertibility of the elements of $\mathcal{B}$ in the algebra $\mathcal{A}$.

Proof. (a) Observe that $G_m(X) \in \{0, 1\}$ in any case. Thus, if $\sigma_{\mathcal{B}}(X) \cap \{0, 1\} = \emptyset$, one-dimensional representations cannot exist.
(b) Let M be a mapping from $\mathbb{C}$ into the set of all subsets of $\mathbb{C}$. Given a sequence $(x_n) \subseteq \mathbb{C}$ with $x_n \to 0$ as $n \to \infty$, we consider the set $L(x_n)$ of all limiting points of all sequences (λ_{x_n}) with $\lambda_{x_n} \in M(x_n)$, and we define the *limiting set* $\lim_{x \to 0} M(x)$ as $\cup L(x_n)$ where the

union is taken over all sequences (x_n) with $x_n \to 0$ but $x_n \neq 0$ for all n. Analogously, we define $\lim_{x \to 1} M(x)$.

The function $x \mapsto F_x(Y)$ is continuous on $\sigma_B(X)$. Due to the continuous dependence of the eigenvalues of a matrix on the matrix itself (see [27], Appendix D), one has

$$\sigma(F_0(Y)) = \lim_{x \to 0} \sigma(F_x(Y))$$

and consequently,

$$\sigma(F_0(Y)) = \lim_{x \to 0} \sigma(\Phi_x(Y)). \tag{28}$$

We claim that

$$\lim_{x \to 0} \sigma(\Phi_x(Y)) \subseteq \sigma(\Phi_0(Y)). \tag{29}$$

To prove (29), we need the following supplement to the local principle. Let the notation be as in Theorem 2.

Proposition 10 *Let $a \in \mathcal{A}$ and suppose $a + I_x$ to be invertible for some x. Then there is an open neighborhood U of x such that the cosets $a + I_y$ are invertible and*

$$\|(a + I_y)^{-1}\| \leq 4\|(a + I_x)^{-1}\| \quad \text{for all } y \in U.$$

Proof of Proposition 10. Set $\phi_x(a) := a + I_x$ and let $\phi_x(a)$ be invertible. Then there is a $b \in \mathcal{A}$ such that $\phi_x(ab - e) = \phi_x(ba - e) = 0$. As shown in [1], or [8], Theorem 1.34, or [24], Theorem 1.5, the mappings

$$y \mapsto \|\phi_y(ab - e)\| \quad \text{and} \quad y \mapsto \|\phi_y(ba - e)\|,$$

defined on the maximal ideal space of $\mathcal{C}$, are upper semi-continuous. Hence,

$$\|\phi_y(ab - e)\| < 1/2 \quad \text{and} \quad \|\phi_y(ba - e)\| < 1/2$$

for all maximal ideals y in a certain neighborhood U' of x. Since

$$\phi_y(a)\phi_y(b) = \phi_y(e) + \phi_y(ab - e) \quad \text{and} \quad \phi_y(b)\phi_y(a) = \phi_y(e) + \phi_y(ba - e),$$

and since $\phi(e)$ is the identity element in $\mathcal{A}/I_y$, this implies (Neumann's series) that $\phi_y(a)$ is invertible in $\mathcal{A}/I_y$ and that

$$\|\phi_y(a)^{-1}\| \leq 2\|\phi_y(b)\| \quad \text{for all } y \in U'.$$

Invoking upper semi-continuity once more we get

$$\|\phi_y(b)\| \leq 2\|\phi_x(b)\| = 2\|\phi_x(a)^{-1}\|$$

for all y in a neighborhood U'' of x, which proves Proposition 10. ∎

Continuation of the proof of Proposition 9. Now, in order to prove our claim (29), assume there are sequences $(x_n) \subseteq \sigma_B(X)$ and (λ_n) with $\lambda_n \in \sigma(\Phi_{x_n}(Y))$ such that $x_n \to$

0, $x_n \neq 0$, $\lambda_n \to \lambda$, but $\lambda \notin \sigma(\Phi_0(Y))$. Then $\Phi_0(Y - \lambda I)$ is invertible and, by Proposition 10, $\Phi_{x_n}(Y - \lambda I)$ is invertible and

$$\|\Phi_{x_n}(Y - \lambda I)^{-1}\| \leq 4\|\Phi_0(Y - \lambda I)^{-1}\|$$

for all n large enough. Thus,

$$\text{dist }(\lambda, \sigma(\Phi_{x_n}(Y))) \geq \frac{1}{4\|\Phi_0(Y - \lambda I)^{-1}\|},$$

which contradicts our assumption since

$$|\lambda - \lambda_{x_n}| \geq \text{dist }(\lambda, \sigma(\Phi_{x_n}(Y))).$$

This proves our claim (29).

From (28) and (29) we see that $\sigma(F_0(Y)) \subseteq \sigma(\Phi_0(Y))$ and, analogously, $\sigma(F_1(Y)) \subseteq \sigma(\Phi_1(Y))$. Hence,

$$\sigma(F_0(Y)) \cup \sigma(F_1(Y)) \subseteq \sigma(\Phi_0(Y)) \cup \sigma(\Phi_1(Y)) \subseteq \sigma_{\mathcal{B}}(Y).$$

Evidently,

$$\sigma(F_0(Y)) \cup \sigma(F_1(Y)) = \{1,\, 2,\, \ldots,\, 4N\},$$

and consequently,

$$\sigma_{\mathcal{B}}(Y) \cap \{1,\, 2,\, \ldots,\, 4N\} = \{1,\, 2,\, \ldots,\, 4N\}.$$

In other words, all possible one-dimensional representations occur.

It remains to observe that, for each $A \in \mathcal{B}$, the matrices $F_0(A)$ and $F_1(A)$ are triangular and that the diagonal of $F_0(A)$ equals $(G_1(A), G_3(A), \ldots, G_{2N-1}(A))$, while the diagonal of $F_1(A)$ is $(G_2(A), G_4(A), \ldots, G_{2N}(A))$.

The proof of assertions (c) and (d) can be given in a completely analogous manner. ∎

5 Examples

5.1 Abstract analogues of singular integral operators

Let T be a non-empty proper subset of $\{1,\, 2,\, \ldots,\, 2N\}$, set $p := \sum_{i \in T} p_i$ and $q := I - p$. Elements of the form $A := pPp + q\ (\in \mathcal{B})$ are called *abstract analogues of singular integrals*. Our first concern is to demonstrate how Theorem 9 can be used to compute the spectrum of abstract singular integrals in case the spectrum of X is known. From Theorem 9 we conclude that this spectrum equals

$$\bigcup_{x \in \sigma_{\mathcal{F}}(X) \backslash \{0,1\}} \sigma(F_x(A)) \cup \bigcup_{m \in \sigma_{\mathcal{F}}(Y) \cap \{1,\ldots,2N\}} \sigma(G_m(A)),$$

where $\mathcal{F} \in \{\mathcal{A}, \mathcal{B}\}$ depends on whether we want to know the spectrum of A in $\mathcal{F} = \mathcal{A}$ or in $\mathcal{F} = \mathcal{B}$. Let us first determine the spectrum of $F_x(A)$ for $x \in \sigma_{\mathcal{F}}(X) \setminus \{0, 1\}$. Let $\lambda \in \mathbb{C}$ and set $D(x) := \det(F_x(A - \lambda I))$. Further, let t, t_o, and t_e refer to the number of the elements of the sets T, $T \cap \{1, 3, \ldots, 2N-1\}$, and $T \cap \{2, 4, \ldots 2N\}$, respectively. Also put $v := t_o - t_e$. Changing the rows and columns of $F_x(A)$ in an appropriate way produces a matrix of the form

$$\begin{pmatrix} F_{11} & 0 \\ 0 & I \end{pmatrix} \tag{30}$$

where F_{11} is a $t \times t$ matrix and I is the $(2N-t) \times (2N-t)$ identity matrix. The determinant $D(x)$ of (30) is a polynomial of first degree in x (see the proof of Proposition 7), and

$$D(0) = (-\lambda)^{t_o}(1-\lambda)^{t_e}(1-\lambda)^{2N-t}, \quad D(1) = (1-\lambda)^{t_o}(-\lambda)^{t_e}(1-\lambda)^{2N-t},$$

the factors $(1-\lambda)^{2N-t}$ coming from the lower right corner in (30) and the other factors resulting from the upper left one. Thus,

$$D(x) = (1-\lambda)^{2N-t}[x(1-\lambda)^{t_o}(-\lambda)^{t_e} + (1-x)(-\lambda)^{t_o}(1-\lambda)^{t_e}].$$

Depending on whether $v > 0$, $v = 0$, or $v < 0$, this equals

$$D(x) = (1-\lambda)^{2N-t}(1-\lambda)^{t_e}(-\lambda)^{t_e}[x(1-\lambda)^v + (1-x)(-\lambda)^v],$$

$$D(x) = (1-\lambda)^{2N-t}(1-\lambda)^{t_o}(-\lambda)^{t_o},$$

$$D(x) = (1-\lambda)^{2N-t}(1-\lambda)^{t_o}(-\lambda)^{t_o}[x(-\lambda)^{|v|} + (1-x)(1-\lambda)^{|v|}],$$

respectively. Thus, if $v = 0$, $\sigma(F_x(A)) = \{0, 1\}$. In case $v > 0$, we have

$$x(1-\lambda)^v + (1-x)(-\lambda)^v = 0 \tag{31}$$

if and only if

$$\left(\frac{\lambda}{\lambda-1}\right)^v = \frac{x}{x-1} \tag{32}$$

(observe that $x \neq 1$ by assumption and that (31) cannot vanish if $\lambda = 1$). Hence, on denoting by $\zeta_0(x), \ldots, \zeta_{v-1}(x)$ the v roots of $x/(x-1)$, we infer from (32) that the spectrum of $F_x(A)$ equals

$$\{0, 1\} \cup \left\{ \frac{\zeta_0(x)}{\zeta_0(x)-1}, \ldots, \frac{\zeta_{v-1}(x)}{\zeta_{v-1}(x)-1} \right\} \quad \text{for} \quad t_e > 0, \tag{33}$$

$$\{1\} \cup \left\{ \frac{\zeta_0(x)}{\zeta_0(x)-1}, \ldots, \frac{\zeta_{v-1}(x)}{\zeta_{v-1}(x)-1} \right\} \quad \text{for} \quad t_e = 0. \tag{34}$$

In the case $v < 0$ we obtain analogously that $\sigma(F_x(A))$ is

$$\{0, 1\} \cup \left\{ \frac{-1}{\zeta_0(x)-1}, \ldots, \frac{-1}{\zeta_{|v|-1}(x)-1} \right\} \quad \text{for} \quad t_o > 0, \tag{35}$$

$$\{1\} \cup \left\{ \frac{-1}{\zeta_0(x)-1}, \ldots, \frac{-1}{\zeta_{|v|-1}(x)-1} \right\} \quad \text{for} \quad t_o = 0. \tag{36}$$

Finally, it is evident that $G_m(A) \in \{0, 1\}$ for all m, and it is clear which value is actually assumed.

The case where $\sigma_{\mathcal{F}}(X)$ is a logarithmic double spiral is of particular interest for applications. For $\delta \in \mathbb{R}$ and $\nu \in (0, 1)$, put $\mathcal{S}_{\delta,\nu} := \{re^{i\delta \log r} e^{2\pi i \nu} : r \in (0, \infty)\}$, and given two distinct numbers $z, w \in \mathbb{C}$, let

$$\mathcal{S}(z, w; \delta; \nu) := \{(w\zeta - z)/(\zeta - 1) : \zeta \in \mathcal{S}_{\delta,\nu}\} \cup \{z, w\}.$$

If $\delta = 0$, then $\mathcal{S}_{\delta,\nu}$ is a ray and hence, $\mathcal{S}(z, w; \delta; \nu)$ is a circular arc between z and w, which degenerates to the line segment $[z, w]$ in case $\nu = 1/2$. If $\delta \neq 0$, then $\mathcal{S}_{\delta,\nu}$ is a logarithmic spiral and therefore $\mathcal{S}(z, w; \delta; \nu)$ is a double spiral wriggling out of z and scrolling up at w. We call a set a *logarithmic double spiral* (between z and w) if it is of the form $\mathcal{S}(z, w; \delta; \nu)$ with some $\delta \in \mathbb{R}$ and $\nu \in (0, 1)$. Notice that segments and circular arcs are logarithmic double spirals in this sense.

Now suppose $\sigma_{\mathcal{F}}(X) = \mathcal{S}(0, 1; \delta; \nu)$ and let $x \in \mathcal{S}(0, 1; \delta; \nu) \setminus \{0, 1\}$. Assume first that $v := t_o - t_e > 0$ and $t_e > 0$. Then $\sigma(F_x(A))$ is given by (33). If $x = re^{i\delta \log r} e^{2\pi i \nu} / (re^{i\delta \log r} e^{2\pi i \nu} - 1)$, then $x/(x-1) = re^{i\delta \log r} e^{2\pi i \nu} = re^{i\delta \log r} e^{2\pi i (\nu + k)}$ and consequently, the v roots of $\zeta^v = x/(x-1)$ are

$$\zeta_k(x) = r^{1/v} e^{i\delta(\log r)/v} e^{2\pi i (\nu + k)/v} = s e^{i\delta \log s} e^{2\pi i (\nu + k)/v}$$

where $s := r^{1/v}$ and $k = 0, \ldots, v - 1$. Thus, if x traces out $\mathcal{S}(0, 1; \delta; \nu) \setminus \{0, 1\}$ then $\zeta_k(x)/(\zeta_k(x) - 1)$ describes the logarithmic double spiral $\mathcal{S}(0, 1; \delta; (\nu + k)/v) \setminus \{0, 1\}$. In the case where $v < 0$ we similarly see that if x ranges over $\mathcal{S}(0, 1; \delta; \nu) \setminus \{0, 1\}$ then $-1/(\zeta_k(x) - 1)$ moves along the logarithmic double spiral $\mathcal{S}(1, 0; \delta; (\nu + k)/|v|) \setminus \{0, 1\}$. Taking into account that spectra are closed we so obtain from (33) – (36) the following result, a concrete version of which was by means of slightly different methods already proved and explicitly stated in [6] (Theorem 2.2.2).

Theorem 10 *Let $\mathcal{F}$ be $\mathcal{A}$ or $\mathcal{B}$. If $\sigma_{\mathcal{F}}(X)$ is the logarithmic double spiral $\mathcal{S}(0, 1; \delta; \nu)$, then the spectrum of $A := pPp + q$ in $\mathcal{F}$ equals*

$$\{0, 1\} \quad for \quad v = 0,$$

$$\bigcup_{k=0}^{v-1} \mathcal{S}(0, 1; \delta; (\nu + k)/v) \quad for \quad v > 0, \tag{37}$$

$$\bigcup_{k=0}^{|v|-1} \mathcal{S}(1, 0; \delta; (\nu + k)/|v|) \quad for \quad v < 0. \tag{38}$$

Clearly, if $\sigma_{\mathcal{F}}(X)$ is a union of logarithmic spirals, $\sigma_{\mathcal{F}}(X) = \bigcup_{\delta \in [\delta_1, \delta_2]} \mathcal{S}(0, 1; \delta; \nu)$, then the conclusion of Theorem 10 remains true with (37) and (38) replaced by

$$\bigcup_{k=0}^{v-1} \bigcup_{\delta \in [\delta_1, \delta_2]} \mathcal{S}(0, 1; \delta; (\nu + k)/v) \quad for \quad v > 0,$$

$$\bigcup_{k=0}^{|v|-1} \bigcup_{\delta \in [\delta_1, \delta_2]} \mathcal{S}(1, 0; \delta; (\nu + k)/|v|) \quad for \quad v < 0.$$

5.2   Applications to singular integral operators

The results of this paper yield a symbol calculus for the closed algebra generated by singular integral operators with piecewise continuous coefficients. This symbol calculus reduces the question of deciding whether an operator is Fredholm to the problem of finding out whether a family of matrix functions consists of invertible matrices only. The simplest nontrivial operator in the algebra mentioned is the Cauchy singular integral operator S_Γ, and we now apply the results of Section 5.1 to the operator S_Γ. To avoid complications that go beyond the scope of this paper, we will not study the problem in full generality.

A *simple arc* is an oriented rectifiable curve in the plane which is homeomorphic to a line segment. The union of finitely many simple arcs each pair of which have at most endpoints in common is called a *composed curve*. If Γ is a composed curve and $z \in \Gamma$, then in a small neighborhood of z the curve is locally comprised by a finite number of simple arcs. This number is referred to as the *multiplicity* of z and is denoted by $t := t(z)$. At a point z of multiplicity t, the curve has $t_o := t_o(z)$ outgoing and $t_e := t_e(z)$ incoming simple arcs, where $t_o \geq 0$, $t_e \geq 0$, and $t_o + t_e = t$. We call $v := v(z) := t_o - t_e$ the *valency* of the point z.

Let Γ be a composed curve. The curve Γ is said to be a *Carleson curve* (or to be *Ahlfors-David regular*) if

$$\sup_{z \in \Gamma} \sup_{\varepsilon > 0} |\Gamma(z, \varepsilon)|/\varepsilon < \infty$$

where $|\Gamma(z, \varepsilon)|$ denotes the (length) measure of the portion $\Gamma(z, \varepsilon) := \{\zeta \in \Gamma : |\zeta - z| < \varepsilon\}$. David [10], [11] proved that the Cauchy singular integral operator S_Γ,

$$(S_\Gamma f)(z) := \lim_{\varepsilon \to 0} \frac{1}{\pi i} \int_{\Gamma \backslash \Gamma(z, \varepsilon)} \frac{f(\zeta)}{\zeta - z} \, d\zeta \qquad (z \in \Gamma),$$

is a well-defined and bounded operator on $L^p(\Gamma)$ $(1 < p < \infty)$ if and only if Γ is a Carleson curve (see also [33] for the "only if" portion). So let us henceforth suppose that Γ is Carleson. Our aim is to determine the essential spectrum of S_Γ on $L^p(\Gamma)$, i.e. to determine the set

$$\sigma_{\text{ess}}(S_\Gamma) := \{\lambda \in \mathbb{C} : S_\Gamma - \lambda I \text{ is not Fredholm on } L^p(\Gamma)\}.$$

Recall that an operator $A \in L(L^p(\Gamma))$ is Fredholm if and only if it is invertible modulo the ideal $K(L^p(\Gamma))$ of the compact operators, that is, if and only if the coset $\pi(A) := A + K(L^p(\Gamma))$ is invertible in the Calkin algebra $L(L^p(\Gamma))/K(L^p(\Gamma))$.

For $a \in L^\infty(\Gamma)$, let $aI : L^p(\Gamma) \to L^p(\Gamma)$ be the multiplication operator $f \mapsto af$. We denote by $C(\Gamma)$ the continuous functions on Γ and by $PC(\Gamma)$ the closure in $L^\infty(\Gamma)$ of all piecewise constant functions on Γ.

An operator $A \in L(L^p(\Gamma))$ is said to be of *local type* if $AcI - cA$ is a compact operator for every $c \in C(\Gamma)$. Clearly, compact operators as well as multiplication operators are of local type. It is well known that S_Γ is also of local type (see [20], Vol. I, Chap. 1, Theorem 4.3 and [3], Lemma 5.1). One can easily see that the set OLT of all operators of local type is a closed subalgebra of $L(L^p(\Gamma))$ and that an operator $A \in$ OLT is Fredholm if and only if the coset $\pi(A)$ is invertible in $\pi(\text{OLT}) := \text{OLT}/K(L^p(\Gamma))$. For $z \in \Gamma$, let J_z

be the smallest closed two-sided ideal of $\pi(\text{OLT})$ containing $\{\pi(cI) : c \in C(\Gamma), c(z) = 0\}$. Put $\mathcal{A}_z := \pi(\text{OLT})/J_z$ and denote the coset $\pi(A) + J_z$ by $\pi_z(A)$. Allan's local principle (Theorem 2) with $\mathcal{A} := \text{OLT}$ and $\mathcal{C} := \{\pi(cI) : c \in C(\Gamma)\}$ so implies that an operator $A \in \text{OLT}$ is Fredholm on $L^p(\Gamma)$ if and only if $\pi_z(A)$ is invertible in $\mathcal{A}_z$ for every $z \in \Gamma$.

The algebra $\mathcal{A}_z$ contains $P := \pi_z(P_\Gamma)$ and $p_j := \pi_z(\chi_j I)$ $(j = 1, \ldots, t)$ where $P_\Gamma = (I + S_\Gamma)/2$ and $\chi_1, \ldots, \chi_t$ are the characteristic functions of the t connected components of $(\Gamma \cap U) \setminus \{z\}$ (U sufficiently small). Let $\mathcal{B}_z$ stand for the closed subalgebra of $\mathcal{A}_z$ which is generated by P, p_1, p_2, $\ldots$, p_t.

Clearly, $\mathcal{B}_z$ is of much better structure than $\mathcal{A}_z$. It is obvious that p_1, p_2, $\ldots$, p_t are idempotents whose sum is the identity and which satisfy $p_i p_j = \delta_{ij} p_i$. Unfortunately, in general P is not an idempotent, by virtue of which Theorem 9 is not immediately applicable. We therefore construct two other "local algebras" $\mathcal{A}_z^* \supset \mathcal{B}_z^*$ and identify $\pi_z(P_\Gamma)$ as an abstract singular integral (in the sense of Section 5.1) in these algebras.

A counter-clockwise oriented curve homeomorphic to a circle is called a *Jordan curve*. A composed curve consisting of a finite number $N \geq 2$ of Jordan curves which have exactly one point in common is referred to as a *flower*. All points of a flower have valency zero, exactly one point, the center of the flower, has multiplicity $2N$, while the remaining points have multiplicity 2.

Suppose Γ^* is both a flower and a Carleson curve. Denote the center of Γ^* by z, and let $\mathcal{A}_z^*$ and $\mathcal{B}_z^*$ be the algebras that arise from the above construction with Γ^* in place of Γ. If $\varepsilon > 0$ is sufficiently small, then the connected component of the portion $\Gamma^*(z, \varepsilon)$ containing z may be written in the form

$$\bigcup_{i=1}^{N} (\Gamma_{2i}^* \cup \Gamma_{2i-1}^*) \tag{39}$$

where Γ_{2i}^* and Γ_{2i-1}^* $(i = 1, \ldots, N)$ are outgoing and incoming simple arcs, respectively. The algebra $\mathcal{B}_z^*$ is generated by $P := \pi_z(P_{\Gamma^*})$ and $p_j := \pi_z(\chi_j I)$ $(j = 1, \ldots, 2N)$ where χ_j is the characteristic function of Γ_j^*. One can show that now P is idempotent and that (3), (4) hold. Thus, Theorem 9 is applicable to the pair of algebras $\mathcal{A} := \mathcal{A}_z^*$, $\mathcal{B} := \mathcal{B}_z^*$, and we may use the results of Section 5.1 to compute the local spectrum of the singular integral operator

$$A := (\sum_{j \in T} \chi_j I) P_{\Gamma^*} (\sum_{j \in T} \chi_j I) + (\sum_{j \notin T} \chi_j I), \tag{40}$$

i.e. the spectrum of the abstract singular integral $\pi_z(A) = pPp + q$, where T is a non-empty proper subset of $\{1, 2, \ldots, 2N\}$ (note that $\sigma_{\mathcal{A}_z^*} \pi_z(A) = \sigma_{\mathcal{B}_z^*} \pi_z(A) = \{0, 1\}$ for $T = \{1, 2, \ldots, 2N\}$).

What we need is the spectrum $\sigma_{\mathcal{A}_z^*}(X)$ of

$$X := \sum_{i=1}^{N} (p_{2i-1} P p_{2i-1} + p_{2i} Q p_{2i}).$$

It is easily seen that

$$\sigma_{\mathcal{A}_z^*}(X) = \bigcup_{i=1}^{N} \sigma_{\mathcal{A}_z^*}(p_{2i-1} P p_{2i-1} + p_{2i} Q p_{2i}),$$

which reduces the problem to finding the local spectrum of singular integral operators with piecewise continuous coefficients on Carleson Jordan curves. These spectra were completely determined in [4]. In order to illustrate the basic phenomena, let us for the sake of simplicity assume that the arcs Γ_j^* of the flower may be parametrized as

$$\Gamma_j^* = \{\zeta = z + re^{i(\phi(r)+b_j(r))} : 0 \leq r < \varepsilon\} \quad (j = 1, \ldots, 2N) \tag{41}$$

where $\varepsilon \in (0, 1)$, ϕ is a real-valued function of the form

$$\phi(r) = h(\log(-\log r))(-\log r)$$

with a function $h \in C^2(\mathbb{R})$ for which h, h', h'' are bounded on $\mathbb{R}$, and b_j are real-valued functions in $C^1[0, \varepsilon]$ such that

$$0 \leq b_1(r) < b_2(r) < \ldots < b_{2N} < 2\pi \quad \text{for} \quad r \in (0, \varepsilon).$$

We remark that the ansatz $h(\log(-\log r))$ guarantees that $r\dot\phi(r)$ is bounded for $r \in (0, 1)$, which in turn implies that Γ^* is a Carleson curve (see e.g. [4]). Clearly, every piecewise C^1 flower can be parametrized in this way with $h = 0$. If h and $b_1, \ldots, b_{2N}$ are constant functions, then Γ locally consists of $2N$ logarithmic spirals scrolling up at z. The choice

$$h(x) = \delta + \mu \sin x, \quad b_j(r) = b_j = \text{constant} \tag{42}$$

gives $2N$ "oscillating spirals" terminating at z. In accordance with [4], the spirality indices δ_z^- and δ_z^+ of Γ^* at z are defined by

$$\delta_z^- := \liminf_{x \to +\infty} (h(x) + h'(x)) \ (= \liminf_{r \to 0} (-r\dot\phi(r))),$$

$$\delta_z^+ := \limsup_{x \to +\infty} (h(x) + h'(x)) \ (= \limsup_{r \to 0} (-r\dot\phi(r))).$$

In case $h = 0$, i.e. for piecewise C^1 flowers, we have $\delta_z^- = \delta_z^+ = 0$. If h is as in (42), then

$$\delta_z^- = \delta - |\mu|\sqrt{2}, \quad \delta_z^+ = \delta + |\mu|\sqrt{2}.$$

The symbol calculus of [4] implies that

$$\sigma_{A_z^*}(p_{2i-1}Pp_{2i-1} + p_{2i}Qp_{2i}) = \bigcup_{\delta \in [\delta_z^-, \delta_z^+]} S(0, 1; \delta; 1/p)$$

for every $i = 1, \ldots, N$, whence

$$\sigma_{A_z^*}(X) = \bigcup_{\delta \in [\delta_z^-, \delta_z^+]} S(0, 1; \delta; 1/p). \tag{43}$$

The set on the right of (43) is a union of logarithmic double spirals; such sets were called skew spiralic horns in [4] and are logarithmic leaves with a separating point in the terminology of [5]. Clearly, for piecewise C^1 flowers or, more generally, for flowers whose spirality indices are both zero, the set (43) is a circular arc.

Since the set (43) does not separate the complex plane (i.e., does not contain "holes"), a standard result from the theory of Banach algebras implies that $\sigma_{A_z^*}(X) = \sigma_{B_z^*}(X)$.

By a *substar* of the flower Γ^* we understand a set Γ of the form $\Gamma = \cup_{j \in T}\Gamma_j^*$ where the simple arcs Γ_j^* are given by (39) and T is a non-empty subset of $\{1, 2, \ldots, 2N\}$. Obviously, the operator $P_\Gamma = (I + S_\Gamma)/2$ may be identified with the singular integral operator (40). Thus, combining Theorem 10 (and the remark after it) with (43) we arrive at the following result for $S_\Gamma = 2P_\Gamma - I$.

Theorem 11 *Let* Γ^* *be a Carleson flower with the center* z *and let* Γ *be a substar of* Γ^*. *Denote the valency of* $z \in \Gamma$ *by* $v(z)$ *and let* δ_z^-, δ_z^+ *be the spirality indices of* z. *Then the local spectra* $\sigma_{B_z}(S_\Gamma)$ *and* $\sigma_{A_z}(S_\Gamma)$ *of* S_Γ *at* z *coincide and are equal to*

$$\{-1, 1\} \quad \text{if} \quad v(z) = 0,$$

$$\bigcup_{k=0}^{v(z)-1} \bigcup_{\delta \in [\delta_z^-, \delta_z^+]} S(-1, 1; \delta; (1/p + k)/v(z)) \quad \text{if} \quad v(z) > 0,$$

$$\bigcup_{k=0}^{|v(z)|-1} \bigcup_{\delta \in [\delta_z^-, \delta_z^+]} S(1, -1; \delta; (1/p + k)/|v(z)|) \quad \text{if} \quad v(z) < 0.$$

We finally return to the case of an arbitrary composed Carleson curve Γ. A point $z \in \Gamma$ is called a *bud* if there exists an $\varepsilon > 0$ such that the connected component of $\Gamma \cap \{\zeta \in \mathbb{C} : |\zeta - z| < \varepsilon\}$ containing z is a substar of some Carleson flower. It is easily seen that composed Carleson curves which may locally be "parametrized by the radius" as in (41) consist entirely of buds. We conjecture that every point of a composed Carleson curve is a bud; three of the authors are planning to devote a forthcoming paper to this problem. The following theorem is immediate from the preceding discussion.

Theorem 12 *Let* Γ *be a composed Carleson curve each point of which is a bud. Then the essential spectrum of* S_Γ *on* $L^p(\Gamma)$ *is*

$$\sigma_{\text{ess}}(S_\Gamma) = \bigcup_{z \in \Gamma} \sigma(S_\Gamma)$$

where $\sigma(S_\Gamma) := \sigma_{A_z}(S_\Gamma) = \sigma_{B_z}(S_\Gamma)$ *is as in Theorem 11.*

6   Miscellanea

6.1   Other partitions of unity into projections

Besides the (obvious) partition of unity into projections (p_i), there are other partitions in $\mathcal{B}$. Set, for example,

$$w_{2i} = (p_{2i} + p_{2i+1})Q \quad \text{and} \quad w_{2i-1} = (p_{2i-1} + p_{2i})P.$$

Proposition 11 *The set* $(w_i)_{i=1}^{2N}$ *is a partition of unity into projections in* $\mathcal{B}$.

The consideration of this partition is motivated by [22]. To get another one, set $a_i = p_{2i-1} + p_{2i}$ and

$$q_i = \begin{cases} a_i P a_i & \text{if} \quad i = 1, \ldots, N \\ a_{2N+1-i} Q a_{2N+1-i} & \text{if} \quad i = N+1, \ldots, 2N. \end{cases}$$

Proposition 12 *The set* $(q_i)_{i=1}^{2N}$ *is a partition of unity into projections in* $\mathcal{B}$.

The proofs of the preceding propositions are straightforward.

Clearly, the use of other partitions of unity into projections than (p_i) yields other descriptions of the local algebras at points $x \in \sigma_B(X) \setminus \{0, 1\}$. We shall illustrate this for the partition (w_i). Here are the analogues of Propositions 2 and 3.

Proposition 13 *Considered as module over its center, the algebra $\mathcal{B}^0$ is generated by the $(2N)^2$ elements $(w_i)_{i=1}^{2N}$ and $(w_i Y w_j)_{i,j=1}^{2N}$ with $i \neq j$. To be more precise, given $A \in \mathcal{B}^0$, there are polynomials R_{ij} in X such that*

$$A = \sum_{i=1}^{2N} R_{ii}(X) w_i + \sum_{\substack{i,j=1 \\ i \neq j}}^{2N} R_{ij}(X) w_i Y w_j. \tag{44}$$

Proposition 14 *(a) If $l > j > i$ or $j > i > l$ or $i > l > j$ then*

$$w_i Y w_j Y w_l = (X - I) \, w_i Y w_l.$$

(b) If $l > i > j$ or $j > l > i$ or $i > j > l$ then

$$w_i Y w p_j Y w_l = X \, w_i Y w_l.$$

(c) If $i = l$ and $i \neq j$ then

$$w_i Y w_j Y w_i = X(X - I) \, w_i.$$

The proofs are omitted.

As in the proof of Proposition 5 one can show that, for $x \in \sigma_B(X) \setminus \{0, 1\}$, the elements

$$b_{ij} = \begin{cases} (x-1)^{-1} \Phi_x(w_i Y w_j) & \text{if } \quad i < j \\ x^{-1} \Phi_x(w_i Y w_j) & \text{if } \quad i > j \\ \Phi_x(w_i) & \text{if } \quad i = j \end{cases}$$

form a basis of the linear space $\mathcal{B}_x$ which, moreover, satisfies $b_{ij} b_{kl} = \delta_{jk} b_{il}$. Thus, there is an algebra homomorphism $\Psi'_x : \mathcal{B}_x \to \mathbb{C}^{2N \times 2N}$ with $\Psi'_x(b_{ij}) = E_{ij}$. Set $H_x := \Psi'_x \circ \Phi_x$. Then

$$H_x(w_i) = \text{diag} \, (0, \ldots, 0, 1, 0, \ldots, 0) \tag{45}$$

the 1 standing at the ith place, and

$$H_x(Y) = \begin{pmatrix}
x & x-1 & x-1 & x-1 & \cdots & x-1 & x-1 \\
x & x & x-1 & x-1 & \cdots & x-1 & x-1 \\
x & x & x & x-1 & \cdots & x-1 & x-1 \\
x & x & x & x & \cdots & x-1 & x-1 \\
\vdots & \vdots & \vdots & \vdots & \ddots & \vdots & \vdots \\
x & x & x & x & \cdots & x & x-1 \\
x & x & x & x & \cdots & x & x
\end{pmatrix}. \tag{46}$$

Here is the analogue of Corollary 1.

Corollary 3 *Let* $x \in \sigma_B(X) \setminus \{0, 1\}$. *Then*

$$H_x(P) = \text{diag } (1, 0, 1, 0, \ldots, 1, 0),$$

and $H_x(p_i)$ *is the matrix with* $(i-1)$*st column*

$$(1 - x, 1 - x, \ldots, 1 - x, -x, -x, \ldots, -x),$$

ith column

$$(x - 1, x - 1, \ldots, x - 1, x, x, \ldots, x)$$

(the entries $1 - x$ *and* $x - 1$ *both appear* $i - 1$ *times), and all other columns are zero.*

The proof is based on checking that $P = \sum_{i=1}^{N} w_{2i-1}$, that

$$p_i = (I - X)w_{i-1} + Xw_i - \sum_{\substack{k=1 \\ k \neq i-1}}^{2N} w_k Y w_{i-1} + \sum_{\substack{k=1 \\ k \neq i}}^{2N} w_k Y w_i,$$

and on employing (45) and (46). ∎

We renounce to give an explicit formulation of Theorem 9 based on the partition (w_i).

6.2   Other indicator elements

The elements X and Y indicate which matrix representations of the algebra B actually appear. While X is distinguished by the fact that it belongs to the center of B, there is some latitude to choose Y. For example, one can show that the element

$$Z := P + \sum_{i=1}^{2N} 2ip_i$$

can play the role of the Y in the determination of all one-dimensional representations. Indeed, consider the mappings K_m given by $K_{2i}(P) = 0$, $K_{2i+1}(P) = 1$,

$$K_{2i}(p_i) = \begin{cases} 1 & \text{if } i = j \\ 0 & \text{if } i \neq j \end{cases}, \qquad K_{2i+1}(p_i) = \begin{cases} 1 & \text{if } i = j \\ 0 & \text{if } i \neq j \end{cases},$$

where $i = 1, \ldots, 1N$. The analogue of Proposition 7 reads as follows.

Proposition 15 *If* $m \in \sigma_B(Z) \cap \{2, 3, 4, \ldots, 4N + 1\}$ *then the complex-valued mapping* K_m *defined on* $\{P, p_1, \ldots, p_{2N}\}$ *extends to an algebra homomorphism from* B *onto* $\mathbb{C}$.

The proof runs as that of Proposition 7.

The following observation is often useful in order to determine the spectrum of X. For $i = 1, 2, \ldots, 2N$ let B_i denote the algebra $p_i B p_i = \{p_i b p_i, b \in B\}$.

Proposition 16 *If* $\{0, 1\} \subseteq \sigma_{B_i}(p_i X p_i)$ *for some* i *then* $\sigma_B(X) = \sigma_{B_i}(p_i X p_i)$.

Proof. Since (p_i) is a partition of unity into projections and X is in the center of $\mathcal{B}$, we have

$$\sigma_\mathcal{B}(X) = \bigcup_{j=1}^{2N} \sigma_{\mathcal{B}_j}(p_j X p_j). \tag{47}$$

We claim that

$$\sigma_{\mathcal{B}_j}(p_j X p_j) \setminus \{0, 1\} = \sigma_{\mathcal{B}_k}(p_k X p_k) \setminus \{0, 1\} \tag{48}$$

for all j, $k = 1, \ldots, 2N$. Indeed, let $\lambda \notin \sigma_{\mathcal{B}_j}(p_j X p_j)$. Then there is an a in $\mathcal{B}$ such that

$$p_j a p_j (p_j X p_j - \lambda p_j) = p_j.$$

Multiplying this identity from the left hand side by $p_k P p_j$ and from the right hand side by $p_j P p_k$ with some $k \neq j$ yields

$$p_k P p_j a p_j (p_j X p_j - \lambda p_j) p_j P p_k = p_k P p_j P p_k$$

and, by Proposition 3,

$$p_k P p_j a p_j P p_k (p_k X p_k - \lambda p_k) = (-1)^{j-k} X(X - I) p_k.$$

The element $p_k X p_k$ lies in the center of the algebra $\mathcal{B}_k$. Thus, localizing $\mathcal{B}_k$ over its smallest closed subalgebra which contains p_k and $p_k X p_k$ via Theorem 2 yields that at the point $\mu \in \sigma_{\mathcal{B}_k}(p_k X p_k)$ (where Ω_μ refers to the canonical homomorphism from $\mathcal{B}_k$ onto its local algebra at μ) the following equality holds:

$$(\mu - \lambda)\Omega_\mu(p_k P p_j a p_j P p_k) = (-1)^{j-k}\mu(\mu - 1)\Omega_\mu(I).$$

Thus, if $\mu \notin \{0, 1\}$ then $\mu - \lambda \neq 0$ and, hence, $\lambda \notin \sigma_{\mathcal{B}_k}(p_k X p_k) \setminus \{0, 1\}$. This gives our claim (48). Clearly, (48) in combination with (47) proves the assertion. ∎

6.3 The two projections theorem

If $N = 1$ in Theorem 9, then the partition (p_i) consists of two elements p_1 and p_2 with $p_2 = I - p_1$. Moreover, the axioms (3) and (4) reduce to $P^2 = P$ and $Q^2 = Q$, respectively. Thus, $\mathcal{B}$ is nothing but the (general) algebra generated by two idempotents (P and p_1) and the identity.

Obviously, there are some differences between the specification of Theorem 9 to the case $N = 1$ and Theorem 1. In case $N = 1$, set $p := p_1$ and $q := P$ in Theorem 1.

The first difference concerns the indicator element for the one-dimensional representations. In Theorem 1, it is the element $p + 2q$, whereas it is

$$Y = pq + (I - p)(I - q) + p + 3(I - p) = 2pq + 4I - 3p - q$$

in Theorem 9, which seems to be much more complicated. But if Y is replaced by the element Z from preceding remark, then

$$Z = q + 2p + 4(I - p) = q - 2p + 4I$$

which is as simple as $P + 2q$.

The second difference concerns the explicit form of the 2×2 matrices. In Theorem 1, the matrix associated with q at the point $x \in \sigma_B(X) \setminus \{0, 1\}$ is

$$\begin{pmatrix} x & \sqrt{x(1-x)} \\ \sqrt{x(1-x)} & 1-x \end{pmatrix}, \tag{49}$$

whereas the corresponding matrix from Theorem 9 is $\begin{pmatrix} x & 1-x \\ x & 1-x \end{pmatrix}$. But, for $x \in \mathbb{C} \setminus \{0, 1\}$,

$$\begin{pmatrix} x & \sqrt{x(1-x)} \\ \sqrt{x(1-x)} & 1-x \end{pmatrix} = \begin{pmatrix} \sqrt[4]{\frac{x}{1-x}} & 0 \\ 0 & \sqrt[4]{\frac{1-x}{x}} \end{pmatrix} \begin{pmatrix} x & 1-x \\ x & 1-x \end{pmatrix} \begin{pmatrix} \sqrt[4]{\frac{1-x}{x}} & 0 \\ 0 & \sqrt[4]{\frac{x}{1-x}} \end{pmatrix}$$

and, moreover,

$$\begin{pmatrix} 1 & 0 \\ 0 & 0 \end{pmatrix} = \begin{pmatrix} \sqrt[4]{\frac{x}{1-x}} & 0 \\ 0 & \sqrt[4]{\frac{1-x}{x}} \end{pmatrix} \begin{pmatrix} 1 & 0 \\ 0 & 0 \end{pmatrix} \begin{pmatrix} \sqrt[4]{\frac{1-x}{x}} & 0 \\ 0 & \sqrt[4]{\frac{x}{1-x}} \end{pmatrix}$$

where $\sqrt[4]{\frac{x}{1-x}}$ is any number with $\left(\sqrt[4]{\frac{x}{1-x}} \right)^4 = \frac{x}{1-x}$, and $\sqrt[4]{\frac{1-x}{x}}$ is $\left(\sqrt[4]{\frac{x}{1-x}} \right)^{-1}$. Thus, both representations are equivalent.

6.4   Symmetric representations in case $N > 1$

In case $N = 1$, the $2N \times 2N$ dimensional representations of p_i and P can be chosen to be symmetric (and even self-adjoint in case $\sigma_B(X) \subseteq \mathbb{R}$, which is of particular interest in many applications) (compare the matrix (49)). This observation suggests the following question: Is there a symmetric representation in case $N > 1$, too? To be more precise, is there an invertible matrix D such that again

$$D^{-1} F_x(p_i) D = \text{diag } (0, \ldots, 0, 1, 0, \ldots, 0) = F_x(p_i) \tag{50}$$

(which is desirable for symmetry) but, moreover,

$$D^{-1} F_x(P) D = (D^{-1} F_x(P) D)^T \tag{51}$$

where the ^{T} marks the transposed matrix?

In general, a matrix D with these properties does not exist in case $N > 1$! Indeed, (50) involves that D itself is a diagonal matrix, say $D = \text{diag } (d_1, d_2, \ldots, d_{2N})$. Then identity (51) yields for the 2,1, the 4,1, and the 4,2 entry

$$d_2 d_1^{-1} x = d_1 d_2^{-1}(1-x), \tag{52}$$

$$d_4 d_1^{-1} x = d_1 d_4^{-1}(1-x), \tag{53}$$

$$d_4 d_2^{-1}(-x) = d_2 d_4^{-1}(1-x), \tag{54}$$

respectively. Identities (52) and (53) imply that $d_2^2 = d_4^2$, which contradicts (54).

6.5   Coefficient algebras

Again let $\mathcal{A}$ be a Banach algebra with identity I, let $(p_i)_{i=1}^{2N}$ be a partition of unity into projections and P be an idempotent in $\mathcal{A}$ such that the axioms (3) and (4) hold. The smallest closed subalgebra of $\mathcal{A}$ containing the partition (p_i) as well as the element P will be denoted by $\mathcal{B}$ again. Suppose $\mathcal{G}$ is a closed subalgebra of $\mathcal{A}$ containing I and having the property that

$$p_i g = g p_i \quad \text{and} \quad gP = Pg \quad \text{for all} \quad i = 1, \ldots, 2N \quad \text{and} \quad g \in \mathcal{G}.$$

The algebra $\mathcal{G}$ is referred to as a *coefficient algebra*. As in [18], one can derive a version of Theorem 9 which provides us with an invertibility symbol for the smallest closed subalgebra $\mathcal{C}$ of $\mathcal{A}$ which contains the partition (p_i), the idempotent P, and the algebra $\mathcal{G}$. Here is the formulation of this version under the stronger condition that $\mathcal{G}$ be a simple algebra.

Theorem 13 *Let $\mathcal{C}$ be as above and let $\mathcal{G}$ be simple.*

(a) If $x \in \sigma_\mathcal{B}(X) \setminus \{0, 1\}$, then the mapping

$$F_x : \{P, p_1, \ldots, p_{2N}\} \cup \mathcal{G} \to \mathcal{G}^{2N \times 2N}$$

given by

$$F_x(p_i) = \operatorname{diag}(0, \ldots, 0, I, 0, \ldots, 0),$$

the I standing at the ith place,

$$F_x(P) = \operatorname{diag}(I, -I, I, -I, \ldots, I, -I) \times$$

$$\times \begin{pmatrix}
x & x-1 & x-1 & x-1 & \cdots & x-1 & x-1 \\
x & x-1 & x-1 & x-1 & \cdots & x-1 & x-1 \\
x & x & x & x-1 & \cdots & x-1 & x-1 \\
x & x & x & x-1 & \cdots & x-1 & x-1 \\
\vdots & \vdots & \vdots & \vdots & \ddots & \vdots & \vdots \\
x & x & x & x & \cdots & x & x-1 \\
x & x & x & x & \cdots & x & x-1
\end{pmatrix},$$

$$F_x(g) = \operatorname{diag}(g, g, \ldots, g),$$

extends to a continuous algebra homomorphism from $\mathcal{C}$ onto $\mathcal{G}^{2N \times 2N}$.

(b) If $m \in \sigma_\mathcal{B}(Y) \cap \{1, \ldots, 4N\}$, then the mapping

$$G_m : \{P, p_1, \ldots, p_{2N}\} \cup \mathcal{G} \to \mathcal{G}$$

given by

$$G_{4m}(p_i) = \begin{cases} I & \text{if} \ \ i = 2m \\ 0 & \text{if} \ \ i \neq 2m \end{cases}, \qquad G_{4m}(P) = 0,$$

$$G_{4m-1}(p_i) = \begin{cases} I & \text{if} \ \ i = 2m \\ 0 & \text{if} \ \ i \neq 2m \end{cases}, \qquad G_{4m-1}(P) = I,$$

$$G_{4m-2}(p_i) = \begin{cases} I & \text{if } i = 2m-1 \\ 0 & \text{if } i \neq 2m-1 \end{cases}, \quad G_{4m-2}(P) = I,$$

$$G_{4m-3}(p_i) = \begin{cases} I & \text{if } i = 2m-1 \\ 0 & \text{if } i \neq 2m-1 \end{cases}, \quad G_{4m-3}(P) = 0,$$

where $m = 1, \ldots, N$, and by $G_m(g) = g$ extends to a continuous algebra homomorphism from $\mathcal{C}$ onto $\mathcal{G}$.

(c) An element $C \in \mathcal{C}$ is invertible in $\mathcal{C}$ if and only if the matrices $F_x(C)$ are invertible for all $x \in \sigma_B(X) \setminus \{0, 1\}$ and the elements $G_m(C)$ are invertible for all $m \in \sigma_B(Y) \cap \{1, \ldots, 4N\}$.

(d) An element $C \in \mathcal{C}$ is invertible in $\mathcal{A}$ if and only if the matrices $F_x(C)$ are invertible for all $x \in \sigma_A(X) \setminus \{0, 1\}$ and the elements $G_m(C)$ are invertible for all $m \in \sigma_A(Y) \cap \{1, \ldots, 4N\}$.

Observe that the conditions of the theorem are satisfied if, for example, $\mathcal{G}$ is the algebra $\mathbb{C}^{n \times n}$ which yields just the matrix version of Theorem 9.

References

[1] G. R. ALLAN, Ideals of vector valued functions. – Proc. London Math. Soc. **18**(1968), 3, 193 – 216.

[2] A. BÖTTCHER, Toeplitz operators with piecewise continuous symbols - a neverending story? – Jahresber. der Deutschen Mathematiker-Vereinigung **97**(1995), 115 – 129.

[3] A. BÖTTCHER, YU.I. KARLOVICH, Toeplitz and singular integral operators on Carleson curves with logarithmic whirl points. – IEOT **22**(1995), 127 – 161.

[4] A. BÖTTCHER, YU.I. KARLOVICH, Toeplitz and singular integral operators on general Carleson curves. – to appear.

[5] A. BÖTTCHER, YU.I. KARLOVICH, Toeplitz operators with PC symbols on general Carleson Jordan curves with arbitrary Muckenhoupt weights. – to appear.

[6] A. BÖTTCHER, YU.I. KARLOVICH, V.S. RABINOVICH, Emergence, persistence, and disappearance of logarithmic spirals in the spectra of singular integral operators. – to appear.

[7] A. BÖTTCHER, S. ROCH, B. SILBERMANN, I. SPITKOVSKY, A Gohberg-Krupnik-Sarason symbol calculus for algebras of Toeplitz, Hankel, Cauchy, and Carleman operators. – Operator Theory: Advances and Applications **48**(1990), 189-234.

[8] A. BÖTTCHER, B. SILBERMANN, Analysis of Toeplitz operators. – Akademie-Verlag, Berlin 1989 and Springer-Verlag, Berlin, Heidelberg, New York 1990.

[9] A. BÖTTCHER, I. SPITKOVSKY, Pseudodifferential operators with heavy spectrum. – IEOT **19**(1994), 251 – 269.

[10] G. DAVID, L'integrale de Cauchy sur les courbes rectifiables. – Prepublication Univ. Paris-Sud, Dept. Math. 82T05, 1982.

[11] G. DAVID, Opérateurs intégraux singuliers sur certaines courbes du plan complexe. – Ann. Sci. École Norm. Super. **17**(1984), 157 – 189.

[12] J. DIEUDONNÉ, Grundzüge der modernen Analysis 2. – Deutscher Verlag der Wissenschaften, Berlin 1975.

[13] R.G. DOUGLAS, Banach algebra techniques in operator theory. – Academic Press, New York 1972.

[14] R.G. DOUGLAS, Local Toeplitz operators. – Proc. London Math. Soc., 3rd Ser., **36**(1978), 234 - 276.

[15] T. EHRHARDT, S. ROCH, B. SILBERMANN, Symbol calculus for singular integrals with operator-valued PQC coefficients. – Proceedings of the German-Israeli Workshop 1995, to appear.

[16] T. EHRHARDT, S. ROCH, B. SILBERMANN, Finite section method for singular integrals with operator-valued PQC coefficients. – Proceedings of the German-Israeli Workshop 1995, to appear.

[17] T. FINCK, S. ROCH, Banach algebras with matrix symbol of bounded order. – IEOT **18**(1994), 427 – 434.

[18] T. FINCK, S. ROCH, B. SILBERMANN, Two projection theorems and symbol calculus for operators with massive local spectra. – Math. Nachr. **162**(1993), 167 – 185.

[19] I. GOHBERG, N. KRUPNIK, On singular integral operators on a composed curve. Soobshch. Akad. Nauk Gruz. SSR **64**(1971), 21-24 (Russian).

[20] I. GOHBERG, N. KRUPNIK, Introduction to the theory of one-dimensional singular integral operators, Vols. I and II. – Birkhäuser Verlag, Basel, Boston, Stuttgart 1992.

[21] I. GOHBERG, N. KRUPNIK, Extension theorems for invertibility symbols in Banach algebras. – IEOT **15**(1992), 991 - 1010.

[22] I. GOHBERG, N. KRUPNIK, Extension theorems for Fredholm and invertibility symbols. – IEOT **16**(1993), 514 – 529.

[23] I. GOHBERG, N. KRUPNIK, I. SPITKOVSKY, Banach algebras of singular integral operators with piecewise continuous coefficients. General contour and weight. – IEOT **17**(1993), 322 - 337.

[24] R. HAGEN, S. ROCH, B. SILBERMANN, Spectral theory of approximation methods for convolution equations. – Birkhäuser Verlag, Basel, Boston, Berlin 1995.

[25] R. HAGEN, B. SILBERMANN, A Banach algebra approach to the stability of projection methods for singular integral equations. – Math. Nachr. **140**(1989), 285 – 297.

[26] P.R. HALMOS, Two subspaces. – Trans. Amer. Math. Soc. **144**(1969), 381 – 389.

[27] R.A. HORN, C.A. JOHNSON, Matrix analysis. – Cambridge University Press, Cambridge 1986.

[28] A.YA. KHELEMSKII, Banach and semi-normed algebras: general theory, representations, homology. – Nauka, Moscow 1989 (Russian).

[29] N. KRUPNIK, Banach algebras with symbol and singular integral operators. – Operator Theory: Advances and Applications, Vol. 26, Birkhäuser Verlag, Basel 1987.

[30] N. KRUPNIK, Minimal number of idempotent generators of matrix algebras over arbitrary field. – Communications in Algebra **20**(1992), 3251 – 3257.

[31] N. KRUPNIK, S. ROCH, B. SILBERMANN, On C^*-algebras generated by idempotents. – J. Funct. Anal. (to appear).

[32] N. KRUPNIK, E. SPIGEL, Invertibility symbols for a Banach algebra generated by two idempotents and a shift. – IEOT **17**(1993), 567 – 578.

[33] V.A. PAATASHVILI, G.A. KHUSKIVADZE, On the boundedness of the Cauchy singular integral on Lebesgue spaces in the case of non-smooth contours. – Trudy Tbilisk. Matem. Inst. AN GSSR **69**(1982), 93 - 107 (Russian).

[34] G.K. PEDERSEN, Measure theory for C^*-algebras, II. – Math. Scand. **22**(1968), 63 – 74.

[35] S.C. POWER, C^*-algebras generated by Hankel operators and Toeplitz operators. – J. Funct. Anal. **31**(1979), 52 - 68.

[36] S.C. POWER, Hankel operators on Hilbert space. – Pitman Research Notes **64**, Pitman, Boston, London, Melbourne 1982.

[37] S.C. POWER, Essential spectra of piecewise continuous Fourier integral operators. – Proc. Royal Ir. Acad., Vol. 81A, 1(1981), 1 - 7.

[38] I. RAEBURN, A.M. SINCLAIR, The C^*-algebra generated by two projections. – Math. Scand. **65**(1989), 278 – 290.

[39] S. ROCH, Spline approximation methods for Wiener-Hopf operators. – Proceedings of the Winnipeg Conference on Operator Theory 1994 (to appear).

[40] S. ROCH, B. SILBERMANN, Algebras generated by idempotents and the symbol calculus for singular integral operators. – IEOT **11**(1988), 385 – 419.

[41] S. ROCH, B. SILBERMANN, Algebras of convolution operators and their image in the Calkin algebra. – Report R-MATH-05/90 des Karl-Weierstrass-Instituts für Mathematik, Berlin 1990, 157 S.

[42] S. ROCH, B. SILBERMANN, A symbol calculus for finite sections of singular integral operators with flip and piecewise continuous coefficients. – J. Funct. Anal. **78**(1988), 2, 365 – 389.

[43] S. ROCH, B. SILBERMANN, Asymptotic Moore-Penrose invertibility of singular integral operators. – to appear.

[44] B. SILBERMANN, The C^*-algebra generated by Toeplitz and Hankel operators with piecewise quasicontinuous symbols. – IEOT **10**(1987), 730-738.

[45] I. SPITKOVSKY, Singular integral operators with PC symbols on the spaces with general weights. – J. Funct. Anal. **105**(1992), 129 – 143.

[46] I. SPITKOVSKY, Once more on algebras generated by two projections. – Linear Algebra Appl. **208/209**(1994), 377 - 395

[47] V.S. SUNDER, N subspaces. – Canad. J. Math. **40**(1988), 38 – 54.

[48] N. VASILEVSKI, I. SPITKOVSKY, On the algebra generated by two projections. – Dokl. Akad. Nauk Ukrain. SSR, Ser. A, **8**(1981), 10 – 13 (Ukrainian).

[49] Y. WEISS, On algebras generated by two idempotents. – Seminar Analysis: Operator Eq. and Numer. Anal. 1987/88, Karl-Weierstrass-Institut für Mathematik, Berlin 1988, 139 – 145.

A. Böttcher, Yu.I. Karlovich, B. Silbermann
Fakultät für Mathematik
TU Chemnitz-Zwickau
09107 Chemnitz, GERMANY

N. Krupnik
Department of Mathematics
Bar-Ilan University
Ramat-Gan 52900, ISRAEL

I. Spitkovsky
Department of Mathematics
The College of William and Mary
Williamsburg, VA 23187-8795, USA

I. Gohberg
School of Mathematical Sciences
Tel Aviv University
Ramat Aviv 69978, ISRAEL

S. Roch
Mathematisches Institut
Universität Leipzig
Augustusplatz 10 – 11
04109 Leipzig, GERMANY

MSC 1991: Primary 46 H 15
    Secondary 16 W 99, 45 E 05, 46 N 20, 47 D 30, 47 G 10

Operator Theory
Advances and Applications, Vol. 90
© 1996 Birkhäuser Verlag Basel/Switzerland

TOEPLITZ OPERATORS WITH DISCONTINUOUS SYMBOLS: PHENOMENA BEYOND PIECEWISE CONTINUITY

A. Böttcher[1] and S.M. Grudsky[2]

This paper is a systematic introduction (with a number of new results) to some aspects of the theory of Toeplitz operators with oscillating symbols on the Hardy space H^2 of the unit circle. What we are interested in is obtaining answers to the question which geometric and/or algebraic properties of the argument of the symbol imply that the operator is normally solvable, semi-Fredholm, Fredholm, or even invertible. Our discussion includes well known results on symbols in $C + H^\infty$, SAP, or PQC and also less known and new insights into operators whose streched symbols have arguments behaving like x^λ, $\exp(x^\lambda)$, $(\log x)^\lambda$, or $\sin(x^\lambda)$ with $\lambda > 0$ at infinity. We also present some new results on the finite section method for Toeplitz operators.

1. Introduction

The present paper concerns invertibility and one-sided invertibility of Toeplitz operators on H^2, i.e. solvability properties of the infinite linear system

$$\begin{pmatrix} a_0 & a_{-1} & a_{-2} & \cdots \\ a_1 & a_0 & a_{-1} & \cdots \\ a_2 & a_1 & a_0 & \cdots \\ \cdots & \cdots & \cdots & \cdots \end{pmatrix} \begin{pmatrix} f_0 \\ f_1 \\ f_2 \\ \vdots \end{pmatrix} = \begin{pmatrix} g_0 \\ g_1 \\ g_2 \\ \vdots \end{pmatrix} \tag{1}$$

with a given sequence $\{a_n\}_{n=-\infty}^{\infty}$ of complex numbers, a given sequence $\{g_n\}_{n=0}^{\infty} \in l^2$, and sequence $\{f_n\}_{n=0}^{\infty}$ which is sought in l^2. A classical result by Otto Toeplitz [34] (also see [4]) says that the matrix on the left-hand side of (1) generates a bounded operator on l^2 if and only if there exists a function $a \in L^\infty(\mathbf{T})$ such that $\{a_n\}_{n=-\infty}^{\infty}$ is the sequence of the Fourier coefficients of a:

$$a_n = \frac{1}{2\pi} \int_0^{2\pi} a(e^{i\theta}) e^{-in\theta} \, d\theta \quad (n \in \mathbf{Z}). \tag{2}$$

[1] Research supported by the Alfried Krupp Förderpreis für junge Hochschullehrer of the Krupp Foundation

[2] Research supported by the Deutsche Forschungsgemeinschaft and in part also by the Russian fund of fundamental investigations (Grant 93-011-28)

If it exists, the function a (or better: the equivalence class in $L^\infty(\mathbf{T})$ containing a) is uniquely determined, the operator induced by the matrix of (1) on l^2 is then denoted by $T(a)$ and referred to as the *Toeplitz operator with the symbol a*.

When is the operator induced by the matrix of (1) invertible? This question may (and must) again be answered in the terms of the symbol $a \in L^\infty(\mathbf{T})$. A by now also classical theorem of Widom [36] and Devinatz [7] (also see [3, Theorem 2.23]) tells us that $T(a)$ is invertible if and only if

$$\operatorname*{ess\,inf}_{t \in \mathbf{T}} |a(t)| > 0 \qquad (3)$$

and

$$a/|a| = e^{i(\tilde{u}+v+c)} \quad \text{a.e. on } \mathbf{T} \qquad (4)$$

where $c \in \mathbf{R}$, u and v are real-valued functions in $L^\infty(\mathbf{T})$, and $\|v\|_\infty < \pi/2$; here $\tilde{u}$ denotes the conjugate function (= Hilbert transform) of u. However, despite the beauty of the Widom/Devinatz criterion, the problem of deciding whether $a/|a|$ admits a representation of the form (4) is no easy task, which motivates the search for effectively verifiable invertibility criteria for *subclasses* of symbols in $L^\infty(\mathbf{T})$.

Suppose a is *continuous* on $\mathbf{T}$, $a \in C(\mathbf{T})$. Then a maps the unit circle $\mathbf{T}$ into a continuous, closed, and naturally oriented curve $a(\mathbf{T})$. In 1952, Gohberg [13] proved that $T(a)$ is invertible if and only if this curve does not pass through the origin and the winding number of that curve about the origin is zero. If a is *piecewise continuous*, $a \in PC(\mathbf{T})$, we denote by $\sigma a(\mathbf{T})$ the continuous, closed, and naturally oriented curve which is obtained from the (essential) range of a by filling in a line segment between the endpoints $a(t-0)$ and $a(t+0)$ of each jump. In the sixties it was observed by several mathematicians, including Widom, Devinatz, Calderón, Spitzer, Simonenko, Gohberg and Krupnik, that $T(a)$ is invertible if and only if $\sigma a(\mathbf{T})$ does not contain the origin and has vanishing winding number about the origin (see e.g. [16] or [3]).

What happens for symbols beyond $PC(\mathbf{T})$? This question has been studied by many authors for the last few decades. In what follows we will survey part of this development. The main purpose of this paper is to discuss the topic in a systematic manner and our guiding problem is the connection between the invertibility behavior of the operator and the geometric/algebraic properties of the argument of the symbol. We will solve some problems which had been open until now and we will push things forward by considering symbol classes which have not yet been treated so far.

This paper is heavily based on ideas of the second author's works [17], [18], [19], [20]. Part of these works have never been published in English and, for lack of space, they partially do not contain detailed proofs. We therefore decided to quote a few key results of these works here again and to provide them with full proofs.

We tried making this paper self-contained. It is organized as follows. Section 2 is a brief introduction to the general theory of Toeplitz operators. Reading of this section should suffice to comprehend the motivation for the further investigations and to understand the statements as well as large parts of the proofs of the theorems in the subsequent sections. We are aware of the circumstance that Section 2 is not sufficient for catching all details of the proofs – this requires a deeper knowledge of Toeplitz and H^∞ theory, to the extent of

Douglas' book [8], of Chapters 1 and 2 of the book [3], and of pieces of Garnett's monograph [12], say.

In Section 3 we introduce the notion of u-periodic symbols. Many questions about Toeplitz operators generated by such symbols can be disposed of rather quickly. The importance of this symbol class rests on the fact that a series of much more complicated symbols may be factored into a product of nice symbols and u-periodic symbols. We will take profit of this observation throughout the remaining sections.

Sections 4 and 5 are devoted to symbols whose argument increases from $-\infty$ to $+\infty$. The delicacy of the problem may be illustrated as follows. Suppose that, after stretching the argument from $(-\pi, \pi)$ to the real line $\mathbf{R}$, the argument is x^3. It will turn out that the corresponding Toeplitz operator is left-invertible. The same is true for all arguments of the form $x^3 + o(1)$ $(|x| \to \infty)$. However, the class of all arguments of the form $x^3 + O(1)$ $(|x| \to \infty)$ already contains arguments of Toeplitz operators which are not even normally solvable! In Section 6 we consider symbols whose argument comes from $+\infty$ and goes to $+\infty$ (example: x^2). Section 7 is concerned with bounded arguments (example: $\sin x$). Although this will be a digression from the main stream of the paper, we will there also present a little tidbit: we prove the existence of invertible Toeplitz operators with periodic arguments to which the so-called finite section method is not applicable.

Finally, in Section 8 we establish some new results for what we call modulated almost periodic symbols. While the argument of an almost periodic function is always of the form $\lambda x + g(x)$ with some almost periodic function g, the arguments of the symbols considered in Section 8 have the form $\lambda \alpha(x) + g(\alpha(x))$ where g is almost periodic and α is a "regular" homeomorphism of $\mathbf{R}$ onto $\mathbf{R}$. It should be noticed that this section is not an appendage to the preceding sections. The class of modulated almost periodic symbols comprises both the almost periodic symbols $(\alpha(x) = x)$ and a major part of the symbols studied in Sections 4 and 5 $(g(x) = 0)$. Consideration of this class is thus a first step to unifying two more or less independent theories. The next step, developing a theory for modulated semi-almost periodic symbols, will be accomplished in a forthcoming publication.

2. Preliminaries

2.1. Toeplitz and singular integral operators. We denote by $L^p := L^p(\mathbf{T})$ $(1 \le p \le \infty)$ the usual Lebesgue spaces on the complex unit circle $\mathbf{T}$, and we let $H^p := H^p(\mathbf{T})$ denote the corresponding Hardy spaces:

$$H^p = \{f \in L^p : f_n = 0 \quad \text{for} \quad n < 0\}.$$

In what follows we will mainly be concerned with the cases $p = 2$ and $p = \infty$. The *Toeplitz operator* $T(a)$ generated by a function $a \in L^\infty$ is defined by

$$T(a): \ H^2 \to H^2, \ f \mapsto P(af)$$

where a stands for the operator of multiplication by a and P is given by

$$P: L^2 \to H^2, \ \sum_{n=-\infty}^{\infty} f_n \chi_n \mapsto \sum_{n=0}^{\infty} f_n \chi_n; \tag{5}$$

here and in the following $\chi_n(t) := t^n$ $(t \in \mathbf{T})$. Notice that P is the orthogonal projection of L^2 onto H^2. The function a is in this context called the *symbol* of $T(a)$.

An orthonormal basis in H^2 is given by $\{\chi_n/\sqrt{2\pi}\}_{n=0}^{\infty}$. The matrix representation of $T(a)$ with respect to this basis is just the Toeplitz matrix in (1), its entries being defined by (2).

Let S be the Cauchy singular integral operator on L^2,

$$(Sf)(t) = \frac{1}{\pi i} \int_{\mathbf{T}} \frac{f(\tau)}{\tau - t} d\tau \quad (t \in \mathbf{T}).$$

It is well known (see e.g. [16]) that the projection P given by (5) may be written as $P = (I + S)/2$. Put $Q = (I - S)/2$. Numerous questions on singular integral operators on L^2 may be reduced to the study of the *simple singular integral operator* $aP+Q$ on L^2. Many properties of the latter operator are in turn equivalent to the corresponding properties of the Toeplitz operator $T(a)$ on H^2.

2.2. Normally solvable operators. Let A be a bounded linear operator on an infinite-dimensional separable Hilbert space $\mathcal{H}$. Denote the adjoint operator by A^* and put

$$\text{Ker } A = \{f \in \mathcal{H} : Af = 0\}, \quad \text{Ker } A^* = \{f \in \mathcal{H} : A^*f = 0\}.$$

The numbers $\alpha(A)$ and $\beta(A)$ in $\{0, 1, 2, \ldots\} \cup \{\infty\}$ defined by

$$\alpha(A) = \dim \text{Ker } A, \quad \beta(A) = \dim \text{Ker } A^*$$

are referred to as the *kernel* and *cokernel dimensions* of A. The operator A is said to be *normally solvable* if its image space (=range)

$$\text{Im } A = \{Af : f \in \mathcal{H}\}$$

is a closed subset of $\mathcal{H}$. One can show that A^* is normally solvable if and only if so is A. If A is normally solvable, then

$$\alpha(A) = \dim(\text{Im } A^*)^{\perp}, \quad \beta(A) = \dim(\text{Im } A)^{\perp}$$

where $X^{\perp}$ denotes the orthogonal complement of X in $\mathcal{H}$.

A normally solvable operator A is called a *semi-Fredholm operator* if $\alpha(A)$ or $\beta(A)$ is finite. In that case the index of A is defined as

$$\text{Ind } A = \alpha(A) - \beta(A).$$

Clearly, $\text{Ind } A \in \mathbf{Z} \cup \{-\infty, +\infty\}$. If A is normally solvable and both $\alpha(A)$ and $\beta(A)$ are finite , then A is called a *Fredholm operator*. The index of a Fredholm operator is always finite.

The operator A is said to be *left (right) invertible* if there is a bounded linear operator B on $\mathcal{H}$ such that $BA = I$ $(AB = I)$. We remark that left invertibility is equivalent to injectivity and normal solvability, while right invertibility is the same as surjectivity. If A is one-sided (i.e. left or right) invertible, then A is necessarily normally solvable. The converse is not true: there are semi-Fredholm and even Fredholm operators which are neither left nor right invertible.

The set of bounded linear operators on the space H^2 may be represented by a pyramid as in Figure 1. The invertible (i.e. bijective) operators form a subset of the Fredholm operators, these in turn are a subset of the semi-Fredholm operators, which in turn constitute a subset of the normally solvable operators.

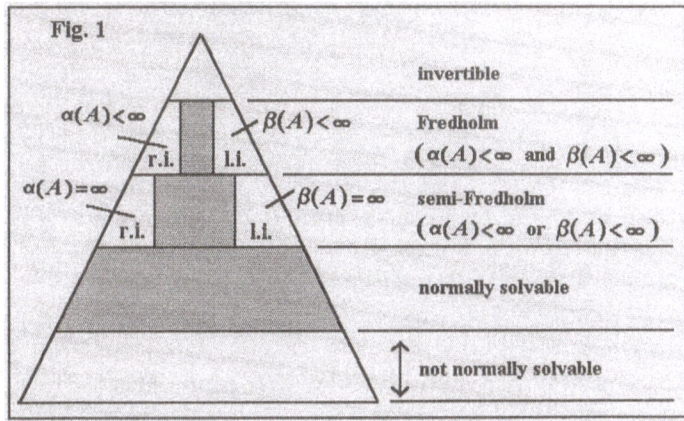

All assertions made in this subsection are generally known. Proofs may be found in [8] or [16], for example.

2.3. The six possibilities for a Toeplitz operator. Now suppose A is the Toeplitz operator $T(a)$ on H^2 generated by a function $a \in L^\infty$. If a is identically zero, then $T(a)$ is the zero operator (and hence normally solvable but not semi-Fredholm). So assume a does not vanish identically. A remarkable theorem by Coburn [5] and Simonenko [33] says that then always $\alpha(T(A)) = 0$ or $\beta(T(a)) = 0$. Consequently, there are no nonzero Toeplitz operators belonging to the dark regions of Figure 1. This pyramid contracts to the pyramid of Figure 2: a nonzero normally solvable Toeplitz operator is automatically left, right, or two-sided invertible ! Thus, given a symbol $a \in L^\infty \setminus \{0\}$, there are six mutually excluding possibilities:

(P_1) $T(a) \in \Phi_0 \Longleftrightarrow T(a)$ is invertible
            $\Longleftrightarrow T(a)$ is Fredholm of index zero;

(P_2) $T(a) \in \Phi(-\mathbf{N}) \Longleftrightarrow T(a)$ is left-invertible and $0 < \beta(T(a)) < \infty$
            $\Longleftrightarrow T(a)$ is Fredholm with negative index;

(P_3) $T(a) \in \Phi(+\mathbf{N}) \Longleftrightarrow T(a)$ is right-invertible and $0 < \alpha(T(a)) < \infty$
            $\Longleftrightarrow T(a)$ is Fredholm with positive index;

(P_4) $T(a) \in \Phi(-\infty) \Longleftrightarrow T(a)$ is left-invertible and $\beta(T(a)) = \infty$;

(P_5) $T(a) \in \Phi(+\infty) \iff T(a)$ is right-invertible and $\alpha(T(a)) = \infty$;
(P_6) $T(a) \in NNS \iff T(a)$ is not normally solvable.

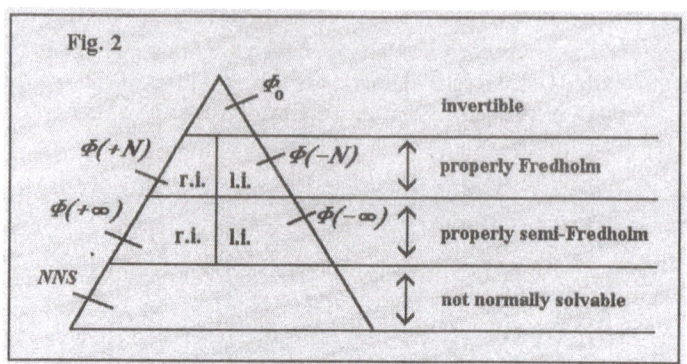

We are now also in a position to make precise the last sentence of Subsection 2.1. Namely, if $a \in L^\infty$, then $aP + Q$ is normally solvable if and only if so is $T(a)$; moreover, one always has $\alpha(aP + Q) = \alpha(T(a))$, $\beta(aP + Q) = \beta(T(a))$.

For proofs of the results of the present subsection see the books [8], [16], [3].

2.4. The Hartman-Wintner-Simonenko theorem. This theorem states that if $a \in L^\infty$ and $T(a)$ is normally solvable, then a is either identically zero or invertible in L^∞. We denote by GL^∞ the set of all functions in L^∞ which are invertible in L^∞, i.e. which satisfy (3). Notice that Hartman and Wintner [21] proved that $a \in GL^\infty$ whenever $T(a)$ is invertible, while Simonenko [33] pointed out that $a \in GL^\infty$ if $T(a)$ is merely known to be normally solvable.

Thus, if $a \in L^\infty \setminus GL^\infty$ then $T(a) \in NNS$. Therefore, *throughout the rest of this article we assume that $a \in GL^\infty$.*

2.5. C and PC. Let $C := C(\mathbf{T})$ stand for the continuous functions on $\mathbf{T}$. If $a \in C \cap GL^\infty$, then $a(\mathbf{T})$ does not contain the origin and it is well known (see [13] or [16]) that $T(a)$ is always Fredholm and that $\operatorname{Ind} T(a)$ is minus the winding number of $a(\mathbf{T})$ with respect to the origin.

Denote by $PC := PC(\mathbf{T})$ the closed algebra of all piecewise continuous functions on $\mathbf{T}$. For $a \in PC$, define $\sigma a(\mathbf{T})$ as in the Introduction. It is again well known (see [16]) that $T(a)$ is Fredholm whenever $0 \notin \sigma a(\mathbf{T})$, in which case $\operatorname{Ind} T(a)$ equals minus the winding number of $\sigma a(\mathbf{T})$ about the origin, and that $T(a)$ is not normally solvable if $0 \in \sigma a(\mathbf{T})$. Notice that if $a \in PC \cap GL^\infty$, then the origin belongs to $\sigma a(\mathbf{T})$ if and only if there is a $t \in \mathbf{T}$ such that $a(t - 0) \neq a(t + 0)$ and $0 \in (a(t - 0), a(t + 0))$.

2.6. Beyond PC. Given $a \in GL^\infty \setminus PC$, which of the six regions of Figure 2 does the operator $T(a)$ belong to ? This question can be decided in a more or less explicit manner for several symbol classes, e.g. for symbols in $H^\infty, \overline{H^\infty}, C + H^\infty, C + \overline{H^\infty}, QC, PQC, AP, SAP$.

Here $\overline{H^\infty}$ is the algebra of the complex conjugates of H^∞ functions, $C + H^\infty$ is the smallest closed subalgebra of L^∞ containing $C \cup H^\infty$, $C + \overline{H^\infty}$ is defined analogously, $QC := (C + H^\infty) \cap (C + \overline{H^\infty})$ is the algebra of quasicontinuous functions, PQC (the piecewise quasicontinuous functions) is the smallest closed subalgebra of L^∞ containing $PC \cup QC$, AP is the algebra of uniformly almost periodic functions, and SAP stands for the algebra of semi-almost periodic functions. We will say more about these symbol classes and their Toeplitz operators in the subsequent sections.

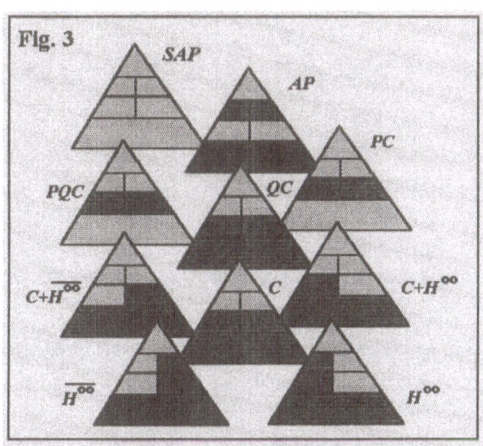

At the present moment we confine ourselves to draw the reader's attention to Figure 3. This figure associates a pyramid with each of the symbol classes mentioned. A light region of the pyramid indicates that there are GL^∞ symbols in the class for which the Toeplitz operators belongs to the corresponding region of Figure 2, while a dark region means that the class does not contain GL^∞ symbols which generate Toeplitz operators belonging to the corresponding region of Figure 2. For example, we infer from Figure 3 that there are $a \in H^\infty \cap GL^\infty$ such that $T(a) \in \Phi_0, T(a) \in \Phi(-\mathbf{N})$, or $T(a) \in \Phi(-\infty)$, but that there are no $a \in H^\infty \cap GL^\infty$ for which $T(a) \in \Phi(+\mathbf{N}), T(a) \in \Phi(+\infty)$, or $T(a) \in NNS$.

We wish to emphasize three observations one may gather from Figure 3:

(O_1) proper semi-Fredholm Toeplitz operators appear only beyond PC
            (or if the reader wants: beyond PQC);

(O_2) the existence of Toeplitz operators which are not normally solvable
            has something to do with "jumps";

(O_3) $SAP \cap GL^\infty$ is the "simplest" class for which everything is possible.

2.7. $H^\infty, \overline{H^\infty}, GH^\infty$. As usual, put

$$H^\infty = \{a \in L^\infty : a_n = 0 \text{ for } n < 0\}, \quad \overline{H^\infty} = \{a \in L^\infty : a_n = 0 \text{ for } n > 0\}.$$

For $a \in L^\infty$ we define $\overline{a} \in L^\infty$ by $\overline{a}(t) = \overline{a(t)} \, (t \in \mathbf{T})$. So $a \in H^\infty$ if and only if $\overline{a} \in \overline{H^\infty}$.

Let $\mathbf{D} := \{z \in \mathbf{C} : |z| < 1\}$ and let $H^\infty(\mathbf{D})$ be the set of all bounded analytic functions in $\mathbf{D}$. Functions in $H^\infty(\mathbf{D})$ have nontangential limits almost everywhere on $\mathbf{T}$, and the map which assigns to a function in $H^\infty(\mathbf{D})$ its boundary function a.e. on $\mathbf{T}$ is a bijection of $H^\infty(\mathbf{D})$ onto H^∞. In what follows we tacitly identify H^∞ and $H^\infty(\mathbf{D})$. Analogously, we identify $\overline{H^\infty}$ with the boundary values of functions which are bounded and analytic in $(\mathbf{C} \cup \{\infty\}) \setminus (\mathbf{D} \cup \mathbf{T})$.

A function $u \in H^\infty$ is called an *inner function* if $|u(t)| = 1$ for almost all $t \in \mathbf{T}$. Examples of inner functions are

$$\chi_n(t) = t^n \ (n \geq 0), \quad S_\lambda(t) = \exp\left(\lambda \frac{t+1}{t-1}\right) (\lambda > 0), \quad b_\alpha(t) = \frac{\alpha - t}{1 - \overline{\alpha}t} \quad (\alpha \in \mathbf{D}).$$

Finite products of inner functions are also inner, and hence so is in particular $\gamma b_{\alpha_1} \ldots b_{\alpha_n}$ $(\alpha_j \in \mathbf{D}, \gamma \in \mathbf{T})$. Functions of the latter form are called *finite Blaschke products*. Note that finite Blaschke products are continuous on $\mathbf{T}$. Let $\{\alpha_j\}_{j=1}^\infty$ be a sequence of complex numbers such that

$$0 < |\alpha_1| \leq |\alpha_2| \leq \ldots < 1, \quad \sum_{j=1}^\infty (1 - |\alpha_j|) < \infty.$$

One can show (see e.g. [12]) that the infinite product

$$b(z) = \prod_{j=1}^\infty \frac{|\alpha_j|}{\alpha_j} \frac{\alpha_j - z}{1 - \overline{\alpha_j}z}$$

converges uniformly in each disk $|z| \leq R < 1$ and that the nontangential limits of b have modulus 1 almost everywhere on $\mathbf{T}$. Thus, b is an inner function. Each α_j is a zero of b (thought of as a function in $H^\infty(\mathbf{D})$) with multiplicity equal to the number of times it occurs in the sequence, and b has no other zeros in $\mathbf{D}$. A function of the form

$$b(z) = \gamma z^m \prod_j \frac{|\alpha_j|}{\alpha_j} \frac{\alpha_j - z}{1 - \overline{\alpha_j}z}$$

is called a *Blaschke product*; here $m \in \{0, 1, 2, \ldots\}, \alpha_j \in \mathbf{D}, \sum_j (1 - |\alpha_j|) < \infty, \gamma \in \mathbf{T}$. The set $\{\alpha_j\}$ may be infinite, finite, or empty. In the first case b is referred to as an *infinite Blaschke product*. In the latter case, $b = \gamma \chi_m$.

An inner function which is not a Blaschke product is called a *singular inner function* (note that some authors also regard constant functions as singular inner). The function S_λ defined above is a singular inner function. A singular inner function u has no zeros in $\mathbf{D}$, but there exists points $t \in \mathbf{T}$ such that the radial limits $\lim_{z \to t} u(z)$ are zero.

The set H^∞ is a closed subalgebra of L^∞. We denote by GH^∞ the functions in H^∞ which are invertible in H^∞. It is well known (see e.g. [8] or [12]) that if $h \in H^\infty$, then

$$h \in GH^\infty \iff \inf_{z \in \mathbf{D}} |h(z)| > 0.$$

The set GH^∞ contains "sufficiently many" functions:

$$\forall a \in GL^\infty \ \exists h \in GH^\infty : \ |a| = |h| \quad \text{a.e. on } \mathbf{T} \tag{6}$$

(see e.g. [8] or [12]). We remark that GH^∞ is also referred to as the collection of all *outer functions* in $GL^\infty \cap H^\infty$. One can show (again see [8] or [12]) that every function $h \in H^\infty \cap GL^\infty$ has a unique factorization $h = BSg$ in which B is a Blaschke product, S is a singular inner function or identically 1, and $g \in GH^\infty$.

If $\bar{g} \in \overline{H^\infty}, a \in L^\infty, h \in H^\infty$, then

$$T(\bar{g}ah) = T(\bar{g})T(a)T(h). \tag{7}$$

This equality is easily verified. It is of great importance throughout the entire theory of Toeplitz operators.

Let $h \in H^\infty \cap GL^\infty$. Then (7) implies that $T(h^{-1})T(h) = T(h^{-1}h) = I$, i.e. $T(h)$ is left-invertible. Moreover, we have

$$T(h) \text{ is invertible} \iff h \in GH^\infty, \tag{8}$$
$$T(h) \text{ is Fredholm} \iff h = Bg \text{ with a finite Blaschke product } B \tag{9}$$
$$\text{and with } g \in GH^\infty.$$

The equivalence (8) is a 1929 result of A. Wintner. It may be easily proved. Indeed, if $h^{-1} \in H^\infty$ then $T(h^{-1})$ is the inverse of $T(h)$ by virtue of (7). On the other hand, if $T(h)$ is invertible then the equation $T(h)f = hf = 1$ has a solution $f \in H^2$, whence $h^{-1} = f \in H^2 \cap L^\infty = H^\infty$. The equivalence (9) is due to Douglas [8, 7.36] (also see [3, 2.64]). The proof of the implication "$\Longrightarrow$" is not trivial. We here confine ourselves to nothing that $\beta(T(B)) = \infty$ if B is an infinite Blaschke product. Let $\alpha \in \mathbf{D}$ be a zero of B. Then the sequence $f_\alpha = \{1, \bar{\alpha}, \bar{\alpha}^2, \ldots\}$ belongs to l^2, and since $T(B)^* = T(\overline{B})$ is represented by the matrix

$$\begin{pmatrix} \overline{B_0} & \overline{B_1} & \overline{B_2} & \cdots \\ 0 & \overline{B_0} & \overline{B_1} & \cdots \\ 0 & 0 & \overline{B_0} & \cdots \\ \cdots & \cdots & \cdots & \cdots \end{pmatrix},$$

a straightforward computation gives that $T(\overline{B})f_\alpha$ is the sequence

$$\{\overline{B(\alpha)}, \bar{\alpha}\overline{B(\alpha)}, \bar{\alpha}^2\overline{B(\alpha)}, \ldots\} = \{0, 0, 0, \ldots\}.$$

Thus, Ker $T(\overline{B})$ contains f_α for each zero α of B and is therefore infinite-dimensional.

2.8. $C + H^\infty$ and $C + \overline{H^\infty}$. Sarason [29] discovered that the smallest closed subalgebra of L^∞ containing both H^∞ and C coincides with the sum

$$C + H^\infty := \{c + h : c \in C, h \in H^\infty\}.$$

The algebra $C + \overline{H^\infty}$ is defined analogously. The theory of Toeplitz operators with symbols in $C + H^\infty$ was worked out by Douglas [8] (also see e.g. [3]). We here only cite the principal results.

If $b \in C + \overline{H^\infty}, a \in L^\infty, c \in C + H^\infty$, then

$$T(bac) - T(b)T(a)T(c) \quad \text{is compact.} \tag{10}$$

Let $a \in (C + H^\infty) \cap GL^\infty$. Then

$$T(a) \in \Phi_0 \cup \Phi(-\mathbf{N}) \cup \Phi(+\mathbf{N}) \cup \Phi(-\infty).$$

One has

$$T(a) \in \Phi_0 \cup \Phi(-\mathbf{N}) \cup \Phi(-\mathbf{N}) \Longleftrightarrow a \in G(C + H^\infty),$$

where $G(C + H^\infty)$ stands for the functions in $C + H^\infty$ which have the inverse in $C + H^\infty$. To decide whether $a \in G(C + H^\infty)$, one may consider the Abel-Poisson means (=harmonic extension) of a. For $0 < r < 1$, define $h_r a \in C$ by

$$(h_r a)(e^{i\theta}) = \sum_{n=-\infty}^{\infty} r^{|n|} a_n e^{in\theta} \quad (e^{i\theta} \in \mathbf{T}).$$

It turns out that $a \in G(C + H^\infty)$ if and only if there is an $r_0 \in (0,1)$ and an $\varepsilon > 0$ such that

$$|(h_r a)(e^{i\theta})| \geq \varepsilon \quad \forall r \in (r_0, 1) \quad \forall e^{i\theta} \in \mathbf{T}.$$

In that case the winding number of $h_r a$ is independent of $r \in (r_0, 1)$ and one can show that $\mathrm{Ind}\, T(a) = -\mathrm{wind}\, h_r a$ where $r \in (r_0, 1)$.

2.9. The role of the argument. The following result was established by Widom [36] and Devinatz [7].

If $a \in GL^\infty$, then there is a function $h \in GH^\infty$ such that

$$T(a/|a|) = T(\overline{h})T(a)T(h) \tag{11}$$

Proof. By virtue of (6), there exists an $h \in GH^\infty$ such that $|a^{-1}|^{1/2} = |h|$. So $a/|a| = \overline{h}ah$, and (11) now follows from (7). ∎

The operators $T(\overline{h})$ and $T(h)$ are invertible due to (8). Thus, the important consequence of (11) is that $T(a)$ and $T(a/|a|)$ belong to one and the same region of Figure 2. In other words, this region is entirely determined by the "frequency modulation" (the "argument") of a and is independent of the "amplitude modulation" (the modulus) of a. Therefore, for our purpose it suffices to consider unimodular symbols only.

In this connection, the following remark might be in order. Toeplitz and singular integral operators may not only be studied on L^2 over the unit circle $\mathbf{T}$ but also on L^2 over an arbitrary rectifiable Jordan curve Γ satisfying the Carleson condition

$$\sup_{t \in \Gamma} \sup_{\varepsilon > 0} |\Gamma(t, \varepsilon)|/\varepsilon < \infty, \tag{12}$$

where $|\Gamma(t, \varepsilon)|$ is the length of the "portion" $\Gamma(t, \varepsilon) := \{\tau \in \Gamma : |\tau - t| < \varepsilon\}$. It was shown by G. David, V.A. Paatashvili, and G.A. Khushkivadze that the Cauchy singular integral operator S_Γ is bounded on $L^2(\Gamma)$ if and only if (12) holds. Hence, in that case every function $a \in L^\infty(\Gamma)$ generates a bounded Toeplitz operator $T_\Gamma(a) : H^2(\Gamma) \to H^2(\Gamma)$, $f \mapsto P_\Gamma(af)$, where $P_\Gamma := (I + S_\Gamma)/2$ and $H^2(\Gamma) := P_\Gamma L^2(\Gamma)$. From the results of Karlovich and one of the authors [2] it follows that there exist Carleson curves Γ and real-valued symbols $a \in PC(\Gamma)$ such that $T_\Gamma(a/|a|)$ is the identity operator whereas $T_\Gamma(a)$ is not normally solvable. Thus, the properties of $T_\Gamma(a)$ and $T_\Gamma(a/|a|)$ are only equivalent if Γ is sufficiently nice !

For another variation of this theme see Subsection 7.8.

2.10. Criteria in analytic language. A function $s \in GL^\infty$ is said to be *sectorial* if the essential range of s is contained in an open half-plane whose boundary passes through the origin. Equivalently, s is sectorial if and only if there exists a complex number $\gamma \in \mathbf{T}$ and a real number $\varepsilon > 0$ such that $\mathrm{Re}\,(\gamma s(t)) \geq \varepsilon$ for almost all $t \in \mathbf{T}$. It is easily seen that sectorial symbols generate invertible Toeplitz operators.

Suppose $a \in GL^\infty$ is unimodular. The following results were essentially established by Widom [36], Devinatz [7], Coburn [5], Douglas and Sarason [10]. Full proofs may also be found in the book [3]..

Given a subset $F \subset L^\infty$ and a function $a \in L^\infty$, we define the distance $\mathrm{dist}_{L^\infty}(a, F)$ between a and F in the usual way: $\mathrm{dist}_{L^\infty}(a, F) = \inf\{\|a - f\|_\infty : f \in F\}$. Notice that since a is required to be unimodular, the functions h and c in the last four of the following criteria automatically belong to GL^∞.

Invertibility (Widom/Devinatz)
$T(a) \in \Phi_0 \Longleftrightarrow \mathrm{dist}_{L^\infty}(a, GH^\infty) < 1$
$\Longleftrightarrow a = hs$ with $h \in GH^\infty$ and a sectorial $s \in GL^\infty$
$\Longleftrightarrow a = e^{i(\tilde{u}+v+c)}$ where $c \in \mathbf{R}$, u and v are real-valued functions in L^∞, $\|v\|_\infty < \pi/2$, and $\tilde{u}$ denotes the conjugate function of u.

Fredholmness (Douglas/Sarason/Coburn)
$T(a) \in \Phi_0 \cup \Phi(-\mathbf{N}) \cup \Phi(+\mathbf{N}) \Longleftrightarrow \mathrm{dist}_{L^\infty}(a, G(C + H^\infty)) < 1$
$\Longleftrightarrow a = cs$ with $c \in G(C + H^\infty)$ and a sectorial $s \in GL^\infty$
$\Longleftrightarrow a = \chi_n b$ with $n \in \mathbf{Z}$ and $T(b) \in \Phi_0$.

Left invertibility (Widom/Devinatz)
$T(a) \in \Phi_0 \cup \Phi(-\mathbf{N}) \cup \Phi(-\infty) \Longleftrightarrow \mathrm{dist}_{L^\infty}(a, H^\infty) < 1$
$\Longleftrightarrow a = hs$ with $h \in H^\infty$ and a sectorial $s \in GL^\infty$.

Right invertibility (Widom/Devinatz)
$T(a) \in \Phi_0 \cup \Phi(+\mathbf{N}) \cup \Phi(+\infty) \Longleftrightarrow \mathrm{dist}_{L^\infty}(a, \overline{H^\infty}) < 1$
$\Longleftrightarrow a = \overline{h}s$ with $h \in H^\infty$ and a sectorial $s \in GL^\infty$.

Left Fredholmness (Douglas/Sarason)
$T(a) \in \Phi_0 \cup \Phi(-\mathbf{N}) \cup \Phi(+\mathbf{N}) \cup \Phi(-\infty) \Longleftrightarrow \mathrm{dist}_{L^\infty}(a, C + H^\infty) < 1$
$\Longleftrightarrow a = cs$ with $c \in C + H^\infty$ and a sectorial $s \in GL^\infty$.

Right Fredholmness (Douglas/Sarason)
$T(a) \in \Phi_0 \cup \Phi(-\mathbf{N}) \cup \Phi(+\mathbf{N}) \cup \Phi(+\infty) \Longleftrightarrow \mathrm{dist}_{L^\infty}(a, C + \overline{H^\infty}) < 1$
$\Longleftrightarrow a = \overline{c}s$ with $c \in C + H^\infty$ and a sectorial $s \in GL^\infty$.

2.11. The stretched symbol. To avoid unnecessary complications, we will mainly consider unimodular symbols $a \in GL^\infty$ with at most one discontinuity on $\mathbf{T}$, at $t = 1$, say. Note that local principles often allow us to reduce the case of several discontinuities to the case of a single discontinuity. Put $\dot{\mathbf{T}} = \mathbf{T} \setminus \{1\}$, let $C(\dot{\mathbf{T}})$ be the collection of all functions on $\mathbf{T}$ which are continuous at every point $t \in \dot{\mathbf{T}}$, and denote by $CU(\dot{\mathbf{T}})$ and $CR(\dot{\mathbf{T}})$ the unimodular and real-valued functions in $C(\dot{\mathbf{T}})$, respectively. Every function $a \in CU(\dot{\mathbf{T}})$ can be written as $a = e^{ib}$ with $b \in CR(\dot{\mathbf{T}})$, and b is uniquely determined up to an additive integral multiple of 2π. We denote by $\arg a \in CR(\dot{\mathbf{T}})$ any such function b; for our purposes the particular choice of $\arg a$ will play no role.

The unimodular and real-valued functions on $\mathbf{R}$ will be denoted by $CU(\mathbf{R})$ and $CR(\mathbf{R})$, respectively. For $a \in L^\infty(\mathbf{T})$, define $a^\# \in L^\infty(\mathbf{R})$ by

$$a^\#(x) = a\left(\frac{x - i}{x + i}\right) \quad (x \in \mathbf{R}).$$

Clearly, the map $a \mapsto a^\#$ is a bijection of $CU(\dot{\mathbf{T}})$ onto $CU(\mathbf{R})$ as well as of $CR(\dot{\mathbf{T}})$ onto $CR(\mathbf{R})$. Working with the "stretched" symbol $a^\# \in CU(\mathbf{R})$ and the "stretched" argument $(\arg a)^\# = \arg(a^\#) \in CR(\mathbf{R})$ is frequently more convenient than considering $a \in CU(\dot{\mathbf{T}})$ and $\arg a \in CR(\dot{\mathbf{T}})$.

If $a \in PC \cap CU(\dot{\mathbf{T}})$ then $(\arg a)^\#$ belongs to $C(\overline{\mathbf{R}})$, i.e. $(\arg a)^\#$ has finite limits at $-\infty$ and $+\infty$. The following sections are devoted to symbols $a \in CU(\dot{\mathbf{T}})$ for which $(\arg a)^\#$ has no finite limits at infinity.

3. Periodic and u-periodic symbols

3.1. Definitions and examples. Let $u \in H^\infty$ be an inner function. If $b \in L^\infty$, then the composition $(b \circ u)(t) = b(u(t))$ is well-defined and is again a function in L^∞. A function $a \in L^\infty$ is said to be u-periodic if there is a function $b \in L^\infty$ such that $a = b \circ u$. We have

$$(b \circ u)^\#(x) = (b \circ u)\left(\frac{x - i}{x + i}\right) = b\left(u\left(\frac{x - i}{x + i}\right)\right) = b(u^\#(x)),$$

i.e. $(b \circ u)^\# = b \circ u^\#$. If

$$u(t) = S_\lambda(t) := \exp\left(\lambda \frac{t + 1}{t - 1}\right) \quad (\lambda > 0), \tag{13}$$

then $u^\#(x) = \exp(i\lambda x)$ and hence

$$(b \circ u)^\#(x) = b(e^{i\lambda x}) \quad (x \in \mathbf{R})$$

is a periodic function with the period $2\pi/\lambda$. Obviously, every function $a \in L^\infty$ for which $a^\#$ is $2\pi/\lambda$-periodic may be represented as $a = b \circ u$ with u given by (13). We call such functions a simply *periodic*.

Let now $b \in C$ be unimodular. Then $b(e^{i\theta}) = e^{i(n\theta + p(\theta))}$ $(0 \le \theta < 2\pi)$ where $n \in \mathbf{Z}$ is the winding number (=index) of the function b with respect to the origin and $p \in CR(\mathbf{R})$ is a 2π-periodic function. Let u be an inner function in $CU(\dot{\mathbf{T}})$. In that case

$$\alpha(x) := \arg u^\#(x) \quad (x \in \mathbf{R})$$

may be taken in $CR(\mathbf{R})$. We have

$$(b \circ u)^{\#}(x) = b(u^{\#}(x)) = b(e^{i\alpha(x)}) = e^{i(n\alpha(x)+p(\alpha(x)))}$$

and thus,

$$\arg(b \circ u)^{\#}(x) = n\alpha(x) + p(\alpha(x)). \tag{14}$$

If u is given by (13), then (14) reads

$$\arg(b \circ u)^{\#}(x) = n\lambda x + p(\lambda x).$$

Figure 4 shows typical plots of such arguments.

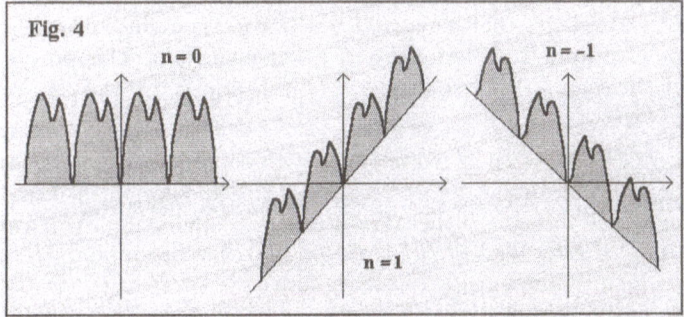

We will see in the next section that infinite Blaschke products $u = B \in CU(\dot{\mathbf{T}})$ produce a variety of different strictly monotonously increasing functions $\alpha = (\arg B)^{\#}$ such that $\alpha(-\infty) = -\infty$ and $\alpha(+\infty) = +\infty$. Typical plots of the argument (14) for this case are shown in Figure 5.

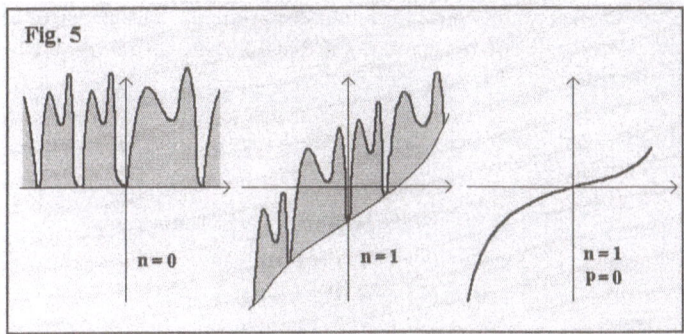

The following result (which actually does not deserve the title "Theorem") is well known; see e.g. [9], [18], [3]. This result tells us a lot about u-periodic symbols and it will be repeatedly employed in the forthcoming sections.

3.2. Theorem. *Let* $u \in H^\infty$ *be a non-constant inner function and let* $b \in GL^\infty$. *Denote by* FB *the set of all finite Blaschke products. Then the following implications hold.*

$$T(b) \text{ invertible} \Longrightarrow T(b \circ u) \text{ invertible}, \tag{15}$$

$$T(b) \text{ left-invertible} \Longrightarrow T(b \circ u) \text{ left-invertible}, \tag{16}$$

$$T(b) \text{ right-invertible} \Longrightarrow T(b \circ u) \text{ right-invertible}, \tag{17}$$

$$T(b) \in \Phi(-\mathbf{N}), u \in FB \Longrightarrow T(b \circ u) \in \Phi(-\mathbf{N}), \tag{18}$$

$$T(b) \in \Phi(-\mathbf{N}), u \notin FB \Longrightarrow T(b \circ u) \in \Phi(-\infty), \tag{19}$$

$$T(b) \in \Phi(+\mathbf{N}), u \in FB \Longrightarrow T(b \circ u) \in \Phi(+\mathbf{N}), \tag{20}$$

$$T(b) \in \Phi(+\mathbf{N}), u \notin FB \Longrightarrow T(b \circ u) \in \Phi(+\infty), \tag{21}$$

Proof. Suppose $T(b)$ is left-invertible. Then, by 2.10, $b = hs$ with $h \in H^\infty$ and a sectorial $s \in GL^\infty$. It follows that $b \circ u = (h \circ u)(s \circ u)$, and since $b \circ u \in H^\infty$ and $s \circ n$ is sectorial, we deduce from 2.10 that $T(b \circ u)$ is left-invertible. This proves (16). Taking into account that the map $h \mapsto h \circ u$ also leaves GH^∞ and $\overline{H^\infty}$ invariant, we get (15) and (17) in a similar way.

Now suppose $T(b) \in \Phi(-\mathbf{N})$. Then $b = \chi_n c$ with some integer $n \geq 1$ and $T(c) \in \Phi_0$. We get $b \circ u = u^n (c \circ u)$ and thus $T(b \circ u) = T(c \circ u)T(u^n)$ in view of (7). From (15) we infer that $T(c \circ u)$ is invertible. If $u \in FB$ then $T(u^n) \in \Phi(-\mathbf{N})$ and hence $T(b \circ u) \in \Phi(-\mathbf{N})$. So let $u \notin FB$. Then $u^n \notin FB$. We have $u^n = BS$ with a Blaschke product B and a function S which is either singular inner or identically 1. Since $u^n \notin FB$, S is non-constant or B is infinite. In either case (9) implies that $T(u^n)$ is not Fredholm. Therefore $T(b \circ u)$ cannot be Fredholm. Since, by (16), $T(b \circ u)$ is left-invertible, it results that $T(b \circ u) \in \Phi(-\infty)$. This completes the proof of (18) and (19). The proof of (20) and (21) is analogous. ∎

We can show that the reverse implications in (15) to (17) are also true, that is,

$$T(b) \text{ is (left/right) invertible} \Longleftrightarrow T(b \circ u) \text{ is (left/right) invertible} \tag{22}$$

for every $b \in L^\infty$ and every non-constant inner function u. The proof of the implication "$\Longleftarrow$" is less trivial and will not be given here. Let us instead fix what Theorem 3.2 says in the case of arguments as in Figures 4 or 5.

3.3. Corollary. *Let* u *be given by (13) or let* $u \in CU(\dot{\mathbf{T}})$ *be an infinite Blaschke product. Then if* $b \in C \cap GL^\infty$,

$$T(b) \text{ is invertible} \Longleftrightarrow T(b \circ u) \text{ is invertible},$$

$$T(b) \in \Phi(-\mathbf{N}) \Longleftrightarrow T(b \circ u) \in \Phi(-\infty),$$

$$T(b) \in \Phi(+\mathbf{N}) \Longleftrightarrow T(b \circ u) \in \Phi(+\infty).$$

Proof. We have either $T(b) \in \Phi_0$ or $T(b) \in \Phi(-\mathbf{N})$ or $T(b) \in \Phi(+\mathbf{N})$. So Theorem 3.2 implies that $T(b \circ u) \in \Phi_0$ or $T(b \circ u) \in \Phi(-\infty)$ or $T(b \circ u) \in \Phi(+\infty)$, respectively, and thus gives the implications "$\Longrightarrow$". Since the sets $\Phi_0, \Phi(-\infty), \Phi(+\infty)$ are pairwise disjoint, we automatically get the implications "$\Longleftarrow$". ∎

The next proposition states an interesting property of u-periodic functions; this property will find application in Section 4.

3.4. Proposition. *Let u and v be inner functions such that $u(0) = v(0)$. Then*

$$\int_{\mathbf{T}} |f(u(t))|^2 \, |dt| = \int_{\mathbf{T}} |f(v(t))|^2 \, |dt|$$

for every function $f \in L^2$.

Proof. Let $f(t) = \sum_k f_k t^k$ $(t \in \mathbf{T})$. Then

$$\int_{\mathbf{T}} |f(u(t))|^2 \, |dt| = \int_{\mathbf{T}} f(u(t)) \overline{f(u(t))} \, |dt| = \sum_{k,l} f_k \overline{f_l} \int_{\mathbf{T}} u(t)^{k-l} \, |dt|$$

and the residue theorem gives

$$\int_{\mathbf{T}} u(t)^{k-l} \, |dt| = \frac{1}{i} \int_{\mathbf{T}} u(t)^{k-l} \frac{dt}{t} = \begin{cases} 2\pi u(0)^{k-l} & \text{for } k - l \geq 0, \\ 2\pi \overline{u(0)}^{l-k} & \text{for } k - l < 0, \end{cases}$$

which depends only on the value of u at 0. ∎

3.5. Digression: composition operators on H^2. At least since Nordgren's paper [25], composition operators on H^2 have gained permanently increasing popularity. If $u \in H^\infty(\mathbf{D})$ is an inner function and $f \in H^2(\mathbf{D})$, then the composition $f \circ u$ is analytic in $\mathbf{D}$. The question is whether $f \circ u$ also belongs to L^2 and thus to H^2, i.e. whether the map $C_u : f \mapsto f \circ u$ is a well-defined and bounded operator on H^2. It turns out that this is always the case ! A really beautiful discussion of this set of problems and several proofs of the boundedness of C_u on H^2 are contained in Joel Shapiro's book [32]. With the help of the previous proposition we can give one more proof of the fact that C_u is bounded on H^2 for every inner function u. Indeed, let v be the inner function given by

$$v(z) = (u(0) - z)/(1 - \overline{u(0)}z).$$

Then $v(0) = u(0)$, and Proposition 3.4 implies that

$$\|C_u f\|_2^2 = \|C_v f\|_2^2 = \int_{\mathbf{T}} \left| f\left(\frac{u(0) - t}{1 - \overline{u(0)}t}\right) \right|^2 |dt| = (1 - |u(0)|^2) \int_{\mathbf{T}} \frac{|f(\tau)|^2 \, |d\tau|}{|1 - \overline{u(0)}\tau|^2}$$

(the last equality resulting from the substitution $\tau = (u(0) - t)/(1 - \overline{u(0)}t)$). Taking into account that $1 - |u(0)| \leq |1 - \overline{u(0)}\tau| \leq 1 + |u(0)|$ for all $\tau \in \mathbf{T}$, we arrive at the estimates

$$\left(\frac{1 - |u(0)|}{1 + |u(0)|}\right)^{1/2} \|f\|_2 \leq \|C_u f\|_2 \leq \left(\frac{1 + |u(0)|}{1 - |u(0)|}\right)^{1/2} \|f\|_2,$$

which, in particular, show that C_u is bounded on H^2.

4. Orientation preserving whirls

4.1. The symbol class. We define an equivalence relation on $CR(\mathbf{R})$ by saying that f and g in $CR(\mathbf{R})$ are equivalent, $f \sim g$, if the difference $f - g$ is bounded, i.e. $f - g \in L^{\infty}(\mathbf{R})$. A function $f \in CR(\mathbf{R})$ is said to be *essentially monotonous* if it is equivalent to a monotonous function in $CR(\mathbf{R})$.

Two functions $f, g \in CR(\dot{\mathbf{T}})$ are called equivalent, $f \sim g$, if $f^{\#} \sim g^{\#}$. In the same vein, we say that $f \in CR(\dot{\mathbf{T}})$ is essentially monotonous if so is $f^{\#} \in CR(\mathbf{R})$.

In this section we mainly focus our attention on symbols $a \in CU(\dot{\mathbf{T}})$ whose argument $\arg a \in CR(\dot{\mathbf{T}})$ is subject to the following conditions:

$$\lim_{t \to 1+0} \arg a(t) = -\infty, \ \lim_{t \to 1-0} \arg a(t) = +\infty, \tag{23}$$

$$\arg a \text{ is essentially monotonous.} \tag{24}$$

Obviously, (23) and (24) are equivalent to the assumptions

$$(\arg a)^{\#}(-\infty) = -\infty, (\arg a)^{\#}(+\infty) = +\infty, \tag{25}$$

$$(\arg a)^{\#} \text{ is essentially monotonous.} \tag{26}$$

Of course, if (25) is required then $(\arg a)^{\#}$ must be essentially monotonously increasing, i.e. $(\arg a)^{\#}$ is equivalent to a monotonously increasing function. Also notice that (23),(24) and (25),(26) are well-posed hypotheses in the sense that they do not depend on the particular choice of the argument $\arg a \in CR(\mathbf{T})$.

4.2. Example: almost periodic functions. An almost periodic polynomial is a function on $\mathbf{R}$ of the form

$$f(x) = \sum_{j} f_j e^{i\lambda_j x} \quad (x \in \mathbf{R})$$

where $f_j \in \mathbf{C}, \lambda_j \in \mathbf{R}$, and the sum is finite. The closure of the set of all almost periodic polynomials in $L^{\infty}(\mathbf{R})$ is denoted by $AP(\mathbf{R})$ and referred to as the *algebra of (uniformly) almost periodic functions*. Clearly, $AP(\mathbf{R}) \subset C(\mathbf{R}) \cap L^{\infty}(\mathbf{R})$.

A well known theorem by Bohr (see e.g. [22] or [24]) states that if $f \in AP(\mathbf{R}) \cap GL^{\infty}(\mathbf{R})$ then

$$f(x) = e^{i(\lambda x + g(x))} \quad (x \in \mathbf{R})$$

with some number $\lambda \in \mathbf{R}$ and some function $g \in AP(\mathbf{R}) \cap L^{\infty}(\mathbf{R})$. Hence, the argument of an (invertible) almost periodic function is always equivalent to a linear function and thus to a monotonous function.

A function $a \in L^{\infty}(\mathbf{T})$ is called *almost periodic*, $a \in AP$, if $a^{\#} \in AP(\mathbf{R})$. If $a \in AP \cap GL^{\infty}$ then $a/|a| \in CU(\dot{\mathbf{T}})$ and $\arg(a/|a|)$ is essentially monotonous.

4.3. Arguments of Blaschke products. Consider the infinite Blaschke product

$$B(t) = \prod_{k=-\infty}^{\infty} \frac{|\alpha_k|}{\alpha_k} \frac{\alpha_k - t}{1 - \overline{\alpha}_k t} \quad (t \in \mathbf{T}) \tag{27}$$

with $|\alpha_k| < 1$ and $\sum_{k=-\infty}^{\infty}(1 - |\alpha_k|) < \infty$; for what follows it will be more convenient to index the zeros α_k by $k \in \mathbf{Z}$ and not by $k \in \mathbf{N}$. The discontinuities of B on $\mathbf{T}$ are located at the cluster points of the zeros α_k. Hence, B may be discontinuous on massive subsets of the unit circle $\mathbf{T}$ (and even on all of $\mathbf{T}$), in which case a reasonable definition of $\arg B$ causes serious problems. For our further purposes it suffices to consider Blaschke products with the property that

$$\lim_{|k| \to \infty} \alpha_k = 1. \tag{28}$$

In that case $B \in CU(\dot{\mathbf{T}})$ and hence we may take $\arg B \in CR(\dot{\mathbf{T}})$.

The following result, which was established in [18], shows that $(\arg B)^{\#}$ is always strictly monotonously increasing. Later we will prove that if a satisfies (23),(24) then there exists a Blaschke product (27),(28) such that $\arg a \sim \arg B$; in other words, every monotonous growth of the argument $\arg a$ from $-\infty$ to $+\infty$ is essentially modelled by the argument of a Blaschke product !

4.4. Lemma. *Let B be an infinite Blaschke product of the form (27),(28) and let $\arg B \in CR(\dot{\mathbf{T}})$ be any argument. Then $\arg B(t)$ is strictly monotonously increasing as t traces out the unit circle counter-clockwise from $1 + 0$ to $1 - 0$, the limits*

$$A_+ := \lim_{t \to 1+0} \arg B(t), \quad A_- := \lim_{t \to 1-0} \arg B(t) \tag{29}$$

exist and at least one of these limits is infinite. There is an argument $\arg B \in CR(\dot{\mathbf{T}})$ such that $A_+ < 0, A_- > 0$, and

$$\arg B(e^{i\theta}) = \begin{cases} -2\pi\,\mathrm{card}\,\{\theta_k > \theta\} - 2\sum_{k=-\infty}^{\infty} \varphi_k(\theta) & \text{for} \quad 0 < \theta < \pi, \\ 2\pi\,\mathrm{card}\,\{\theta_k < \theta\} - 2\sum_{k=-\infty}^{\infty} \varphi_k(\theta) & \text{for} \quad -\pi < \theta < 0, \end{cases}$$

$$= \begin{cases} -2\Big(\sum_{\theta_k > \theta}(\pi + \varphi_k(\theta)) + \sum_{\theta_k \le \theta} \varphi_k(\theta)\Big) & \text{for} \quad 0 < \theta < \pi, \\ 2\Big(\sum_{\theta_k < \theta}(\pi - \varphi_k(\theta)) - \sum_{\theta_k \ge \theta} \varphi_k(\theta)\Big) & \text{for} \quad -\pi < \theta < 0, \end{cases}$$

where

$$\alpha_k = r_k e^{i\theta_k}, \ 0 < r_k < 1, -\pi < \theta_k \le \pi, \ \theta_k \to 0 \quad as \quad |k| \to \infty,$$

$$\varphi_k(\theta) := \arctan\left(\varepsilon_k \cot\left(\frac{\theta - \theta_k}{2}\right)\right), \quad \varepsilon_k := \frac{1 - r_k}{1 + r_k}, \tag{30}$$

$\arctan x \in (-\pi/2, \pi/2)$ for $x \in \mathbf{R}$, $\varphi_k(\theta_k) := \pi/2$, and $\mathrm{card}\,\{\theta_k > 0\}, \mathrm{card}\,\{\theta_k < 0\}$ denotes the number of $k \in \mathbf{Z}$ for which $\theta_k > 0, \theta_k < 0$, respectively.

Proof. Elementary plane geometry shows that if $r \in (0,1)$ and $\theta \in (-\pi, 0) \cup (0, \pi]$, then

$$(r - e^{i\theta})/(1 - re^{i\theta}) = e^{-2i\varphi(\theta)} \tag{31}$$

with

$$\varphi(\theta) := \arctan\left(\frac{1 - r}{1 + r} \cot\frac{\theta}{2}\right). \tag{32}$$

The function φ is continuous and strictly monotonously decreasing on $(-\pi, 0)$ and $(0, \pi]$, and we have

$$\varphi(-\pi + 0) = 0, \quad \varphi(0 - 0) = -\frac{\pi}{2}, \quad \varphi(0 + 0) = \frac{\pi}{2}, \quad \varphi(\pi) = 0.$$

Let B_k denote the kth factor in (27),

$$B_k(e^{i\theta}) = (r_k - e^{i(\theta - \theta_k)})/(1 - r_k e^{i(\theta - \theta_k)}),$$

and define $\varphi_k(\theta)$ by (30). From (31),(32) we infer that

$$B_k(e^{i\theta}) = e^{-2i\varphi_k(\theta)}.$$

The function φ_k is also continuous and strictly monotonously decreasing on $(-\pi, 0)$ and $(0, \pi]$. Furthermore,

$$\varphi(\theta_k - 0) = -\frac{\pi}{2}, \quad \varphi_k(\theta_k + 0) = \frac{\pi}{2}, \quad \varphi_k(-\pi + 0) = \varphi_k(\pi).$$

Thus, $e^{i\theta} \mapsto -2\varphi_k(\theta)$ is not a function in $CR(\dot{\mathbf{T}})$ unless $\theta_k = 0$. In case $\theta_k = 0$, we put

$$\arg B_k(e^{i\theta}) := \psi_k(\theta) := -2\varphi_k(\theta) \quad \text{for} \quad \theta \in (-\pi, 0) \cup (0, \pi].$$

Note that then $\arg B_k \in CR(\dot{\mathbf{T}})$ is monotonously increasing.

Let $\theta_k > 0$. In this case put

$$\psi_k(\theta) := \begin{cases} -2\varphi_k(\theta) & \text{for} \quad \theta \in (-\pi, 0) \cup [\theta_k, \pi], \\ -2\varphi_k(\theta) - 2\pi & \text{for} \quad \theta \in (0, \theta_k). \end{cases}$$

Clearly,

$$B_k(e^{i\theta}) = e^{-2i\varphi_k(\theta)} = e^{i\psi_k(\theta)} \quad \text{for} \quad \theta \in (-\pi, 0) \cup (0, \pi],$$

and since $\psi_k(\theta_k - 0) = -\pi = \psi_k(\theta_k + 0)$, it follows that ψ_k is continuous and strictly monotonously increasing $(-\pi, 0)$ and on $(0, \pi]$. Thus, if we set

$$\arg B_k(e^{i\theta}) := \psi_k(\theta) \quad \text{for} \quad \theta \in (-\pi, 0) \cup (0, \pi],$$

then $\arg B_k \in CR(\dot{\mathbf{T}})$ is a strictly monotonously increasing argument of B_k.

Analogously, if $\theta_k < 0$ we let

$$\psi_k(\theta) := \begin{cases} -2\varphi_k(\theta) & \text{for} \quad \theta \in (-\pi, \theta_k] \cup (0, \pi], \\ -2\varphi_k(\theta) + 2\pi & \text{for} \quad \theta \in (\theta_k, 0). \end{cases}$$

Then $\psi_k(\theta_k - 0) = \pi = \psi_k(\theta_k + 0)$ and hence,

$$\arg B_k(e^{i\theta}) := \psi_k(\theta) \quad \text{for} \quad \theta \in (-\pi, 0) \cup (0, \pi]$$

defines a strictly monotonously increasing argument of B_k belonging to $CR(\dot{\mathbf{T}})$.

An argument $\arg B \in CR(\dot{\mathbf{T}})$ is now given by

$$\arg B(e^{i\theta}) = \sum_{k=-\infty}^{\infty} \arg B_k(e^{i\theta}) = \sum_{k=-\infty}^{\infty} \psi_k(\theta). \tag{33}$$

If $\theta \in (0, \pi)$, then the right-hand side of (33) equals

$$\sum_{-\pi < \theta_k < 0} (-2\varphi_k(\theta)) + \sum_{0 \le \theta_k \le \theta} (-2\varphi_k(\theta)) + \sum_{\theta < \theta_k \le \pi} (-2\varphi_k(\theta) - 2\pi)$$

$$= -2\pi \operatorname{card} \{\theta_k > \theta\} - 2 \sum_{k=-\infty}^{\infty} \varphi_k(\theta)$$

$$= -2 \left(\sum_{\theta_k > \theta} (\pi - \varphi_k(\theta)) + \sum_{\theta_k \le \theta} \varphi_k(\theta) \right),$$

while for $\theta \in (-\pi, 0)$ the right-hand side of (33) is

$$\sum_{-\pi < \theta_k < \theta} (-2\varphi_k(\theta) + 2\pi) + \sum_{\theta \le \theta_k < 0} (-2\varphi_k(\theta)) + \sum_{0 \le \theta_k < \pi} (-2\varphi_k(\theta))$$

$$= 2\pi \operatorname{card} \{\theta_k < \theta\} - 2 \sum_{k=-\infty}^{\infty} \varphi_k(\theta)$$

$$= 2 \left(\sum_{\theta_k < \theta} (\pi - \varphi_k(\theta)) - \sum_{\theta_k \ge \theta} \varphi_k(\theta) \right).$$

Obviously, $\arg B$ is strictly monotonously increasing. Since $\psi_k(0+0) \le 0$ and $\psi_k(0-0) \ge 0$ for all k, it follows that for our argument the limits (29) exist and that $A_+ \le 0$ and $A_- \ge 0$. Because $\psi_k(0+0) < 0$ and $\psi_k(0-0) > 0$ whenever $\theta_k \ne \pi$, we conclude that actually $A_+ < 0$ and $A_- > 0$.

We now show that at least one of the numbers A_+ and A_- is infinite. Suppose $\theta_k > 0$ for infinitely many k. Then

$$\sum_{\theta_k > \theta} (\pi + \varphi_k(\theta)) \ge \sum_{\theta_k > \theta} \left(\pi - \frac{\pi}{2} \right) \to +\infty \quad \text{as} \quad \theta \to 0 + 0,$$

and since $\varphi_k(\theta_k) \ge 0$ for $\theta_k \le \theta$, we see that $A_+ = -\infty$. In the same way we get $A_- = +\infty$ if $\theta_k < 0$ for infinitely many k. So we are left with the case where $\theta_k = 0$ for all but finitely many k. As

$$-2 \sum_{\theta_k=0} \varphi_k(0 \pm 0) = \sum_{\theta_k=0} (\mp \pi) = \mp\infty$$

in this case and both $\operatorname{card} \{\theta_k > 0\}$ and $\operatorname{card} \{\theta_k < 0\}$ remain finite as $\theta \to 0 + 0$ and $\theta \to 0 - 0$, respectively, it follows that $A_+ = -\infty$ and $A_- = +\infty$.

Since any argument $\arg B \in CR(\dot{\mathbf{T}})$ differs from the argument constructed above only by an integral multiple of 2π, we deduce that any such argument is strictly monotonously increasing and satisfies (29) with $A_+ = -\infty$ or $A_- = +\infty$. ∎

In [1] (also see [3, 2.26]) it was shown that if the assumptions of the next theorem are satisfied, then the operator $T(a)$ cannot be Fredholm. The problem whether such operators can belong to $\Phi(+\infty)$ had been open since then. The following theorem tells us that this cannot happen. Notice that the hypotheses of this theorem are obviously satisfied if (23),(24) or (25),(26) are in force.

4.5. Theorem. *Let* $\arg a \in CR(\dot{\mathbf{T}})$, *suppose* $(\arg a)^{\#}(+\infty) = +\infty$ *and* $(\arg a)^{\#}$ *is bounded from above at* $-\infty$, *which means that there exist* $M \in \mathbf{R}$ *and* $x_0 \in \mathbf{R}$ *such that* $(\arg a)^{\#}(x) < M$ *for all* $x < x_0$. *Then*

$$T(a) \in \Phi(-\infty) \cup NNS. \tag{34}$$

Proof. Assume (34) is not true. Then $T(a)$ is right-Fredholm and hence, by 2.10, we have $a = cs$ with $c \in C + \overline{H^{\infty}}$ and a sectorial function $s \in GL^{\infty}$. The function c may be written as $c = b\overline{h}$ with $b \in C \cap GL^{\infty}$ and $h \in H^{\infty}$ (see [24, 165.52.2]) and h in turn may be factored in the form $h = ug$ with an inner function u and a function $g \in GH^{\infty}$ (recall 2.7). Thus $a = b\overline{u}\,\overline{g}s$.

The set of all Blaschke products is uniformly dense in the set of all inner functions (see [12, II.6.5]). Hence, given $\varepsilon > 0$, there exists a Blaschke product B such that $\|u - B\|_{\infty} < \varepsilon$. It follows that $\|1 - \overline{B}u\|_{\infty} = \|u - B\|_{\infty} < \varepsilon$, hence $\overline{B}u = 1 + \delta$ and thus $u = B(1 + \delta)$ with $\|\delta\|_{\infty} < \varepsilon$. In summary, $a = bs(1 + \delta)\overline{B}\overline{g}$. We may choose ε small enough, so that $r := s(1 + \delta)$ is also sectorial. Consequently, $Bg = \overline{a}\overline{b}^{-1}\overline{r}^{-1}$ with $a \in CU(\dot{\mathbf{T}})$, $b \in C \cap GL^{\infty}$, and a sectorial function $r \in GL^{\infty}$. If $t_0 \in \dot{\mathbf{T}}$, then $\overline{a}\overline{b}^{-1}$ is continuous at t_0, which implies that there exists a number $\gamma \in \mathbf{T}$ such that $\mathrm{Re}\,(\gamma\overline{a}\overline{b}^{-1}\overline{r}^{-1}) \geq \eta > 0$ a.e. on some open arc $I \subset \mathbf{T}$ containing t_0. It results that $\mathrm{Re}\,(\gamma Bg) \geq \eta > 0$ a.e. on I and therefore B is analytic on I (see [12, Chap. II, Exercise 14]). As $t_0 \in \dot{\mathbf{T}}$ may be picked arbitrarily, we conclude that $B \in CU(\dot{\mathbf{T}})$ and hence we may choose $\arg B$ in $CR(\dot{\mathbf{T}})$.

Because $g \in GH^{\infty}$, the function $w := \log|g|$ belongs to L^{∞}. Let $\tilde{w}$ denote the conjugate function of w, normalized so that $g = \exp(w + i\tilde{w})$. Since each $t_0 \in \dot{\mathbf{T}}$ is contained in some open subarc I of $\mathbf{T}$ such that $\mathrm{Re}\,(\gamma Bg) \geq \eta > 0$ a.e. on I and since B is continuous on I, it follows that g is sectorial on some open arc J such that $t_0 \in J \subset I$. Hence, there is a constant $\lambda_J \in \mathbf{R}$ such that $|\tilde{w} - \lambda_J| < \pi/2$ a.e. on J.

We may write $b = |b|e^{id}$ and $r = |r|e^{i(-v-\mu)}$ with bounded real-valued functions d and v on $\mathbf{T}$, $\mu \in \mathbf{R}$, and $\|v\|_{\infty} < \pi/2$. We so obtain

$$a = |b|e^{id}|r|e^{i(-v-\mu)}e^{-i\arg B}e^{w-i\tilde{w}},$$

whence

$$e^{i\arg a} = a/|a| = e^{id}e^{i(-v-\mu)}e^{-i\arg B}e^{-i\tilde{w}},$$

and thus

$$e^{i(v+\tilde{w}+\mu)} = e^{i(d-\arg a-\arg B)}.$$

It follows that

$$d - \arg a - \arg B = v + \tilde{w} + \mu + 2\pi k \tag{35}$$

where k is a function on $\mathbf{T}$ assuming only integral values. Fix $t_0 \in \dot{\mathbf{T}}$. Since $\varphi := d - \arg a - \arg B - \mu$ is continuous at t_0, $\|v\|_{\infty} < \pi/2$, and

$$\underset{t \in J}{\mathrm{ess\,sup}}\,|\tilde{w}(t) - \lambda_J| < \pi/2$$

for some open arc $J \subset \mathbf{T}$ containing t_0, it follows that

$$2\pi \operatorname*{ess\,sup}_{t \in J} |k(t) - k(t_0)|$$

$$\leq \sup_{t \in J} |\varphi(t) - \varphi(t_0)| + \operatorname*{ess\,sup}_{t \in J} |v(t) - v(t_0)| + \operatorname*{ess\,sup}_{t \in J} |\tilde{w}(t) - \tilde{w}(t_0)|$$

$$\leq \sup_{t \in J} |\varphi(t) - \varphi(t_0)| + 2\|v\|_\infty + 2 \operatorname*{ess\,sup}_{t \in J} |\tilde{w}(t) - \lambda_J|,$$

and this is less than 2π if only J is sufficiently small. Hence, k is locally and thus globally constant. Consequently, (35) holds with some integer $k \in \mathbf{Z}$.

Because v and w belong to L^∞, we deduce from a theorem by Fefferman (see [12, VI.1.5]) that the right-hand side of (35) belongs to $BMO(\mathbf{T})$. The same is therefore true for the left-hand side $\psi := d - \arg a - \arg B$. It follows that $\psi^\# \in BMO(\mathbf{R})$ (see [12, VI.1.3]). The boundedness of d, the assumption made in the present theorem for a, and Lemma 4.4 for B imply that $\psi^\#(+\infty) = -\infty$ and that $\psi^\#$ is bounded from below at $-\infty$. We claim that such a function cannot belong to $BMO(\mathbf{R})$.

If $\psi^\# \in BMO(\mathbf{R})$, then the function $f(x) := (\psi^\#(-x) - \psi^\#(x))/2$ also belongs to $BMO(\mathbf{R})$. Because f is odd, the mean value $f_I := (1/|I|) \int_I f(\xi) \, d\xi$ is zero for every I of the form $I = (-x, x)$. So the BMO condition gives

$$\sup_{x > 0} \frac{1}{x} \int_0^x |f(\xi)| \, d\xi =: N < \infty. \tag{36}$$

Put $F(x) = \int_0^x |f(\xi)| \, d\xi$. Obviously, $F(0) = 0$, $F(x) \geq 0$ for $x \geq 0$, and (36) says that

$$F(x) \leq Nx \quad \text{for} \quad x > 0. \tag{37}$$

It is precisely the conditions that $\psi^\#(+\infty) = -\infty$ and that $\psi^\#$ be bounded from below at $-\infty$ which imply that $f(+\infty) = +\infty$. Thus, by the mean value theorem, there exists an $x_0 > 0$ such that

$$F(2x_0) - F(x_0) = \int_{x_0}^{2x_0} |f(\xi)| \, d\xi \geq (2N + 1)x_0,$$

whence

$$\frac{F(2x_0)}{2x_0} = \frac{F(x_0) + (2N + 1)x_0}{2x_0} \geq \frac{(2N + 1)x_0}{2x_0} = N + \frac{1}{2},$$

which contradicts (37) and completes the proof. ∎

If $B \in CU(\dot{\mathbf{T}})$ is an infinite Blaschke product, then $T(B) \in \Phi(-\infty)$. To see this notice first that $T(\overline{B})$ is a left inverse of $T(B)$. We so have at least two possibilities to conclude that $T(B) \in \Phi(-\infty)$: we may have recourse to 2.7, where we showed that $\dim \operatorname{Ker} T(\overline{B}) = \infty$, or we may combine Lemma 4.4 and Theorem 4.5 to see that $T(B) \in \Phi(-\infty)$. The following theorem was established in [18].

4.6. Theorem. *Let* $\arg a \in CR(\dot{\mathbf{T}})$ *and suppose*

the limits $(\arg a)^\#(\pm\infty)$ *exist and at least one of them is infinite,* $\qquad$ (38)

$(\arg a)^\#$ *is essentially monotonously increasing.* $\qquad$ (39)

Then there exists an infinite Blaschke product B of the form (27),(28) such that $\arg a \sim \arg B$. In particular, each function $\arg a$ satisfying (38) and (39) is equivalent to the argument of a symbol $B \in CU(\dot{\mathbf{T}})$ for which $T(B)$ is in $\Phi(-\infty)$.

Proof. Put $f(\theta) = \arg a(e^{i\theta})$ for $\theta \in (-\pi, 0) \cup (0, \pi]$. For the sake of definiteness, suppose

$$f(0 + 0) = (\arg a)^{\#}(-\infty) = -\infty,$$

and without loss of generality assume $f(\pi) = 0$ and f is strictly monotonously increasing.

Define numbers $\theta_1, \theta_2, \ldots \in (0, \pi)$ by $f(\theta_k) = -(2k - 1)\pi$. Obviously, θ_k goes monotonously to zero as $k \to \infty$. Let k_0 be a positive integer such that $\cot((\theta_k - \theta_{k+1})/2) > 1$ for all $k \geq k_0$. Choose any numbers r_k for $1 \leq k \leq k_0$ and define r_k for $k > k_0$ by the condition

$$\max\left\{\varepsilon_k \cot \frac{\theta_k - \theta_{k+1}}{2}, \ \varepsilon_k \cot \frac{\theta_{k-1} - \theta_k}{2}\right\} = \frac{1}{2^k} \tag{40}$$

where $\varepsilon_k := (1 - r_k)/(1 + r_k)$. Then put $\alpha_k = r_k e^{i\theta_k}$. We have

$$\sum_{k > k_0} (1 - |\alpha_k|) = \sum_{k > k_0} (1 - r_k) \leq 2 \sum_{k > k_0} \varepsilon_k \leq 2 \sum_{k > k_0} \frac{1}{2^k} < \infty.$$

Let B_+ be the infinite Blaschke product with the zeros $\alpha_1, \alpha_2, \ldots$.

If $n > k_0$ then, by Lemma 4.4,

$$\begin{aligned}
&|\arg B_+(e^{i\theta_n}) - f(\theta_n)| \\
&= 2\left|\sum_{k=1}^{n-1}(\pi - \varphi_k(\theta_n)) + \varphi_n(\theta_n) + \sum_{k=n+1}^{\infty} \varphi_k(\theta_n) + \frac{f(\theta_n)}{2}\right| \\
&= 2\left|\sum_{k \in \mathbb{N} \setminus \{n\}} \varphi_k(\theta_n) + (n - 1)\pi + \frac{\pi}{2} + \frac{f(\theta_n)}{2}\right| \\
&= 2\left|\sum_{k \in \mathbb{N} \setminus \{n\}} \varphi_k(\theta_n)\right| \leq 2 \sum_{k \in \mathbb{N} \setminus \{n\}} |\varphi_k(\theta_n)|
\end{aligned}$$

and since

$$|\varphi_k(\theta_n)| \leq \varepsilon_k \cot \frac{|\theta_n - \theta_k|}{2} \leq \begin{cases} \varepsilon_k \cot((\theta_k - \theta_{k+1})/2) & \text{for} \quad k < n, \\ \varepsilon_k \cot((\theta_{k-1} - \theta_k)/2) & \text{for} \quad k > n, \end{cases}$$

it follows that

$$|\arg B_+(e^{i\theta_n}) - f(\theta_n)| \leq 2 \sum_{k=1}^{k_0} \varepsilon_k \cot \frac{\theta_k - \theta_{k+1}}{2} + 2 \sum_{k > k_0} \frac{1}{2^k} =: M.$$

Notice that M is independent of n. Let now $\theta \in (\theta_{n-1}, \theta_n)$. Then

$$\begin{aligned}
|\arg B_+(e^{i\theta}) - f(\theta)| &\leq |\arg B_+(e^{i\theta}) - \arg B_+(e^{i\theta_n})| \\
&+ |\arg B_+(e^{i\theta_n}) - f(\theta_n)| + |f(\theta) - f(\theta_n)|,
\end{aligned} \tag{41}$$

and since $\arg B_+(e^{i\theta})$ and $f(\theta)$ are monotonous, we see that (41) is at most

$$
\begin{aligned}
&|\arg B_+(e^{i\theta_{n-1}}) - \arg B_+(e^{i\theta_n})| + M + |f(\theta_{n-1}) - f(\theta_n)| \\
&\leq |\arg B_+(e^{i\theta_{n-1}}) - f(\theta_{n-1})| + |f(\theta_{n-1}) - f(\theta_n)| \\
&+|f(\theta_n) - \arg B_+(e^{i\theta_n})| + M + |f(\theta_{n-1}) - f(\theta_n)| \\
&\leq M + 2\pi + M + M + 2\pi = 3M + 4\pi.
\end{aligned}
$$

Thus, $\arg B_+(e^{i\theta}) - f(\theta)$ is bounded on $(0, \pi)$.

If $\theta \in (-\pi, 0)$, we obtain from Lemma 4.4 and (40) that

$$
\arg B_+(e^{-i\pi}) \leq \arg B_+(e^{i\theta}) = -2 \sum_{k=1}^{\infty} \varphi_k(\theta)
$$

$$
\leq -2 \sum_{k=1}^{\infty} \varphi_k(0) = 2 \sum_{k=1}^{\infty} |\varphi_k(0)|
$$

$$
\leq 2 \sum_{k=1}^{\infty} \varepsilon_k \cot \frac{\theta_k}{2} \leq 2 \sum_{k=1}^{\infty} \varepsilon_k \cot \frac{\theta_k - \theta_{k+1}}{2} \leq \sum_{k=1}^{\infty} \frac{1}{2^k} = 2.
$$

Consequently, if $f(0-0) = (\arg a)^{\#}(+\infty)$ is finite, we have $f \sim \arg B_+$. If $f(0-0) = (\arg a)^{\#}(+\infty) = +\infty$, we may in a similar way construct an infinite Blaschke product B_- such that

$$
\begin{aligned}
|\arg B_-(e^{i\theta}) - f(\theta)| &\leq M \quad \text{for} \quad \theta \in (-\pi, 0), \\
|\arg B_-(e^{i\theta})| &\leq M \quad \text{for} \quad \theta \in (0, \pi].
\end{aligned}
$$

Then $B := B_- B_+$ is a Blaschke product satisfying $\arg B \sim \arg a$. ∎

Our next objective is to show the following counterpart of Theorem 4.6: if $\arg a \in CR(\dot{\mathbf{T}})$ satisfies (38) and (39), then there exists a function $b \in CU(\dot{\mathbf{T}})$ such that $\arg a \sim \arg b$ but $T(b) \in NNS$. The problem whether such b exist had been open for a long time; it was, for example, posed by I. Spitkovsky to one of the authors already 10 years ago. Unfortunately, we do not yet have a sufficiently simple proof of Theorem 4.10, which will give the solution to this problem. The following three subsections serve the preparation of the proof of Theorem 4.10.

4.7. Lemma. *Let $B \in CU(\dot{\mathbf{T}})$ be an infinite Blaschke product and let $\arg B \in CR(\dot{\mathbf{T}})$ be any argument of B. Put $\varphi(\theta) = (\arg B)(e^{i\theta})$. Then for every $\alpha = e^{i\theta_0} \in \dot{\mathbf{T}}$ and every $f \in L^2$,*

$$
\int_{\mathbf{T}} |f(t)|^2 |dt| = \frac{1}{\varphi'(\theta_0)} \int_{\mathbf{T}} |f(B(t))|^2 \left| \frac{B(t) - B(\alpha)}{t - \alpha} \right|^2 |dt|. \tag{42}
$$

Moreover, if $g \in H^2$ and $k \in \overline{H^2}$, then

$$
g(B(t)) \frac{B(t) - B(\alpha)}{t - \alpha} \quad \text{and} \quad k(B(t)) \frac{B(t) - B(\alpha)}{t - \alpha} \overline{B(t)} \tag{43}
$$

are functions in H^2 and $\chi_{-1}\overline{H^2}$, respectively.

Proof. Let first $z \in \mathbf{D}$ and put $c = (1 - |z|^2)/(1 - |B(z)|^2)$. We claim that

$$\int_{\mathbf{T}} |f(t)|^2 |dt| = c \int_{\mathbf{T}} |f(B(t))|^2 \left| \frac{B(t) - B(z)}{t - z} \right|^2 |dt|. \tag{44}$$

The substitution

$$t = \frac{z - \tau}{1 - \overline{z}\tau}, \quad dt = -\frac{1 - |z|^2}{(1 - \overline{z}\tau)^2} d\tau$$

transforms the integral on the right of (44) into

$$\frac{c}{1 - |z|^2} \int_{\mathbf{T}} |f(B_z(\tau))|^2 |B_z(\tau) - B(z)|^2 |d\tau| \tag{45}$$

where $B_z(\tau) := B((z - \tau)/(1 - \overline{z}\tau))$. Proposition 3.4 with

$$u(\tau) = B_z(\tau) \quad \text{and} \quad v(\tau) = (B(z) - \tau)/(1 - \overline{B(z)}\tau)$$

shows that (45) equals

$$\frac{c}{1 - |z|^2} \int_{\mathbf{T}} \left| f\left(\frac{B(z) - \tau}{1 - \overline{B(z)}\tau} \right) \right|^2 \left| \frac{B(z) - \tau}{1 - \overline{B(z)}\tau} - B(z) \right|^2 |d\tau|. \tag{46}$$

Substituting

$$t = \frac{B(z) - \tau}{1 - \overline{B(z)}\tau}, \quad \tau = \frac{B(z) - t}{1 - \overline{B(z)}t}, \quad d\tau = -\frac{1 - |B(z)|^2}{(1 - \overline{B(z)}t)^2} dt$$

finally yields that (46) is equal to the left-hand side of (44). This proves our claim.

Now write down (44) with $z = r\alpha$ and let r go to 1. Because

$$\lim_{r \to 1} \frac{1 - |r\alpha|^2}{1 - |B(r\alpha)|^2} = \lim_{r \to 1} \frac{1 - r^2}{1 - |B(\alpha) + B'(\alpha)\alpha(r - 1) + O((1 - r)^2)|^2}$$

$$= \lim_{r \to 1} \frac{1 - r^2}{2\mathrm{Re}\,(B'(\alpha)\overline{B(\alpha)}\alpha)(1 - r) + O((1 - r)^2)} = \frac{1}{\mathrm{Re}\,(B'(\alpha)\overline{B(\alpha)}\alpha)}$$

$$= \lim_{\theta \to \theta_0} 1 \Big/ \left(\mathrm{Re}\,\left(\frac{e^{i\varphi(\theta)} - e^{i\varphi(\theta_0)}}{e^{i\theta} - e^{i\theta_0}} \frac{e^{-i\varphi(\theta_0)}}{e^{-i\theta_0}} \right) \right) = 1 \Big/ \left(\mathrm{Re}\,\frac{i\varphi'(\theta_0)}{i} \right) = 1/\varphi'(\theta_0),$$

we arrive at (42).

Obviously, $h(t) = (B(t) - B(\alpha))/(t - \alpha)$ is an H^∞ function, which implies that the first function in (43) belongs to H^2 whenever $g \in H^2$ (also recall 3.5). Let now $k \in \overline{H^2}$. Then $k(B(t))$ is a function in $\overline{H^2}$ (again by 3.5). For $t \in \mathbf{T}$, we have

$$\frac{B(t) - B(\alpha)}{t - \alpha} \overline{B}(t) = t^{-1} \frac{1 - B(\alpha)\overline{B(t)}}{1 - \alpha\overline{t}} \tag{47}$$

and it is clear that (47) is a function in $\chi_{-1}\overline{H^\infty}$. ∎

4.8. The construction. Let $f(\theta) = \arg a(e^{i\theta})$ be an essentially monotonous function for which

$$\lim_{\theta\to 0+0} f(\theta) = -\infty, \quad \lim_{\theta\to 2\pi-0} f(\theta) = A_- \in \mathbf{R} \cup \{+\infty\}.$$

By Theorem 4.6, we may find an infinite Blaschke product $B \in CU(\dot{\mathbf{T}})$ such that if we define $\varphi(\theta) = \arg B(e^{i\theta})$, then $\varphi \sim f$ and

$$\lim_{\theta\to 0+0} \varphi(\theta) = -\infty, \quad \lim_{\theta\to 2\pi-0} \varphi(\theta) = A_-.$$

Without loss of generality assume $A_- > 0$. Define points $\{\theta_j\}_{j=-M}^{\infty}$ on $(0, 2\pi)$ by $\varphi(\theta_j) = -2\pi j$; in case $A_- = +\infty$ we have $M = \infty$.

In what follows we denote by z^γ ($\gamma \in \mathbf{R}$) the branch of the function which is analytic in $\mathbf{C} \setminus [0, +\infty)$ and takes the value $e^{i\pi\gamma}$ at $z = -1$. For $e^{i\theta} \in \mathbf{T}$, put $B_0(e^{i\theta}) = (B(e^{i\theta}))^{1/2}$. Then $B_0(e^{i\theta}) = e^{i\varphi_0(\theta)}$ where

$$\varphi_0(\theta) = -\pi j + \frac{1}{2}\varphi(\theta) \quad \text{for} \quad \theta \in [\theta_j, \theta_{j-1})$$

(see Figure 6). Choose points $\xi_j \in (\theta_j, \theta_{j-1})$, define ψ on $[\theta_j, \xi_j]$ by $\psi(\theta) = \varphi_0(\theta)$ and let $\psi(\theta)$ join $\psi(\xi_j)$ and $\varphi_0(\theta_{j-1})$ linearly as θ goes from ξ_j to θ_{j-1}. Then ψ is a continuous and real-valued function on $(0, 2\pi)$. Finally, put

$$b(e^{i\theta}) = e^{i\psi(\theta)} \quad \text{for} \quad \theta \in (0, 2\pi).$$

Obviously, $\arg b = \psi \sim \varphi_0 \sim \varphi \sim f$.

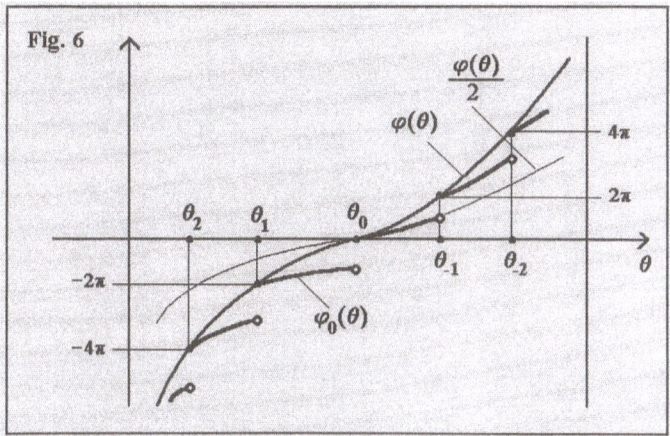

Fig. 6

4.9. Lemma. *There is a choice of the numbers $\{\xi_j\}_{j=-M}^{\infty}$ such that $T(b) \in NNS$.*

Proof. For $\delta \in (0, 1)$ and $j = -(M-1), \ldots, 0, 1, \ldots$ put

$$x_\delta(t) = \delta^{1/2}\left((t-1)^{-1/2+\delta/2} - (t-1)^{-1/2+\delta}\right)$$

and
$$y_{\delta,j}(t) = x_\delta(B(t))d_j(B(t) - 1)/(t - e^{i\theta_j}),$$

where $d_j := 1/\sqrt{\varphi'(\theta_j)}$. Clearly, x_δ is in H^2 and $y_{\delta,j}$ is analytic in $\mathbf{D}$. Lemma 4.7 shows that $\|y_{\delta,j}\|_2 = \|x_\delta\|_2$, and hence $y_{\delta,j} \in H^2$. Furthermore,

$$
\begin{aligned}
\|x_\delta\|_2^2 &= \int_0^{2\pi} |x_\delta(e^{i\theta})|^2 d\theta \geq \int_0^1 |x_\delta(e^{i\theta})|^2 d\theta \\
&= \delta \int_0^1 |e^{i\theta} - 1|^{-1+\delta} |1 - (e^{i\theta} - 1)^{\delta/2}|^2 d\theta,
\end{aligned}
$$

and since $|e^{i\theta} - 1| \leq \theta$, it follows that

$$\|x_\delta\|_2^2 \geq \delta \int_0^1 \theta^{-1+\delta}(1 - \theta^{\delta/2}) d\theta = \delta \frac{1}{6\delta} = \frac{1}{6}.$$

Thus, because $\|y_{\delta,j}\|_2 = \|x_\delta\|_2$,

$$\|y_{\delta,j}\|_2 \geq 1/\sqrt{6} \quad \forall \delta \quad \forall j. \tag{48}$$

We now show that
$$\lim_{\delta \to 0} \|T(B_0)y_{\delta,j}\|_2 = 0 \text{ uniformly in } j. \tag{49}$$

We have

$$
\begin{aligned}
&t^{1/2}(t - 1)^{-1/2+\delta/2} - t^{1/2}(t - 1)^{-1/2+\delta} \\
&= t^{1/2-\delta/2}(t - 1)^{-1/2+\delta/2} t^{\delta/2} - t^{1/2-\delta}(t - 1)^{-1/2+\delta} t^\delta \\
&= \left(\frac{t}{t - 1}\right)^{1/2-\delta/2} t^{\delta/2} - \left(\frac{t}{t - 1}\right)^{1/2-\delta} t^\delta \\
&= \left(\frac{t}{t - 1}\right)^{1/2-\delta/2} - \left(\frac{t}{t - 1}\right)^{1/2-\delta} \\
&\quad - \left(\frac{t}{t - 1}\right)^{1/2-\delta/2}(1 - t^{\delta/2}) + \left(\frac{t}{t - 1}\right)^{1/2-\delta}(1 - t^\delta)
\end{aligned}
$$

and therefore,

$$
\begin{aligned}
(T(B_0)y_{\delta,j})(t) &= P\left(B(t)^{1/2} y_{\delta,j}(t)\right) \\
&= d_j \delta^{1/2} P\left\{\left(\left(\frac{B(t)}{B(t) - 1}\right)^{1/2-\delta/2} - \left(\frac{B(t)}{B(t) - 1}\right)^{1/2-\delta}\right) \frac{B(t) - 1}{t - e^{i\theta_j}}\right\} \\
&\quad - d_j \delta^{1/2} P\left\{\left(\frac{B(t)}{B(t) - 1}\right)^{1/2-\delta/2}(1 - B(t)^{\delta/2}) \frac{B(t) - 1}{t - e^{i\theta_j}}\right\} \\
&\quad + d_j \delta^{1/2} P\left\{\left(\frac{B(t)}{B(t) - 1}\right)^{1/2-\delta}(1 - B(t)^\delta) \frac{B(t) - 1}{t - e^{i\theta_j}}\right\}. \tag{50}
\end{aligned}
$$

Since

$$k(t) = \left(\left(\frac{t}{t-1}\right)^{1/2-\delta/2} - \left(\frac{t}{t-1}\right)^{1/2-\delta}\right)t$$

is a function in $\overline{H^2}$, we deduce from Lemma 4.7 that the expression in the braces of the first term on the right of (50) is a function in $\chi_{-1}\overline{H^2}$. Hence this term vanishes. Also by Lemma 4.7, the squared L^2 norm of the second term on the right of (50) is not greater than

$$\delta \int_{\mathbf{T}} \left|\frac{t}{t-1}\right|^{1-\delta} |1 - t^{\delta/2}|^2 |dt|. \tag{51}$$

Because $|e^{i\theta} - 1| = 2\left|\sin\frac{\theta}{2}\right|$ and $|1 - e^{i\theta\delta/2}| \le |\theta|\delta/2$, it follows that (51) may be estimated from above by

$$\delta \int_{-\pi}^{\pi} \frac{|\theta|^2\delta^2/4}{\left(2\left|\sin\frac{\theta}{2}\right|\right)^{1-\delta}} d\theta = \frac{\delta^3}{2} \int_0^{\pi} \frac{\theta^2}{\left(2\sin\frac{\theta}{2}\right)^{1-\delta}} d\theta$$

$$\le \frac{\delta^3}{2} \int_0^{\pi} \frac{\theta^2}{(2\theta/\pi)^{1-\delta}} d\theta = \frac{\delta^3}{2}\left(\frac{\pi}{2}\right)^{1-\delta} \frac{\pi^{2+\delta}}{2+\delta} = o(1) \quad (\delta \to 0).$$

In the same way we may estimate the third term on the right of (50). This completes the proof of (49).

We now specify the numbers ξ_j. As θ moves from θ_j to θ_{j-1}, the value of $\varphi(\theta)$ increases from $-2\pi j$ to $-2\pi j + 2\pi$. For fixed $\delta > 0$, define $\eta_{\delta,j} \in (\theta_j, \theta_{j+1})$ by

$$\varphi(\eta_{\delta,j}) = -2\pi j + 2\pi - \exp(-1/\delta^2).$$

Notice that $\exp(-1/\delta^2)$ is very small if δ is a small number, so that $\eta_{\delta,j}$ lies very close to θ_{j-1}. Then choose a sequence $\{\delta_n\}_{n=1}^{\infty}$ such that

$$\delta_n \to 0 \quad \text{and} \quad \delta_n/\varphi'(\theta_n) \to 0 \quad \text{as} \quad n \to \infty.$$

Finally, define $\xi_j \in (\theta_j, \theta_{j-1})$ by

$$|\xi_j - \theta_{j-1}| = \min\left\{|\eta_{\delta,j} - \theta_{j-1}|, \frac{|\theta_j - \theta_{j-1}|^2}{j^2 + 1 + |\theta_j - \theta_{j-1}|}, \frac{|\theta_j - \theta_{j+1}|}{j^2 + 1}\right\}. \tag{52}$$

Put $z_n = y_{\delta_n,j}/\|y_{\delta_{n,n}}\|_2$. Then $\|z_n\|_2 = 1$. We claim that

$$\|T(b)z_n\|_2 \to 0 \quad \text{as} \quad n \to \infty. \tag{53}$$

From Theorem 4.5 we know that $T(b)$ is either left-invertible or not normally solvable. If (53) holds, then $T(b)$ cannot be left-invertible, so $T(b) \in NNS$. Hence, our proof will be complete as soon as (53) is shown. We have

$$\|T(b)z_n\|_2 \le \|T(B_0)z_n\|_2 + \|T(b - B_0)z_n\|_2,$$

and from (48) and (49) we infer that

$$\|T(B_0)z_n\|_2 = \|T(B_0)y_{\delta_n,n}\|_2/\|y_{\delta_n,n}\|_2 \leq \sqrt{6}\|T(B_0)y_{\delta_n,n}\|_2$$

goes to zero as $n \to \infty$. Furthermore,

$$\|T(b - B_0)z_n\|_2^2 \leq \|(b - B_0)z_n\|_2^2$$

$$= \sum_{j=-(M-1)}^{\infty} \int_{\xi_j}^{\theta_j-1} |b(e^{i\theta}) - B_0(e^{i\theta})|^2 |z_n(e^{i\theta})|^2 \, d\theta$$

$$\leq 4\left(\sum_{j=-(M-1)}^{n-1} \int_{\xi_j}^{\theta_j-1} + \int_{\xi_n}^{\theta_{n-1}} + \sum_{j=n+1}^{\infty} \int_{\xi_j}^{\theta_j-1} \right)|z_n(e^{i\theta})|^2 \, d\theta. \tag{54}$$

Due to (52) the middle integral in (54) is not greater than

$$\int_{\eta_{\delta_n,n}}^{\theta_{n-1}} |z_n(e^{i\theta})|^2 \, d\theta \leq \sum_{j=(M-1)}^{\infty} \int_{\eta_{\delta_n,j}}^{\theta_j-1} |z_n(e^{i\theta})|^2 \, d\theta. \tag{55}$$

Since $z_n(t) = x_{\delta_n}(B(t))d_n(B(t) - 1)/((t - e^{i\theta_n})\|y_{\delta_n,n}\|_2)$, we may conclude from Lemma 4.7 that the right-hand side of (55) equals

$$\int_{-\exp(-1/\delta_n^2)}^{0} \left(|x_{\delta_n}(e^{i\theta})|^2/\|y_{\delta_n,n}\|_2^2 \right) d\theta$$

$$\leq 6\,\delta_n \int_{-\exp(-1/\delta_n^2)}^{0} |(e^{i\theta} - 1)^{-1/2+\delta_n/2} - (e^{i\theta} - 1)^{1/2+\delta_n}|^2 \, d\theta$$

$$= 6\,\delta_n \int_{-\exp(-1/\delta_n^2)}^{0} |e^{i\theta} - 1|^{-1+\delta_n} |1 - (e^{i\theta} - 1)^{\delta_n/2}|^2 \, d\theta$$

$$\leq 6\,\delta_n(1 + 2^{\delta_n/2})^2 \int_{-\exp(-1/\delta_n^2)}^{0} |e^{i\theta} - 1|^{-1+\delta_n} \, d\theta$$

$$\leq \text{const}\,\delta_n \int_{0}^{\exp(-1/\delta_n^2)} \theta^{-1+\delta_n} \, d\theta$$

$$= \text{const}\,\exp(-1/\delta_n) = o(1) \quad \text{as} \quad n \to \infty.$$

The first sum in (54) is not greater than

$$4 \cdot 6 \frac{\delta_n}{\varphi'(\theta_n)} \sum_{j=-(M-1)}^{n-1} \int_{\xi_j}^{\theta_j-1} |(B(e^{i\theta}) - 1)^{-1/2+\delta_n/2} - (B(e^{i\theta}) - 1)^{-1/2+\delta_n}|^2 \left|\frac{B(e^{i\theta}) - 1}{e^{i\theta} - e^{i\theta_0}}\right|^2 \, d\theta$$

$$\leq \text{const}\, \frac{\delta_n}{\varphi'(\theta_n)} \sum_{j=-(M-1)}^{n-1} \int_{\xi_j}^{\theta_j-1} \frac{d\theta}{|e^{i\theta} - e^{i\theta_n}|^2}$$

$$\leq \text{const} \ \frac{\delta_n}{\varphi'(\theta_n)} \sum_{j=-(M-1)}^{n-1} \frac{|\xi_j - \theta_{j-1}|}{|\theta_{j-1} - \theta_n| \, |\xi_j - \theta_n|}$$

$$\leq \text{const} \ \frac{\delta_n}{\varphi'(\theta_n)} \sum_{j=-(M-1)}^{n-1} \frac{|\xi_j - \theta_{j-1}|}{|\theta_{j-1} - \theta_j| \, |\xi_j - \theta_j|}$$

$$\leq \text{const} \ \frac{\delta_n}{\varphi'(\theta_n)} \sum_{j=-(M-1)}^{n-1} \frac{|\theta_j - \theta_{j-1}|}{(j^2 + 1 + |\theta_j - \theta_{j-1}|)(|\theta_j - \theta_{j-1}| - |\xi_j - \theta_{j-1}|)}$$

(here we used (52))

$$\leq \text{const} \ \frac{\delta_n}{\varphi'(\theta_n)} \sum_{j=-(M-1)}^{n-1} \frac{|\theta_j - \theta_{j-1}|}{(j^2 + 1 + |\theta_j - \theta_{j-1}|)\left(|\theta_j - \theta_{j-1}| - \frac{|\xi_j - \theta_{j-1}|}{j^2 + 1 + |\theta_j - \theta_{j-1}|}\right)}$$

(again we had recourse to (52))

$$= \text{const} \ \frac{\delta_n}{\varphi'(\theta_n)} \sum_{j=-(M-1)}^{n-1} \frac{1}{j^2 + 1} \leq \text{const} \ \frac{\delta_n}{\varphi'(\theta_n)} = o(1) \quad \text{as} \quad n \to \infty.$$

In a similar way one can show that the third term on the right of (54) goes to zero as n tends to infinity. ∎

4.10. Theorem. *Let* $\arg a \in CR(\dot{\mathbf{T}})$ *and suppose* (38) *and* (39) *are satisfied. Then there exists a function* $b \in CU(\dot{\mathbf{T}})$ *such that* $\arg a \sim \arg b$ *and* $T(b) \in NNS$.

Proof. In 4.8 and 4.9 we constructed such a b in case $(\arg a)^{\#}(-\infty) = -\infty$. An analogous construction gives the desired b if $(\arg a)^{\#}(+\infty) = +\infty$. ∎

5. Concrete orientation preserving whirls

5.1. The symbol class. The results of the preceding section show that at a finite distance of an argument satisfying (23), (24) (or equivalently: (25), (26)) there exist both arguments of normally solvable and arguments of not normally solvable Toeplitz operators. In this section we describe a class of arguments subject to the conditions

$$(\arg a)^{\#}(-\infty) = -\infty, \ (\arg a)^{\#}(+\infty) = +\infty, \tag{56}$$
$$(\arg a)^{\#} \text{ is strictly monotonous on } \mathbf{R} \tag{57}$$

for which $T(a)$ is normally solvable and hence, by Theorem 4.5, belongs to $\Phi(-\infty)$.

Let $a \in CU(\dot{\mathbf{T}})$, choose $\arg a \in CR(\dot{\mathbf{T}})$, and suppose

$$(\arg a)(e^{i\theta}) = 2\pi f(\theta) + o(1) \quad \text{as} \quad \theta \to 0 \tag{58}$$

where

$$f \text{ is continuous and strictly monotonous on } (-\pi, 0) \text{ and } (0, \pi), \tag{59}$$
$$f(0 - 0) = +\infty, \ f(0 + 0) = -\infty. \tag{60}$$

Clearly, in this case (56) and (57) are satisfied $((\arg a)^{\#}(\pm\infty) = f(0 \mp 0))$. Without loss of generality assume also that $f(-\pi + 0) = f(\pi - 0) = 0$, i.e. that $(\arg a)^{\#}(0) = 0$.

Define $\theta(x)$ for $x \in \mathbf{R} \setminus \{0\}$ by $\theta(x) = f^{-1}(-x)$. Notice that

$$\theta(-\infty) = \theta(+\infty) = 0, \quad \theta(0-0) = -\pi, \quad \theta(0+0) = \pi$$

and that θ is strictly monotonously decreasing on $(-\infty, 0)$ and on $(0, +\infty)$. Figure 7 shows plots of $f^{\#}, f$, and θ.

Put $\theta(0) = \pi$,

$$\Delta(n) = \begin{cases} \theta(n) - \theta(n+1) & \text{for} \quad n = 0, 1, 2, \ldots \\ \theta(n-1) - \theta(n) & \text{for} \quad n = -1, -2, -3, \ldots \end{cases}$$

and consider the functions $\psi_n \in C[-1/2, 1/2]$ given for $n \in \mathbf{Z} \setminus \{0\}$ by

$$\psi_n(s) = \frac{\theta(n) - \theta(n+s)}{\Delta(n)} \quad (-1/2 \leq s \leq 1/2).$$

Suppose

$$\sup_{n \in \mathbf{Z}} \sum_{j \in \mathbf{Z} \setminus \{n\}} \left(\frac{\theta(j) - \theta(j+1)}{\theta(j) - \theta(n)} \right)^2 < \infty \tag{61}$$

and

there exists a function $\psi \in C[-1/2, 1/2]$ satisfying $\psi(-1/2) < 0$ and $\psi(1/2) > 0$ such that $\psi_n(s)$ converges uniformly to $\psi(s)$ and $-\psi(-s)$ as $n \to +\infty$ and $n \to -\infty$, respectively. $\tag{62}$

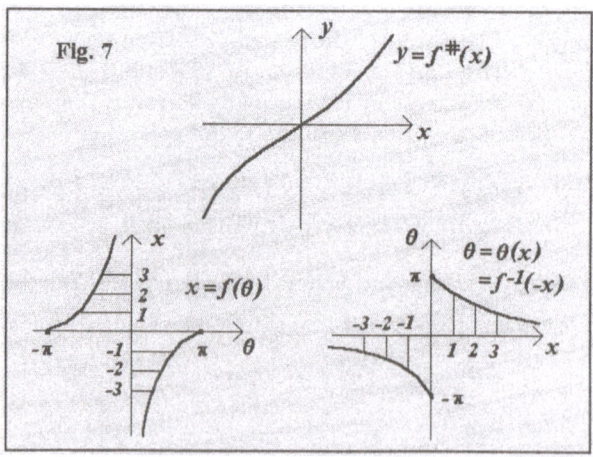

Fig. 7

We understand that conditions (61) and (62) are difficult to comprehend and give rise to at least three questions.

(Q_1) Why just these conditions ? Answer: Because we can prove the following Theorem 5.2 under just these conditions.

(Q_2) Are these conditions effectively verifiable in concrete cases ? We will show in Subsections 5.3–5.9 that (61) and (62) are satisfied for a series of interesting classes of arguments.

(Q_3) What do these conditions mean ? This is, of course, the most difficult question. First of all, we remark that (61) is a condition on the behavior of $\theta(x)$ at integral values of x while (62) governs the behavior of $\theta(x)$ between the integers. It is easily seen that (61) implies that

$$\sum_{j \in \mathbf{Z} \backslash \{0\}} \left(1 - \theta(j+1)/\theta(j)\right)^2 < \infty. \tag{63}$$

Thus, $\xi_j := \theta(j+1)/\theta(j)$ converges sufficiently fast to 1 as $|j| \to \infty$. Since

$$\lim_{n \to \infty} \prod_{j=1}^{n} \xi_j = \lim_{n \to \infty} \frac{\theta(n+1)}{\theta(1)} = 0$$

and, analogously, $\prod_{j=-1}^{n} \xi_j \to 0$ as $n \to \infty$, we also obtain that

$$\sum_{j \in \mathbf{Z} \backslash \{0\}} |1 - \xi_j| = \sum_{j=\mathbf{Z} \backslash \{0\}} |1 - \theta(j+1)/\theta(j)| = \infty, \tag{64}$$

which says that ξ_j cannot approach 1 too fast as $|j| \to \infty$. Further, from the definition of $\psi_n(s)$ we have

$$\theta(n+s) = \theta(n) - \psi_n(s)\Delta(n)$$

and hence, for $n \geq 1$,

$$\theta(n) - \psi_n\left(\frac{1}{2}\right)\Delta(n) = \theta\left(n + \frac{1}{2}\right) = \theta\left(n + 1 - \frac{1}{2}\right) = \theta(n+1) - \psi_{n+1}\left(-\frac{1}{2}\right)\Delta(n+1),$$

which gives that

$$\frac{\Delta(n+1)}{\Delta(n)} = -\frac{1 - \psi_n(1/2)}{\psi_{n+1}(-1/2)}.$$

Now condition (62) shows that

$$\lim_{n \to \infty} \frac{\Delta(n+1)}{\Delta(n)} = -\frac{1 - \psi(1/2)}{\psi(-1/2)} =: c \geq 0 \tag{65}$$

Because $\xi_n \to 1$ as $n \to +\infty$, we also have

$$c = \lim_{n \to +\infty} \frac{\Delta(n+1)}{\Delta(n)} = \lim_{n \to \infty} \frac{\xi_n - \xi_{n+1}\xi_n}{1 - \xi_n} = \lim_{n \to +\infty} \frac{1 - \xi_{n+1}}{1 - \xi_n}.$$

Thus, in the case $0 \leq c < 1$ the series (64) would converge, whereas for $c > 1$ the series (63) would diverge. Hence $c = 1$. A similar argument for $n \to -\infty$ gives

$$\lim_{|n| \to \infty} \frac{\Delta(n+1)}{\Delta(n)} = 1 \tag{66}$$

and consequently, by (65),

$$\psi(1/2) - \psi(-1/2) = 1. \tag{67}$$

The following theorem was essentially established in [18], [20].

5.2. Theorem. *Suppose* $a \in CU(\dot{\mathbf{T}})$ *satisfies the conditions* (58)–(62). *Then the operator* $T(a)$ *belongs to* $\Phi(-\infty)$.

Proof. For $M > 0$ and $k \in \mathbf{Z} \setminus \{0\}$, put

$$r_k = (M - \Delta(k))/(M + \Delta(k)), \ \alpha_k = r_k e^{i\theta(k)}.$$

We have

$$\sum_{k \in \mathbf{Z} \setminus \{0\}} (1 - r_k) \leq \frac{2}{M} \sum_{k \in \mathbf{Z} \setminus \{0\}} \Delta(k) = \frac{2}{M}(\theta(1) - \theta(0 - 0)) \leq \frac{4\pi}{M} < \infty.$$

Let $B_M \in CU(\dot{\mathbf{T}})$ denote the Blaschke product with the zeros $\{\alpha_k\}_{k \in \mathbf{Z}}$. Note that, up to the missing zero α_0, B_M is of the form (27),(28). We show that if M is sufficiently large, then $a = B_M b$ with some function $b \in GL^\infty$ which factors into a product of nice functions and a B_M-periodic function, so that $T(b)$ is invertible. This implies that $T(a) \in \Phi(-\infty)$.

For $x \in \mathbf{R} \setminus \{0\}$, set $f_M(x) = \arg B_M(e^{i\theta(x)})$. Let first $x > 0$ and write $x = n + s$ with $n \in \mathbf{Z}$ and $s \in [-1/2, 1/2]$. Since $\theta(x) \in (0, \pi)$, we obtain from Lemma 4.4 that

$$f_M(x) = \quad -2\pi(n - 1) - 2 \sum_{k=1}^{n-1} \varphi_k(\theta(x)) - 2\tilde{\varphi}_n(\theta(x))$$

$$-2 \sum_{k=n+1}^{\infty} \varphi_k(\theta(x)) - 2 \sum_{k=-\infty}^{-1} \varphi_k(\theta(x)) \tag{68}$$

where

$$\varphi_k(\theta(x)) = \arctan\left(\varepsilon_k \cot \frac{\theta(x) - \theta(k)}{2}\right), \ \varepsilon_k = \frac{1 - r_k}{1 + r_k} = \frac{\Delta(k)}{M},$$

$$\tilde{\varphi}_n(\theta(x)) = \begin{cases} \varphi_n(\theta(x)) & \text{for} \quad s \leq 0, \\ \pi + \varphi_n(\theta(x)) & \text{for} \quad s > 0. \end{cases}$$

Our aim is to show that for sufficiently large M the quantity $\sup_{x>0} |\sum_{k \neq n} \varphi_k(\theta(x))|$ is small while $\tilde{\varphi}_n(\theta(x))$ is close to a periodic function.

In what follows we write $g(x, n, s, M) = O(1/M)$ if there exists a constant $C > 0$ independent on x, n, s, M such that $|g(x, n, s, M)| \leq C/M$ for all x, n, s, M.

We claim that

$$\sum_{k=1}^{n-2} \varphi_k(\theta(x)) = \sum_{k=1}^{n-2} \arctan\left(\frac{2}{M} \frac{\Delta(k)}{\theta(x) - \theta(k)}\right) + O\left(\frac{1}{M}\right). \tag{69}$$

Indeed, taking into account the inequalities

$$\left|\cot u - \frac{1}{u}\right| \le \text{const } |u| \quad \text{for} \quad |u| \le |u_0| < \frac{\pi}{2} \tag{70}$$

and

$$|\arctan u - \arctan v| \le |\arctan(u - v)| \quad \text{for} \quad u, v < 0$$

we get

$$\left|\sum_{k=1}^{n-2} \arctan\left(\varepsilon_k \cot \frac{\theta(x) - \theta(k)}{2}\right) - \arctan\left(\frac{2}{M}\frac{\Delta(k)}{\theta(x) - \theta(k)}\right)\right|$$

$$\le \sum_{k=1}^{n-2}\left|\arctan\left(\varepsilon_k \cot\left(\frac{\theta(x) - \theta(k)}{2}\right) - \frac{2}{M}\frac{\Delta(k)}{\theta(x) - \theta(k)}\right)\right|$$

$$= \sum_{k=1}^{n-2}\left|\arctan\left(\frac{\Delta(k)}{M}\left(\cot\frac{\theta(x) - \theta(k)}{2} - \frac{2}{\theta(x) - \theta(k)}\right)\right)\right|$$

$$\le \sum_{k=1}^{n-2}\frac{\Delta(k)}{M}\text{const}\frac{|\theta(x) - \theta(k)|}{2} = \sum_{k=1}^{n-2}\frac{\Delta(k)}{M}\text{const}\frac{\theta(x) - \theta(k)}{2}$$

$$\le \frac{\text{const}}{M}\sum_{k=1}^{n-2}\Delta(k) = \frac{\text{const}}{M}(\theta(1) - \theta(n - 1)) \le \frac{\text{const}}{M},$$

which proves (69).

We next show that

$$\sum_{k=1}^{n-2}\arctan\left(\frac{2}{M}\frac{\Delta(k)}{\theta(x) - \theta(k)}\right) = \frac{2}{M}\sum_{k=1}^{n-2}\log\left(1 + \frac{\Delta(k)}{\theta(x) - \theta(k)}\right) + O\left(\frac{1}{M}\right). \tag{71}$$

Using the inequality

$$\left|\arctan\left(\frac{2}{M}u\right) - \frac{2}{M}\log(1 + u)\right| \le \text{const }\frac{u^2}{M} \quad \text{for} \quad u \ge u_0 > -1 \tag{72}$$

we obtain

$$\left|\sum_{k=1}^{n-2}\left(\arctan\frac{2}{M}\frac{\Delta(k)}{\theta(x) - \theta(k)}\right) - \frac{2}{M}\log\left(1 + \frac{\Delta(k)}{\theta(x) - \theta(k)}\right)\right|$$

$$\le \frac{\text{const}}{M}\sum_{k=1}^{n-2}\left(\frac{\Delta(k)}{\theta(x) - \theta(k)}\right)^2 \le \frac{\text{const}}{M}\sum_{k=1}^{n-2}\left(\frac{\theta(k) - \theta(k + 1)}{\theta(k) - \theta(n - 1)}\right)^2,$$

and the latter term is $O(1/M)$ by virtue of (61); we remark that application of (72) may be justified as follows: from (66) we infer that $\Delta(k + 1)/\Delta(k) \ge \Delta > 0$ for all k, from (62) we deduce that $\psi_n(1/2) \ge \varepsilon > 0$ for all n, and thus,

$$\left|\frac{\Delta(k)}{\theta(x) - \theta(k)}\right| = \frac{1}{1 + \frac{\theta(k+1) - \theta(x)}{\Delta(k)}} \le \frac{1}{1 + \frac{\theta(k+1) - \theta(k+3/2)}{\Delta(k)}}$$

$$= \frac{1}{1 + \frac{\Delta(k+1)}{\Delta(k)}\psi_{k+1}(1/2)} \le |u_0| < 1.$$

So the proof of (71) is complete.

For the sum on the right of (71) we have

$$\frac{2}{M}\sum_{k=1}^{n-2}\log\left(1+\frac{\Delta(k)}{\theta(x)-\theta(k)}\right)=\frac{2}{M}\log\prod_{k=1}^{n-2}\frac{\theta(x)-\theta(k+1)}{\theta(x)-\theta(k)}$$
$$=\frac{2}{M}\log\left(\frac{\theta(x)-\theta(n-1)}{\theta(x)-\theta(1)}\right).$$

This together with (69) and (71) gives

$$\sum_{k=1}^{n-2}\varphi_k(\theta(x))=\frac{2}{M}\log\left(\frac{\theta(x)-\theta(n-1)}{\theta(x)-\theta(1)}\right)+O\left(\frac{1}{M}\right). \tag{73}$$

Proceeding in a similar way one gets

$$\sum_{k=n+2}^{\infty}\varphi_k(\theta(x))=\frac{2}{M}\log\left(\frac{\theta(x)}{\theta(x)-\theta(n+2)}\right)+O\left(\frac{1}{M}\right), \tag{74}$$

$$\sum_{k=-\infty}^{-1}\varphi_k(\theta(x))=\frac{2}{M}\log\left(\frac{\theta(x)+\pi}{\theta(x)}\right)+O\left(\frac{1}{M}\right). \tag{75}$$

Finally, we have

$$|\varphi_{n-1}(\theta(x))|\leq\varepsilon_{n-1}\left|\cot\frac{\theta(x)-\theta(n-1)}{2}\right|\leq\varepsilon_{n-1}\frac{2}{\theta(n-1)-\theta(x)}$$
$$=\frac{2}{M}\frac{\Delta(n-1)}{\theta(n-1)-\theta(x)}\leq\frac{2}{M}\frac{\Delta(n-1)}{\theta(n-1)-\theta(n-1/2)}=\frac{2}{M}\frac{1}{\psi_{n-1}(1/2)}=O\left(\frac{1}{M}\right) \tag{76}$$

and, analogously,

$$|\varphi_{n+1}(\theta(x))|=O(1/M). \tag{77}$$

Adding (73)–(76) we arrive at the equality

$$\sum_{k\neq n}\varphi_k(\theta(x))=\frac{2}{M}\log\left(\frac{\theta(x)+\pi}{\theta(1)-\theta(x)}\right)+\frac{2}{M}\log\left(\frac{\theta(n-1)-\theta(x)}{\theta(x)-\theta(n+2)}\right)+O\left(\frac{1}{M}\right). \tag{78}$$

Clearly, the first term on the right of (78) is $O(1/M)$. We claim that the same is also true for the second term. To see this notice first that

$$\frac{\theta(n-1)-\theta(n-1/2)}{\theta(n-1/2)-\theta(n+2)}\leq\frac{\theta(n-1)-\theta(x)}{\theta(x)-\theta(n+2)}\leq\frac{\theta(n-1)-\theta(n+1/2)}{\theta(n+1/2)-\theta(n+1)},$$

hence

$$\frac{\psi_{n-1}(1/2)\Delta(n-1)}{\theta(n-1/2)-\theta(n+2)}\leq\frac{\theta(n-1)-\theta(x)}{\theta(x)-\theta(n+2)}\leq-\frac{\theta(n-1)-\theta(n+1/2)}{\psi_{n+1}(-1/2)\Delta(n+1)},$$

and thus

$$\frac{\psi_{n-1}(1/2)}{-\psi_{n-1}(1/2)+1+\frac{\Delta(n)}{\Delta(n-1)}+\frac{\Delta(n+1)}{\Delta(n-1)}}\leq\frac{\theta(n-1)-\theta(x)}{\theta(x)-\theta(n+2)}$$
$$\leq-\frac{\frac{\Delta(n-1)}{\Delta(n+1)}+\frac{\Delta(n)}{\Delta(n+1)}+\psi_{n+1}(-1/2)}{\psi_{n+1}(-1/2)}. \tag{79}$$

Taking into account (66) and (67) we obtain that $\psi_{n-1}(1/2) < 1$, $\psi_{n+1}(-1/2) > -1$,

$$\Delta(n)/\Delta(n-1) \geq \Delta, \quad \Delta(n+1)/\Delta(n-1) \geq \Delta^2$$
$$\Delta(n)/\Delta(n+1) \leq 1/\Delta, \quad \Delta(n-1)/\Delta(n+1) \leq 1/\Delta^2$$

with some $\Delta \in (0,1)$. Since, by (62), the numbers $|\psi_n(\pm 1/2)|$ are bounded away from zero, the estimate (79) gives

$$\left| \log \left(\frac{\theta(n-1) - \theta(x)}{\theta(x) - \theta(n+2)} \right) \right| \leq C$$

with some constant C independent of x and n, which proves our claim. Thus, (78) reduces to

$$\sum_{k \neq n} \varphi_k(\theta(x)) = O(1/M). \tag{80}$$

Let us now look at the term $\tilde{\varphi}_n(\theta(x))$ in (68). From (70) we get without difficulty that

$$\tilde{\varphi}_n(\theta(x)) = \begin{cases} -\arctan \frac{2}{M\psi_n(s)} + O\left(\frac{1}{M}\right) & \text{for} \quad -1/2 \leq s \leq 0, \\ \pi - \arctan \frac{2}{M\psi_n(s)} + O\left(\frac{1}{M}\right) & \text{for} \quad 0 < s < 1/2 \end{cases}$$

(recall that $x = n + s$.) Define $\alpha_0 \in C[-1/2, 1/2]$ by

$$\alpha_0(s) = \begin{cases} -\arctan \frac{2}{M\psi(s)} & \text{for} \quad -1/2 \leq s \leq 0, \\ \pi - \arctan \frac{2}{M\psi(s)} & \text{for} \quad 0 < s < 1/2 \end{cases}$$

and put

$$m(x) = -\arctan \frac{2}{M\psi_n(s)} + \arctan \frac{2}{M\psi(s)}. \tag{81}$$

Then

$$\tilde{\varphi}_n(\theta(x)) = \alpha_0(s) + m(x) + O(1/M). \tag{82}$$

The identity $\arctan u + \arctan u^{-1} = \pi/2$ gives

$$|m(x)| \leq |\arctan(M\psi_n(s)/2) - \arctan(M\psi(s)/2)|$$
$$\leq |\arctan(M(\psi_n(s) - \psi(s))/2)|$$

and hence, by (62), $m(x) \to 0$ as $x = n + s \to +\infty$.

The function m has jumps at the points $n + 1/2$. From (81) and (62) we infer that

$$\left| m\left(n \pm \frac{1}{2}\right) \right| \leq \arctan \left(\frac{2}{M} \left| \frac{1}{\psi_n(\pm 1/2)} - \frac{1}{\psi(\pm 1/2)} \right| \right) = O\left(\frac{1}{M}\right),$$

and therefore (82) may be written in the form

$$\tilde{\varphi}_n(\theta(x)) = \alpha_0(s) + c_0(x) + O(1/M) \tag{83}$$

where c_0 is continuous on $(0, +\infty)$ and satisfies

$$c_0(+\infty) = 0. \tag{84}$$

Combining (68), (80), (83) we arrive at the representation

$$\begin{aligned} f_M(x) &= -2\pi(n-1) - 2\alpha_0(s) - 2c_0(x) + O(1/M) \\ &= -2\pi x + \alpha(x) + c(x) + O(1/M) \end{aligned} \tag{85}$$

with $c(x) = -2c_0(x)$ and

$$\alpha(x) = -2\alpha_0(x - [x + 1/2]) + 2\pi(x - [x + 1/2]) + 2\pi.$$

Note that α is a 1-periodic function on $\mathbf{R}$. Moreover, taking into consideration the concrete form of α_0 one can easily see that $F(x) := -2\pi x + \alpha(x)$ is strictly monotonously decreasing and that

$$\alpha\left(n + \frac{1}{2} \pm 0\right) = \pi + 2\arctan\frac{2}{M\psi(\pm 1/2)}.$$

The sizes of these jumps are $O(1/M)$ and therefore we may rewrite (85) in the form

$$f_M(x) = -2\pi x + \beta(x) + c(x) + O(1/M) \tag{86}$$

with a continuous 1-periodic function β on $\mathbf{R}$ such that $G(x) := -2\pi x + \beta(x)$ is strictly monotonously decreasing.

The representation (86) holds for $x > 0$. We may repeat the above reasoning with B_M replaced by the Blaschke product $\tilde{B}_M := 1/B_M(t^{-1})$. The zeros of $\tilde{B}_M$ are the complex conjugates of the zeros of B_M and we have

$$\arg\tilde{B}_M(e^{i\theta}) = -\arg B_M(e^{-i\theta}).$$

Taking into account (62) we arrive at the conclusion that the equality (86) is also valid for $x < 0$.

As already said, the function $y = G(x) = -2\pi x + \beta(x)$ is strictly monotonously decreasing. Let $x = G^{-1}(y)$ be the inverse function. Clearly, $G^{-1}(y) = -y/(2\pi) + \gamma(y)$ with some continuous 2π-periodic function γ on $\mathbf{R}$. From (86) we get

$$G^{-1}(f_M(x)) = x + \delta_1(x) + \delta_2(x, M)$$

with

$$\begin{aligned} \delta_1(x) &= G^{-1}(-2\pi x + \beta(x) + c(x)) - G^{-1}(-2\pi x + \beta(x)), \\ \delta_2(x, M) &= G^{-1}(-2\pi x + \beta(x) + c(x) + O(1/M)) - G^{-1}(-2\pi x + \beta(x) + c(x)). \end{aligned}$$

Since G^{-1} is uniformly continuous, it follows that

$$\sup_{x\in\mathbf{R}} |\delta_2(x, M)| \to 0 \text{ as } M \to \infty \tag{87}$$

and (also recall (85))

$$\delta_1(x) \to 0 \text{ as } |x| \to \infty. \tag{88}$$

Because $G^{-1}(y) = -y/(2\pi) + \gamma(y)$ and $f_M(x) = \arg B_M(e^{i\theta(x)})$, we obtain

$$-\frac{\arg B_M(e^{i\theta(x)})}{2\pi} + \gamma(\arg B_M(e^{i\theta(x)})) = x + \delta_1(x) + \delta_2(x, M). \tag{89}$$

Now write $t = e^{i\theta(x)}$. Multiplying (89) by $-2\pi i$ and taking exponentials gives

$$e^{i\arg B_M(t)} e^{-2\pi i\gamma(\arg B_M(t))} = e^{-2\pi i x} e^{-2\pi i\delta_1(x)} e^{-2\pi\delta_2(x,M)}.$$

Clearly, $e^{i\arg B_M(t)} = B_M(t)$. Since $f(\theta(x)) = -x$, we conclude from (58) that

$$e^{-2\pi i x} = e^{2\pi i f(\theta(x))} = e^{i\arg a(e^{i\theta(x)}) - i\delta_3(x)}$$

with $\delta_3 \in CR(\mathbf{R})$ satisfying

$$\delta_3(x) \to 0 \quad \text{as } |x| \to \infty. \tag{90}$$

For $t = e^{i\theta(x)}$, put

$$b(t) = e^{-2\pi i\gamma(\arg t)}, \quad d_1(t) = e^{2\pi i\delta_1(x)},$$
$$d_2(t, M) = e^{2\pi i\delta_2(x,M)}, \quad d_3(t) = e^{i\delta_3(x)}.$$

We so have

$$B_M(t)b(B_M(t)) = a(t)d_3^{-1}(t)d_1^{-1}(t)d_2^{-1}(t, M),$$

i.e.

$$a(t) = b(B_M(t))d_1(t)d_2(t, M)d_3(t)B_M(t). \tag{91}$$

Since $\delta_3 \in CR(\mathbf{R})$ and because of (90), the function d_3 belongs to $C(\mathbf{T})$, has no zeros on $\mathbf{T}$ and has winding number zero. In the same way we obtain from (88) that $d_1 \in C(\mathbf{T}) \cap GL^\infty$ and wind $d_1 = 0$. Thus, we may write $d_1 d_3 = p(1+\varepsilon_1)$ where p is a trigonometric polynomial, $p(t) \neq 0$ for $t \in \mathbf{T}$, wind $p = 0$, and $\varepsilon_1 \in L^\infty$ has arbitrarily small norm. Since $\gamma \in CR(\mathbf{R})$ is 2π-periodic, the function b also belongs to $C(\mathbf{T}) \cap GL^\infty$ and has winding number zero. Consequently, $b = q(1 + \varepsilon_2)$ where q is a trigonometric polynomial, $q(t) \neq 0$ for $t \in \mathbf{T}$, wind $q = 0$, and ε_2 is a function in L^∞ whose norm can be made as small as desired. We may in particular achieve that

$$\| \arg(1 + \varepsilon_1)\|_\infty + \| \arg(1 + \varepsilon_2)\|_\infty < \pi/4.$$

From (87) we infer that $\| \arg \delta_2(\cdot, M)\|_\infty < \pi/4$ whenever M is sufficiently large. For such M the function

$$s(t) := (1 + \varepsilon_1(t))(1 + \varepsilon_2(B_M(t)))\delta_2(t, M)$$

is sectorial. Finally, we have $p = p_-p_+$ and $q = q_-q_+$ with

$$p_- = \exp(Q \log p) \in G\overline{H^\infty}, \quad p_+ = \exp(P \log p) \in GH^\infty,$$
$$q_- = \exp(Q \log q) \in G\overline{H^\infty}, \quad q_+ = \exp(P \log q) \in GH^\infty$$

("classical" Wiener-Hopf factorization) and it is clear that $q_- \circ B_M \in G\overline{H^\infty}$ and $q_+ \circ B_M \in GH^\infty$. Since

$$a(t) = p_-(B_M(t))s(t)p_+(t)q_+(B_M(t))B_M(t),$$

it follows from (7) that

$$T(a) = T(p_-)T(q_- \circ B_M)T(s)T(p_+)T(q_+ \circ B_M)T(B_M)$$

where $T(p_\pm), T(q_\pm \circ B_M), T(s)$ are invertible and $T(B_M) \in \Phi(-\infty)$. Consequently, $T(a) \in \Phi(-\infty)$. ∎

The following proposition provides simple sufficient conditions for (61) and (62) to hold.

5.3. Proposition. *Let θ be continuous and strictly monotonously decreasing on $(-\infty, 0]$ and on $[0, +\infty)$, let $\theta(x) < 0$ for $x \in (-\infty, 0]$ and $\theta(x) > 0$ for $x \in [0, +\infty)$, $\theta(-\infty) = 0$, $\theta(+\infty) = 0$. Suppose θ is differentiable on $\mathbf{R} \setminus \{0\}$, θ' is continuous and monotonously decreasing on $(-\infty, 0]$ and continuous and monotonously increasing on $[0, +\infty)$. Also suppose $\theta'(-\infty) = 0$ and $\theta'(+\infty) = 0$. If*

$$\lim_{|x| \to \infty} \frac{\theta'(x+1)}{\theta'(x)} = 1, \tag{92}$$

then (62) holds. If, in addition,

$$\sup_{x \in \mathbf{R} \setminus \{0\}} \left| \frac{\theta'(x)}{\theta'(2x)} \right| < \infty, \tag{93}$$

then (61) is true. In particular, under the above hypotheses $T(a) \in \Phi(-\infty)$.

Proof. Let $n > 0$. By the mean value theorem,

$$\psi_n(s) = \frac{\theta(n) - \theta(n+s)}{\theta(n) - \theta(n+1)} = s \frac{\theta'(\xi)}{\theta'(\eta)}$$

with $\xi \in (n, n+s)$ and $\eta \in (n, n+1)$. Taking into account the monotonity of θ', we get

$$\left| \frac{\theta'(n+1)}{\theta'(n)} \right| \le \left| \frac{\theta'(\xi)}{\theta'(\eta)} \right| = \frac{\theta'(\xi)}{\theta'(\eta)} \le \left| \frac{\theta'(n)}{\theta'(n+1)} \right|,$$

and hence (92) implies that $\psi_n(s) \to s$ uniformly as $n \to +\infty$. In the same way one may verify that $\psi_s(s)$ converges uniformly to $-(-s) = s$ as $n \to -\infty$.

To prove (61), suppose $n > 0$. Using the monotonity of θ and θ' we obtain

$$\sum_{j \ne n, j \ne 0} \left(\frac{\theta(j) - \theta(j+1)}{\theta(j) - \theta(n)} \right)^2 = \sum_{j=-\infty}^{-1} + \sum_{j=1}^{[n/2]} + \sum_{j=[n/2]+1}^{n-1} + \sum_{j=n+1}^{\infty}$$

$$\le \sum_{j=-\infty}^{-1} \left(\frac{\theta(j) - \theta(j+1)}{\theta(j) - \theta(2j)} \right)^2 + \sum_{j=1}^{[n/2]} \left(\frac{\theta(j) - \theta(j+1)}{\theta(j) - \theta(2j)} \right)^2$$

$$+ \sum_{j=[n/2]+1}^{n-1} \left(\frac{\theta(j) - \theta(j+1)}{\theta(j) - \theta(n)} \right)^2 + \sum_{j=n+1}^{\infty} \left(\frac{\theta(j) - \theta(j+1)}{\theta(n) - \theta(j)} \right)^2$$

$$\le \sum_{j=-\infty}^{-1} \left(\frac{\theta'(j+1)}{\theta'(2j)} \right)^2 \frac{1}{j^2} + \sum_{j=1}^{[n/2]} \left(\frac{\theta'(j)}{\theta'(2j)} \right)^2 \frac{1}{j^2}$$

$$+ \sum_{j=[n/2]+1}^{n-1} \left(\frac{\theta'(j)}{\theta'(n)} \right)^2 \frac{1}{(j-n)^2} + \sum_{j=n+1}^{\infty} \left(\frac{\theta'(j)}{\theta'(j)} \right)^2 \frac{1}{(n-j)^2} \le C \sum_{j=1}^{\infty} \frac{1}{j^2}$$

with some constant $C > 0$ independent of n. For $n < 0$ one may proceed analogously. ∎

5.4. Remark. The membership of $T(a)$ in $\Phi(-\infty)$ or NNS is not affected by replacing a with ac (and thus $\arg a$ with $\arg a + \arg c$) where $c \in C \cap GL^\infty$. This implies that Theorem 5.2 also holds with (61) replaced by the condition

$$\limsup_{|n| \to \infty} \sum_{j \in \mathbb{Z} \setminus \{n\}} \left(\frac{\theta(j) - \theta(j+1)}{\theta(j) - \theta(n)} \right)^2 < \infty. \tag{94}$$

Moreover, it follows that $T(a) \in \Phi(-\infty)$ if the assumptions of Proposition 5.3 are satisfied only for $|x| > x_0$ with some $x_0 > 0$.

5.5. Example: power-like growth. Suppose

$$f(\theta) = \begin{cases} -\alpha|\theta|^{-\mu} & \text{for } \theta > 0, \\ \beta|\theta|^{-\lambda} & \text{for } \theta < 0, \end{cases} \tag{95}$$

where $\alpha, \beta > 0$ and $\mu, \lambda > 0$. In that case

$$\theta(x) = \begin{cases} (x/\alpha)^{-1/\mu} & \text{for } x > 0, \\ -(|x|/\beta)^{-1/\lambda} & \text{for } x < 0. \end{cases}$$

It is easily seen that the hypothesis of Proposition 5.3 are satisfied. Thus, if $a \in C\dot{U}(\dot{\mathbb{T}})$ is given by (58) and (95), then $T(a) \in \Phi(-\infty)$. We remark that this result was first proved in [17] and, by different techniques, also in [23].

5.6. Example: power-logarithmic-like growth. Let now

$$f(\theta) = \begin{cases} -\alpha|\theta|^{-\mu}(\log|\theta|)^\gamma & \text{for } \theta > 0, \\ \beta|\theta|^{-\lambda}(\log|\theta|)^\delta & \text{for } \theta < 0, \end{cases} \tag{96}$$

where $\alpha, \beta > 0$, $\mu, \lambda > 0$, $\gamma, \delta \in \mathbb{R}$. Using standard asymptotic analysis (see e.g. [26]) we get

$$\theta(x) = \varrho x^{-1/\mu}(\log x)^{\gamma/\mu}(1 + O(\log\log x/\log x)),$$
$$\theta'(x) = -\frac{\varrho}{\mu} x^{-1/\mu-1}(\log x)^{\gamma/\mu}(1 + O(\log\log x/\log x)),$$
$$\theta''(x) = \frac{\varrho}{\mu}\left(\frac{1}{\mu}+1\right) x^{-1/\mu-2}(\log x)^{\gamma/\mu}(1 + O(\log\log x/\log x)),$$
$$\theta'''(x) = -\frac{\varrho}{\mu}\left(\frac{1}{\mu}+1\right)\left(\frac{1}{\mu}+2\right) x^{-1/\mu-3}(\log x)^{\gamma/\mu}(1 + \mathcal{O}(\log\log x/\log x))$$

as $x \to +\infty$; here $\varrho = \alpha^{-1/\mu}\mu^{-\gamma/\mu}$. Since $\theta''(x) > 0$ for all sufficiently large x, the function $\theta'(x)$ is eventually monotonously increasing. Clearly, $\theta'(+\infty) = 0$. Because $\theta'''(x) < 0$ if x is large enough, the function $\theta''(x)$ is monotonously decreasing for these x. Hence

$$\lim_{x \to +\infty} \frac{\theta'(x+1)}{\theta'(x)} = 1 + \lim_{x \to +\infty} \frac{\theta'(x+1) - \theta'(x)}{\theta'(x)} = 1 + \lim_{x \to +\infty} \frac{\theta''(\xi(x))}{\theta'(x)},$$

the latter limit being included between

$$\lim_{x \to +\infty} \frac{\theta''(x)}{\theta'(x)} \quad \text{and} \quad \lim_{x \to +\infty} \frac{\theta''(x+1)}{\theta'(x)}.$$

But these two limits are easily seen to be zero. It can be straightforwardly verified that

$$\sup_{x>0} |\theta'(x)/\theta'(2x)| < \infty.$$

The situation is analogous for $x < 0$. Consequently, the assumptions of Proposition 5.3 are fulfilled in the sense of Remark 5.4. Therefore $T(a) \in \Phi(-\infty)$ for the symbol $a \in CU(\dot{\mathbb{T}})$ given by (58) and (95).

5.7. Example: exponential and hyperexponential growth. Assume

$$f(\theta) = \begin{cases} -\alpha \exp(|\theta|^{-\mu}) & \text{for } \theta > 0, \\ \beta \exp(|\theta|^{-\lambda}) & \text{for } \theta < 0, \end{cases} \tag{97}$$

where $\alpha, \beta > 0$ and $\mu, \lambda > 0$, or let

$$f(\theta) = \begin{cases} -\alpha \exp(\exp(|\theta|^{-\mu})) & \text{for } \theta > 0, \\ \beta \exp(\exp(|\theta|^{-\lambda})) & \text{for } \theta < 0, \end{cases} \tag{98}$$

with $\alpha, \beta > 0$ and $\mu, \lambda > 0$. In both case f can be explicitly inverted and the conditions of Proposition 5.3/Remark 5.4 can be easily checked to be true for sufficiently large $|x|$. Hence, if arg is determined by (58) and (97) or (98), then $T(a) \in \Phi(-\infty)$.

5.8. Example: logarithmic growth. Now suppose

$$f(\theta) = \begin{cases} -\alpha(-\log|\theta|)^{\mu} & \text{for } \theta > 0, \\ \beta(-\log|\theta|)^{\lambda} & \text{for } \theta < 0, \end{cases} \tag{99}$$

where $\alpha, \beta > 0$ and $\mu, \lambda > 0$. For $x > 0$ we get

$$\theta(x) = \exp(-(x/\alpha)^{1/\mu}).$$

It can be verified straightforwardly that $\theta'(x)$ converges monotonously to zero as $x \to +\infty$. We have

$$\lim_{x \to +\infty} \frac{\theta'(x+1)}{\theta'(x)} = \lim_{x \to +\infty} \exp\left(\left(\frac{x}{\alpha}\right)^{1/\mu} - \left(\frac{x+1}{\alpha}\right)^{1/\mu}\right)\left(\frac{x+1}{x}\right)^{1/\mu-1}$$

and this is 1 for $\mu > 1$, is $e^{-1/\alpha}$ for $\mu = 1$, and is $+\infty$ for $\mu < 1$. Thus, (92) holds only for $\mu > 1$. Still worse,

$$\lim_{x \to +\infty} \frac{\theta'(x)}{\theta'(2x)} = \lim_{x \to +\infty} \left(\frac{1}{2}\right)^{1/\mu-1} \exp\left(\frac{2^{1/\mu}-1}{2^{1/\mu}}x^{1/\mu}\right) = +\infty$$

for every $\mu > 0$, telling us that (93) is never satisfied in the case at hand.

Although Proposition 5.3 does not work in the present situation, it turns out that Theorem 5.2 does its job if $\mu > 2$ and $\lambda > 2$. Indeed, since in this case (92) is valid, we see that (62) is satisfied. In order to check (61) directly, let $n > 0$ and write

$$\sum_{j \neq n, j > 0} \left(\frac{\theta(j) - \theta(j+1)}{\theta(j) - \theta(n)} \right)^2 = \left(\sum_{j=1}^{[n/2]} + \sum_{j=[n/2]+1}^{n-1} + \sum_{j=n+1}^{\infty} \right) \left(\frac{\theta(j) - \theta(j+1)}{\theta(j) - \theta(n)} \right)^2.$$

There are constants $q < 1$, $C > 0$, $\gamma > 0$ independent of j and n such that if $1 \leq j \leq [n/2]$, then

$$\frac{\theta(n)}{\theta(j)} = \exp\left(\left(\frac{j}{\alpha} \right)^{1/\mu} - \left(\frac{n}{\alpha} \right)^{1/\mu} \right) \leq q,$$

if $j \geq 1$ and $\mu > 2$, then

$$1 - \frac{\theta(j+1)}{\theta(j)} = 1 - \exp\left(\left(\frac{j}{\alpha} \right)^{1/\mu} \left(1 - \left(1 + \frac{1}{j} \right)^{1/\mu} \right) \right) \leq C j^{1/\mu - 1},$$

if $[n/2] \leq j \leq n - 1$ and $\mu > 2$, then

$$1 - \frac{\theta(n)}{\theta(j)} = 1 - \exp\left(\left(\frac{j}{\alpha} \right)^{1/\mu} \left(1 - \left(1 + \frac{n-j}{j} \right)^{1/\mu} \right) \right) \geq \gamma(n-j) j^{1/\mu - 1}.$$

Thus,

$$\sum_{j \neq n, j > 0} \left(\frac{\theta(j) - \theta(j+1)}{\theta(j) - \theta(n)} \right)^2 \leq \sum_{j=1}^{[n/2]} \left(\frac{\theta(j) - \theta(j+1)}{\theta(j)(1-q)} \right)^2$$

$$+ \sum_{j=[n/2]+1}^{n-1} \left(\frac{\theta(j) - \theta(j+1)}{\theta(j) - \theta(n)} \right)^2 + \sum_{j=n+1}^{\infty} \left(\frac{\theta'(j)}{\theta'(n)} \right)^2 \frac{1}{(j-n)^2}$$

$$\leq \frac{C^2}{(1-q)^2} \sum_{j=1}^{\infty} j^{2(1/\mu - 1)} + C^2 \gamma^2 \sum_{j=[n/2]+1}^{n-1} \frac{1}{(n-j)^2} + \sum_{j=n+1}^{\infty} \frac{1}{(j-n)^2} \leq M$$

with some constant M which does not depend on n. In the case where $n < 0$ or $j < 0$ we may estimate in a similar way.

To summarize, if $a \in CU(\dot{\mathbf{T}})$ is given by (58) and (99) with $\mu > 2$ and $\lambda > 2$, then $T(a) \in \Phi(-\infty)$.

5.9. Example: mixed growth. Apart from the requirement that $\psi_n(s) \to \psi(s)$ as $n \to +\infty$ and $\psi_n(s) \to -\psi(-s)$ as $n \to -\infty$, conditions (61) and (62) do not demand a relation between the behavior of $f(\theta)$ on different sides of the origin. Under the assumption of Proposition 5.3 and in the Examples 5.4–5.8, the function $\psi_n(s)$ converges to s as $n \to +\infty$ and to $s = -(-s)$ as $n \to +\infty$. Conclusion: if $\arg a$ is defined by (58) and $f(\theta)$ is any of the functions (95)–(99) for $\theta > 0$ and any (possibly other) of these functions for $\theta < 0$, then $T(a) \in \Phi(-\infty)$; in case one of the functions (99) participates, we have of course to require that $\mu, \lambda > 2$.

5.10. Stretched arguments. Assume we are given the stretched argument $(\arg a)^\#$ and suppose (56), (57) hold. Let

$$(\arg a)^\#(x) = F(x) + o(1) \quad (|x| \to \infty)$$

with some continuously differentiable and strictly monotonously increasing function $y = F(x)$ such that (without loss of generality) $F(0) = 0$.

In order to get the $f(\theta)$ in (58) we write

$$(x - i)/(x + i) = e^{i\theta}, \text{ i.e. } x = -1/\tan\frac{\theta}{2}$$

and so have

$$\frac{1}{2\pi}\arg a(e^{i\theta}) = \frac{1}{2\pi}F\left(-1/\tan\frac{\theta}{2}\right) + o(1) \quad (\theta \to 0).$$

The function $\theta(x)$ we have worked with so far is the solution of the equation $f(\theta(x)) = -x$, which now assumes the form

$$\frac{1}{2\pi}F\left(-1/\tan\frac{\theta(x)}{2}\right) = -x. \tag{100}$$

Let $x = G(y)$ denote the inverse function of $y = F(x)$. Then (100) is equivalent to

$$\begin{aligned}\theta(x) &= -2\arctan(1/G(-2\pi x)) \\ &= \begin{cases} 2\arctan G(-2\pi x) - \pi & \text{for } x < 0, \\ 2\arctan G(-2\pi x) + \pi & \text{for } x > 0 \end{cases}\end{aligned}$$

(recall that $F(0) = G(0) = 0$). Since

$$\theta'(x) = -4\pi G'(-2\pi x)/(1 + G(-2\pi x)^2),$$

we may rephrase Proposition 5.3 in terms of the stretched argument as follows.

If $G'(y)/(1 + G(y^2))$ vanishes at $y = \pm\infty$, is monotonously increasing on $(-\infty - y_0)$ and monotonously decreasing on $(y_0, +\infty)$ for some $y_0 > 0$, and if

$$\frac{G'(y + 1)}{G'(y)}\frac{1 + G(y)^2}{1 + G(y + 1)^2} \to 1 \quad \text{as } |y| \to \infty,$$

$$\frac{G'(y)}{G'(2y)}\frac{1 + G(2y)^2}{1 + G(y)^2} = O(1) \quad \text{as } |y| \to \infty,$$

then $T(a) \in \Phi(-\infty)$.

This result (along with some straightforward but tedious computations) implies in particular that the following functions $F(x)$ generate Toeplitz operators in $\Phi(-\infty)$:

$$F(x) = \begin{cases} \alpha|x|^\mu & \text{for } x > 0 \\ -\beta|x|^\lambda & \text{for } x < 0 \end{cases} \tag{101}$$

with $\alpha, \beta > 0$ and $\mu, \lambda > 0$;

$$F(x) = \begin{cases} \alpha|x|^\mu(\log|x|)^\gamma & \text{for } x > 1 \\ -\beta|x|^\lambda(\log|x|)^\delta & \text{for } x < -1 \end{cases} \tag{102}$$

with $\alpha, \beta > 0$, $\mu, \lambda > 0$, $\gamma, \delta \in \mathbf{R}$; .

$$F(x) = \begin{cases} \alpha\exp(\mu|x|^\gamma) & \text{for } x > 0 \\ -\beta\exp(\lambda|x|^\delta) & \text{for } x < 0 \end{cases} \tag{103}$$

with $\alpha, \beta > 0$, $\mu, \lambda > 0$, $\gamma, \delta > 0$.

5.11. SO, QC, PSO, PQC. Theorem 5.2 and Proposition 5.3 (or 5.10) often show that $T(a) \in \Phi(-\infty)$ if $(\arg a)^\#$ increases sufficiently fast. The most critical arguments are those of "moderate growth", such as $(\log x)^\lambda$ for $\lambda \in [1, 2]$ (recall Example 5.8). We will say more about such arguments later. Our next concern is arguments which increase "very slowly". Such arguments frequently fall into the class PSO.

For a function $F \in CR(\mathbf{R})$, the *oscillation* osc (F, I) on a set $I \subset \mathbf{R}$ is defined as

$$\text{osc } (F, I) = \sup\{|F(s) - F(t)| : s, t \in I\}.$$

A function $F \in CR(\mathbf{R})$ is said to be *slowly oscillating*, $F \in SO$, if

$$\text{osc } (F, [-2x, -x] \cup [x, 2x]) \to 0 \text{ as } x \to \infty. \tag{104}$$

Clearly, (104) relates $F(x)$ and $F(-x)$ in a fairly strong way, and (104) is never satisfied if $F(-\infty) = -\infty$ and $F(+\infty) = +\infty$. We say that a function $F \in CR(\mathbf{R})$ is *piecewise slowly oscillating*, $F \in PSO$, if

$$\text{osc } (F, [x, 2x]) \to 0 \text{ as } |x| \to \infty. \tag{105}$$

For example, if

$$F(x) = \begin{cases} \alpha(\log|x|)^\lambda & \text{for } x > 1 \\ -\beta(\log|x|)^\mu & \text{for } x < -1 \end{cases} \tag{106}$$

with $\alpha, \beta \in \mathbf{R}$ and $\mu, \lambda \in (0, 1)$, then $F \in PSO$. Indeed, if e.g. $x > 1$, then

$$\text{osc } (F, [x, 2x]) = F(2x) - F(x) = xF'(\xi(x)) = x\alpha\lambda(\log\xi(x))^{\lambda-1}\xi(x)^{-1}$$

with $\xi(x) \in (x, 2x)$ and hence osc $(F, [x, 2x]) \to 0$ as $x \to +\infty$.

Let us turn over to the unit circle. A function $f \in L^\infty(\mathbf{T})$ is said to be *quasicontinuous*, $f \in QC$, if both f and $\overline{f}$ (the complex conjugate of f) belong to $C + H^\infty$:

$$QC = (C + H^\infty) \cap (C + \overline{H^\infty}).$$

The algebra PQC, which was inroduced by Sarason [31], is defined as the smallest closed subalgebra of $L^\infty(\mathbf{T})$ containing PC and QC. The functions in PQC are referred to as the *piecewise quasicontinuous functions*. Power [28] showed the following implications:

$$a \in CU(\dot{\mathbf{T}}), \ (\arg a)^\# \in SO \Longrightarrow a \in QC, \tag{107}$$
$$a \in CU(\dot{\mathbf{T}}), \ (\arg a)^\# \in PSO \Longrightarrow a \in PQC \tag{108}$$

From 2.8 we deduce that $T(a)$ is always Fredholm if $a \in QC \cap GL^\infty$, because QC is a C^*-subalgebra of L^∞ and hence $a^{-1} \in QC$ in that case. Fredholm criteria and index formulas for symbols in PQC were established Sarason [31] (also see [3]). Using (108) and Sarason's PQC results, Power [27] found a nice Fredholm criterion for Toeplitz operators whose stretched argument is in PSO. The following theorem is essentially Power's.

5.12. Theorem. *Suppose* $a \in CU(\dot{\mathbf{T}})$ *and* $(\arg a)^{\#} \in PSO$. *Denote by*

$$d(x) = \mathrm{dist}\,(0, [a^{\#}(-x), a^{\#}(x)])$$

the distance between the line segment $[a^{\#}(-x), a^{\#}(x)]$ *and the origin. Then*

$$T(a) \text{ is Fredholm} \iff \liminf_{|x| \to \infty} d(x) > 0 \tag{109}$$

and

$$T(a) \in NNS \iff \liminf_{|x| \to \infty} d(x) = 0. \tag{110}$$

Proof. The equivalence (109) is explicitly in Power's paper [27] and implicitly in Sarason's article [31]. We therefore restrict ourselves to proving (110).

What we must show is that if $(\arg a)^{\#} \in PSO$ and $T(a)$ is normally solvable, then $T(a)$ is automatically Fredholm. So suppose $T(a)$ is normally solvable. Then, by 2.3,

$$T(a) \in \Phi_0 \cup \Phi(-\mathbf{N}) \cup \Phi(+\mathbf{N}) \cup \Phi(-\infty)$$

or

$$T(a) \in \Phi_0 \cup \Phi(-\mathbf{N}) \cup \Phi(+\mathbf{N}) \cup \Phi(+\infty),$$

i.e. $T(a)$ is left- or right-Fredholm. Equivalently, the coset $T(a) + \mathcal{K}(l^2)$ is left- or right-invertible in the Calkin algebra $\mathcal{L}(l^2)/\mathcal{K}(l^2)$. For the sake of definiteness, assume $T(a) + \mathcal{K}(l^2)$ is left-invertible. In every unital C^*-algebra the equivalences

$$A \text{ is left-invertible} \iff A^*A \text{ is invertible},$$
$$A \text{ is right-invertible} \iff AA^* \text{ is invertible}$$

hold. Hence, we know that

$$T(\bar{a})T(a) + \mathcal{K}(l^2) \text{ is invertible in } \mathcal{L}(l^2)/\mathcal{K}(l^2).$$

From (108) we infer that a and $\bar{a}$ belong to PQC. Since Toeplitz operators with symbols in PQC commute modulo compact operators (see [31] or [3, Proposition 4.83]), it follows that

$$T(a)T(\bar{a}) + \mathcal{K}(l^2) \text{ is invertible in } \mathcal{L}(l^2)/\mathcal{K}(l^2)$$

and consequently, $T(a)$ is right-Fredholm and thus Fredholm. ∎

5.13. Corollary. *Let* $a \in CU(\dot{\mathbf{T}})$ *and* $(\arg a)^{\#} \in PSO$. *If* $(\arg a)^{\#}(+\infty) = +\infty$ *and* $(\arg a)^{\#}$ *is bounded from above at* $-\infty$, *then* $T(a) \in NNS$.

Proof. We have

$$d(x) = 0 \iff 0 \in [e^{i(\arg a)^{\#}(-x)}, e^{i(\arg a)^{\#}(x)}]$$
$$\iff e^{i(\arg a)^{\#}(x)} = e^{i(\arg a)^{\#}(-x)}$$
$$\iff (\arg a)^{\#}(x) - (\arg a)^{\#}(-x) = (2k+1)\pi.$$

Our assumptions imply that $(\arg a)^{\#}(x) - (\arg a)^{\#}(-x)$ goes to $+\infty$ as $x \to +\infty$, and hence there is a sequence $x_k \to +\infty$ such that

$$(\arg a)^{\#}(x_k) - (\arg a)^{\#}(-x_k) = (2k+1)\pi.$$

Since $d(x_k) = 0$, we deduce from Theorem 5.12 that $T(a) \in NNS$. ∎

5.14. Example. If $(\arg a)^{\#} \in PSO$ and (56) holds, then the previous corollary implies that $T(a) \in NNS$. In particular, if $(\arg a)^{\#}(x) = F(x) + o(1)$ as $|x| \to \infty$ where $F(x)$ is given by (106) with $\alpha, \beta > 0$ and $\mu, \lambda \in (0,1)$, then $T(a) \in NNS$.

The following theorem was announced in [20]. It is applicable to symbols whose arguments increase moderately provided $f(\theta)$ and $f(-\theta)$ are related in a quite strong manner.

5.15 Theorem *Let $a \in CU(\dot{\mathbf{T}})$ satisfy (58)–(60) and (62). Instead of (61) assume that*

$$\theta(-k) = -c\theta(k) \quad \text{for all sufficiently large } k \geq 1 \tag{111}$$

with some $c > 0$ and that

$$\limsup_{k \to \infty}(\theta(k+1)/\theta(k)) =: q < 1. \tag{112}$$

Then $T(a) \in \Phi(-\infty)$.

Proof. By virtue of Remark 5.4 we may without loss of generality assume that (111) holds for all integers $k \geq 1$ and that $\sup_{k \geq 1}(\theta(k+1)/\theta(k)) \leq q$.

Define the Blaschke product B_M as in the proof of Theorem 5.2. Then (68) holds. Our aim is to show (80) under the hypotheses of the present theorem.

As in the proof of Theorem 5.2 we see that (69) is true. Hence

$$\left| \sum_{k=1}^{n-2} (\varphi_k(\theta(x)) + \varphi_{-k}(\theta(x))) \right|$$

$$= \left| \sum_{k=1}^{n-2} \left(\arctan\left(\frac{2}{M} \frac{\Delta(k)}{\theta(x) - \theta(k)} \right) + \arctan\left(\frac{2}{M} \frac{\Delta(-k)}{\theta(x) - \theta(-k)} \right) \right) \right| + o\left(\frac{1}{M} \right)$$

$$= \left| \sum_{k=1}^{n-2} \left(-\arctan\left(\frac{2}{M} \frac{\Delta(k)}{\theta(k) - \theta(x)} \right) + \arctan\left(\frac{2}{M} \frac{\Delta(-k)}{\theta(x) - \theta(-k)} \right) \right) \right| + o\left(\frac{1}{M} \right)$$

$$\leq \frac{\text{const}}{M} \sum_{k=1}^{n-2} \left| \frac{\Delta(k)}{\theta(k) - \theta(x)} - \frac{\Delta(-k)}{\theta(x) - \theta(-k)} \right| + o\left(\frac{1}{M} \right). \tag{113}$$

From (111) we infer that

$$\frac{\Delta(k)}{\theta(k) - \theta(x)} > \frac{\Delta(k)}{\theta(k)} = \frac{(1/c)\Delta(-k)}{(-1/c)\theta(-k)} > \frac{\Delta(-k)}{\theta(x) - \theta(-k)}$$

and therefore the sum in (113) equals

$$\sum_{k=1}^{n-2} \left(\frac{\Delta(k)}{\theta(k) - \theta(x)} - \frac{\Delta(-k)}{\theta(x) - \theta(k)} \right). \tag{114}$$

The inequality

$$u - v \leq \log((1 - v)/(1 - u)) \text{ for } 0 \leq v \leq u \leq 1$$

gives that (114) is at most

$$\sum_{k=1}^{n-2} \log \left(\frac{\theta(x) - \theta(-k-1)}{\theta(x) - \theta(-k)} \cdot \frac{\theta(k) - \theta(x)}{\theta(k+1) - \theta(x)} \right)$$

$$= \log \left(\frac{\theta(x) - \theta(-n+1)}{\theta(x) - \theta(-1)} \cdot \frac{\theta(1) - \theta(x)}{\theta(n-1) - \theta(x)} \right)$$

$$\leq \text{const} + \log \left(\frac{\theta(x) - \theta(-n+1)}{\theta(n-1) - \theta(x)} \right)$$

$$\leq \text{const} + \log \left(\frac{\theta(n-1) + c\theta(n-1)}{\theta(n-1) - \theta(n-1/2)} \right)$$

$$\leq \text{const} + \log \frac{\theta(n-1)}{\theta(n-1) - \theta(n-1/2)}$$

$$= \text{const} + \log \frac{\Delta(n-1)}{\theta(n-1) - \theta(n-1/2)} + \log \frac{\theta(n-1)}{\Delta(n-1)}$$

$$= \text{const} + \log(1/\psi_{n-1}(1/2)) + \log \left(1 \Big/ \left(1 - \frac{\theta(n)}{\theta(n-1)} \right) \right) \leq \text{const},$$

the latter estimate resulting from (62) and (112). To summarize, we have shown that

$$\left| \sum_{k=1}^{n-2} \left(\varphi_k(\theta(x)) + \varphi_{-k}(\theta(x)) \right) \right| = O\left(\frac{1}{M}\right). \tag{115}$$

Further, by (70),

$$\left| \sum_{k=n+2}^{\infty} \varphi_k(\theta(x)) \right| \leq \frac{\text{const}}{M} \sum_{k=n+2}^{\infty} \frac{\Delta(k)}{\theta(x) - \theta(k)}$$

$$\leq \frac{\text{const}}{M} \sum_{k=n+2}^{\infty} \frac{\theta(k)}{\theta(n+1) - \theta(k)} = \frac{\text{const}}{M} \sum_{k=n+2}^{\infty} 1 \Big/ \left(\frac{\theta(n+1)}{\theta(k)} - 1 \right).$$

Condition (112) implies that $\theta(n+1)/\theta(k) \geq (1/q)^{k-n-1}$ and hence,

$$\left| \sum_{k=n+2}^{\infty} \varphi_k(\theta(x)) \right| \leq \frac{\text{const}}{M} \sum_{k=n+2}^{\infty} 1 \Big/ \left(\left(\frac{1}{q}\right)^{k-n-1} - 1 \right) = O\left(\frac{1}{M}\right). \tag{116}$$

In a similar fashion one obtains

$$\left| \sum_{k=-\infty}^{-n+1} \varphi_k(\theta(x)) \right| = O\left(\frac{1}{M}\right). \tag{117}$$

Finally, by virtue of (70) and (62),

$$|\varphi_{n\pm1}(\theta(x))| \leq \frac{\text{const}}{M} \frac{\Delta(n\pm1)}{|\theta(n\pm1) - \theta(x)|}$$

$$\leq \frac{\text{const}}{M} \frac{\Delta(n\pm1)}{|\theta(n\pm1) - \theta(n\pm1/2)|} = \frac{\text{const}}{M} \frac{1}{|\psi_{n\pm1}(\mp1/2)|} = O\left(\frac{1}{M}\right). \tag{118}$$

Adding (115)-(118) we arrive at the estimate

$$\sum_{k\neq n} \varphi_k(\theta(x)) = O\left(\frac{1}{M}\right).$$

The rest of the proof is literally the part of the proof of Theorem 5.2 following after (80). ∎

5.16. Remark. Fix any $r \in (q, 1)$. Then $\theta(k+1)/\theta(k) < r$ for all $k \geq k_0$ due to condition (112), which implies that

$$\theta(n) < \theta(k_0)r^{n-k_0} = O(e^{-n\log(1/r)}) \quad \text{as} \quad n \to \infty. \tag{119}$$

Thus, $\theta(n)$ has nevertheless to go to zero sufficiently rapidly.

As pointed out in 5.1, $\theta(n+1)/(\theta(n)) \to 1$ as $n \to +\infty$ under the hypotheses of Theorem 5.2. One can show that under the assumptions of Theorem 5.15 the sequence $\{\theta(n+1)/\theta(n)\}$ actually has a limit, i.e. $\lim_{n\to\infty} \theta(n+1)/\theta(n) = q$.

5.17. Example: the pure logarithm. Let f be given by (99) with $\alpha, \beta > 0$. We know from 5.8 that $T(a) \in \Phi(-\infty)$ if $\mu > 2$ and $\lambda > 2$ and we showed in 5.14 that $T(a) \in NNS$ if $\mu < 1$ and $\lambda < 1$. Suppose now that

$$f(\theta) = \begin{cases} -\alpha(-\log|\theta|)^\mu & \text{for } \theta > 0 \\ \alpha(-\log|\theta|)^\mu & \text{for } \theta < 0 \end{cases} \tag{120}$$

with $\alpha > 0$ and $\mu > 0$. Then

$$\theta(x) = \begin{cases} \exp(-(|x|/\alpha)^{1/\mu} & \text{for } x > 0 \\ -\exp(-(|x|/\alpha)^{1/\mu} & \text{for } x < 0 \end{cases}$$

and thus (111) is satisfied. From (119) we deduce that (112) cannot hold in case $\mu > 1$. Since $T(a) \in NNS$ if $\mu < 1$, there must be a condition of Theorem 5.15 which is violated for $\mu < 1$. This is condition (62): if $\mu < 1$, then $\psi_n(s) \to 1$ for $s > 0$, $\psi_n(s) \to 0$ for $s = 0$, $\psi_n(s) \to -\infty$ for $s < 0$, so that condition (62) is indeed not fulfilled and Theorem 5.15 not applicable to this case. In the case $\mu = 1$ Theorem 5.15 does its job, since then

$$\lim_{k\to\infty} \theta(k+1)/\theta(k) = e^{-1/\alpha} =: q < 1$$

and $\psi_n(s)$ converges uniformly to $\psi(s) = (1 - q^s)/(1 - q)$ as $n \to +\infty$ and to $-\psi(-s)$ as $n \to -\infty$. Consequently, if f is given by (120) with $\alpha > 0$ and $\mu = 1$, then $T(a) \in \Phi(-\infty)$. In 5.2 we proved that $T(a) \in \Phi(-\infty)$ for $\mu > 2$. The remaining gap is the values μ satisfying $1 < \mu \leq 2$. We can show that $T(a) \in \Phi(-\infty)$ in this case, too. For lack of space, we omit the proof here.

6. Orientation changing whirls

6.1. The symbol class. This section is concerned with symbols $a \in CU(\dot{\mathbf{T}})$ for which

$$\lim_{t \to 1+0} \arg a(t) = +\infty, \quad \lim_{t \to 1-0} \arg a(t) = +\infty$$

or equivalently,

$$(\arg a)^{\#}(-\infty) = +\infty, \quad (\arg a)^{\#}(+\infty) = +\infty. \tag{121}$$

Furthermore, we require that

$$(\arg a)^{\#} \text{ is essentially monotonous on } (-\infty, 0) \text{ and on } (0, +\infty), \tag{122}$$

which means that there exists a function $f \in CR(\mathbf{R})$ which is monotonously decreasing on $(-\infty, 0)$ and monotonously increasing on $(0, +\infty)$ such that $(\arg a)^{\#} \sim f$, i.e. such that $(\arg a)^{\#} - f \in L^\infty(\mathbf{R})$.

6.2. Example: semi-almost periodic functions. Let $C(\overline{\mathbf{R}})$ denote the set of continuous functions on $\mathbf{R}$ which have finite (but not necessarily coinciding) limits at $-\infty$ and $+\infty$. The algebra $SAP(\mathbf{R})$ of all *semi-almost periodic functions on $\mathbf{R}$* is defined as the smallest closed subalgebra of $L^\infty(\mathbf{R})$ containing $C(\overline{\mathbf{R}})$ and $AP(\mathbf{R})$. A function $a \in L^\infty(\mathbf{T})$ is said to be *semi-almost* periodic, $a \in SAP$, if $a^{\#} \in SAP(\mathbf{R})$. If

$$(\arg a)^{\#}(x) = \begin{cases} \lambda x + c_0(x) + p(x) & \text{for } x > 0 \\ \mu x + c_0(x) + q(x) & \text{for } x < 0 \end{cases} \tag{123}$$

·where $\lambda, \mu \in \mathbf{R}$, $c_0 \in C(\overline{\mathbf{R}})$, $c_0(\pm\infty) = 0$, and $p, q \in AP(\mathbf{R})$, then $a \in SAP$. In particular, in case $\lambda > 0, \mu < 0$, this function satisfies (121) and (122).

6.3. Each of the six possibilities occurs. If a is as in Subsection 4.1, then there are only two possibilities: either $T(a) \in \Phi(-\infty)$ or $T(a) \in NNS$ (Theorem 4.5). In contrast to this, everything is possible for symbols subject to (121) and (122).

Let $(\arg a)^{\#}(x) = (\log |x|)^\lambda$ $(0 < \lambda < 1)$ for $|x| > 1$. Then $(\arg a)^{\#} \in SO$ (recall 5.12), so $a \in QC$ (by (107)), and since $a \in GL^\infty$, it follows that $T(a)$ is Fredholm. Consequently, the set $\{T(\chi_n a) : n \in \mathbf{Z}\}$ contains Fredholm operators of every (finite) index and, in particular, an invertible operator. Obviously, all symbols $\chi_n a$ satisfy (121) and (122).

Now take the same $a \in QC$ as in the preceding paragraph and let $B \in CU(\dot{\mathbf{T}})$ be an infinite Blaschke product such that $(\arg B)^{\#}(-\infty)$ is finite and $(\arg B)^{\#}(+\infty) = +\infty$ (recall Lemma 4.4). Then $T(aB) = T(a)T(B)$, and since $T(a)$ is Fredholm and $T(B) \in \Phi(-\infty)$, we deduce that $T(aB) \in \Phi(-\infty)$. Clearly, aB fulfils (121) and (122).

To get an operator in $\Phi(+\infty)$, let $a \in QC$ be as before but choose an infinite Blaschke product B so that $(\arg B)^\#(-\infty)$ is finite, $(\arg B)^\#(+\infty) = +\infty$, $(\arg a - \arg B)^\#(+\infty) = +\infty$ (this is again possible due to Lemma 4.4). The function $a\overline{B}$ satisfies (121) and (122), $T(a)$ is Fredholm, $T(\overline{B}) \in \Phi(+\infty)$, and hence, $T(a\overline{B}) = T(\overline{B})T(a) \in \Phi(+\infty)$.

Finally, let $p \in PC$ be a symbol such that $T(p) \in NNS$ and let $a \in QC$ be as above. Then (121) and (122) hold for ap. If $T(pa)$ were normally solvable, then $T(pa)$ were left- or right-Fredholm. Since $T(pa)$ equals both $T(p)T(a)$ and $T(a)T(p)$ modulo compact operators, it would follow that $T(p)$ were left- or right-Fredholm and thus normally solvable. This shows that $T(ap) \in NNS$.

The following theorem, which was established in [1] (also see [3, 2.26]), shows that $(\arg a)^\#$ cannot grow too fast if $T(a)$ is Fredholm and at the same time (121) and (122) hold. Notice that it is again the arguments of "moderate growth" which cause problems.

6.4. Theorem. *Let $a \in CU(\dot{\mathbf{T}})$ and suppose (121) and (122) are satisfied. If $T(a)$ is Fredholm then necessarily*

$$(\arg a)^\#(x) = O(\log |x|) \ \text{as} \ |x| \to \infty. \tag{124}$$

Proof. If $T(a)$ is Fredholm, then there is an $n \in \mathbf{Z}$ such that $T(\chi_n a)$ is invertible. Clearly, $(\arg \chi_n a)^\# \sim (\arg a)^\#$. Hence it suffices to show (124) for the case where $T(a)$ is invertible. So assume $T(a)$ is invertible.

By the Widom/Devinatz criterion cited in 2.10, $a = e^{i(\tilde{u}+v+c)}$ with $c \in \mathbf{R}$ and real-valued functions $u, v \in L^\infty$ such that $\|v\|_\infty < \pi/2$. Since a is continuous on $\dot{\mathbf{T}}$, it follows that $e^{i\tilde{u}} = ae^{-i(v+c)}$ is locally sectorial on $\dot{\mathbf{T}}$. In other words, for each point $t_0 \in \dot{\mathbf{T}}$ there exists an open arc $I \subset \dot{\mathbf{T}}$ containing t_0 and a number $\lambda_I \in \mathbf{R}$ such that

$$\text{ess} \sup_{t \in I} |\tilde{u}(t) - \lambda_I| < \pi/2.$$

We may therefore proceed as in the proof of Theorem 4.5 to conclude that

$$\arg a = \tilde{u} + v + c + 2k\pi \ \text{a.e. on} \ \mathbf{T}$$

with some $k \in \mathbf{Z}$.

Also as in the proof of Theorem 4.5 we see that $(\arg a)^\# \in BMO(\mathbf{R})$. We have $(\arg a)^\# = f + \delta$ where $\delta \in L^\infty(\mathbf{R})$, $f(-\infty) = f(+\infty) = +\infty$, $f(0) = 0$, f is strictly monotonously decreasing on $(-\infty, 0)$ and strictly monotonously increasing on $(0, +\infty)$. Since $L^\infty(\mathbf{R}) \subset BMO(\mathbf{R})$, we obtain that $f \in BMO(\mathbf{R})$. Put $g(x) = f(1/x)$ for $x \neq 0$. Because BMO is conformally invariant (see [12, VI.1.3]), g also belongs to $BMO(\mathbf{R})$.

There is an $x_0 > 0$ such that $g(x) > 0$ for $|x| < x_0$. Denote the mean value $g_{(-x_0, x_0)}$ by g_0. The John-Nirenberg theorem (see [3, VI.2.1]) implies that

$$|\{x \in (-x_0, x_0) : |g(x) - g_0| > \lambda\}| \leq Ce^{-\gamma\lambda} \quad \forall \lambda > 0$$

where $C > 0$ and $\gamma > 0$ are constants independent of λ. Hence, if we define $x_1(\lambda) \in (0, x_0)$ and $x_2(\lambda) \in (0, x_0)$ by $g(x_1(\lambda)) = g(-x_2(\lambda)) = g_0 + \lambda$, then $x_1(\lambda) + x_2(\lambda) \le Ce^{-\gamma\lambda}$. So $x_i(\lambda) \le Ce^{-\gamma\lambda}$, whence $\log x_i(\lambda) \le \log C - \gamma\lambda$ and thus,

$$g(x_1(\lambda)) = g_0 + \lambda \le g_0 + (1/\gamma)\log c - (1/\gamma)\log x_1(\lambda),$$
$$g(-x_2(\lambda)) = g_0 + \lambda \le g_0 + (1/\gamma)\log c - (1/\gamma)\log x_2(\lambda).$$

On replacing $x_i(\lambda)$ by x we get $g(x) \le A(-\log|x|)$ for all x in a sufficiently small neighborhood of $x = 0$. Thus, $f(x) = O(\log|x|)$. ∎

7. Undecided symbols with bounded arguments

7.1. The symbol class. This section is devoted to arguments in the "zeroth" equivalence class of $CR(\mathbf{R})$ with respect to the equivalence relation "$\sim$" introduced in 4.1. Thus, suppose $a \in CU(\dot{\mathbf{T}})$ and

$$(\arg a)^{\#} \in CR(\mathbf{R}) \cap L^{\infty}(\mathbf{R}). \tag{125}$$

If $a \in PC \cap CU(\dot{\mathbf{T}})$ then (125) holds. Toeplitz operators with PC symbols realize four of the six possibilities of 2.3: they may be not normally solvable, may be Fredholm of every index and may be invertible. If $(\arg a)^{\#}$ is periodic, then $T(a)$ is invertible due to Corollary 3.3. We also know from Corollary 3.3 that $T(a)$ is invertible if $(\arg a)^{\#}(x) = p(\alpha(x))$ where p is 2π-periodic and $\alpha(x)$ is the argument of an infinite Blaschke product in $CU(\dot{\mathbf{T}})$. To have one more example, note that $a^{\#}(x) = \sin(\log|x|)^{\lambda}$ $(0 < \lambda < 1)$ is a function in SO for which $T(a)$ is invertible.

The following theorem shows that two of the six possibilities of 2.3 do never occur for the symbols considered here.

7.2. Theorem. *If $a \in CU(\dot{\mathbf{T}})$ and (125) holds, then*

$$T(a) \notin \Phi(-\infty) \cup \Phi(+\infty).$$

Proof. For the sake of definiteness, assume $T(a) \in \Phi(-\infty)$. Then $T(a)$ is right-invertible and as in the proof of Theorem 4.5 we may conclude that

$$T(e^{i(d-\arg a - \arg B)}) = T(e^{id}B^{-1}a^{-1}) \tag{126}$$

is invertible, where $e^{id} \in C \cap GL^{\infty}$ and $B \in CU(\dot{\mathbf{T}})$ is a Blaschke product. By the Widom/Devinatz criterion, the invertibility of the operator (126) is equivalent to the invertibility of the operator

$$T(e^{i(-d+\arg a + \arg B)}) = T(e^{-id}Ba). \tag{127}$$

If B is an infinite Blaschke product, then the operator (127) cannot be invertible due to Lemma 4.4 and Theorem 4.5. Therefore B is a finite Blaschke product and thus $B \in C \cap GL^{\infty}$. It follows from (10) that both

$$T(e^{-id}Ba) - T(e^{-id}B)T(a) \text{ and } T(e^{-id}Ba) - T(a)T(e^{-id}B)$$

are compact and hence, since (127) is invertible, $T(a)$ is Fredholm. This contradicts the assumption that $T(a) \in \Phi(-\infty)$. ∎

7.3. AP and SAP. If $a \in AP$, then either (125) holds or a satisfies (23),(24) or $\tilde{a}$ satisfies (23),(24). If $a \in SAP$ then, in addition to these cases, a may satisfy (121),(122) or $\tilde{a}$ may fulfil (121),(122). Thus, the results of Sections 4 and 6 in conjunction with Theorem 7.2 tells us a lot (but not all) about Toeplitz operators with symbols in SAP ($\supset AP$).

In fact, with respect to the questions we study in this paper, everything is known for SAP symbols. The following results were established by Gohberg and Feldman [14], Coburn and Douglas [6], and Sarason [30]; full proofs may also be found in the book [3].

If $a \in AP \cap GL^\infty$, then $(\arg a)^\#(x) = \lambda x + g(x)$ with $\lambda \in \mathbf{R}$ and $g \in AP(\mathbf{R})$ by Bohr's theorem. One can show that $T(a) \in \Phi(-\infty)$ if $\lambda > 0$, $T(a) \in \Phi(+\infty)$ if $\lambda < 0$, and that $T(a)$ is invertible if $\lambda = 0$.

If $a \in SAP \cap GL^\infty$, then $(\arg a)^\#$ may be written in the form (123). Equivalently,

$$(\arg a)^\#(x) = u(x)(\lambda x + p(x)) + v(x)(\mu x + q(x)) + c(x)$$

where u, v, c are real-valued functions in $CR(\mathbf{R})$ such that $u(-\infty) = 1$, $u(+\infty) = 0$, $v(-\infty) = 0$, $v(+\infty) = 1$, $c(-\infty) = c(+\infty) = 0$, $\lambda \in \mathbf{R}$, $\mu \in \mathbf{R}$, $p \in AP(\mathbf{R})$, $q \in AP(\mathbf{R})$. Put

$$m(p) = \lim_{T \to -\infty} \frac{1}{T} \int_T^{2T} p(x)\, dx, \quad m(q) = \lim_{T \to +\infty} \frac{1}{T} \int_T^{2T} q(x)\, dx$$

and define $a_0 \in PC \cap CU(\dot{\mathbf{T}})$ by

$$(\arg a_0)^\#(x) = m(p)u(x) + m(q)v(x) + c(x).$$

It can be shown that $T(a) \in NNS$ if $\lambda\mu < 0$, that $T(a) \in \Phi(-\infty)$ if $\lambda \geq 0$, $\mu \geq 0$ and $\lambda^2 + \mu^2 > 0$, that $T(a) \in \Phi(+\infty)$ if $\lambda \leq 0$, $\mu \leq 0$ and $\lambda^2 + \mu^2 > 0$, that $T(a) \in NNS$ if $\lambda = \mu = 0$ and $T(a_0) \in NNS$, and that $T(a)$ is Fredholm of index $\varkappa$ if $\lambda = \mu = 0$ and $T(a_0)$ is Fredholm of index $\varkappa$.

7.4. The finite section method. Let $a \in L^\infty$ and consider the infinite linear system (1). In order to solve this system approximately, we may replace it by the truncated system

$$
\begin{pmatrix}
a_0 & a_{-1} & \cdots & a_{-n} \\
a_1 & a_0 & \cdots & a_{-n+1} \\
\cdots & \cdots & \cdots & \cdots \\
a_n & a_{n-1} & \cdots & a_0
\end{pmatrix}
\begin{pmatrix}
f_0^{(n)} \\
f_1^{(n)} \\
\cdots \\
f_n^{(n)}
\end{pmatrix}
=
\begin{pmatrix}
g_0 \\
g_1 \\
\cdots \\
g_n
\end{pmatrix}
\tag{128}
$$

One says that the *finite section method is applicable* to the operator $T(a)$ if there is an $n_0 \geq 0$ such that the systems (128) have a unique solution for all $n \geq n_0$ and all $g \in l^2$ and if

$$\{f_0^{(n)}, f_1^{(n)}, \ldots, f_n^{(n)}, 0, 0, \ldots\} \in l^2$$

converges in l^2 to a solution $f \in l^2$ of the system (1). In this case one writes $T(a) \in \Pi\{P_n\}$.

Denote the matrix in (128) by $T_n(a)$ and let P_n stand for the projection on l^2 defined by

$$P_n : \{f_0, f_1, f_2, \ldots\} \mapsto \{f_0, f_1, \ldots, f_n, 0, 0, \ldots\}.$$

It is well known (see e.g. [15] or [3]) that $T(a) \in \Pi\{P_n\}$ if and only if $T(a)$ is invertible, if $T_n(a)$ is an invertible matrix for all sufficiently large n, for $n \geq n_0$ say, and if $\sup_{n \geq n_0} \|T_n^{-1}(a)P_n\| < \infty$. In particular, the implication

$$T(a) \in \Pi\{P_n\} \implies T(a) \text{ is invertible} \tag{129}$$

is true. The reverse implication,

$$T(a) \text{ is invertible} \implies T(a) \in \Pi\{P_n\} \tag{130}$$

has been studied by many mathematicians, including Baxter, Reich, Gohberg, Feldman, Ambartsumyan, Widom, Silbermann, and it was shown to be valid if

$$a \in (C + H^\infty) \cup (C + \overline{H^\infty}) \cup PQC;$$

see the books [15], [3] for more about this topic. Until 1987, it had not been clear whether (130) holds for every $a \in L^\infty$, and only in 1987 Treil [35] was able to find a function $a \in L^\infty$ such that $T(a)$ is invertible but $T(a) \notin \Pi\{P_n\}$. Treil's symbol belongs to $CU(\dot{\mathbf{T}})$ but is rather complicated. The purpose of the rest of this section is to show that *we need not look for such symbols in the abyss of L^∞*. Theorem 7.6 will provide a sufficiently simple class of such symbols. For instance, this theorem shows that there are $a \in CU(\mathbf{T}) \cap AP$ such that $T(a)$ is invertible but $T(a) \notin \Pi\{P_n\}$. Notice, however, that our proof of Theorem 7.6 is essentially based on the proof of Treil and is thus not much simpler than his proof.

7.5. Lemma. *There exist universal constants γ and Γ with the following property. Given $\varepsilon \in (0,1)$ and $n \geq \gamma\varepsilon^{-5/4}$, one can find a trigonometric polynomial*

$$T_n(\theta) = c_0 + \sum_{k=1}^{n}(c_k \sin(k\theta) + d_k \cos(k\theta))$$

such that

$$\|T_n\|_{L^2(-\pi,\pi)} = 1 \tag{131}$$

and

$$\|T_n\|_{L^2((-\pi,\pi)\backslash(-\varepsilon,\varepsilon))} \leq \Gamma n^{-2}\varepsilon^{-5/2}. \tag{132}$$

Proof. Fix a twice continuously differentiable function f on $(-\pi, \pi)$ such that $\|f\|_{L^2(-\pi,\pi)} = 1$ and $f^{(j)}(-\pi) = f^{(j)}(\pi) = 0$ for $j = 0, 1, 2$. Define f_ε by

$$f_\varepsilon(\theta) = \begin{cases} (\pi/\varepsilon)^{1/2}f(\pi\theta/\varepsilon) & \text{for } |\theta| \leq \varepsilon, \\ 0 & \text{for } |\theta| > \varepsilon. \end{cases}$$

Obviously, $\|f\|_{L^2(-\pi,\pi)} = 1$. Jackson's theorem (see e.g. [11, p. 205]) implies the existence of a trigonometric polynomial $\tilde{T}_n$ of degree n such that

$$|f_\varepsilon(\theta) - \tilde{T}_n(\theta)| \leq \frac{144}{n}\omega\left(f'_\varepsilon, \frac{1}{n}\right)$$

where
$$\omega(f,\delta) := \sup\left\{|f(\theta+\delta) - f(\theta)| : \theta \in (-\pi, \pi),\ \theta + \delta \in (-\pi, \pi)\right\}$$
is the continuity modulus of f. Since
$$\omega\left(f'_\varepsilon, \frac{1}{n}\right) \le \|f''_\varepsilon\|_{L^\infty(-\pi,\pi)}\frac{1}{n} \le \pi^{5/2}\|f''\|_{L^\infty(-\pi,\pi)}\varepsilon^{-5/2}\frac{1}{n},$$
we get
$$|f_\varepsilon(\theta) - \tilde{T}_n(\theta)| \le 144\,\pi^{5/2}\|f''\|_{L^\infty(-\pi,\pi)}n^{-2}\varepsilon^{-5/2} =: qn^{-2}\varepsilon^{-5/2}.$$
Now put $T_n = \tilde{T}_n/\|\tilde{T}_n\|_{L^2(-\pi,\pi)}$. Then $\|T_n\|_{L^2(-\pi,\pi)} = 1$. Furthermore,
$$\|T_n\|_{L^2((-\pi,\pi)\backslash(-\varepsilon,\varepsilon))} = \|\tilde{T}_n\|_{L^2((-\pi,\pi)\backslash(-\varepsilon,\varepsilon))}\|\tilde{T}_n\|^{-1}_{L^2(-\pi,\pi)}$$
$$= \|\tilde{T}_n - f_\varepsilon\|_{L^2((-\pi,\pi)\backslash(-\varepsilon,\varepsilon))}\|\tilde{T}_n\|^{-1}_{L^2(-\pi,\pi)}$$
$$\le \|\tilde{T}_n - f_\varepsilon\|_{L^2(-\pi,\pi)}\|\tilde{T}_n\|^{-1}_{L^2(-\pi,\pi)}$$
$$\le q(2\pi)^{1/2}\|\tilde{T}_n\|^{-1}_{L^2(-\pi,\pi)}n^{-2}\varepsilon^{-5/2}.$$

Let $\gamma := (2\pi)^{1/4}\,2^{1/2}q^{1/2}$. If $n \ge \gamma\varepsilon^{-5/4}$, then
$$\|f_\varepsilon - \tilde{T}_n\|_{L^2(-\pi,\pi)} \le (2\pi)^{1/2}\|f_\varepsilon - \tilde{T}_n\|_{L^\infty(-\pi,\pi)} \le (2\pi)^{1/2}qn^{-2}\varepsilon^{-5/2} < 1/2,$$
whence
$$\|\tilde{T}_n\|_{L^2(-\pi,\pi)} \ge \|f_\varepsilon\|_{L^2(-\pi,\pi)} - \|f_\varepsilon - \tilde{T}_n\|_{L^2(-\pi,\pi)} \ge 1 - \frac{1}{2} = \frac{1}{2},$$
and thus
$$\|T_n\|_{L^2((-\pi,\pi)\backslash(-\varepsilon,\varepsilon))} \le q(2\pi)^{1/2}2n^{-2}\varepsilon^{-5/2} =: \Gamma n^{-2}\varepsilon^{-5/2}.\ \blacksquare$$

7.6. Theorem. *Let $b \in C$ be any function such that $T(b)$ is invertible and such that the zeroth Fourier coefficient of b is zero, $b_0 = 0$. Let u be the inner function*
$$u(t) = \exp\left(\frac{t+1}{t-1}\right) \quad (t \in \dot{\mathbf{T}})$$
and put $a(t) = b(u(t))$ for $t \in \dot{\mathbf{T}}$. Then $T(a)$ is invertible but $T(a) \notin \Pi\{P_n\}$.

Proof. The invertibility of $T(a)$ results from Corollary 3.3. In order to show that $T(a) \notin \Pi\{P_n\}$, it suffices to find a sequence $\{f_{n(l)}\}_{l=1}^\infty$ of
$$f_{n(l)} = \sum_{j=0}^{2n(l)} \alpha_{j,l}\,t^j \quad (t \in \mathbf{T}) \tag{133}$$
such that $n(l) \to \infty$ as $l \to \infty$, $\|f_{n(l)}\|_2 = 1$, and
$$\delta(l) := \|P_{2n(l)}T(a)f_{n(l)}\|_2 \to 0 \quad \text{as}\ l \to \infty. \tag{134}$$
Put $t = e^{i\theta}$ with $\theta \in (-\pi, \pi]$. We then have
$$u(e^{i\theta}) = \exp((e^{i\theta} + 1)/(e^{i\theta} - 1)) = \exp(-i\cot(\theta/2))$$

and we write $A(\theta) := -\cot(\theta/2)$. For $j \in \mathbf{Z}$, define $\theta_j \in (-\pi, \pi)$ by $A(\theta_j) = (2j-1)\pi$. Put $I_j = (\theta_j, \theta_{j+1})$ for $j \in \mathbf{Z} \setminus \{0\}$ and define $I_0 = (\theta_0 \cup \pi] \cup [-\pi, \theta_1)$. For $\theta \in I_j$, set $A_j(\varphi) = A(\varphi) - 2j\pi$. Then A_j is a bijective map of I_j onto $(-\pi, \pi)$. Further, put $t_j = e^{i\theta_j}$ and denote by γ_j the arc of $\mathbf{T}$ between t_j and t_{j-1}. Let u_j be the restriction of u to γ_j. Then u_j maps γ_j bijectively onto $\mathbf{T}$. See Figure 8. Clearly,

$$A_j'(\theta) = 1/(2\sin^2(\theta/2)). \tag{135}$$

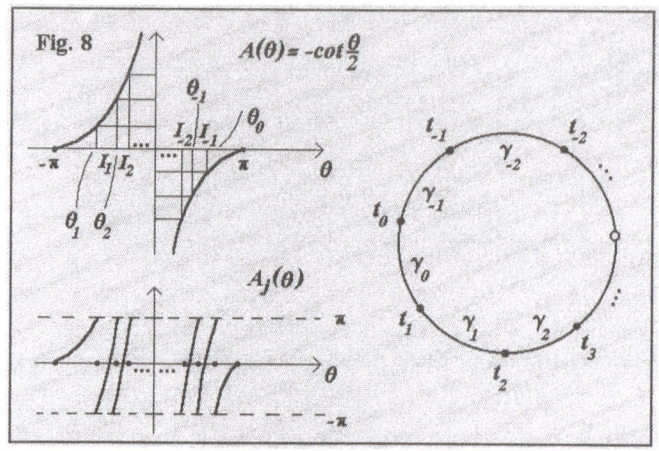

Fig. 8

There is an l_0 such that

$$(-\varepsilon, \varepsilon) = \bigcup_{|j| \ge l} I_j \subset (-1, 1) \tag{136}$$

for all $l \ge l_0$. For $l \ge l_0$, define ε by (136). Since $\gamma\varepsilon^{-4/3} > \gamma\varepsilon^{-5/4}$, Lemma 7.5 guarantees the existence of a trigonometric polynomial $T_{n(l)}$ of degree

$$n = n(l) \in [\gamma\varepsilon^{-4/3}, \gamma\varepsilon^{-4/3} + 1] \tag{137}$$

such that (131) and (132) hold. Put

$$f_{n(l)}(e^{i\theta}) = e^{in(l)\theta} T_n(\theta).$$

Since $1/(Cl) \le |\theta_l| = \varepsilon \le C/l$ with some constant $C > 0$ independent of l, we infer from (137) that

$$l^{4/3}/M \le n(l) \le Ml^{4/3} \tag{138}$$

with some constant $M > 0$ which does not depend on n.

Let $\mathcal{P}_{2n(l)}$ stand for the set of all polynomials of the form (133). Clearly, $f_{n(l)} \in \mathcal{P}_{2n(l)}$. Denote the Fourier coefficients of $a f_{n(l)}$ by φ_j ($j \in \mathbf{Z}$). We have

$$\delta(l)^2 = \sum_{j=0}^{2n(l)} |\varphi_j|^2 = \max\left\{ \left| \sum_{j=0}^{2n(l)} \varphi_j \overline{q_j} \right| : q \in \mathcal{P}_{2n(l)}, \|q\|_2 = 1 \right\}$$

$$= \max\left\{ \left| \frac{1}{2\pi} \int_{\mathbf{T}} \left(\sum_{j=-\infty}^{\infty} \varphi_j t^j \right) \left(\sum_{k=0}^{2n(l)} \overline{q_k} t^{-k} \right) dt \right| : q \in \mathcal{P}_{2n(l)}, \|q\|_2 = 1 \right\}$$

and consequently, there is a polynomial $q_{n(l)} \in \mathcal{P}_{2n(l)}$ such that $\|q_{n(l)}\|_2 = 1$ and

$$\delta(l)^2 = \frac{1}{2\pi}\left|\int_{\mathbb{T}} b(u(t))f_{n(l)}(t)\overline{q_{n(l)}(t)}\,|dt|\right| = \frac{1}{2\pi}\left|\int_{-\pi}^{\pi} b(u(e^{i\theta}))f_{n(l)}(e^{i\theta})\overline{q_{n(l)}(e^{i\theta})}\,d\theta\right|.$$

Decompose the latter integral into the sum

$$\int_{-\pi}^{\pi} = \int_{-\varepsilon}^{\varepsilon} + \int_{(-\pi,\pi)\backslash(-\varepsilon,\varepsilon)}.$$

By the construction of $f_{n(l)}$ and $q_{n(l)}$,

$$\left|\int_{(-\pi,\pi)\backslash(-\varepsilon,\varepsilon)} b(u(e^{i\theta}))f_{n(l)}(e^{i\theta})\overline{q_{n(l)}(e^{i\theta})}\,d\theta\right|$$

$$\leq \|b\|_\infty\|T_n\|_{L^2((-\pi,\pi)\backslash(-\varepsilon,\varepsilon))}\|q_n\|_{L^2((-\pi,\pi)\backslash(-\varepsilon,\varepsilon))} \leq \|b\|_\infty \Gamma n(l)^{-2}\varepsilon^{-5/2}.$$

For the first integral we have

$$\left|\int_{-\varepsilon}^{\varepsilon} b(u(e^{i\theta}))f_{n(l)}(e^{i\theta})\overline{q_{n(l)}(e^{i\theta})}\,d\theta\right|$$

$$\leq \sum_{|j|\geq l}\left|\int_{I_j} b(u(e^{i\theta}))f_{n(l)}(e^{i\theta})\overline{q_{n(l)}(e^{i\theta})}\,d\theta\right| = \sum_{|j|\geq l}\left|\int_{I_j} b(u(e^{i\theta}))f_{n(l)}(e^{i\theta})\overline{q_{n(l)}(e^{i\theta})}\frac{dA_j(\theta)}{A'_j(\theta)}\right|.$$

Denote by $g_{l,j}$ the function

$$g_{l,j}(\theta) = 2f_{n(l)}(e^{i\theta})\overline{q_{n(l)}(e^{i\theta})}\sin^2(\theta/2)$$

(recall (135)). If $\psi_j \in I_j$, then

$$|g_{l,j}(\theta) - g_{l,j}(\psi_j)| \leq \|g'_{l,j}\|_{L^\infty(I_j)}|\theta - \psi_j|.$$

For an arbitrary trigonometric polynomial S_n of degree n we have Jackson's inequality (see e.g. [11, p. 116])

$$\|S_n\|_{L^\infty(-\pi,\pi)} \leq \sqrt{n}\|S_n\|_{L^2(-\pi,\pi)}$$

and Bernstein's inequality (see e.g. [11, p. 246])

$$\|S'_n\|_{L^\infty(-\pi,\pi)} \leq n\|S_n\|_{L^\infty(-\pi,\pi)}.$$

Consequently,

$$|g'_{l,j}(\theta)| \leq 2|f'_{n(l)}(e^{i\theta})|\,|q_{n(l)}(e^{i\theta})|\sin^2(\theta/2)$$
$$+2|f_{n(l)}(e^{i\theta})|\,|q'_{n(l)}(e^{i\theta})|\sin^2(\theta/2) + |f_{n(l)}(e^{i\theta})|\,|q_{n(l)}(e^{i\theta})|\,\sin\theta|$$
$$\leq 2\cdot 2n(l)\sqrt{2n(l)}\|f_{n(l)}\|_2\sqrt{2n(l)}\|q_{n(l)}\|_2 \sin^2(\theta/2)$$
$$+2\sqrt{2n(l)}\|f_{n(l)}\|_2\cdot 2n(l)\sqrt{2n(l)}\|q_{n(l)}\|_2 \sin^2(\theta/2)$$
$$+\sqrt{2n(l)}\|f_{n(l)}\|_2\sqrt{2n(l)}\|q_{n(l)}\|_2|\sin\theta|$$
$$= 16n(l)^2\sin^2(\theta/2) + 2n(l)|\sin\theta|. \tag{139}$$

If $\theta \in I_j$ then

$$\sin^2(\theta/2) \le C/|j|^2, \quad |\sin\theta| \le C/|j| \tag{140}$$

with some constant C independent of j. The sum (139) equals

$$\sum_{|j|\ge l} \left| \int_{I_j} b(u(e^{i\theta})) g_{l,j}(\theta) \, dA_j(\theta) \right|$$

$$\le \sum_{|j|\ge l} |g_{l,j}(\psi_j)| \left| \int_{I_j} b(u(e^{i\theta})) \, dA_j(\theta) \right|$$

$$+ \sum_{|j|\ge l} \|g'_{l,j}\|_{L^\infty(I_j)} \int_{I_j} |b(u(e^{i\theta}))| |\theta - \psi_j| \, dA_j(\theta). \tag{141}$$

We have

$$\int_{I_j} b(u(e^{i\theta})) \, dA_j(\theta) = \int_{-\pi}^{\pi} b(e^{i\varphi}) \, d\varphi = 2\pi b_0 = 0,$$

and therefore the first sum in (141) vanishes. By (139) and (140), the second sum in (141) is not larger than

$$\sum_{|j|\ge l} \left(\frac{16Cn(l)^2}{|j|^2} + \frac{2Cn(l)}{|j|} \right) \|b\|_\infty |I_j| \int_{I_j} dA_j(\theta). \tag{142}$$

The integral in (142) is equal to

$$A_j(\theta_{j+1}) - A_j(\theta_j) = 2\pi.$$

In summary, we have

$$\delta(l)^2 \le \|b\|_\infty \Gamma n(l)^{-2} \varepsilon^{-5/2} + \|b\|_\infty \sum_{|j|\ge l} \left(\frac{16Cn(l)^2}{|j|^2} + \frac{2Cn(l)}{|j|} \right) 2\pi |I_j|.$$

Because $|I_j| \le c/|j|^2$ with some constant $c > 0$ independent of j and $\varepsilon = |\theta_l| \ge 1/(Cl)$, it follows that

$$\delta(l)^2 \le D\left(\frac{l^{5/4}}{n(l)} \right)^2 + E\left(\left(\frac{n(l)}{l^{3/2}} \right)^2 + \frac{n(l)}{l^2} \right)$$

with certain constants D and E independent of l. The estimates (138) finally imply that

$$\delta(l)^2 \le D\left(\frac{Ml^{5/4}}{l^{4/3}} \right)^2 + E\left(\left(\frac{Ml^{4/3}}{l^{3/2}} \right)^2 + \frac{Ml^{4/3}}{l^2} \right) = o(1) \quad \text{as } l \to \infty,$$

as desired. ∎

7.7. Examples. Let g be a real-valued function in $C[0, \pi/2]$ and suppose $g(0) = 0$, $g(\pi/2) = \pi$. Define f on $[0, \pi]$ by

$$f(x) = \begin{cases} g(x) & \text{for } x \in [0, \pi/2] \\ \pi + g(x - \pi/2) & \text{for } x \in [\pi/2, \pi] \end{cases}$$

and extend f to an even and 2π-periodic function on all of $\mathbf{R}$. Obviously, $f \in CR(\mathbf{R})$. Then define $b \in C(\mathbf{T})$ by $b(e^{ix}) = e^{if(x)}$. If e^{ix} once runs through $\mathbf{T}$, $b(e^{ix})$ twices traces out $\mathbf{T}$, once in the positive and once in the negative direction. Hence, $b(t) \neq 0$ for $t \in \mathbf{T}$ and wind $b = 0$. We have

$$b_0 = \int_{-\pi}^{\pi} e^{if(x)}\, dx = 2\int_0^\pi e^{if(x)}\, dx = 2\int_0^{\pi/2} e^{ig(x)}\, dx + 2\int_0^{\pi/2} e^{i(\pi+g(x))}\, dx = 0$$

and thus the hypotheses of Theorem 7.6 are satisfied. If u is the inner function in that theorem, then $u^{\#}(x) = e^{ix}$ and hence for $a(t) = b(u(t))$ we have

$$a^{\#}(x) = b(u^{\#}(x)) = b(e^{ix}) = e^{if(x)},$$

i.e. $(\arg a)^{\#}(x) = f(x)$. In particular, if $(\arg a)^{\#}$ is as in Figures 9 and 10, then $T(a)$ is invertible but $T(a) \notin \Pi\{P_n\}$. Clearly, the symbols constructed in the above way belong to $CU(\dot{\mathbf{T}}) \cap AP$.

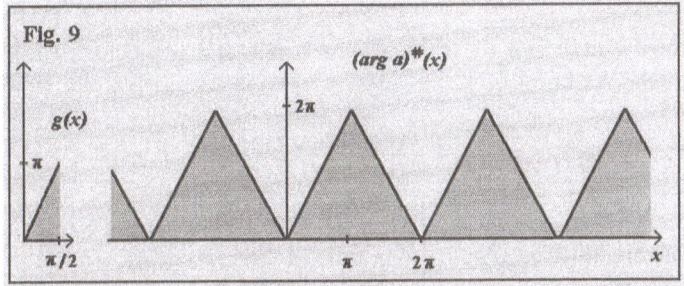

Fig. 9

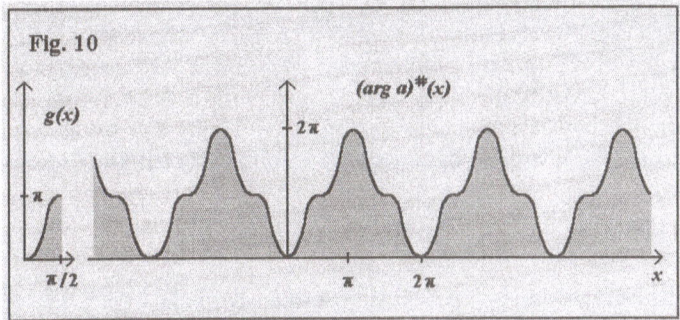

Fig. 10

In absolutely explicit language: the argument plotted in Figure 9 is given by

$$(\arg a)^{\#}(x) = \pi - \frac{8}{\pi}\left(\cos x + \frac{1}{3^2}\cos 3x + \frac{1}{5^2}\cos 5x + \dots\right), \tag{143}$$

so $a \in AP$, the operator $T(a)$ is invertible, and the finite section method is not applicable to $T(a)$.

7.8. One more example or once again on the role of amplitude modulation.
Now define $b \in C(\mathbf{T})$ by

$$b(t) = \left(\frac{\sqrt{3} - t}{\sqrt{3} - \bar{t}}\right)^2 \quad (t \in \mathbf{T}).$$

Clearly, $b(t) \neq 0$ for $t \in \mathbf{T}$ and wind $b = 0$ (note that wind $(\sqrt{3} - t) = $ wind $(\sqrt{3} - \bar{t}) = 0$).
Furthermore, by the residue theorem we have

$$
\begin{aligned}
b_0 &= \frac{1}{2\pi} \int_{\mathbf{T}} b(t) \, |dt| = \frac{1}{2\pi i} \int_{\mathbf{T}} b(t) \frac{dt}{t} = \frac{1}{2\pi i} \frac{1}{3} \int_{\mathbf{T}} \left(\frac{\sqrt{3} - t}{t - 1/\sqrt{3}}\right)^2 t \, dt \\
&= \frac{1}{3} \operatorname{Res} \left(\left(\frac{\sqrt{3} - t}{t - 1/\sqrt{3}}\right)^2 t, \ 1/\sqrt{3} \right) = \frac{1}{3} \left[(\sqrt{3} - t)^2 t \right]'_{t = 1/\sqrt{3}} = 0.
\end{aligned}
$$

Thus, if we put $a(t) = b(u(t))$, then $T(a)$ is invertible but $T(a) \notin \Pi\{P_n\}$. Explicitly:

$$a(t) = \left(\frac{\sqrt{3} - u(t)}{\sqrt{3} - \overline{u(t)}}\right)^2 \quad (t \in \mathbf{T}), \qquad a^{\#}(x) = \left(\frac{\sqrt{3} - e^{ix}}{\sqrt{3} - e^{-ix}}\right)^2 \quad (x \in \mathbf{R}). \qquad (144)$$

Put $h(t) = (\sqrt{3} - u(t))^4$ for $t \in \dot{\mathbf{T}}$. Then $h \in GH^{\infty}$ and consequently, $T(h)$ is invertible
and $T(h) \in \Pi\{P_n\}$. We know from 2.9 that $T(h/|h|)$ is also invertible. However, since

$$\frac{h(t)}{|h(t)|} = \frac{(\sqrt{3} - u(t))^4}{|\sqrt{3} - u(t)|^4} = \frac{(\sqrt{3} - u(t))^2}{(\sqrt{3} - \overline{u(t)})^2} = a(t),$$

we arrive at the conclusion that $T(h/|h|) \notin \Pi\{P_n\}$. Moral: while amplitude modulation
does not affect invertibility of Toeplitz operators, it affects applicability of the finite section
method.

Finally, we remark that symbols as in Theorem 7.6 (and in particular the symbols
given by (143) or (144)) own another interesting property: although $T(a)$ is invertible, the
harmonic extension of a into $\mathbf{D}$ is not bounded away from zero near $\partial \mathbf{D} = \mathbf{T}$ (see [9, pp.
13–14] or [3, p. 211]). Crazy question: is it true that for $a \in L^{\infty}$ (or at least for $a \in AP$)
the finite section method is applicable to $T(a)$ if and only if $T(a)$ is invertible and the
harmonic extension of a into $\mathbf{D}$ is bounded away from zero near $\partial \mathbf{D}$?

8. Modulated almost periodic symbols

8.1. The symbol class. In this section we consider symbols $a \in CU(\dot{\mathbf{T}})$ whose stretched
argument is of the form

$$(\arg a)^{\#}(x) = \lambda \alpha(x) + p(\alpha(x)) \qquad (145)$$

where $\lambda \in \mathbf{R}$, $p \in AP(\mathbf{R})$, $\alpha \in CR(\mathbf{R})$ is a strictly monotonously increasing function
satisfying

$$\alpha(-\infty) = -\infty \quad \text{and} \quad \alpha(+\infty) = +\infty.$$

The case in which λ is an integer, p is 2π-periodic and α is known to be the stretched
argument of an inner function in $CU(\dot{\mathbf{T}})$ was already tackled in Section 3. Notice also that

if $\lambda \neq 0$, then the symbols of the above form fall into the class considered in Section 4, while for $\lambda = 0$ we have symbols as in Section 7.

The main ingredient in (145) is the function $\alpha(x)$. We call $\alpha(x)$ *regular* if the function a whose stretched function $a^\#$ is given by $a^\#(x) = e^{i\alpha(x)}$ (i.e. the function given by (145) with $\lambda = 1$ and $p(x) = 0$) satisfies (58)–(62) and, in addition,

$$\psi(s) = ds + O(s^2) \quad (s \to 0) \quad \text{with some } d > 0. \tag{146}$$

If $\alpha(x) = F(x)$ is any of the functions (101)–(103) then $\alpha(x)$ is regular, because, as shown in 5.5–5.7, in these cases the hypotheses of Proposition 5.3 are satisfied and therefore $\psi(s) = s$.

8.2. Lemma. *Let* $a^\#(x) = e^{i\alpha(x)}$ *with a regular function* $\alpha(x)$. *Then for each* $\varepsilon > 0$ *there exist an inner function* $u_\varepsilon \in CU(\mathbf{T})$, *a function* $\varphi_\varepsilon \in C \cap GL^\infty$, *and a function* $s_\varepsilon \in L^\infty$ *with* $\|s_\varepsilon\|_\infty < \varepsilon$ *such that*

$$a(t) = u_\varepsilon(t)\varphi_\varepsilon(t)(1 + s_\varepsilon(t)) \quad \text{a.e. on } \mathbf{T}. \tag{147}$$

Proof. For $M > 0$, define the Blaschke product B_M as in the proof of Theorem 5.2 and put

$$\sigma = \frac{Md - 2\pi}{Md + 2\pi}, \quad u_\varepsilon(t) = \frac{\sigma - B_M(t)}{1 - \sigma B_M(t)},$$

where d is from (146). Clearly, u_ε is inner. For the function $f_M(x)$ given by (68) we have the representation (86):

$$f_M(x) = -2\pi x + \beta(x) + c(x) + O(1/M), \tag{148}$$

where $\beta \in CR(\mathbf{R})$ is 1-periodic, $c \in CR(\mathbf{R})$ vanishes at $\pm\infty$, $G(x) := -2\pi x + \beta(x)$ is strictly monotonously decreasing, and $\beta(x) = \beta_0(x) + O(1/M)$ with

$$\beta_0(x) = \begin{cases} 2\pi s + 2\pi + 2\arctan\frac{2}{M\psi(s)} & \text{for } s \in (-1/2, 0) \\ 2\pi s + 2\arctan\frac{2}{M\psi(s)} & \text{for } s \in (0, 1/2] \end{cases}$$

where $s := x - [x + 1/2]$ (note that the β_0 is the α of the proof of Theorem 5.2). If (146) holds, then $\beta(x) = \gamma(x) + O(1/M)$ with

$$\gamma(x) = \begin{cases} 2\pi s + 2\pi + 2\arctan\left(\frac{2\pi}{Md}\cot \pi s\right) & \text{for } s \in (-1/2, 0) \\ 2\pi s + 2\arctan\left(\frac{2\pi}{Md}\cot \pi s\right) & \text{for } s \in (0, 1/2] \end{cases}$$

where again $s = x - [x + 1/2]$. Indeed, we have

$$|\beta_0(s) - \gamma(s)| = 2\left|\arctan\frac{2}{M\psi(s)} - \arctan\left(\frac{2\pi}{Md}\cot \pi s\right)\right|$$

$$\leq \frac{\text{const}}{M}\left|\frac{1}{\psi(s)} - \frac{\pi}{d\tan \pi s}\right| = \frac{\text{const}}{M}\left|\frac{d\tan \pi s - \pi\psi(s)}{d\psi(s)\tan \pi s}\right|$$

$$= \frac{\text{const}}{M}\left|\frac{(d\pi s + O(s^3)) - (\pi ds + O(s^2))}{d(ds + O(s^2))(\pi s + O(s^3))}\right|$$

$$\leq \frac{\text{const}}{M}\frac{s^2}{s^2|1 + O(s)||1 + O(s^2)|} \leq \frac{\text{const}}{M}.$$

Thus, from (148) we get

$$f_M(x) = -2\pi x + \gamma(x) + c(x) + O(1/M). \tag{149}$$

Clearly, γ is continuous and 1-periodic and the function $H(x) := -2\pi x + \gamma(x)$ is strictly monotonously decreasing. The inverse function H^{-1} is of the form $H^{-1}(y) = -y/(2\pi) + \delta(y)$ with

$$\delta(y) = \frac{u}{2\pi} + \frac{1}{\pi}\arctan\left(\frac{2\pi}{Md}\cot\frac{u}{2}\right), \quad u = y - 2\pi\left[\frac{y}{2\pi}\right] \in [0, 2\pi). \tag{150}$$

From (149) and the uniform continuity of H^{-1} we obtain

$$-\frac{f_M(x)}{2\pi} + \delta(f_M(x)) = x + g(x) + O(1/M) \tag{151}$$

with some function $g \in CR(\mathbf{R})$ vanishing at $\pm\infty$.

We now compute the argument of the inner function u_ε. We have

$$\arg u_\varepsilon(e^{i\theta(x)}) = \arg\left(\frac{\sigma - \exp(if_M(x))}{1 - \sigma\exp(if_M(x))}\right)$$

and (31), (32) so imply that

$$\arg u_\varepsilon(e^{i\theta(x)}) = -2\arctan\left(\frac{1-\sigma}{1+\sigma}\cot\frac{f_M(x)}{2}\right) + 2\pi\left[\frac{f_M(x)}{2\pi}\right]$$
$$= -2\arctan\left(\frac{2\pi}{Md}\cot\frac{f_M(x)}{2}\right) + 2\pi\left[\frac{f_M(x)}{2\pi}\right]$$
$$= -2\pi\delta(f_M(x)) + f_M(x) \quad \text{(by(150))}$$
$$- -2\pi x - 2\pi g(x) + O(1/M) \quad \text{(by(151))}.$$

Taking into account that $x = -f(\theta)$ we arrive at the formula

$$e^{2\pi if(\theta)} = u_\varepsilon(e^{i\theta})e^{2\pi ig(x)}e^{iO(1/M)}.$$

If M is sufficiently large then $\|e^{iO(1/M)} - 1\|_\infty \le \varepsilon$, which completes the proof. ∎

8.3. Theorem. *Let $a \in CU(\dot{\mathbf{T}})$ be given by (145) and suppose $\mu\alpha(x)$ is regular for every $\mu > 0$. Then $T(a)$ is Fredholm if $\lambda = 0$, $T(a) \in \Phi(-\infty)$ if $\lambda > 0$, and $T(a) \in \Phi(+\infty)$ if $\lambda < 0$.*

Proof. Define $a_\mu \in CU(\dot{\mathbf{T}})$ by $a_\mu^\#(x) = e^{i\mu\alpha(x)}$. From Lemma 8.2 we know that for $\varepsilon > 0$ there is a factorization

$$a_\mu = u_{\mu,\varepsilon}\varphi_{\mu,\varepsilon}(1 + s_{\mu,\varepsilon}) \tag{152}$$

where $u_{\mu,\varepsilon} \in H^\infty$ is inner, $\varphi_{\mu,\varepsilon} \in C \cap GL^\infty$, $s_{\mu,\varepsilon} \in L^\infty$, $\|s_{\mu,\varepsilon}\|_\infty < \varepsilon$. From (152) and the closedness of $C + H^\infty$ we infer that

$$a_\mu \in C + H^\infty \quad \text{and} \quad a_{-\mu} = a_\mu^{-1} \in C + \overline{H^\infty} \tag{153}$$

for all $\mu > 0$.

Assume first that $p \in AP(\mathbf{R})$ belongs to $AP_W(\mathbf{R})$, which means that p is of the form

$$p(x) = \sum_j p_j e^{i\mu_j x} \text{ with } \sum_j |p_j| < \infty.$$

Put

$$p_-(x) = \sum_{\mu_j < 0} p_j e^{i\mu_j x}, \quad p_+(x) = \sum_{\mu_j \geq 0} p_j e^{i\mu_j x}.$$

We then have

$$a^\#(x) = e^{ip_-(\alpha(x))} e^{i\lambda\alpha(x)} e^{ip_+(\alpha(x))} \tag{154}$$

and from (153) we deduce that

$$p_- \circ \alpha = \sum_{\mu_j < 0} p_j a_{\mu_j} \in C + \overline{H^\infty}, \quad p_+ \circ \alpha = \sum_{\mu_j \geq 0} p_j a_{\mu_j} \in C + H^\infty,$$

whence

$$e^{i(p-\circ\alpha)} \in G(C + \overline{H^\infty}), \quad e^{i(p+\circ\alpha)} \in G(C + H^\infty),$$

which implies that $T(e^{i(p-\circ\alpha)})$ and $T(e^{i(p+\circ\alpha)})$ are Fredholm. From (154) in conjunction with (10) we obtain

$$T(a) = T(e^{i(p-\circ\alpha)})T(a_\lambda)T(e^{i(p+\circ\alpha)}) + K \tag{155}$$

with some compact operator K. Thus, $T(a)$ is Fredholm if and only if so is $T(a_\lambda)$, which, by Theorem 5.2, happens if and only if $\lambda = 0$. If $\lambda > 0$, we have, by (152) and (10),

$$T(a_\lambda) = T(\varphi_{\lambda,\varepsilon})T(1 + s_{\lambda,\varepsilon})T(u_{\lambda,\varepsilon}) + L_{\lambda,\varepsilon}$$

with some compact operator $L_{\lambda,\varepsilon}$. Because $T(\varphi_{\lambda,\varepsilon})$ is Fredholm, $T(1+s_{\lambda,\varepsilon})$ is invertible and $T(u_{\lambda,\varepsilon})$ is left-invertible, it follows that $T(a_\lambda)$ is left-Fredholm. So (155) implies that $T(a)$ is also left-Fredholm, and since $T(a)$ is Fredholm only for $\lambda = 0$, we see that $T(a) \in \Phi(-\infty)$. Considering the adjoint of (155) gives that $T(a) \in \Phi(+\infty)$ whenever $\lambda < 0$.

Finally, if p is an arbitrary function in $AP(\mathbf{R})$, we may write $p = q + \delta$ with $q \in AP_W(\mathbf{R})$ and a function $\delta \in AP(\mathbf{R})$ for which $\|\delta\|_\infty < \pi/2$. We obtain

$$T(a) = T(e^{i(q-\circ\alpha)})T(e^{i\delta})T(a_\lambda)T(e^{i(q+\circ\alpha)}) + K_1$$

for $\lambda \leq 0$ and

$$T(a) = T(e^{i(q-\circ\alpha)})T(a_\lambda)T(e^{i\delta})T(e^{i(q+\circ\alpha)}) + K_2$$

for $\lambda \geq 0$, where K_1 and K_2 are compact. Taking into account that $T(e^{i\delta})$ is invertible, we get the assertion as above. ∎

8.4. Examples. As already said in 8.1, if $\alpha(x)$ is any of the functions $F(x)$ given by (101)–(103), then $\alpha(x)$ is regular. Since the classes (101)–(103) are invariant under multiplication by a positive constant, it follows that $\mu\alpha(x)$ is regular for all $\mu > 0$. Thus, Theorem 8.3 implies in particular that if $\mu > 0$ and $(\arg a)^\#(x)$ is an odd function which for $x \to +\infty$ is of the form

$$\sin(x^\mu), \sin(x^\mu \log x), \sin(e^{\mu x}), \ldots$$

then $T(a)$ is Fredholm.

References

[1] A Böttcher: *On Toeplitz operators generated by symbols with three essential cluster points*. Preprint P-Math-04/86, Akad. Wiss. DDR, Inst. Math., Berlin 1986.

[2] A. Böttcher and Yu.I. Karlovich: *Toeplitz and singular integral operators on Carleson curves with logarithmic whirl points*. Integral Equations and Operator Theory **22** (1995), 127-161.

[3] A. Böttcher and B. Silbermann: *Analysis of Toeplitz Operators*. Akademie-Verlag, Berlin 1989 and Springer-Verlag, Berlin, Heidelberg, New York 1990.

[4] A. Brown and P. Halmos: *Algebraic properties of Toeplitz operators*. J. reine angew. Math. **231** (1963), 89–102.

[5] L.A. Coburn: *Weyl's theorem for non-normal operators*. Michigan Math. J. **13** (1966), 285–286.

[6] L.A. Coburn and R.G. Douglas: *Translation operators on a half-line*. Proc. Nat. Acad. Sci. USA **62** (1969), 1010–1013.

[7] A. Devinatz: *Toeplitz operators on H^2 spaces*. Trans. Amer. Math. Soc. **112** (1964), 304–317.

[8] R.G. Douglas: *Banach Algebra Techniques in Operator Theory*. Academic Press, New York 1972.

[9] R.G. Douglas: *Banach Algebra Techniques in the Theory of Toeplitz Operators*. CBMS Lecture Notes **15**, Amer. Math. Soc., Providence, RI, 1973.

[10] R.G. Douglas and D. Sarason: *Fredholm Toeplitz Operators*. Proc. Amer. Math. Soc. **26** (1970), 117–120.

[11] V.K. Dzyadyk: *Introduction to the Theory of Uniform Approximation by Polynomials*. Nauka, Moscow 1971 [Russian].

[12] J.B. Garnett: *Bounded Analytic Functions*. Academic Press, New York 1981.

[13] I. Gohberg: *On an application of the theory of normed rings to singular integral equations*. Uspehi Matem. Nauk **7** (1952), 149–156 [Russian].

[14] I. Gohberg and I.A. Feldman: *On Wiener–Hopf integro-difference equations*. Soviet. Math. Dokl. **9** (1968), 1312–1316.

[15] I. Gohberg and I.A. Feldman: *Convolution Equations and Projection Methods for Their Solution*. Amer. Math. Soc. Transl. of Math. Monographs, Vol. **41**, Providence, RI, 1974 [Russian original: Nauka, Moscow 1967].

[16] I. Gohberg and N. Krupnik: *One-dimensional Linear Singular Integral Equations*. Vols. I and II. Birkhäuser Verlag, Basel, Boston, Berlin 1992 [Russian original: Shtiintsa, Kishinev 1973].

[17] S.M. Grudsky: *Singular integral equations and the Riemann boundary value problem with infinite index in the space $L_p(\Gamma, \varrho)$.* Izv. Akad. Nauk. SSSR **49** (1985), 55–80 [Russian].

[18] S.M. Grudsky: *Singular integral operators with infinite index and Blaschke products.* Math. Nachr. **129** (1986), 313–331 [Russian].

[19] S.M. Grudsky: *Factorization of u-periodic matrix-valued functions and problems with infinite index.* Soviet Math. Dokl. **36** (1988), 180–183.

[20] S.M. Grudsky: *u-factorization and Toeplitz operators with infinite index.* In: Problems and Methods in Mathematical Physics, (L. Jentsch, F. Tröltzsch, eds.), pp. 59–70, Teubner-Text zur Math., Vol. **134**, Teubner, Stuttgart and Leipzig 1994.

[21] P. Hartman and A. Wintner: *The spectra of Toeplitz's matrices.* Amer. J. Math. **76** (1954), 867–882.

[22] B.M. Levitan: *Almost Periodic Functions.* Moscow 1953 [Russian].

[23] V.N. Monahov and E.V. Semenko: *Classes of well-posed boundary value problems for conjugate analytic functions with infinite index.* Dokl. Akad. Nauk SSSR **286** (1986), 27–30 [Russian].

[24] N.K. Nikolski: *Treatise on the Shift Operator.* Springer-Verlag, Berlin and Heidelberg 1993.

[25] E. Nordgren: *Composition operators.* Canadian J. Math. **20** (1968), 442–449.

[26] F.W.J. Olver: *Asymptotics and Special Functions.* Academic Press, New York 1974.

[27] S.C. Power: *Fredholm Toeplitz operators and slow oscillation.* Can. J. Math. **32** (1980), 1058–1071.

[28] S.C. Power: *Hankel Operators on Hilbert Space.* Pitman Research Notes, No. **64**, Pitman, Boston, London, Melbourne 1982.

[29] D. Sarason: *Generalized interpolation in H^∞.* Trans. Amer. Math. Soc. **127** (1967), 179–203.

[30] D. Sarason: *Toeplitz operators with semi-almost periodic symbols.* Duke Math. J. **44** (1977), 357–364.

[31] D. Sarason: *Toeplitz operators with piecewise quasicontinuous symbols.* Indiana Univ. Math. J. **26** (1977), 817–838.

[32] J. Shapiro: *Composition Operators and Classical Function Theory.* Springer-Verlag, Berlin, Heidelberg, New York 1993.

[33] I.B. Simonenko: *Some general questions of the theory of the Riemann boundary value problem.* Math. USSR Izv. **2** (1968), 1091–1099.

[34] O. Toeplitz: *Zur Theorie der quadratischen und bilinearen Formen von unendlich vielen Veränderlichen.* Math. Annalen **70** (1911), 351–376.

[35] S. Treil: *Invertibility of Toeplitz operators does not imply applicability of the finite section method.* Dokl. Akad. Nauk SSSR **292** (1987), 563–567 [Russian].

[36] H. Widom: *Inversion of Toeplitz matrices, III.* Notices Amer. Math. Soc. **7** (1960), p. 63.

Albrecht Böttcher   Sergei M. Grudsky
TU Chemnitz–Zwickau   Rostov-on-Don State University
Fakultät für Mathematik   Faculty of Mechanics and Mathematics
09107 Chemnitz, Germany   Ul. Bolshaya Sadovaya 105
                                           344 711 Rostov-on-Don, Russia

MSC 1991:   Primary 47B35
            Secondary 30D50, 30D55, 42A75, 45E05, 47A53

Operator Theory
Advances and Applications, Vol. 90
© 1996 Birkhäuser Verlag Basel/Switzerland

TOEPLITZ AND SINGULAR INTEGRAL OPERATORS
ON GENERAL CARLESON JORDAN CURVES

A. Böttcher[1] and Yu.I. Karlovich[2]

This paper is concerned with the spectra of Toeplitz operators with piecewise continuous symbols and with the symbol calculus for singular integral operators with piecewise continuous coefficients on $L^p(\Gamma)$ where $1 < p < \infty$ and Γ is a Carleson Jordan curve. It is well known that piecewise smooth curves lead to the appearance of circular arcs in the essential spectra of Toeplitz operators, and only recently the authors discovered that certain Carleson curves metamorphose these circular arcs into logarithmic double-spirals. In the present paper we dispose of the matter by determining the local spectra produced by a general Carleson curve. These spectra are of a qualitatively new type and may, in particular, be heavy sets - until now such a phenomenon has only be observed for spaces with general Muckenhoupt weights.

1. Introduction

Let Γ be a rectifiable (closed) Jordan curve in the complex plane and equip Γ with Lebesgue length measure $|d\tau|$. A function $w : \Gamma \to [0, \infty]$ is referred to as a weight if w is measurable and $w^{-1}(\{0\})$ as well as $w^{-1}(\{\infty\})$ have measure zero. We denote by $L^p(\Gamma)$ $(1 \leq p \leq \infty)$ the usual Lebesgue spaces on Γ, and given a weight w on Γ, we let $L^p(\Gamma, w)$ $(1 < p < \infty)$ stand for the Lebesgue space with the norm

$$\|f\|_{p,w} := \left(\int_\Gamma |f(\tau)|^p w(\tau)^p \, |d\tau| \right)^{1/p}.$$

Obviously, if $w \in L^p(\Gamma)$ and $w^{-1} \in L^q(\Gamma)$ $(1/p + 1/q = 1)$, then

$$L^\infty(\Gamma) \subset L^p(\Gamma, w) \subset L^1(\Gamma).$$

The curve Γ divides the plane into a bounded connected component D_+ and an unbounded connected component D_-. Without loss of generality assume that D_+ contains the origin. We provide Γ with counter-clockwise orientation, i.e. we require that D_+ stays on the left of Γ if the curve is traced out in the positive direction.

---

[1] Research supported by the Alfried Krupp Förderpreis für junge Hochschullehrer of the Krupp Foundation and in part also by NATO Collaborative Research Grant CRG 950332

[2] Research supported by NATO Collaborative Research Grant CRG 950332

The Cauchy singular integral of a function $f \in L^1(\Gamma)$ is defined by

$$(Sf)(t) := \lim_{\varepsilon \to 0} \frac{1}{\pi i} \int_{\Gamma \setminus \Gamma(t,\varepsilon)} \frac{f(\tau)}{\tau - t} d\tau \quad (t \in \Gamma), \tag{1}$$

where $\Gamma(t,\varepsilon) := \{\tau \in \Gamma : |\tau - t| < \varepsilon\}$ is the portion of Γ in the disk of radius ε centered at t. A nice discussion of the problem concerning the existence of the Cauchy singular integral is in Dynkin's survey [7] (see Theorem 2.22 and the discussion following it). There it is shown that Calderón's 1977 paper [4] together with a 1964 result by Havin [15] imply that the limit (1) exists almost everywhere on Γ whenever Γ is rectifiable and f is a function in $L^1(\Gamma)$. The operator S is said to be bounded on $L^p(\Gamma, w)$ if $L^p(\Gamma, w) \cap L^1(\Gamma)$ is dense in $L^p(\Gamma, w)$ and there is a constant $M > 0$ such that

$$\|Sf\|_{p,w} \leq M\|f\|_{p,w} \text{ for all } f \in L^p(\Gamma, w) \cap L^1(\Gamma).$$

The problem of finding conditions ensuring the boundedness of S on $L^p(\Gamma, w)$ has been studied by many people for a long time. Here is the final result, which is essentially due do Hunt, Muckenhoupt, Wheeden [17], David [5], [6], Paatashvili and Khuskivadze [19] (also see e.g. [8], [12], [2]).

Theorem 1.1. *Let $1 < p < \infty$ and let w be a weight on the rectifiable Jordan curve Γ. The operator S is bounded on $L^p(\Gamma, w)$ if and only if*

$$w \in L^p(\Gamma), \quad w^{-1} \in L^q(\Gamma) \quad (1/p + 1/q = 1) \tag{2}$$

and

$$\sup_{t \in \Gamma} \sup_{\varepsilon > 0} \left(\frac{1}{\varepsilon} \int_{\Gamma(t,\varepsilon)} w(\tau)^p \, |d\tau| \right)^{1/p} \left(\frac{1}{\varepsilon} \int_{\Gamma(t,\varepsilon)} w(\tau)^{-q} \, |d\tau| \right)^{1/q} < \infty. \tag{3}$$

The set of all weights w satisfying (2) and (3) is usually referred to as the Muckenhoupt class and is denoted by $A_p(\Gamma)$. If $w \in A_p(\Gamma)$ then Hölder's inequality implies that

$$C_\Gamma := \sup_{t \in \Gamma} \sup_{\varepsilon > 0} \frac{1}{\varepsilon} |\Gamma(t,\varepsilon)| < \infty, \tag{4}$$

where $|\Gamma(t,\varepsilon)|$ is the length of the portion $\Gamma(t,\varepsilon)$. Condition (4) is a condition for solely the curve Γ, and curves Γ subject to this condition are commonly called Carleson curves. Although (4) is contained in (2) and (3), Theorem 1.1 is by psychological and historical reasons often phrased in the following form: *S is bounded on $L^p(\Gamma, w)$ $(1 < p < \infty)$ if and only if Γ is a Carleson curve and w is a Muckenhoupt weight.*

From now on we always suppose that $1 < p < \infty$, that Γ is a Carleson Jordan curve, and that $w \in A_p(\Gamma)$. So S is bounded on $L^p(\Gamma, w)$, and one can show that $S^2 = I$ (see [12] and [2]). Hence, $P := (I + S)/2$ and $Q := (I - S)/2$ are bounded and complementary projections on $L^p(\Gamma, w)$. We define

$$L^p_+(\Gamma, w) := PL^p(\Gamma, w), \quad \dot{L}^p_-(\Gamma, w) := QL^p(\Gamma, w), \quad L^p_-(\Gamma, w) := \dot{L}^p_-(\Gamma, w) + \mathbf{C},$$

where $\mathbf{C}$ stands for the constant functions on Γ. Since $\operatorname{Im} P = \operatorname{Ker} Q$ and $\operatorname{Im} Q = \operatorname{Ker} P$, the spaces $L^p_+(\Gamma, w)$, $L^p_-(\Gamma, w)$, $\overset{\circ}{L}{}^p_-(\Gamma, w)$ are closed subspaces of $L^p(\Gamma, w)$. The space $L^p_+(\Gamma, w)$ is frequently called the pth Hardy space of Γ and w.

The Toeplitz operator $T(a)$ induced by a function $a \in L^\infty(\Gamma)$ is the bounded operator

$$T(a) : L^p_+(\Gamma, w) \to L^p_+(\Gamma, w), \ f \mapsto P(af).$$

The function a is in this context referred to as the symbol of the operator $T(a)$. Our main concern is the description of the spectrum of $T(a)$, i.e. of the set

$$\operatorname{sp} T(a) := \{\lambda \in \mathbf{C} : T(a) - \lambda I \text{ is not invertible on } L^p_+(\Gamma, w)\}.$$

Since $T(a) - \lambda I = T(a - \lambda)$, our problem amounts to finding invertibility criteria for Toeplitz operators.

A bounded linear operator A on a Banach space X is said to be Fredholm if its image (= range) $\operatorname{Im} A$ is closed and its cokernel $\operatorname{Coker} A := X/\operatorname{Im} A$ and its kernel $\operatorname{Ker} A := \{f \in X : Af = 0\}$ have finite dimension. In that case the index of A is defined by $\operatorname{Ind} A := \dim \operatorname{Ker} A - \dim \operatorname{Coker} A$. Let $\mathcal{L}(X)$ be the Banach algebra of all bounded linear operators on X and let $\mathcal{K}(X)$ stand for the ideal of compact linear operators on X. It is well known that $A \in \mathcal{L}(X)$ is Fredholm if and only if the coset $A + \mathcal{K}(X)$ is invertible in the quotient algebra $\mathcal{L}(X)/\mathcal{K}(X)$.

It turns out ("Coburn's lemma") that a Toeplitz operator is invertible if and only if it is Fredholm of index zero; see e.g. [22] or [12]. Thus, the problem of describing $\operatorname{sp} T(a)$ may be split into the problem of finding the essential spectrum of $T(a)$,

$$\operatorname{sp}_{\operatorname{ess}} T(a) := \{\lambda \in \mathbf{C} : T(a) - \lambda I \text{ is not Fredholm on } L^p_+(\Gamma, w)\},$$

and the problem of establishing index formulas for Toeplitz operators.

The following theorem by Simonenko [21], [22], [23] (also see [12] and [2]) provides a general Fredholm criterion.

Theorem 1.2. *Let $a \in L^\infty(\Gamma)$. Then $T(a)$ is Fredholm on $L^p_+(\Gamma, w)$ if and only if $a^{-1} \in L^\infty(\Gamma)$ and a can be factored in the form*

$$a(\tau) = a_-(\tau)\tau^\varkappa a_+(\tau) \ \text{for almost all } \tau \in \Gamma \tag{5}$$

where $\varkappa$ is an integer, $\varkappa \in \mathbf{Z}$, and the functions $a_\pm$ have the following properties:

$$(i) \ a_- \in L^p_-(\Gamma, w), \ a_-^{-1} \in L^q_-(\Gamma, w^{-1}), \ a_+ \in L^q_+(\Gamma, w^{-1}), \ a_+^{-1} \in L^p_+(\Gamma, w);$$
$$(ii) \ |a_+^{-1}|w \in A_p(\Gamma).$$

In that case $\operatorname{Ind} T(a) = -\varkappa$.

We remark that condition (ii) may be replaced by the requirement that $|a_-|w \in A_p(\Gamma)$. A factorization (5) with the properties (i) and (ii) is usually called a (generalized) Wiener-Hopf factorization in $L^p(\Gamma, w)$.

Despite the great generality of Simonenko's theorem, it is by no means easy to check in concrete situations whether $a \in L^\infty(\Gamma)$ admits a Wiener-Hopf factorization in $L^p(\Gamma, w)$. The purpose of what follows is to establish effectively verifiable Fredholm criteria and index formulas for Toeplitz operators with piecewise continuous symbols. Notice that in the case of continuous symbols we have the following result, which goes back to Gohberg's 1952 paper [10] and was for general Carleson curves and general Muckenhoupt weights only recently proved in [2].

Theorem 1.3. *Let $a \in C(\Gamma)$. Then $T(a)$ is Fredholm on $L^p_+(\Gamma, w)$ if and only if $a(\tau) \neq 0$ for all $\tau \in \Gamma$. In that case $\operatorname{Ind} T(a) = -\operatorname{wind} a$, where $\operatorname{wind} a$ denotes the winding number of the (naturally oriented) curve $a(\Gamma)$ about the origin. In other terms, $\operatorname{sp_{ess}} T(a) = a(\Gamma)$ and*

$$\operatorname{sp} T(a) = a(\Gamma) \cup \{\lambda \in \mathbf{C} \setminus a(\Gamma) : \operatorname{wind}(a - \lambda) \neq 0\}.$$

The paper is organized as follows. Section 2 is a brief report about the history of the problem and about former results on the topic. Our main result is stated in Section 3 and examples are given in Section 4. The proof of the main theorem occupies Sections 5 to 9. In Section 10 we formulate the consequences of our results on the essential spectra of Toeplitz operators for the symbol calculus of singular integral operators.

2. The first three metamorphoses

We denote by $PC(\Gamma)$ the C^*- algebra of all piecewise continuous functions on Γ: a function $a \in L^\infty(\Gamma)$ belongs to $PC(\Gamma)$ if and only if the one-sided limits $a(t \pm 0) := \lim_{\tau \to t \pm 0} a(\tau)$ exist for every $t \in \Gamma$. Here $\tau \to t - 0$ means that τ approaches t following the positive orientation of Γ, while $\tau \to t + 0$ says that τ goes to t in the opposite direction. For $a \in PC(\Gamma)$, let Λ_a stand for the points at which a has a jump,

$$\Lambda_a := \{t \in \Gamma : a(t - 0) \neq a(t + 0)\},$$

and let $\mathcal{R}(a)$ denote the essential range of a,

$$\mathcal{R}(a) := \bigcup_{t \in \Gamma} \{a(t - 0), a(t + 0)\} = a(\Gamma) \cup \bigcup_{t \in \Lambda_a} \{a(t - 0), a(t + 0)\}.$$

We remark that the set Λ_a is at most countable and that for each $\delta > 0$ the set

$$\{t \in \Gamma : |a(t + 0) - a(t - 0)| > \delta\}$$

is finite.

The story of describing $\operatorname{sp_{ess}} T(a)$ for $a \in PC(\Gamma)$ has its beginning in the sixties, when several mathematicians, including I.B. Simonenko, A. Calderón, F. Spitzer, H. Widom, A. Devinatz, I. Gohberg, and N. Krupnik, observed that if Γ is a piecewise smooth curve, w

is identically 1, and $p = 2$, then $\mathrm{sp}_{\mathrm{ess}} T(a)$ is the closed continuous curve resulting from the essential range of a by filling in a line segment between the endpoints of each jump:

$$\mathrm{sp}_{\mathrm{ess}} T(a) = \mathcal{R}(a) \cup \bigcup_{t \in \Lambda_a} [a(t-0), a(t+0)].$$

First metamorphosis: circular arcs. The first surprise came with Gohberg and Krupnik's paper [11] (but also see Widom's 1960 paper [26]). They assumed that Γ is piecewise smooth, w is identically 1, but allowed p to take on values between 1 and ∞. Their result says that $\mathrm{sp}_{\mathrm{ess}} T(a)$ is obtained from the essential range of a by joining $a(t-0)$ and $a(t+0)$ by a certain circular arc for each jump. To be more precise, given $z, w \in \mathbf{C}$ and $\nu \in (0,1)$, define

$$\mathcal{A}(z, w; \nu) := \left\{ \xi \in \mathbf{C} \setminus \{z, w\} : \arg \frac{\xi - z}{\xi - w} \in 2\pi\nu + 2\pi\mathbf{Z} \right\} \cup \{z, w\}.$$

A little thought reveals that $\mathcal{A}(z, w; \nu)$ is a circular arc between z and w whose shape is determined by ν. The Gohberg/Krupnik/Widom result may be stated in the form

$$\mathrm{sp}_{\mathrm{ess}} T(a) = \mathcal{R}(a) \cup \bigcup_{t \in \Lambda_a} \mathcal{A}(a(t-0), a(t+0); 1/p).$$

Subsequently, Gohberg and Krupnik (see e.g. [12]) also studied spaces with so-called power weights, i.e. weights of the form

$$w(\tau) = \prod_{j=1}^{n} |\tau - t_j|^{\mu_j} \quad (\tau \in \Gamma), \tag{6}$$

where $t_1, \ldots, t_n$ are distinct points on Γ and $\mu_1, \ldots, \mu_n$ are nonzero real numbers. The weight (6) belongs to $A_p(\Gamma)$ if and only if $-1/p < \mu_j < 1/q$ for all j. They showed that in the presence of a power weight

$$\mathrm{sp}_{\mathrm{ess}} T(a) = \mathcal{R}(a) \cup \bigcup_{t \in \Lambda_a} \mathcal{A}(a(t-0), a(t+0); \nu_t)$$

where $\nu_t = 1/p$ for $t \notin \{t_1, \ldots, t_n\}$ and $\nu_{t_j} = 1/p + \mu_j$ for $j = 1, \ldots, n$. Thus, although now the circular arcs participating in the spectrum may have different shape, they nevertheless remain circular arcs.

Second metamorphosis: horns. The development had paused many years until 1990, when Spitkovsky [25] made a spectacular discovery. He considered again the case of a piecewise smooth curve Γ, but he admitted arbitrary Muckenhoupt weights $w \in A_p(\Gamma)$ $(1 < p < \infty)$. His result tells us that the presence of Muckenhoupt weights may metamorphose the circular arcs into so-called horns. A horn is a set of the form

$$\mathcal{H}(z, w; \nu_1, \nu_2) := \left\{ \xi \in \mathbf{C} \setminus \{z, w\} : \arg \frac{\xi - z}{\xi - w} \in 2\pi[\nu_1, \nu_2] + 2\pi\mathbf{Z} \right\} \cup \{z, w\}$$

$$= \bigcup_{\nu \in [\nu_1, \nu_2]} \mathcal{A}(z, w; \nu),$$

where $z, w \in \mathbf{C}$ and $\nu_1, \nu_2 \in (0,1)$ satisfy $\nu_1 \leq \nu_2$; in words, a horn is a closed subset of $\mathbf{C}$ which is bounded by two circular arcs. Spitkovsky associated two numbers ν_t^- and ν_t^+ with each point $t \in \Gamma$ which, in a sense, measure the "powerlikeness" of the weight w in a neighborhood of t and proved that

$$\text{sp}_{\text{ess}} T(a) = \mathcal{R}(a) \cup \bigcup_{t \in \Lambda_a} \mathcal{H}(a(t-0), a(t+0); \nu_t^-, \nu_t^+).$$

Third metamorphosis: spiralic horns. In 1994, the authors turned attention to the case where Γ is no longer supposed to be piecewise smooth but is allowed to be a more complicated Carleson Jordan curve [2]. The class of Carleson curves considered in [2] is as follows. Fix a point $t \in \Gamma$. We then have $\tau - t = |\tau - t| e^{i \arg(\tau - t)}$ for $\tau \in \Gamma \setminus \{t\}$, and the argument $\arg(\tau - t)$ may be chosen so that it is a continuous function on $\Gamma \setminus \{t\}$. Seifullayev [20] showed that for an arbitrary Carleson curve the estimate

$$\arg(\tau - t) = O(-\log|\tau - t|) \text{ as } \tau \to t \tag{7}$$

holds. In [2], we studied Carleson curves subject to the stronger condition

$$\arg(\tau - t) = -\delta_t \log|\tau - t| + O(1) \text{ as } \tau \to t \tag{8}$$

with some $\delta_t \in \mathbf{R}$. Such curves may scroll up like two (in view of the $O(1)$ possibly slightly perturbed) logarithmic spirals.

We proved that if Γ is a Carleson Jordan curve satisfying (8) for each $t \in \Gamma$ and if w is an arbitrary Muckenhoupt weight in $A_p(\Gamma)$ $(1 < p < \infty)$, then $\text{sp}_{\text{ess}} T(a)$ arises from $\mathcal{R}(a)$ by filling in a so-called spiralic horn between the endpoints of each jump. A spiralic horn between $z, w \in \mathbf{C}$ is a set of the form

$$\mathcal{S}(z, w; \delta; \nu_1, \nu_2)$$
$$:= \left\{ \xi \in \mathbf{C} \setminus \{z, w\} : \arg \frac{\xi - z}{\xi - w} - \delta \log \left| \frac{\xi - z}{\xi - w} \right| \in 2\pi[\nu_1, \nu_2] + 2\pi \mathbf{Z} \right\} \cup \{z, w\} \tag{9}$$

where $\delta \in \mathbf{R}, \nu_1, \nu_2 \in (0,1)$, and $\nu_1 \leq \nu_2$. Various spiralic horns are plotted in [2]. We measured again the "powerlikeness" of the weight w in a neighborhood of $t \in \Gamma$ by two numbers ν_t^- and ν_t^+ and showed that

$$\text{sp}_{\text{ess}} T(a) = \mathcal{R}(a) \cup \bigcup_{t \in \Lambda_a} \mathcal{S}(a(t-0), a(t+0); \delta_t; \nu_t^-, \nu_t^+)$$

where δ_t is given by (8). Since $\mathcal{S}(z, w; 0; \nu_1, \nu_2) = \mathcal{H}(z, w; \nu_1, \nu_2)$, spiralic horns become usual horns at points $t \in \Gamma$ where (8) holds with $\delta_t = 0$ ("nonhelical" points). If w is a power weight (and, in particular, if $w \equiv 1$), then $\nu_t^- = \nu_t^+$ and the spiralic horns have "thickness" zero, that is, they degenerate to a logarithmic double-spiral.

The question we are left with is what happens if condition (8) is not satisfied for some $t \in \Gamma$. There are two conflicting suggestions. First, the plots contained in [2] indicate that the set we have to fill in between $a(t-0)$ and $a(t+0)$ mimics the shape of the curve near t. So if the curve is more complicated, if it scrolls up like two "oscillating spirals", say, is it an oscillating double-spiral we have to fill in ? We will prove that the answer to this question is NO. Secondly, even if due to the $O(1)$ in (8) the two spirals of the curve scrolling up at t are no longer "pure" logarithmic spirals, the boundaries of the spiralic horn between $a(t-0)$ and $a(t+0)$ are nevertheless "pure" logarithmic double-spirals. So is in the general case the part of the spectrum joining $a(t-0)$ and $a(t+0)$ always bounded by "pure" logarithmic double-spirals ? Our main result says that, at least in the absence of a weight, the answer to this question is YES !

3. Main result

We describe $\mathrm{sp}_{\mathrm{ess}}\, T(a)$ for $a \in PC(\Gamma)$ on $L^p_+(\Gamma)$ in case $p \in (1,\infty)$ and Γ is an arbitrary Carleson Jordan curve. From now on we do not include weights into the consideration in order to make crystal-clear that the phenomena we will encounter in the following, and in particular the appearance of the heavy spectra, are exclusively caused by the behavior of the curve and in no way tied in with the presence of a weight. The subject of this paper is to dispose of the fourth spectral metamorphosis for Toeplitz operators. We will embark on spaces with general Muckenhoupt weights (and thus, on the fifth and last spectral metamorphosis) in a forthcoming article.

So let Γ be a Carleson Jordan curve and suppose $1 < p < \infty$. Fix $t \in \Gamma$. The curve Γ may be given by

$$\tau - t = |\tau - t|e^{i\arg(\tau-t)} \quad (\tau \in \Gamma \setminus \{t\})$$

with some continuous argument $\Gamma \setminus \{t\} \to \mathbf{R}$, $\tau \mapsto \arg(\tau - t)$. Put $d_t := \max_{\tau \in \Gamma}|\tau - t|$. For $R \in (0, d_t]$, define $\partial\Gamma(t, R) := \{\tau \in \Gamma : |\tau - t| = R\}$; notice that in general $\partial\Gamma(t, R)$ is not the boundary of the portion $\Gamma(t, R) := \{\tau \in \Gamma : |\tau - t| < R\}$. For $x \in (0, \infty)$, put

$$\tilde{\varrho}_t(x) := \limsup_{R \to 0} \frac{\max\limits_{\tau \in \partial\Gamma(t,xR)} e^{-\arg(\tau-t)}}{\min\limits_{\tau \in \partial\Gamma(t,R)} e^{-\arg(\tau-t)}}, \tag{10}$$

and define $\delta_t^-, \delta_t^+ \in [-\infty, +\infty]$ by

$$\delta_t^- := \limsup_{x \to 0} \frac{\log \tilde{\varrho}_t(x)}{\log x}, \quad \delta_t^+ := \limsup_{x \to \infty} \frac{\log \tilde{\varrho}_t(x)}{\log x}. \tag{11}$$

Finally, given $z, w \in \mathbf{C}$, $\delta_1, \delta_2 \in \mathbf{R}$, $\nu_1, \nu_2 \in (0, 1)$ such that $\delta_1 \le \delta_2$ and $\nu_1 \le \nu_2$, we define the skew spiralic horn $\mathcal{S}(z, w; \delta_1, \delta_2; \nu_1, \nu_2)$ from z to w as the set

$$\left\{\xi \in \mathbf{C} \setminus \{z, w\} : \; \arg\frac{\xi - z}{\xi - w} - \delta\log\left|\frac{\xi - z}{\xi - w}\right| \in 2\pi[\nu_1, \nu_2] + 2\pi\mathbf{Z}\right.$$

$$\left. \text{for some } \delta \in [\delta_1, \delta_2]\right\} \cup \{z, w\}.$$

Obviously,

$$\mathcal{S}(z, w; \delta_1, \delta_2; \nu_1, \nu_2) = \bigcup_{\delta \in [\delta_1, \delta_2]} \mathcal{S}(z, w; \delta; \nu_1, \nu_2)$$

where $\mathcal{S}(z, w; \delta; \nu_1, \nu_2)$ is the spiralic horn (9).

Main theorem. *Let Γ be a Carleson Jordan curve, $1 < p < \infty$, and $a \in PC(\Gamma)$. Then $-\infty < \delta_t^- \leq \delta_t^+ < +\infty$ for every $t \in \Gamma$, and the essential spectrum of the Toeplitz operator $T(a)$ on $L_+^p(\Gamma)$ is given by*

$$\mathrm{sp}_{\mathrm{ess}}\, T(a) = \mathcal{R}(a) \cup \bigcup_{t \in \Lambda_a} \mathcal{S}\left(a(t-0), a(t+0); \delta_t^-, \delta_t^+; \frac{1}{p}, \frac{1}{p}\right). \qquad (12)$$

Plots of several skew spiralic horns are shown in Figures 1 - 4. It is easily seen that a skew spiralic horn is always bounded by pieces of at most two logarithmic double-spirals. Notice that the fourth spectral metamorphosis leads to heavy parts of the essential spectrum, for example to the "butterfly" in Figure 1 or the "dino coming out from an egg" in Figure 4. Also notice that the points 0 and 1 are not only cluster points of the set $\mathcal{S}(0, 1; \delta_t^-, \delta_t^+; 1/p, 1/p)$, as this is the case if (8) is satisfied and thus $\delta_t^- = \delta_t^+ = \delta_t$, but that in the general case 0 and 1 may even be interior points of this set.

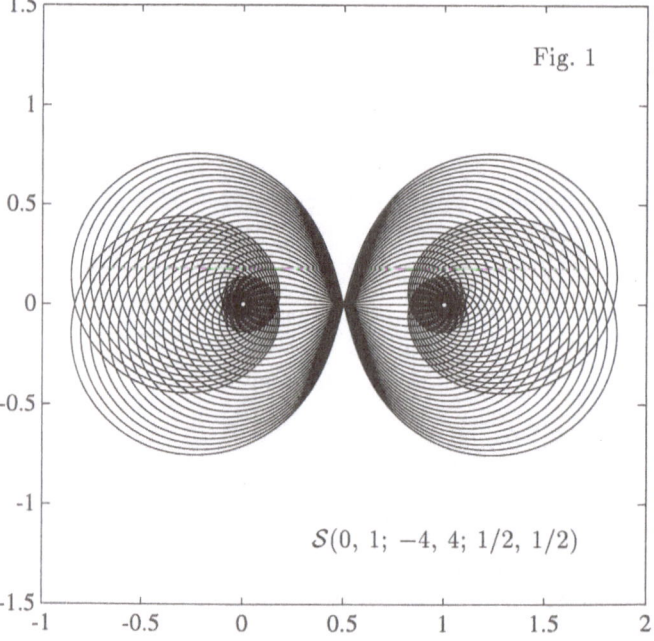

Fig. 1

$\mathcal{S}(0,\ 1;\ -4,\ 4;\ 1/2,\ 1/2)$

In a sense, the numbers δ_t^- and δ_t^+ measure the "logarithmic spirallikeness" of the curve at t, i.e. they measure how far the curve differs from the curve given by (7) in a neighborhood of t. One might conjecture that for spaces $L^p(\Gamma, w)$ with $w \in A_p(\Gamma)$ the equality (11) holds with $\frac{1}{p}, \frac{1}{p}$ replaced by ν_t^-, ν_t^+ where ν_t^- and ν_t^+ measure the "powerlikeness" of the weight w at the point t. However, things seem to be more complicated: there may arise some kind of interference between the oscillation of the curve and the oscillation of the weight, as the result of which the set we have to fill in between $a(t-0)$ and $a(t+0)$ may deviate from a skew spiralic horn ! As already said, this problem will be studied in a forthcoming paper.

4. Examples

To get a feeling for the main theorem let us consider a few examples before proceeding to its proof.

Let $\varphi \in C(0,1] \cap C^1(0,1)$ and $b \in C[0,1] \cap C^1(0,1)$ be real-valued functions and assume

$$b(1) \in \{0, 2\pi\}, \ 0 < b(r) < 2\pi \text{ for } r \in (0,1).$$

Pick a point $t \in \mathbf{C}$ and put

$$\Gamma_+ := \{\tau = t + re^{-i\varphi(r)} : r \in (0,1]\}, \quad \Gamma_- := \{\tau = t + re^{-i(\varphi(r)-b(r))} : r \in (0,1]\}.$$

Then $\Gamma := \{t\} \cup \Gamma_+ \cup \Gamma_-$ is a Jordan curve.

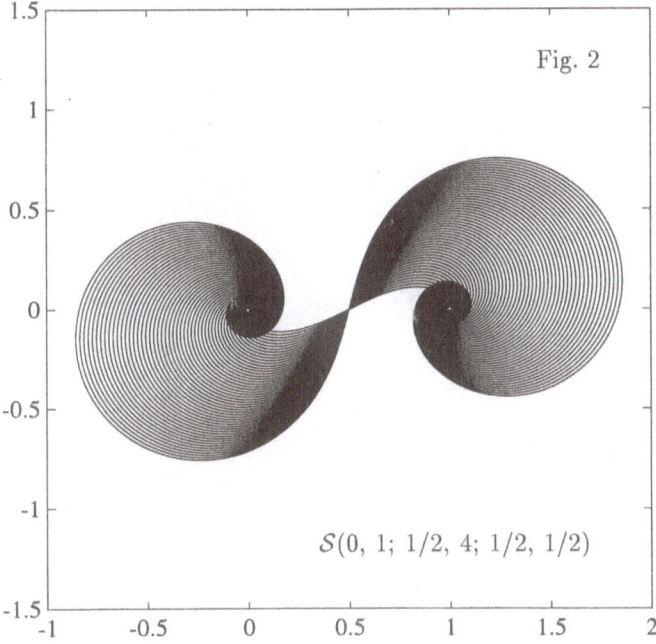

Fig. 2

$S(0, 1; 1/2, 4; 1/2, 1/2)$

In what follows we denote by a dot the derivative d/dr of a function of the variable r.

Lemma 4.1. Γ *is a Carleson curve if the functions*

$$r\dot{\varphi}(r) \quad and \quad r\dot{b}(r) \quad are \ bounded \ for \ r \in (0,1). \tag{13}$$

Proof. By (13), there is an $M > 0$ such that

$$|r\dot\varphi(r)| \le M, \quad |r(\dot\varphi(r) - \dot b(r))| \le M \quad \text{for} \quad r \in (0,1). \tag{14}$$

We have

$$|d\tau| = \sqrt{1 + r^2\dot\varphi(r)^2}\, dr \text{ on } \Gamma_+, \quad |d\tau| = \sqrt{1 + r^2(\dot\varphi(r) - \dot b(r))^2}\, dr \text{ on } \Gamma_-.$$

Hence,

$$\frac{|\Gamma(t,\varepsilon)|}{\varepsilon} \doteq \frac{1}{\varepsilon} \int_0^\varepsilon \left(\sqrt{1 + r^2\dot\varphi(r)^2} + \sqrt{1 + r^2(\dot\varphi(r) - \dot b(r))^2} \right) dr$$

$$\le 2\sqrt{1 + M^2}\frac{1}{\varepsilon}\varepsilon = 2\sqrt{1 + M^2}$$

for every $\varepsilon \in (0,1)$.

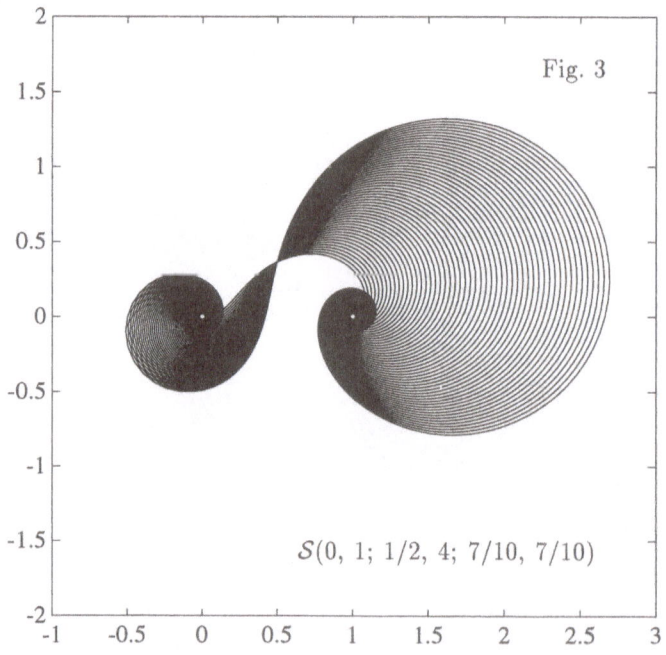

Fig. 3

$\mathcal{S}(0, 1; 1/2, 4; 7/10, 7/10)$

Now let $t_0 \in \Gamma \setminus \{t\}$ and put $r_0 = |t_0 - t|$. If $\varepsilon < r_0$, then

$$\frac{|\Gamma(t_0,\varepsilon)|}{\varepsilon} \le \frac{|\Gamma(t,r_0+\varepsilon) \setminus \Gamma(t,r_0-\varepsilon)|}{\varepsilon}$$

$$= \frac{1}{\varepsilon} \int\limits_{r_0 - \varepsilon}^{\min(r_0 + \varepsilon, 1)} \left(\sqrt{1 + r^2 \dot{\varphi}(r)^2} + \sqrt{1 + r^2 (\dot{\varphi}(r) - \dot{b}(r))^2} \right) dr$$

$$\leq 2\sqrt{1 + M^2} \, \frac{1}{\varepsilon} (\min(r_0 + \varepsilon, 1) - (r_0 - \varepsilon)) \leq 4\sqrt{1 + M^2},$$

while if $\varepsilon \geq r_0$, we get

$$\frac{|\Gamma(t_0, \varepsilon)|}{\varepsilon} \leq \frac{|\Gamma(t, r_0 + \varepsilon)|}{\varepsilon} \leq 2\sqrt{1 + M^2} \frac{r_0 + \varepsilon}{\varepsilon} \leq 2\sqrt{1 + M^2}(\varepsilon + \varepsilon)/\varepsilon = 4\sqrt{1 + M^2}.$$

This proves that (4) is satisfied with $C_\Gamma \leq 4\sqrt{1 + M^2}$. ■

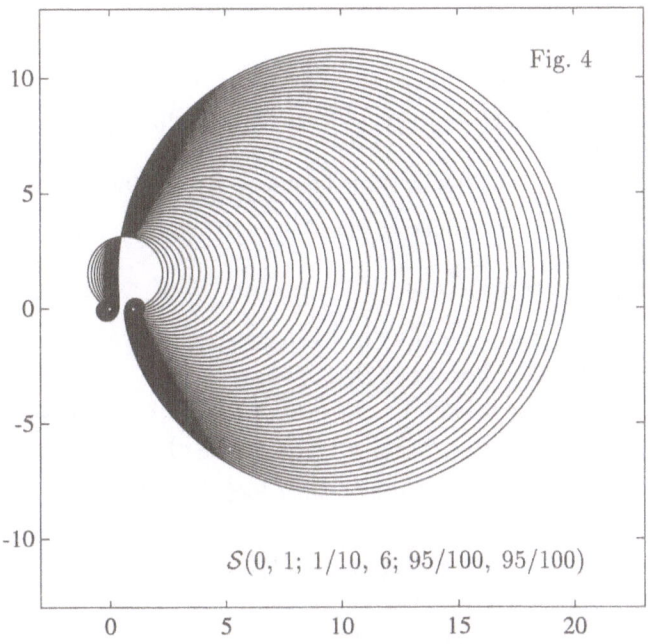

Fig. 4

$$\mathcal{S}(0, 1; 1/10, 6; 95/100, 95/100)$$

So suppose (13) is in force. We have

$$\arg(\tau - t) = -\varphi(|\tau - t|) \text{ for } \tau \in \Gamma_+,$$
$$\arg(\tau - t) = -\varphi(|\tau - t|) + b(|\tau - t|) \text{ for } \tau \in \Gamma_-,$$

hence for xR and R in $(0, d_t)$,

$$\max_{\tau \in \partial\Gamma(t, xR)} e^{-\arg(\tau - t)} = \max \left\{ e^{\varphi(xR)}, e^{\varphi(xR) - b(xR)} \right\} = e^{\varphi(xR)},$$

$$\min_{\tau \in \partial\Gamma(t, R)} e^{-\arg(\tau - t)} = \min \left\{ e^{\varphi(R)}, e^{\varphi(R) - b(R)} \right\} = e^{\varphi(R) - b(R)},$$

and thus, in accordance with (10),

$$\tilde{\varrho}_t(x) = \limsup_{R \to 0} e^{\varphi(xR) - \varphi(R)} e^{b(R)} = \limsup_{R \to 0} e^{\varphi(xR) - \varphi(R)}. \tag{15}$$

Lemma 4.2. *Let $h \in C^2(\mathbf{R})$ and suppose the functions*

$$h(x), h'(x), h''(x) \quad \text{are bounded for } x \in \mathbf{R}. \tag{16}$$

Put

$$\varphi(r) = h(\log(-\log r))(-\log r) \quad \text{for } r \in (0,1) \tag{17}$$

and $\varphi(1) = 0$. Then $r\dot{\varphi}(r)$ is bounded for $r \in (0,1)$ and the numbers δ_t^-, δ_t^+ defined by (11) and (15) are given by

$$\delta_t^- = \liminf_{r \to 0} r\dot{\varphi}(r), \quad \delta_t^+ = \limsup_{r \to 0} r\dot{\varphi}(r). \tag{18}$$

Proof. Since

$$r\dot{\varphi}(r) = rh'(\log(-\log r))\frac{-\log r}{r \log r} + rh(\log(-\log r))\left(-\frac{1}{r}\right), \tag{19}$$

we infer from (16) that the function $r\dot{\varphi}(r)$ is bounded. To compute

$$\limsup_{R \to 0}(\varphi(xR) - \varphi(R)) \tag{20}$$

put $y = -\log R$ and $\varepsilon = -\log x$. Then

$$\begin{aligned}
\varphi(xR) - \varphi(R) &= h(\log(y + \varepsilon))(y + \varepsilon) - h(\log y)y \\
&= h(\log y + \log(1 + \varepsilon/y))(y + \varepsilon) - h(\log y)y \\
&= (h(\log y + \log(1 + \varepsilon/y)) - h(\log y))(y + \varepsilon) + \varepsilon h(\log y) \\
&= h'(\xi(y))\log(1 + \varepsilon/y)(y + \varepsilon) + \varepsilon h(\log y)
\end{aligned}$$

with some $\xi(y) \in (\log y, \log y + \log(1 + \varepsilon/y))$. Since $\log(1 + \varepsilon/y) = \varepsilon/y + O(1/y^2)$, it follows that

$$\begin{aligned}
\varphi(xR) - \varphi(R) &= h'(\xi(y))\varepsilon + h(\log y)\varepsilon + O(1/y) \\
&= (h'(\log y + O(1/y)) + h(\log y))\varepsilon + O(1/y) \\
&= h''(\eta(y))O(1/y) + (h'(\log y) + h(\log y))\varepsilon + O(1/y)
\end{aligned}$$

with $\eta(y) \in (\log y, \log y + O(1/y))$. Taking into account (16) we so see that (20) equals

$$\begin{aligned}
&\limsup_{y \to \infty}[(h'(\log y) + h(\log y))\varepsilon] \\
&= \limsup_{r \to 0}[(h'(\log(-\log r)) + h(\log(-\log r)))(-\log x)] \\
&= \limsup_{r \to 0}[(-r\dot{\varphi}(r))(-\log x)] \quad \text{by (19)} \\
&= \limsup_{r \to 0}[r\dot{\varphi}(r)\log x] \\
&= \begin{cases} (\liminf_{r \to 0} r\dot{\varphi}(r))\log x & \text{for } x \in (0,1) \; (\Leftrightarrow \log x < 0) \\ (\limsup_{r \to 0} r\dot{\varphi}(r))\log x & \text{for } x \in (1, \infty) \; (\Leftrightarrow \log x > 0). \end{cases}
\end{aligned}$$

Thus

$$\tilde{\varrho}_t(x) = \begin{cases} \exp\left\{(\liminf_{r\to 0} r\dot\varphi(r))\log x\right\} & \text{for } x \in (0,1) \\ \exp\left\{(\limsup_{r\to 0} r\dot\varphi(r))\log x\right\} & \text{for } x \in (1,\infty), \end{cases}$$

implying that

$$\delta_t^- := \limsup_{x\to 0} \frac{\log \tilde{\varrho}_t(x)}{\log x} = \liminf_{r\to 0} r\dot\varphi(r),$$

$$\delta_t^+ := \limsup_{x\to\infty} \frac{\log \tilde{\varrho}_t(x)}{\log x} = \limsup_{r\to 0} r\dot\varphi(r). \;\blacksquare$$

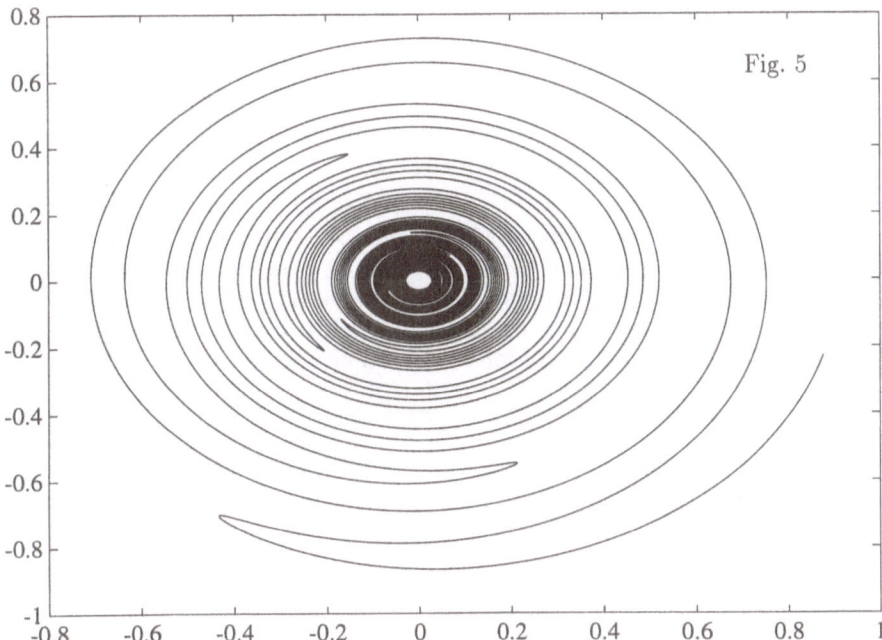

Fig. 5

Via (16), (17) we may construct a lot of Carleson curves for which the numbers δ_t^-, δ_t^+ are available. To have a concrete example, choose

$$h(x) = \delta + \mu \sin \lambda x \qquad (21)$$

with $\delta, \mu \in \mathbf{R}$ and $\lambda \in \mathbf{R} \setminus \{0\}$. Notice that condition (8) is satisfied if and only if $\mu = 0$. By (19),

$$\begin{aligned} r\dot\varphi(r) &= -h'(\log(-\log r)) - h(\log(-\log r)) \\ &= -\mu\lambda\cos(\lambda\log(-\log r)) - \delta - \mu\sin(\lambda\log(-\log r)), \end{aligned}$$

whence

$$\delta_t^- = \liminf_{r \to 0} r\dot\varphi(r) = -\delta - |\mu| \max_{x \in \mathbb{R}} (\lambda \cos \lambda x + \sin \lambda x) = -\delta - |\mu|\sqrt{1 + \lambda^2},$$

$$\delta_t^+ = \limsup_{r \to 0} r\dot\varphi(r) = -\delta + |\mu| \max_{x \in \mathbb{R}} (\lambda \cos \lambda x + \sin \lambda x) = -\delta + |\mu|\sqrt{1 + \lambda^2}.$$

In particular, the choice (17), (21) with appropriate values of δ and μ (and $\lambda = 1$) provides us with Carleson curves with arbitrarily prescribed segments

$$[\delta_t^-, \delta_t^+] = [-\delta - |\mu|\sqrt{2}, -\delta + |\mu|\sqrt{2}].$$

Figure 5 shows the part Γ_+ of the Carleson curve produced by $h(x) = 2 + 5\sin(9x)$; in order to make visible this curve, which approaches the point t extremely rapidly, we chose a logarithmic scale of the radius.

5. Plan of the proof of the main theorem

Localization techniques reduce the problem to determining $\text{sp}_{\text{ess}} T(g_t)$ for some "canonical" piecewise continuous function g_t with a jump at $t \in \Gamma$.

The function g_t is constructed as follows. We may conformally map the interior and the exterior of the complex unit circle onto D_+ and D_-, respectively, so that the point 1 is mapped to t and the points 0 and ∞ are left fixed. Denote by Λ_0 and Λ_∞ the images of $[0,1]$ and $(1, \infty) \cup \{\infty\}$ under this map. The curve $\Lambda_0 \cup \Lambda_\infty$ joins $0 \in D_+$ with the point $\infty \in D_-$ and meets Γ at exactly one point, namely t. Let $\arg z$ be any continuous branch of the argument in $\mathbb{C} \setminus (\Lambda_0 \cup \Lambda_\infty)$ and for $\gamma \in \mathbb{C}$, define

$$z^\gamma := |z|^\gamma e^{i\gamma \arg z} \text{ for } z \in \mathbb{C} \setminus (\Lambda_0 \cup \Lambda_\infty).$$

Then z^γ is an analytic function in $\mathbb{C} \setminus (\Lambda_0 \cup \Lambda_\infty)$. The restriction of z^γ to $\Gamma \setminus \{t\}$ will be denoted by $g_t(\tau) = \tau^\gamma$. Clearly, $g_t \in PC(\Gamma)$, $g_t^{-1} \in L^\infty(\Gamma)$, and $g_t(t+0)/g_t(t-0) = e^{-2\pi i\gamma}$.

What we are primarily interested in is the set

$$\Phi_t := \Phi_t(\Gamma, p) := \{\gamma \in \mathbb{C} : T(g_t) \text{ is Fredholm on } L_+^p(\Gamma)\}; \tag{22}$$

notice that the dependence on γ is suppressed in the notation g_t. The function $g_t(\tau) = \tau^\gamma$ admits the factorization

$$\tau^\gamma = (1 - t/\tau)^{\varkappa - \gamma} \tau^\varkappa (\tau - t)^{\gamma - \varkappa} \quad (\tau \in \Gamma \setminus \{t\}) \tag{23}$$

with appropriately chosen branches of $(1 - t/\tau)^{\varkappa - \gamma}$ and $(\tau - t)^{\gamma - \varkappa}$: define $\arg z$ for $z \in \mathbb{C} \setminus (\Lambda_0 \cup \Lambda_\infty)$ as above, take any continuous branch of $\arg(z - t)$ for $z \in \mathbb{C} \setminus \Lambda_\infty$, define

$$\arg(1 - t/z) = \arg[(z - t)/z] := \arg(z - t) - \arg z$$

for $z \in \mathbb{C} \setminus (\Lambda_0 \cup \Lambda_\infty)$, and then put

$$(z - t)^\alpha := |z - t|^\alpha e^{i\alpha \arg(z-t)}, \quad (1 - t/z)^\alpha := |1 - t/z|^\alpha e^{i\alpha \arg(1-t/z)}$$

for $\alpha \in \mathbf{C}$. The functions $(z - t)^\alpha$ and $(1 - t/z)^\alpha$ are analytic and nonzero in $\mathbf{C} \setminus \Lambda_\infty$ and $\mathbf{C} \setminus \Lambda_0$, respectively, and are continuous on $D_+ \cup (\Gamma \setminus \{t\})$ and $D_- \cup (\Gamma \setminus \{t\})$, respectively.

A suitable choice of the integer $\varkappa$ guarantees that the factorization (23) satisfies condition (i) of Theorem 1.2. In order to deal with condition (ii) of that theorem, we have to study the set

$$N_t := N_t(\Gamma, p) := \{\gamma \in \mathbf{C} : |(\tau - t)^\gamma| \in A_p(\Gamma)\} \tag{24}$$

and are thus led to the question of deciding whether a continuous function $\psi : \Gamma \setminus \{t\} \to (0, \infty)$ satisfies the Muckenhoupt condition. To tackle the latter question, we associate a function $\varrho_t : (0, \infty) \to (0, \infty)$ with ψ which turns out to be submultiplicative and whose asymptotic behavior near 0 and ∞ may therefore be characterized by two numbers, the so-called lower and upper indices α_t and β_t. We then prove that if $\log \varrho_t$ is bounded in a neighborhood of the point 1, then $\psi \in A_p(\Gamma)$ if and only if both α_t and β_t belong to the interval $(-1/p, 1/q)$. In the case we are interested in, that is, in the case where $\psi(\tau) = |(\tau - t)^\gamma|$, the corresponding function $\log \varrho_t$ can indeed be shown to be bounded near 1. This provides us with the information we need to understand N_t and thus Φ_t.

6. Submultiplicative functions

In imitation of [15, Chapter 2] and [1, p. 146] we call a function $\varrho : (0, \infty) \to (0, \infty]$ submultiplicative if

$$\varrho(x_1 x_2) \leq \varrho(x_1)\varrho(x_2) \quad \text{for all } x_1, x_2 \in (0, \infty). \tag{25}$$

First of all we remark that ϱ is bounded and bounded away from zero in a neighborhood of the point 1 if and only if $\log \varrho$ is bounded in that neighborhood.

Lemma 6.1. *Let $\varrho : (0, \infty) \to (0, \infty]$ be submultiplicative.*

(a) If $\log \varrho$ is bounded in a neighborhood of the point 1, then $\log \varrho$ is bounded on every closed interval $[a, b] \subset (0, \infty)$.

(b) If ϱ is (Lebesgue) measurable and $\log \varrho(x) \in (0, \infty)$ for all $x \in (0, \infty)$, then $\log \varrho$ is bounded on every closed interval $[a, b] \subset (0, \infty)$.

Proof. (a) Obvious.

(b) Apply Theorem 7.4.1 of [16] to $v(x) := \log \varrho(e^x)$. ∎

If $\log \varrho$ is bounded near 1, then the asymptotic behavior of a submultiplicative function ϱ near 0 and ∞ is characterized by the two numbers

$$\alpha_\varrho := \sup_{x \in (0,1)} \frac{\log \varrho(x)}{\log x}, \quad \beta_\varrho := \inf_{x \in (1,\infty)} \frac{\log \varrho(x)}{\log x},$$

which are referred to as the lower and upper indices of ϱ, respectively. Clearly, $\alpha_\varrho > -\infty$ and $\beta_\varrho < +\infty$.

Theorem 6.2. *Let $\varrho : (0, \infty) \to (0, \infty)$ be submultiplicative and suppose $\log \varrho$ is bounded in a neighborhood of the point 1. Then*

(a) $-\infty < \alpha_\varrho \le \beta_\varrho < +\infty$;

(b) $\alpha_\varrho = \lim\limits_{x\to 0} \frac{\log \varrho(x)}{\log x}$, $\beta_\varrho = \lim\limits_{x\to\infty} \frac{\log \varrho(x)}{\log x}$;

(c) $\varrho(x) \ge x^{\alpha_\varrho}$ for $x \in (0,1)$, $\varrho(x) \ge x^{\beta_\varrho}$ for $x \in (1,\infty)$;

(d) given $\varepsilon > 0$, there exists an $x_0 = x_0(\varepsilon) \in (0,1)$ such that $\varrho(x) \le x^{\alpha_\varrho - \varepsilon}$ for $x \in (0, x_0)$ and $\varrho(x) \le x^{\beta_\varrho + \varepsilon}$ for $x \in (x_0^{-1}, \infty)$;

(e) for every $\mu > 0$ the function ϱ^μ is submultiplicative and its lower and upper indices are $\mu\alpha_\varrho$ and $\mu\beta_\varrho$, respectively.

Proof. Theorem 7.6.2 of [16] and Theorem 1.3 of Chapter 2 of [18]. ∎

Lemma 6.3. *Let Γ be a Carleson Jordan curve, fix $t \in \Gamma$, and suppose $\psi : \Gamma \setminus \{t\} \to (0,\infty)$ is continuous. For $R_1, R_2 \in (0, d_t]$, define*

$$\Delta(R_1, R_2, \psi) := \frac{\max\limits_{\tau \in \partial\Gamma(t, R_1)} \psi(\tau)}{\min\limits_{\tau \in \partial\Gamma(t, R_2)} \psi(\tau)} \qquad (26)$$

and put

$$\varrho_t(x) := \begin{cases} \sup\limits_{R \in (0, d_t]} \Delta(xR, R, \psi) & \text{for } x \in (0,1], \\ \sup\limits_{R \in (0, d_t]} \Delta(R, x^{-1}R, \psi) & \text{for } x \in [1, \infty). \end{cases} \qquad (27)$$

Then $\varrho_t : (0,\infty) \to (0,\infty]$ is submultiplicative.

Proof. It is readily seen that (25) holds if $x_1, x_2 \in (0,1]$ or $x_1, x_2 \in [1,\infty)$. So assume $x = x_1 x_2$, $x_1 \in (0,1]$, $x_2 \in (1,\infty)$, $x \in (0,1]$. Then

$$\varrho_t(x) = \sup\limits_{R \in (0, d_t]} \Delta(xR, R, \psi) \le \sup\limits_{R \in (0, d_t]} \Delta(xR, x_1 R, \psi) \sup\limits_{R \in (0, d_t]} \Delta(x_1 R, R, \psi)$$

$$= \sup\limits_{R' \in (0, x d_t]} \Delta(R', x_2^{-1} R', \psi) \sup\limits_{R \in (0, d_t]} \Delta(x_1 R, R, \psi)$$

$$\le \sup\limits_{R \in (0, d_t]} \Delta(R, x_2^{-1} R, \psi) \sup\limits_{R \in (0, d_t]} \Delta(x_1 R, R, \psi) = \varrho_t(x_2)\varrho_t(x_1).$$

In a similar way one can verify (25) in case $x = x_1 x_2$, $x_1 \in (0,1]$, $x_2 \in (1,\infty)$, $x \in (1,\infty)$. ∎

The following result is an analogue of Lemma 2(a) of [3] (see also Theorem 4.8.18 of [1]).

Lemma 6.4. *Let Γ be a Carleson Jordan curve, let $t \in \Gamma$, and let $\psi : \Gamma \setminus \{t\} \to (0,\infty)$ be continuous. Define the function ϱ_t by (26) and suppose $\log \varrho_t$ is bounded in a neighborhood of the point 1. Put*

$$\tilde{\varrho}_t(x) := \limsup\limits_{R \to 0} \Delta(xR, R, \psi) \quad \text{for } x \in (0, \infty). \qquad (28)$$

Then $\tilde{\varrho}_t : (0,\infty) \to (0,\infty)$ is submultiplicative and $\log \tilde{\varrho}_t$ is bounded in a neighborhood of the point 1. Moreover,

$$\alpha_t = \tilde{\alpha}_t \quad \text{and} \quad \beta_t = \tilde{\beta}_t, \qquad (29)$$

where α_t and β_t are the lower and upper indices of ϱ_t, and $\tilde{\alpha}_t$ and $\tilde{\beta}_t$ stand for the lower and upper indices of $\tilde{\varrho}_t$.

Proof. The inequality $\tilde{\varrho}_t(x_1 x_2) \leq \tilde{\varrho}_t(x_1)\tilde{\varrho}_t(x_2)$ is easily verified if $\tilde{\varrho}_t(x) > 0$ for all $x \in (0, \infty)$. Since clearly $\tilde{\varrho}_t(x) \leq \varrho_t(x)$ for all $x \in (0, \infty)$, the function $\tilde{\varrho}_t$ is finite on $(0, \infty)$ and bounded near the point 1.

By Theorem 6.2(c),(d), there is a function $\varepsilon : (0, 1) \to [0, \infty)$ such that $\varepsilon(x) \to 0$ as $x \to 0$ and

$$x^{\alpha_t} \leq \varrho_t(x) \leq x^{\alpha_t - \varepsilon(x)} \text{ for } x \in (0, 1). \tag{30}$$

Fix an $x \in (0, 1)$. Then for every integer $m \geq 1$ there exists a number $R_m \in (0, d_t]$ such that

$$\varrho_t(x^m) \leq \Delta(x^m R_m, R_m, \psi)x^{-1}.$$

Hence, by (30),

$$\Delta(x^m R_m, R_m, \psi) \geq x^{m\alpha_t + 1}. \tag{31}$$

On the other hand, we infer from (30) that

$$\Delta(x^m R_{2m}, R_{2m}, \psi) \leq \varrho_t(x^m) \leq x^{m\alpha_t - m\varepsilon(x^m)}. \tag{32}$$

We have

$$\prod_{n=m}^{2m-1} \Delta(x^{n+1} R_{2m}, x^n R_{2m}, \psi) \geq \Delta(x^{2m} R_{2m}, x^m R_{2m}, \psi)$$
$$\geq \Delta(x^{2m} R_{2m}, R_{2m}, \psi)/\Delta(x^m R_{2m}, R_{2m}, \psi)$$

and hence, by (31) and (32),

$$\prod_{n=m}^{2m-1} \Delta(x^{n+1} R_{2m}, x^n R_{2m}, \psi) \geq x^{2m\alpha_t + 1}/x^{m\alpha_t - m\varepsilon(x^m)} = x^{m\alpha_t + 1 + m\varepsilon(x^m)}. \tag{33}$$

Choose $n(m)$ so that

$$\Delta(x^{n(m)+1} R_{2m}, x^{n(m)} R_{2m}, \psi) = \max\left\{\Delta(x^{n+1} R_{2m}, x^n R_{2m}, \psi) : m \leq n \leq 2m - 1\right\}.$$

Then

$$\left[\Delta(x^{n(m)+1} R_{2m}, x^{n(m)} R_{2m}, \psi)\right]^m \geq \prod_{n=m}^{2m-1} \Delta(x^{n+1} R_{2m}, x^n R_{2m}, \psi)$$

and (33) implies that

$$\Delta(x^{n(m)+1} R_{2m}, x^{n(m)} R_{2m}, \psi) \geq x^{\alpha_t + 1/m + \varepsilon(x^m)}.$$

Because $x^{n(m)} R_{2m} \leq x^m d_t = o(1)$ and $\varepsilon(x^m) = o(1)$ as $m \to \infty$, it follows that

$$\tilde{\varrho}_t(x) = \limsup_{R \to 0} \Delta(xR, R, \psi) \geq \limsup_{m \to \infty} \Delta(x^{n(m)+1} R_{2m}, x^{n(m)} R_{2m}, \psi) \geq x^{\alpha_t} \tag{34}$$

for every $x \in (0, 1)$. In particular, $\tilde{\varrho}_t(x) > 0$ for $x \in (0, 1)$ and $\tilde{\varrho}_t(x)$ is bounded away from zero for all $x \in (0, 1)$ close enough to 1. At the point 1 itself we obviously have $\tilde{\varrho}_t(1) \geq 1$.

From (30) we also obtain that

$$\tilde{\varrho}_t(x) \leq \varrho_t(x) \leq x^{\alpha_t - \varepsilon(x)} \text{ for } x \in (0,1).$$ (35)

Combining (34) and (35) we get

$$x^{\alpha_t} \leq \tilde{\varrho}_t(x) \leq x^{\alpha_t - \varepsilon(x)} \text{ for } x \in (0,1).$$

Thus,

$$\alpha_t \geq \tilde{\alpha}_t := \sup_{x \in (0,1)} \frac{\log \tilde{\varrho}_t(x)}{\log x} \geq \sup_{x \in (0,1)} (\alpha_t - \varepsilon(x)) = \alpha_t,$$

which proves that $\alpha_t = \tilde{\alpha}_t$.

In a similar way one can show that $\tilde{\varrho}_t(x) > 0$ for $x > 1$ and that $\tilde{\varrho}_t(x)$ is bounded away from zero in a right neighborhood of the point 1 (including 1 itself) and that the equality $\beta_t = \tilde{\beta}_t$ holds. ∎

We finally specify $\psi(\tau)$ to $\psi_t(\tau) := e^{-\arg(\tau - t)}$ and maintain the notation $\varrho_t(x)$ and $\tilde{\varrho}_t(x)$ for the functions defined by (27) and (28) with ψ_t in place of ψ. Our aim is to show that in this case $\log \varrho_t$ is bounded in a neighborhood of the point 1.

A connected subset of $\Gamma \setminus \{t\}$ will be referred to as an arc of Γ. Given an arc $\gamma \subset \Gamma$, we define $\{\arg(\tau - t)\}_{\tau \in \gamma}$ as the increment of $\arg(\tau - t)$ as τ traces out γ following the orientation of Γ.

Lemma 6.5. *Let Γ be a Carleson Jordan curve and fix $t \in \Gamma$. Let γ be any arc of Γ whose endpoints lie on the concentric circles $\{z \in \mathbf{C} : |z - t| = r\}$ and $\{z \in \mathbf{C} : |z - t| = R\}$ with $0 < r \leq R$. Then*

$$\left| \left\{ \arg(\tau - t) \right\}_{\tau \in \gamma} \right| \leq 2\pi C_\Gamma R/r$$

where C_Γ is given by (4).

Proof. Put $K(y) = \{z \in \mathbf{C} : |z - t| = y\}$ and denote by $M(y)$ the set of the boundary points of the set $\Gamma \cap K(y)$. The set $\gamma \setminus (M(r) \cup M(R))$ consists of at most countably many open arcs $\gamma_j \subset \gamma$. We divide the collection of these arcs γ_j into three pairwise disjoint classes. Let t'_j and t''_j denote the endpoints of the arc γ_j. Put

$$\mathcal{N}_1 := \left\{ \gamma_j : \{t'_j, t''_j\} \subset K(r) \text{ and } \gamma_j \subset \Gamma(t, r) \right\},$$
$$\mathcal{N}_2 := \left\{ \gamma_j : \{t'_j, t''_j\} \subset K(r) \cup K(R) \text{ and } \gamma_j \subset \Gamma(t, 2R) \setminus \Gamma(t, r) \right\},$$
$$\mathcal{N}_3 := \left\{ \gamma_j : \{t'_j, t''_j\} \subset K(R) \text{ and } \gamma_j \not\subset \Gamma(t, 2R) \right\}.$$

Let $\gamma_j \in \mathcal{N}_1$. We denote by $\tilde{\gamma}_j$ the arc of $K(r)$ whose endpoints are t'_j and t''_j and which is uniquely determined by the requirement that t belongs to the unbounded component of $\mathbf{C} \setminus (\gamma_j \cup \tilde{\gamma}_j)$. Clearly,

$$0 < \left| \left\{ \arg(\tau - t) \right\}_{\tau \in \gamma_j} \right| < 2\pi.$$ (36)

If $0 < |\{\arg(\tau - t)\}_{\tau \in \gamma_j}| < \pi$ then

$$|\gamma_j| \geq |t_j'' - t_j'| \geq \frac{2}{\pi}|\tilde{\gamma}_j| = \frac{2}{\pi}r|\{\arg(\tau - t)\}_{\tau \in \gamma_j}|,$$

while if $\pi \leq |\{\arg(\tau - t)\}_{\tau \in \gamma_j}| < 2\pi$, we have

$$|\gamma_j| \geq 2r = \frac{1}{\pi}r2\pi > \frac{1}{\pi}r|\{\arg(\tau - t)\}_{\tau \in \gamma_j}|.$$

Thus, for every $\gamma_j \in \mathcal{N}_1$ the inequality

$$|\gamma_j| > \frac{1}{\pi}r|\{\arg(\tau - t)\}_{\tau \in \gamma_j}| \tag{37}$$

holds.

If $\gamma_j \in \mathcal{N}_2$ then $\gamma_j \subset \Gamma(t, 2R) \setminus \Gamma(t, r)$, whence

$$|\gamma_j| \geq r|\{\arg(\tau - t)\}_{\tau \in \gamma_j}| > \frac{1}{\pi}r|\{\arg(\tau - t)\}_{\tau \in \gamma_j}|. \tag{38}$$

If $\gamma_j \in \mathcal{N}_3$, then the endpoints of γ_j lie on $K(R)$ and γ_j is not a subset of $\Gamma(t, 2R)$. This implies that $|\gamma_j \cap \Gamma(t, 2R)| \geq 2R$. On the other hand, again (36) is valid. Therefore,

$$|\gamma_j \cap \Gamma(t, 2R)| \geq \frac{1}{\pi}R2\pi > \frac{1}{\pi}r|\{\arg(\tau - t)\}_{\tau \in \gamma_j}|. \tag{39}$$

Since $\gamma_j \subset \Gamma(t, 2R)$ for $\gamma_j \in \mathcal{N}_1 \cup \mathcal{N}_2$, we obtain

$$|\Gamma(t, 2R)| \geq \sum_{\gamma_j \in \mathcal{N}_1} |\gamma_j| + \sum_{\gamma_j \in \mathcal{N}_2} |\gamma_j| + \sum_{\gamma_j \in \mathcal{N}_3} |\gamma_j \cap \Gamma(t, 2R)|,$$

and taking into account (37)–(39) we so get

$$|\Gamma(t, 2R)| > \frac{1}{\pi}r \sum_{\gamma_j \in \mathcal{N}_1 \cup \mathcal{N}_2 \cup \mathcal{N}_3} |\{\arg(\tau - t)\}_{\tau \in \gamma_j}| \geq \frac{1}{\pi}r|\{\arg(\tau - t)\}_{\tau \in \gamma}|. \tag{40}$$

From (4) we know that $|\Gamma(t, 2R)| \leq C_\Gamma 2R$, and thus (40) gives

$$|\{\arg(\tau - t)\}_{\tau \in \gamma}| \leq \frac{\pi}{r}C_\Gamma 2R = 2\pi C_\Gamma R/r. \blacksquare$$

Lemma 6.6. *Let Γ be a Carleson Jordan curve and $t \in \Gamma$. Define ϱ_t by (27) with $\psi(\tau) := \psi_t(\tau) = e^{-\arg(\tau - t)}$. Then*

$$e^{-4\pi C_\Gamma} \leq \varrho_t(x) \leq e^{4\pi C_\Gamma} \quad \text{for } x \in [1/2, 2].$$

Proof. Let $x \in [1/2, 1]$ and $R \in (0, d_t]$.

Let $\tau' \in \partial\Gamma(t, xR)$ and $\tau'' \in \partial\Gamma(t, R)$ be the points which are determined by

$$\psi_t(\tau') = \max_{\tau \in \partial\Gamma(t, xR)} \psi_t(\tau), \quad \psi_t(\tau'') = \min_{\tau \in \partial\Gamma(t, R)} \psi_t(\tau).$$

Then

$$|\log \Delta(xR, R, \psi_t)| = \left|\log \frac{\psi_t(\tau')}{\psi_t(\tau'')}\right| = |-\log \psi_t(\tau'') + \log \psi_t(\tau')|$$
$$= |\arg(\tau'' - t) - \arg(\tau' - t)| = \left|\left\{\arg(\tau - t)\right\}_{\tau \in \gamma}\right|,$$

where γ is the arc of Γ between τ' and τ''. Consequently, by Lemma 6.5,

$$|\log \Delta(xR, R, \psi_t)| \leq 2\pi C_\Gamma R/(xR) = 2\pi C_\Gamma x^{-1} \leq 4\pi C_\Gamma,$$

whence

$$e^{-4\pi C_\Gamma} \leq \Delta(xR, R, \psi_t) \leq e^{4\pi C_\Gamma}$$

for all $x \in [1/2, 1]$ and all $R \in (0, d_t]$. In a similar way one can show that

$$e^{-4\pi C_\Gamma} \leq \Delta(R, x^{-1}R, \psi_t) \leq e^{4\pi C_\Gamma}$$

for $x \in [1, 2]$ and $R \in (0, d_t]$. ∎

Corollary 6.7. *Let Γ be a Carleson Jordan curve, $t \in \Gamma$, and define ϱ_t and $\tilde{\varrho}_t$ by (27) and (28)=(10) with $\psi(\tau) := e^{-\arg(\tau-t)}$. Then*

$$\alpha_t = \tilde{\alpha}_t = \delta_t^- \quad \text{and} \quad \beta_t = \tilde{\beta}_t = \delta_t^+,$$

where $\alpha_t, \tilde{\alpha}_t, \beta_t, \tilde{\beta}_t$ are as in Lemma 6.4 and δ_t^-, δ_t^+ are the numbers given by (11).

Proof. Since $\log \varrho_t$ is bounded near the point 1 (Lemma 6.6), we may apply Lemma 6.4 to deduce that $\tilde{\varrho}_t$ is submultiplicative, that $\log \tilde{\varrho}_t$ is bounded in a neighborhood of the point 1, and that $\alpha_t = \tilde{\alpha}_t$ and $\beta_t = \tilde{\beta}_t$. So Theorem 6.2(b) shows that in (11) the "limsup" may actually be replaced by "lim" and that $\delta_t^- = \tilde{\alpha}_t$ and $\delta_t^+ = \tilde{\beta}_t$. ∎

7. Description of the set N_t

Whether a function ψ satisfies the Muckenhoupt condition (2) is often not easy to decide by directly verifying the inequality (2). The following theorem provides us with a more effective criterion for functions ψ which have at most a single discontinuity on Γ. This criterion is in terms of the indices of a submultiplicative function associated with ψ. Since we have not been able to find such a criterion in the literature, we will give a full proof.

Theorem 7.1. *Let Γ be a Carleson Jordan curve, let $t \in \Gamma$, and let $\psi : \Gamma \setminus \{t\} \to (0, \infty)$ be a continuous function. Define ϱ_t by (26) and suppose $\log \varrho_t$ is bounded in a neighborhood of the point 1 and let α_t and β_t denote the lower and upper indices of ϱ_t, respectively. Then $\psi \in A_p(\Gamma)$ if and only if*

$$-1/p < \alpha_t \leq \beta_t < 1/q, \tag{41}$$

where $1 < p < \infty$ *and* $1/p + 1/q = 1$.

Proof. Suppose (41) holds. Then there is an $\varepsilon > 0$ such that

$$- 1/p < \alpha_t - \varepsilon < \beta_t + \varepsilon < 1/q. \tag{42}$$

Since $\log \varrho_t$ is bounded near 1, Theorem 6.2(d) and Lemma 6.1(a) imply that there are $x_0 \in (0,1)$ and $C_t > 0$ such that

$$\varrho_t(x) \le x^{\alpha_t - \varepsilon} \text{ for } x \in (0, x_0), \quad \varrho_t(x) \le x^{\beta_t + \varepsilon} \text{ for } x \in (x_0^{-1}, \infty), \tag{43}$$
$$\varrho_t(x) \le C_t \text{ for } x \in [x_0, x_0^{-1}]. \tag{44}$$

From the definition (27) we see that

$$\sup_{R \in (0, d_t]} \Delta(xR, R, \psi^{-1}) = \sup_{R \in (0, d_t]} \Delta(R, xR, \psi) = \varrho_t(x^{-1}) \tag{45}$$

for $x \in (0,1)$. Hence, by (27) and (45), for every $x \in (0,1)$ and every $\tau \in \partial\Gamma(t, xR)$ we have

$$\psi(\tau) \le \max_{\tau \in \partial\Gamma(t, xR)} \psi(\tau) \le \varrho_t(x) \min_{\tau \in \partial\Gamma(t, R)} \psi(\tau), \tag{46}$$
$$\psi^{-1}(\tau) \le \max_{\tau \in \partial\Gamma(t, xR)} \psi^{-1}(\tau) \le \varrho_t(x^{-1}) \min_{\tau \in \partial\Gamma(t, R)} \psi^{-1}(\tau). \tag{47}$$

Using (43) we also obtain for $n > 1$ that

$$\sup_{x \in [x_0^n, x_0^{n-1})} \varrho_t(x) \le \sup_{x \in [x_0^n, x_0^{n-1})} x^{\alpha_t - \varepsilon}$$
$$\le \max\left\{ x_0^{n(\alpha_t - \varepsilon)}, x_0^{(n-1)(\alpha_t - \varepsilon)} \right\} = c_1 x_0^{(n-1)(\alpha_t - \varepsilon)}, \tag{48}$$
$$\sup_{x \in [x_0^n, x_0^{n-1})} \varrho_t(x^{-1}) \le \sup_{x \in [x_0^n, x_0^{n-1})} x^{-(\beta_t + \varepsilon)}$$
$$\le \max\left\{ x_0^{-n(\beta_t + \varepsilon)}, x_0^{-(n-1)(\beta_t + \varepsilon)} \right\} = c_2 x_0^{-(n-1)(\beta_t + \varepsilon)}, \tag{49}$$

where $c_1 := \max\{1, x_0^{\alpha_t - \varepsilon}\}$ and $c_2 := \max\{1, x_0^{-(\beta_t + \varepsilon)}\}$.

By virtue of (42), $1 + p(\alpha_t - \varepsilon) > 0$ and $1 - q(\beta_t + \varepsilon) > 0$. Consequently,

$$0 < x_0^{1 + p(\alpha_t - \varepsilon)} < 1, \quad 0 < x_0^{1 - q(\beta_t + \varepsilon)} < 1. \tag{50}$$

Further, by the Carleson condition (4),

$$|\Gamma(t, x_0^{n-1} R)| - |\Gamma(t, x_0^n R)| \le C_\Gamma x_0^{n-1} R - x_0^n R = c_0 x_0^{n-1} R \tag{51}$$

with $c_0 := C_\Gamma - x_0 > 0$.

Taking into account (46), (51), (44), (48) we get

$$\int_{\Gamma(t, R)} \psi^p(\tau) |d\tau| = \sum_{n=1}^{\infty} \int_{\Gamma(t, x_0^{n-1} R) \setminus \Gamma(t, x_0^n R)} \psi^p(\tau) |d\tau|$$

$$\leq \sum_{n=1}^{\infty} \sup_{x \in [x_0^n, x_0^{n-1})} \varrho_t^p(x) \Big(\min_{\tau \in \partial \Gamma(t,R)} \psi^p(\tau) \Big) \Big(|\Gamma(t, x_0^{n-1}R| - |\Gamma(t, x_0^n R)| \Big)$$

$$\leq \Big(C_t^p c_0 R + \sum_{n=2}^{\infty} c_1^p x_0^{p(n-1)(\alpha_t - \varepsilon)} c_0 x_0^{n-1} R \Big) \min_{\tau \in \partial \Gamma(t,R)} \psi^p(\tau)$$

$$= c_0 \Big(C_t^p + c_1^p \frac{x_0^{1+p(\alpha_t - \varepsilon)}}{1 - x_0^{1+p(\alpha_t - \varepsilon)}} \Big) R \min_{\tau \in \partial \Gamma(t,R)} \psi^p(\tau), \tag{52}$$

while (47), (51), (44), (49) yield

$$\int_{\Gamma(t,R)} \psi^{-q}(\tau)\, |d\tau| = \sum_{n=1}^{\infty} \int_{\Gamma(t, x_0^{n-1}R) \setminus \Gamma(t, x_0^n R)} \psi^{-q}(\tau)\, |d\tau|$$

$$\leq \sum_{n=1}^{\infty} \sup_{x \in [x_0^n, x_0^{n-1})} \varrho_t^q(x^{-1}) \Big(\min_{\tau \in \partial \Gamma(t,R)} \psi^{-q}(\tau) \Big) \Big(|\Gamma(t, x_0^{n-1} R| - |\Gamma(t, x_0^n R| \Big)$$

$$\leq \Big(C_t^q c_0 R + \sum_{n=2}^{\infty} c_2^q x_0^{-q(n-1)(\beta_t + \varepsilon)} c_0 x_0^{n-1} R \Big) \min_{\tau \in \partial \Gamma(t,R)} \psi^{-q}(\tau)$$

$$= c_0 \Big(C_t^q + c_2^q \frac{x_0^{1-q(\beta_t + \varepsilon)}}{1 - x_0^{1-q(\beta_t + \varepsilon)}} \Big) R \min_{\tau \in \partial \Gamma(t,R)} \psi^{-q}(\tau). \tag{53}$$

Put

$$c_3 := c_0 \Big(C_t^p + c_1^p \frac{x_0^{1+p(\alpha_t - \varepsilon)}}{1 - x_0^{1+p(\alpha_t - \varepsilon)}} \Big), \quad c_4 := c_0 \Big(C_t^q + c_2^q \frac{x_0^{1-q(\beta_t + \varepsilon)}}{1 - x_0^{1-q(\beta_t + \varepsilon)}} \Big).$$

Then (52) and (53) read

$$\Big(\frac{1}{R} \int_{\Gamma(t,R)} \psi^p(\tau)\, |d\tau| \Big)^{1/p} \leq c_3^{1/p} \min_{\tau \in \partial \Gamma(t,R)} \psi(\tau),$$

$$\Big(\frac{1}{R} \int_{\Gamma(t,R)} \psi^{-q}(\tau)\, |d\tau| \Big)^{1/q} \leq c_4^{1/q} \min_{\tau \in \partial \Gamma(t,R)} \psi^{-1}(\tau) \leq c_4^{1/q} \Big(\min_{\tau \in \partial \Gamma(t,R)} \psi(\tau) \Big)^{-1},$$

and multiplication of these two inequalities gives

$$B_t := \sup_{R>0} \Big(\frac{1}{R} \int_{\Gamma(t,R)} \psi^p(\tau)\, |d\tau| \Big)^{1/p} \Big(\frac{1}{R} \int_{\Gamma(t,R)} \psi^{-q}(\tau)\, |d\tau| \Big)^{1/q} \leq c_3^{1/p} c_4^{1/q} < \infty. \tag{54}$$

Let now $t_0 \in \Gamma \setminus \{t\}$. Suppose first that $R \geq |t - t_0|/2$. Then $(R + |t - t_0|)/R \leq 3$, and since always $\Gamma(t_0, R) \subset \Gamma(t, R + |t - t_0|)$, we obtain

$$\frac{1}{R} \int_{\Gamma(t_0,R)} \psi^p(\tau)\, |d\tau| \leq \frac{R + |t - t_0|}{R} \frac{1}{R + |t - t_0|} \int_{\Gamma(t, R+|t-t_0|)} \psi^p(\tau)\, |d\tau|$$

$$\leq \frac{3}{R + |t - t_0|} \int_{\Gamma(t, R+|t-t_0|)} \psi^p(\tau)\, |d\tau|,$$

$$\frac{1}{R} \int\limits_{\Gamma(t_0,R)} \psi^{-q}(\tau) \, |d\tau| \leq \frac{R + |t - t_0|}{R} \frac{1}{R + |t - t_0|} \int\limits_{\Gamma(t,R+|t-t_0|)} \psi^{-q}(\tau) \, |d\tau|$$

$$\leq \frac{3}{R + |t - t_0|} \int\limits_{\Gamma(t,R+|t-t_0|)} \psi^{-q}(\tau) \, |d\tau|.$$

Consequently, by (54),

$$\sup_{R \geq |t-t_0|/2} \left(\frac{1}{R} \int\limits_{\Gamma(t_0,R)} \psi^p(\tau) \, |d\tau| \right)^{1/p} \left(\frac{1}{R} \int\limits_{\Gamma(t_0,R)} \psi^{-q}(\tau) \, |d\tau| \right)^{1/q}$$

$$\leq 3 \sup_{\varepsilon > 0} \left(\frac{1}{\varepsilon} \int\limits_{\Gamma(t,\varepsilon)} \psi^p(\tau) \, |d\tau| \right)^{1/p} \left(\frac{1}{\varepsilon} \int\limits_{\Gamma(t,\varepsilon)} \psi^{-q}(\tau) \, |d\tau| \right)^{1/q} = 3B_t < \infty. \tag{55}$$

Now suppose $0 < R < |t - t_0|/2$. Put

$$R_0 := \min \{R, d_t - |t - t_0|\}, \quad R_1 := \min \{|t - t_0| + R, d_t\}.$$

The estimates (46) and (47) show that

$$\left(\frac{1}{R} \int\limits_{\Gamma(t_0,R)} \psi^p(\tau) \, |d\tau| \right)^{1/p} \leq C_\Gamma^{1/p} \sup_{\tau \in \Gamma(t_0,R)} \psi(\tau)$$

$$\leq C_\Gamma^{1/p} \sup_{x \in [-R,R_0]} \max_{\tau \in \partial\Gamma(t,|t-t_0|+x)} \psi(\tau)$$

$$\leq C_\Gamma^{1/p} \sup_{x \in [-R,R_0]} \varrho_t \left(\frac{|t - t_0| + x}{R_1} \right) \min_{\tau \in \partial\Gamma(t,R_1)} \psi(\tau),$$

$$\left(\frac{1}{R} \int\limits_{\Gamma(t_0,R)} \psi^{-q}(\tau) \, |d\tau| \right)^{1/q} \leq C_\Gamma^{1/q} \sup_{\tau \in \Gamma(t_0,R)} \psi^{-1}(\tau)$$

$$\leq C_\Gamma^{1/q} \sup_{x \in [-R,R_0]} \max_{\tau \in \partial\Gamma(t,|t-t_0|+x)} \psi^{-1}(\tau)$$

$$\leq C_\Gamma^{1/q} \sup_{x \in [-R,R_0]} \varrho_t \left(\frac{R_1}{|t - t_0| + x} \right) \min_{\tau \in \partial\Gamma(t,R_1)} \psi^{-1}(\tau)$$

$$\leq C_\Gamma^{1/q} \sup_{x \in [-R,R_0]} \varrho_t \left(\frac{R_1}{|t - t_0| + x} \right) \left(\min_{\tau \in \partial\Gamma(t,R_1)} \psi(\tau) \right)^{-1}.$$

Multiplying the latter two inequalities we arrive at the estimate

$$\sup_{0 < R < |t-t_0|/2} \left(\frac{1}{R} \int\limits_{\Gamma(t_0,R)} \psi^p(\tau) \, |d\tau| \right)^{1/p} \left(\frac{1}{R} \int\limits_{\Gamma(t_0,R)} \psi^{-q}(\tau) \, |d\tau| \right)^{1/q}$$

$$\leq C_\Gamma \sup_{0 < R < |t-t_0|/2} \sup_{x \in [-R,R_0]} \varrho_t \left(\frac{|t - t_0| + x}{R_1} \right) \sup_{x \in [-R,R_0]} \varrho_t \left(\frac{R_1}{|t - t_0| + x} \right). \tag{56}$$

We now consider

$$\frac{|t - t_0| + x}{R_1} = \frac{|t - t_0| + x}{\min\{|t - t_0| + R, d_t\}}$$

for $x \in [-R, R_0] = [-R, \min\{R, d_t - |t - t_0|\}]$ and $0 < R < |t - t_0|/2$. If $R \leq d_t - |t - t_0|$, then

$$1 \geq \frac{|t - t_0| + x}{\min\{|t - t_0| + R, d_t\}} \geq \frac{|t - t_0| - R}{|t - t_0| + R} \geq \frac{|t - t_0| - |t - t_0|/2}{|t - t_0| + |t - t_0|/2} = \frac{1}{3},$$

while if $R > d_t - |t - t_0|$, we have $d_t - |t - t_0| < R < |t - t_0|/2$, whence $d_t < (3/2)|t - t_0|$ and thus,

$$1 \geq \frac{|t - t_0| + x}{\min\{|t - t_0| + R, d_t\}} \geq \frac{|t - t_0| - R}{d_t} \geq \frac{|t - t_0|/2}{(3/2)|t - t_0|} = \frac{1}{3}.$$

Consequently, in either case

$$\frac{1}{3} \leq \frac{|t - t_0| + x}{\min\{|t - t_0| + R, d_t\}} \leq 1. \tag{57}$$

Combining (56) and (57) we obtain

$$\sup_{0 < R < |t - t_0|/2} \left(\frac{1}{R} \int_{\Gamma(t_0, R)} \psi^p(\tau) |d\tau| \right)^{1/p} \left(\frac{1}{R} \int_{\Gamma(t_0, R)} \psi^{-q}(\tau) |d\tau| \right)^{1/q}$$

$$\leq C_\Gamma \sup_{y \in [1/3, 3]} \varrho_t^2(y) < \infty, \tag{58}$$

the last inequality resulting from (43), (44). Putting (55) and (58) together we arrive at the estimate

$$B_{t_0} := \sup_{R > 0} \left(\frac{1}{R} \int_{\Gamma(t_0, R)} \psi^p(\tau) |d\tau| \right)^{1/p} \left(\frac{1}{R} \int_{\Gamma(t_0, R)} \psi^{-q}(\tau) |d\tau| \right)^{1/q}$$

$$\leq \max \left\{ 3B_t, C_\Gamma \sup_{y \in [1/3, 3]} \varrho_t^2(y) \right\} < \infty \tag{59}$$

for $t_0 \in \Gamma \setminus \{t\}$. From (54) and (59) we finally infer that $\psi \in A_p(\Gamma)$, which proves the "if" portion of the theorem.

To show the "only if" part, assume that $\alpha_t < -1/p$. Since $\log \varrho_t$ is bounded near 1, Theorem 6.2(c) yields that

$$\varrho_t(x) \geq x^{\alpha_t} \text{ for } x \in (0, 1). \tag{60}$$

From the definition (27) we infer that if $c \in (0, 1]$ and $R \in (0, d_t]$, then

$$\min_{\tau \in \partial \Gamma(t, cR)} \psi(\tau) = (\Delta(R, cR, \psi))^{-1} \max_{\tau \in \partial \Gamma(t, R)} \psi(\tau) \geq \frac{1}{\varrho_t(c^{-1})} \max_{\tau \in \partial \Gamma(t, R)} \psi(\tau), \tag{61}$$

$$\min_{\tau \in \partial \Gamma(t, cR)} \psi^{-1}(\tau) = (\Delta(R, cR, \psi^{-1}))^{-1} \max_{\tau \in \partial \Gamma(t, R)} \psi^{-1}(\tau)$$

$$= (\Delta(cR, R, \psi))^{-1} \max_{\tau \in \partial \Gamma(t, R)} \psi^{-1}(\tau) \geq \frac{1}{\varrho_t(c)} \max_{\tau \in \partial \Gamma(t, R)} \psi^{-1}(\tau). \tag{62}$$

Fix $\varepsilon \in (0, 1)$ and $\kappa \in (0, 1)$. Since ϱ_t is positive on $(0, \infty)$, we see from definition (27) that for each integer $n \geq 1$ there is an $R_n \in (0, d_t]$ such that

$$\max_{\tau \in \partial \Gamma(t, \kappa^n R_n)} \psi(\tau) \geq (1 - \varepsilon) \varrho_t(\kappa^n) \min_{\tau \in \partial \Gamma(t, R_n)} \psi(\tau). \tag{63}$$

If $c \in [\kappa, 1)$, we obtain from (61), (63), (60) that

$$
\min_{\tau \in \partial \Gamma(t, c\kappa^n R_n)} \psi(\tau) \geq \frac{1}{\varrho_t(c^{-1})} \max_{\tau \in \partial \Gamma(t, \kappa^n R_n)} \psi(\tau)
$$

$$
\geq \frac{1-\varepsilon}{\varrho_t(c^{-1})} \varrho_t(\kappa^n) \min_{\tau \in \partial \Gamma(t, R_n)} \psi(\tau) \geq \frac{1-\varepsilon}{\varrho_t(c^{-1})} \kappa^{n\alpha_t} \min_{\tau \in \partial \Gamma(t, R_n)} \psi(\tau). \tag{64}
$$

By (62), we have for $c \in [\kappa, 1)$

$$
\min_{\tau \in \partial \Gamma(t, c R_n)} \psi^{-1}(\tau) \geq \frac{1}{\varrho_t(c)} \max_{\tau \in \partial \Gamma(t, R_n)} \psi^{-1}(\tau). \tag{65}
$$

Put $\Gamma_{m,n} := \Gamma(t, \kappa^m R_n) \setminus \Gamma(t, \kappa^{m+1} R_n)$ for $m \geq 0$. Clearly, $|\Gamma_{m,n}| \geq (1-\kappa)\kappa^m R_n$ for $m \geq 0$ and $n \geq 1$. This together with (64) and (65) implies that

$$
\left(\frac{1}{R_n} \int_{\Gamma(t, R_n)} \psi^p(\tau) \, |d\tau| \right)^{1/p} \geq \left(\frac{|\Gamma_{n,n}|}{R_n} \frac{1}{|\Gamma_{n,n}|} \int_{\Gamma_{n,n}} \psi^p(\tau) \, |d\tau| \right)^{1/p}
$$

$$
\geq \left(\frac{(1-\kappa)\kappa^n R_n}{R_n} \right)^{1/p} \inf_{c \in [\kappa, 1)} \min_{\tau \in \partial \Gamma(t, c\kappa^n R_n)} \psi(\tau)
$$

$$
\geq (1-\kappa)^{1/p} \kappa^{n/p} \cdot \frac{1-\varepsilon}{\sup_{c \in [\kappa, 1)} \varrho_t(c^{-1})} \kappa^{n\alpha_t} \min_{\tau \in \partial \Gamma(t, R_n)} \psi(\tau), \tag{66}
$$

$$
\left(\frac{1}{R_n} \int_{\Gamma(t, R_n)} \psi^{-q}(\tau) \, |d\tau| \right)^{1/q} \geq \left(\frac{|\Gamma_{0,n}|}{R_n} \frac{1}{|\Gamma_{0,n}|} \int_{\Gamma_{0,n}} \psi^{-q}(\tau) \, |d\tau| \right)^{1/q}
$$

$$
\geq \left(\frac{(1-\kappa)R_n}{R_n} \right)^{1/q} \inf_{c \in [\kappa, 1)} \min_{\tau \in \partial \Gamma(t, c R_n)} \psi^{-1}(\tau)
$$

$$
\geq (1-\kappa)^{1/q} \frac{1}{\sup_{c \in [\kappa, 1)} \varrho_t(c)} \max_{\tau \in \partial \Gamma(t, R_n)} \psi^{-1}(\tau). \tag{67}
$$

Taking into account that

$$
\tilde{c}_0 := \sup_{c \in [\kappa, 1)} \varrho_t(c) \sup_{c \in [\kappa, 1)} \varrho_t(c^{-1}) < \infty,
$$

we conclude from (66) and (67) that

$$
B_t := \sup_{R > 0} \left(\frac{1}{R} \int_{\Gamma(t, R)} \psi^p(\tau) \, |d\tau| \right)^{1/p} \left(\frac{1}{R} \int_{\Gamma(t, R)} \psi^{-q}(\tau) \, |d\tau| \right)^{1/q}
$$

$$
\geq \left(\frac{1}{R_n} \int_{\Gamma(t, R_n)} \psi^p(\tau) \, |d\tau| \right)^{1/p} \left(\frac{1}{R_n} \int_{\Gamma(t, R_n)} \psi^{-q}(\tau) \, |d\tau| \right)^{1/q}
$$

$$
\geq (1-\kappa)(1-\varepsilon)\tilde{c}_0^{-1} \kappa^{n(1/p+\alpha_t)}.
$$

Since $\kappa \in (0, 1)$ and $1/p + \alpha_t < 0$, we see that $\kappa^{n(1/p+\alpha_t)} \to \infty$ as $n \to \infty$. Hence $B_t = \infty$, which tells us that ψ cannot belong to $A_p(\Gamma)$.

It can be shown analogously that $B_t = \infty$ and thus $\psi \notin A_p(\Gamma)$ in case $\beta_t > 1/q$.

Suppose finally that $\alpha_t = -1/p$ or $\beta_t = 1/q$ and assume $\psi \in A_p(\Gamma)$. Then, by a theorem of Simonenko [24], there exists an $\varepsilon > 0$ such that $\hat{\psi} := \psi^{1+\varepsilon} \in A_p(\Gamma)$. Define $\hat{\varrho}_t$ by (27) with $\hat{\psi}$ in place of ψ. Due to Theorem 6.2 (e), the lower and upper indices of $\hat{\varrho}_t$ are given by $\hat{\alpha}_t = (1+\varepsilon)\alpha_t$, $\hat{\beta}_t = (1+\varepsilon)\beta_t$. Consequently, we have

$$\hat{\alpha}_t = (1+\varepsilon)\alpha_t = -(1+\varepsilon)(1/p) < -1/p$$

or

$$\hat{\beta}_t = (1+\varepsilon)\beta_t = (1+\varepsilon)(1/q) > 1/q.$$

By what was shown above, this is impossible if $\hat{\psi} \in A_p(\Gamma)$. This contradiction completes the proof. ∎

We are now in a position to identify the set N_t.

Theorem 7.2. *Let Γ be a Carleson Jordan curve, fix $t \in \Gamma$, and define N_t by (24). Then*

$$N_t = \{\gamma \in \mathbf{C} : -1/p < \operatorname{Re}\gamma + \delta \operatorname{Im}\gamma < 1/q \quad \forall \delta \in [\delta_t^-, \delta_t^+]\},$$

where δ_t^- and δ_t^+ are given by (11).

Proof. Put $\varphi_t(\tau) := |(\tau - t)^\gamma|$ and $\psi_t(\tau) := e^{-\arg(\tau - t)}$ for $\tau \in \Gamma \setminus \{t\}$. We have

$$\varphi_t(\tau) = |\tau - t|^{\operatorname{Re}\gamma} e^{-\operatorname{Im}\gamma \arg(\tau - t)} = |\tau - t|^{\operatorname{Re}\gamma} \left(\psi_t(\tau)^{\operatorname{sign}\operatorname{Im}\gamma}\right)^{|\operatorname{Im}\gamma|}. \tag{68}$$

Define ϱ_t and $\hat{\varrho}_t$ by (27) with ψ replaced by ψ_t and φ_t, respectively. Since

$$\varrho_t(x^{-1}) = \begin{cases} \sup\limits_{R \in (0, d_t]} \Delta(R, xR, \psi_t) & \text{for } x \in (0,1] \\ \sup\limits_{R \in (0, d_t]} \Delta(x^{-1}R, R, \psi_t) & \text{for } x \in [1, \infty) \end{cases}$$

$$= \begin{cases} \sup\limits_{R \in (0, d_t]} \Delta(xR, R, \psi_t^{-1}) & \text{for } x \in (0,1] \\ \sup\limits_{R \in (0, d_t]} \Delta(R, x^{-1}R, \psi_t^{-1}) & \text{for } x \in [1, \infty) \end{cases}$$

we infer from (68) that

$$\hat{\varrho}_t(x) = x^{\operatorname{Re}\gamma} \left(\varrho_t(x^{\operatorname{sign}\operatorname{Im}\gamma})\right)^{|\operatorname{Im}\gamma|} \quad \text{for } x \in (0, \infty). \tag{69}$$

We know from Lemma 6.6 that $\log \varrho_t$ is bounded in a neighborhood of the point 1. Hence, by (69), so also is $\log \hat{\varrho}_t$. We may therefore invoke Theorem 6.2(b) to compute the lower and upper indices $\hat{\alpha}_t$ and $\hat{\beta}_t$ of $\hat{\varrho}_t$:

$$\hat{\alpha}_t = \lim_{x \to 0} \frac{\log \hat{\varrho}_t(x)}{\log x} = \operatorname{Re}\gamma + |\operatorname{Im}\gamma| \lim_{x \to 0} \frac{\log \varrho_t(x^{\operatorname{sign}\operatorname{Im}\gamma})}{\log x}$$

$$= \begin{cases} \operatorname{Re} \gamma + \operatorname{Im} \gamma \lim\limits_{x \to 0} \frac{\log \varrho_t(x)}{\log x} & \text{for } \operatorname{Im} \gamma \geq 0 \\ \operatorname{Re} \gamma - \operatorname{Im} \gamma \lim\limits_{x \to \infty} \frac{\log \varrho_t(x)}{-\log x} & \text{for } \operatorname{Im} \gamma < 0 \end{cases}$$

$$= \begin{cases} \operatorname{Re} \gamma + \alpha_t \operatorname{Im} \gamma & \text{for } \operatorname{Im} \gamma \geq 0 \\ \operatorname{Re} \gamma + \beta_t \operatorname{Im} \gamma & \text{for } \operatorname{Im} \gamma < 0, \end{cases}$$

and for $\hat{\beta}_t$ we analogously get

$$\hat{\beta}_t = \begin{cases} \operatorname{Re} \gamma + \beta_t \operatorname{Im} \gamma & \text{for } \operatorname{Im} \gamma \geq 0 \\ \operatorname{Re} \gamma + \alpha_t \operatorname{Im} \gamma & \text{for } \operatorname{Im} \gamma < 0. \end{cases}$$

Theorem 7.1 tells us that $\varphi_t \in A_p(\Gamma)$ if and only if $-1/p < \hat{\alpha}_t \leq \hat{\beta}_t < 1/q$, which happens if and only if

$$-1/p < \operatorname{Re} \gamma + \delta \operatorname{Im} \gamma < 1/q \text{ for all } \delta \in [\alpha_t, \beta_t].$$

Because $\alpha_t = \delta_t^-$ and $\beta_t = \delta_t^+$ by virtue of Corollary 6.7, we arrive at the assertion. ∎

8. Description of the set Φ_t

In order to verify condition (i) of Theorem 1.2 the following lemma often proves useful. For values $\mu \in (0,1)$ this lemma is well known (see e.g. [10, Vol. I, Chap. 2, Theorem 4.8] or [2]). For general $\mu > 0$, both this lemma and the proof given below are due to Grudsky [14].

Lemma 8.1. *Let Γ be a Carleson Jordan curve, let $1 < p < \infty$ and $w \in A_p(\Gamma)$. Suppose $f_\pm$ is analytic in $D_\pm$ and continuous on $D_\pm \cup \Gamma$ with the possible exception of finitely many points $t_1, \ldots, t_n \in \Gamma$. If $f_\pm \in L^p(\Gamma, w)$ and $f_\pm$ admits the estimate*

$$|f_\pm(z)| \leq M/|z - t_k|^\mu \qquad (k = 1, \ldots, n)$$

with $M > 0$ and $\mu > 0$ for all $z \in D_\pm$ sufficiently close to t_k, then $f_\pm|\Gamma \in L^p_\pm(\Gamma, w)$.

Proof. For the sake of definiteness, let us consider the $L^p_-(\Gamma, w)$ case only.

Put $\phi_+ = Pf_-$ and $\phi_- = Qf_-$. Then $\phi_+ = f_- - \phi_-$. Multiply the latter equality by the polynomial g_1 given by

$$g_1(t) = \prod_{k=1}^n (t - t_k)^{[\mu]+1}$$

where $[\mu]$ is the integral part of μ. The function $g_1(f_- - \phi_-)$ may have at most a pole of order $\deg g_1$ at infinity and hence, there is a polynomial g_2 of degree $\deg g_2 \leq \deg g_1$ such that $d := g_1(f_- - \phi_-) - g_2$ is analytic throughout D_-. We also have $d = g_1\phi_+ - g_2$, and since $\phi_+ \in L^p_+(\Gamma, w)$ and g_1, g_2 are polynomials, the function d also belongs to $L^p_+(\Gamma, w)$.

On the other hand, there is a polynomial h_1 of degree $\deg h_1 \leq \deg g_1$ such that $g_1 f_- - h_1$ is analytic in D_-. Put $h_2 = g_2 - h_1$. Then

$$d = (g_1 f_- - h_1) - (g_1\phi_- + h_2) =: \psi_1 - \psi_2.$$

Because d and $\psi_1 := g_1 f_- - h_1$ are analytic in D_-, so also is $\psi_2 := g_1 \phi_- + h_2$. As ϕ_- vanishes at infinity, we have $\deg h_2 \leq \deg g_1 - 1$. The function ψ_1 is analytic in D_- and continuous on $D_- \cup \Gamma$ (recall the choice of g_1), whence $\psi_1 \in L^p_-(\Gamma, w)$. Furthermore, since $\phi_- \in L^p_-(\Gamma, w)$ and $g_1 \in L^\infty(\Gamma)$, we have

$$\psi_2 = g_1 \phi_- + h_2 \in L^p_-(\Gamma, w) + \operatorname{span}\{\chi_1, \chi_2, \ldots, \chi_{\deg g_1 - 1}\},$$

where $\chi_n(t) := t^n$. But ψ_2 has no pole at infinity, so that actually $\psi_2 \in L^p_-(\Gamma, w)$. Consequently, $d = \psi_1 - \psi_2 \in L^p_-(\Gamma, w)$.

In summary, $d \in L^p_+(\Gamma, w) \cap L^p_-(\Gamma, w) = \mathbf{C}$, i. e. $d(t) = c \in \mathbf{C}$ for all $t \in \Gamma$. It follows that the rational function $(g_2 + c)/g_1$ equals ϕ_+ in D_+, is equal to $f_- - \phi_-$ in D_-, and is analytic in $D_+ \cup D_-$. Since $(g_2 + c)/g_1 \in L^p(\Gamma, w) \subset L^1(\Gamma)$, the rational function $(g_2 + c)/g_1$ must not have poles on Γ and is therefore analytic in $D_+ \cup \Gamma \cup D_- = \mathbf{C} \cup \{\infty\}$. Thus, by Liouville's theorem, $(g_2 + c)/g_1$ is a constant function. It results that ϕ_+ is some constant c_0 and hence $f_- = \phi_- + c_0 \in L^p_-(\Gamma, w)$. ∎

Theorem 8.2. *Let Γ be a Carleson Jordan curve, fix $t \in \Gamma$, and define Φ_t by (22). Then*

$$\Phi_t = \{\gamma \in \mathbf{C} : 1/p - \operatorname{Re}\gamma - \delta \operatorname{Im}\gamma \notin \mathbf{Z} \quad \forall \delta \in [\delta_t^-, \delta_t^+]\},$$

where δ_t^-, δ_t^+ are the numbers given by (11). If

$$\varkappa(\delta) := 1/p - \operatorname{Re}\gamma - \delta \operatorname{Im}\gamma \notin \quad \text{for all } \delta \in [\delta_t^-, \delta_t^+],$$

then $[\varkappa(\delta)]$ does not depend on δ and $\operatorname{Ind} T(g_t) = [\varkappa(\delta)]$.

Proof. Let $\gamma \in \mathbf{C}$ and suppose

$$\varkappa(\delta) = 1/p - \operatorname{Re}\gamma - \delta \operatorname{Im}\gamma \notin \mathbf{Z} \quad \forall \delta \in [\delta_t^-, \delta_t^+].$$

Then, obviously, $[\varkappa(\delta)]$ is independent of δ. Put $\varkappa := -[\varkappa(\delta)]$. We have the factorization

$$\tau^\gamma = (1 - t/\tau)^{\varkappa - \gamma} \tau^\varkappa (\tau - t)^{\gamma - \varkappa} \tag{70}$$

(recall Section 5). By the definition of $\varkappa$,

$$-\varkappa < 1/p - \operatorname{Re}\gamma - \delta \operatorname{Im}\gamma < -\varkappa + 1 \quad \forall \delta \in [\delta_t^-, \delta_t^+],$$

whence

$$-1/p < \varkappa - \operatorname{Re}\gamma - \delta \operatorname{Im}\gamma < 1/q \quad \forall \delta \in [\delta_t^-, \delta_t^+]$$

and thus,

$$-1/p < \operatorname{Re}(\varkappa - \gamma) + \delta \operatorname{Im}(\varkappa - \gamma) < 1/q \quad \forall \delta \in [\delta_t^-, \delta_t^+],$$

i. e. $\varkappa - \gamma \in N_t$ due to Theorem 7.2. Consequently, by the definition (24) of N_t,

$$|(\tau - t)^{\varkappa - \gamma}| \in A_p(\Gamma). \tag{71}$$

We have $|(z - t)^{\varkappa - \gamma}| = |z - t|^{\varkappa - \operatorname{Re}\gamma - \theta_t(z)\operatorname{Im}\gamma}$ with

$$\theta_t(z) = \arg(z - t)/(-\log|z - t|) \quad (z \in D_+),$$

and from (7) we infer that $\theta_t(z)$ is bounded for all z in some neighborhood of t. For these z, $|(z-t)^{\varkappa-\gamma}| \leq M/|z-t|^\mu$ with some $\mu > 0$. Since $|(\tau-t)^{\varkappa-\gamma}| \in A_p(\Gamma) \subset L^p(\Gamma)$, we deduce from Lemma 8.1 that $(\tau-t)^{\varkappa-\gamma} \in L_+^q(\Gamma)$. Analogously one can check that

$$(\tau-t)^{\gamma-\varkappa} \in L_+^q(\Gamma), \quad (1-t/\tau)^{\varkappa-\gamma} \in L_-^p(\Gamma), \quad (1-t/\tau)^{\gamma-\varkappa} \in L_-^q(\Gamma).$$

This together with (71) shows that (70) is a Wiener–Hopf factorization of $g_t(\tau) = \tau^\gamma$ in $L^p(\Gamma)$. Thus, by Theorem 1.2, $T(g_t)$ is Fredholm on $L^p(\Gamma)$ and therefore, $\gamma \in \Phi_t$. Moreover, $\operatorname{Ind} T(g_t) - \varkappa = [\varkappa(\delta)]$.

Conversely, suppose now $\gamma \in \Phi_t$, i.e. suppose $T(g_t)$ is Fredholm on $L^p(\Gamma)$. Denote the index of $T(g_t)$ by $-\varkappa$. Then, again by Theorem 1.2, there is a factorization

$$\tau^\gamma = a_-(\tau)\tau^\varkappa a_+(\tau)$$

subject to the conditions (i) and (ii). We also have

$$\tau^\gamma = (1-t/\tau)^{-k-\gamma}\tau^{-k}(\tau-t)^{\gamma+k}$$

for every positive integer k. Hence,

$$a_+^{-1}(\tau)(\tau-t)^{\gamma+k} = a_-(\tau)(1-t/\tau)^{\gamma+k}\tau^{\varkappa+k}. \tag{72}$$

Since $|(\tau-t)^{\gamma+k}| = |\tau-t|^{k+\operatorname{Re}\gamma+\theta_t(\tau)\operatorname{Im}\gamma}$ with $\theta_t(\tau) = \arg(\tau-t)/(-\log|\tau-t|) = O(1)$ (recall Seifullayev's estimate (7)), it follows that

$$(\tau-t)^{\gamma+k} \in L^\infty(\Gamma), \quad (1-t/\tau)^{\gamma+k} \in L^\infty(\Gamma)$$

whenever k is sufficiently large. For such k we have

$$a_+^{-1}(\tau)(\tau-t)^{\gamma+k} \in L_+^p(\Gamma) \subset L_+^1(\Gamma), \quad a_-(\tau)(1-t/\tau)^{\gamma+k} \in L_-^p(\Gamma) \subset L_-^1(\Gamma)$$

and therefore (72) implies that

$$h(\tau) := a_+^{-1}(\tau)(\tau-t)^{\gamma+k} = a_-(\tau)(1-t/\tau)^{\gamma+k}\tau^{\varkappa+k}$$

is a polynomial in τ of degree $s \leq \varkappa + k$. Since $a_+(z)$ is finite in D_+, we see that $h(z) \neq 0$ for $z \in D_+$, and since $a_-(z) \neq 0$ for $z \in D_-$, we conclude that $h(z) \neq 0$ for $z \in D_-$. Finally, as

$$(\tau-t)^{\gamma+k}/h(\tau) = a_+(\tau) \in L^1(\Gamma),$$

the polynomial $h(\tau)$ cannot possess zeros on $\Gamma \setminus \{t\}$. Consequently, $h(\tau) = (\tau-t)^s$ and

$$a_+(\tau) = (\tau-t)^{\gamma+k-s}. \tag{73}$$

Since, by assumption, $|a_+^{-1}(\tau)| \in A_p(\Gamma)$, we deduce from (73) and (24) that $s - k - \gamma \in N_t$, whence, by Theorem 7.2,

$$-1/p < s - k - \operatorname{Re}\gamma - \delta\operatorname{Im}\gamma < 1/q \qquad \forall \delta \in [\delta_t^-, \delta_t^+]. \tag{74}$$

Obviously, (74) may be rewritten as

$$k - s < 1/p - \operatorname{Re}\gamma - \delta \operatorname{Im}\gamma < k - s + 1 \qquad \forall \delta \in [\delta_t^-, \delta_t^+],$$

which implies that $1/p - \operatorname{Re}\gamma - \delta \operatorname{Im}\gamma \notin \mathbf{Z}$ for $\delta \in [\delta_t^-, \delta_t^+]$. ∎

9. Spectra and essential spectra of Toeplitz operators

Once Theorem 8.2 is established, the main result stated in Section 3 can be proved by standard methods.

Proposition 9.1. *Let* Γ *be a Carleson Jordan curve,* $1 < p < \infty$, $a \in PC(\Gamma)$. *The operator* $T(a)$ *is Fredholm in* $L_+^p(\Gamma)$ *if and only if* $a^{-1} \in L^\infty(\Gamma)$ *and for every* $t \in \Gamma$

$$\varkappa_t(\delta) := \frac{1}{p} - \frac{1}{2\pi}\left(\arg \frac{a(t-0)}{a(t+0)} - \delta \log \left|\frac{a(t-0)}{a(t+0)}\right|\right) \notin \mathbf{Z} \qquad \forall \delta \in [\delta_t^-, \delta_t^+]. \tag{75}$$

If $T(a)$ *is Fredholm and the set* Λ_a *of the points at which* a *has a jump is finite, then*

$$\operatorname{Ind} T(a) = -\frac{1}{2\pi}\sum_\gamma \{\arg a\}_\gamma + \sum_{t \in \Lambda_a}\left([\varkappa_t(\delta)] + \frac{1}{2\pi}\arg \frac{a(t-0)}{a(t+0)}\right), \tag{76}$$

where γ *ranges over the connected components of* $\Gamma \setminus \Lambda_a$, $\{\arg a\}_\gamma$ *denotes the increment of* $\arg a$ *along* γ, *and* $[\varkappa_t(\delta)]$ *is the integral part of* $\varkappa_t(\delta)$ *(which does actually not depend on* δ).

Proof. For $t \in \Gamma$, choose a number $\gamma = \gamma_t$ so that

$$a(t+0)/a(t-0) = e^{-2\pi i \gamma}$$

(notice that γ is uniquely determined up to a summand $2k\pi$ with $k \in \mathbf{Z}$). Then define $g_t \in PC(\Gamma)$ as in Section 5. We have

$$\operatorname{Re}\gamma + \delta \operatorname{Im}\gamma - \frac{1}{2\pi}\left(\arg \frac{a(t-0)}{a(t+0)} - \delta \log \left|\frac{a(t-0)}{a(t+0)}\right|\right) \in \mathbf{Z}$$

and therefore Theorem 8.2 implies that $T(g_t)$ is Fredholm on $L_+^p(\Gamma)$ if and only if (75) holds. The rest of the proof is as in [2]: using a local principle one show that $T(a)$ is Fredholm if and only if $T(g_t)$ is Fredholm for every $t \in \Gamma$, and invoking a principle of separation of the singularities, one can derive the index formula (76) from the equality $\operatorname{Ind} T(g_t) = [\varkappa_t(\delta)]$ (which is part of Theorem 8.2) and the index formula contained in Theorem 1.3. ∎

For $a \in PC(\Gamma)$, denote by $a_{\Gamma,p}$ the set on the right hand–side of (12) and let $a_{\Gamma,p}^\#$ stand for the closed continuous naturally oriented curve which results from the essential range $\mathcal{R}(a)$ by filling in the curve

$$\mathcal{S}\left(a(t-0), a(t+0); \frac{\delta_t^- + \delta_t^+}{2}, \frac{\delta_t^- + \delta_t^+}{2}; \frac{1}{p}, \frac{1}{p}\right)$$

between the endpoints of each jump (note that we could replace $(\delta_t^- + \delta_t^+)/2$ by any other number between δ_t^- and δ_t^+). The following theorem is the main result stated in Section 3 supplemented by an index formula.

Theorem 9.2 *Let Γ be a Carleson Jourdan curve, $1 < p < \infty$, and $a \in PC(\Gamma)$. The operator $T(a)$ is Fredholm on $L_+^p(\Gamma)$ if and only if $0 \notin a_{\Gamma,p}$. In that case $\operatorname{Ind} T(a) = -\operatorname{wind} a_{\Gamma,p}^{\#}$. In terms of spectra, we have*

$$\operatorname{sp}_{\mathrm{ess}} T(a) = a_{\Gamma,p},$$
$$\operatorname{sp} T(a) = a_{\Gamma,p} \cup \{\lambda \in \mathbb{C} \setminus a_{\Gamma,p} : \operatorname{wind}(a_{\Gamma,p}^{\#} - \lambda) \neq 0\}.$$

Proof. Taking into account that (75) is the same as saying that

$$0 \notin \mathcal{S}(a(t-0), a(t+0); \delta_t^-, \delta_t^+; 1/p, 1/p),$$

this theorem can be obtained from Proposition 9.1 exactly as in the proofs of Theorems 7.1 and 7.2 of [2]. ∎

10. The algebra of singular integral operators

We finally state a Fredholm criterion for operators in the smallest closed subalgebra of $\mathcal{L}(L^p(\Gamma))$ containing all operators of the form $aP + bQ$ with a and b in $PC(\Gamma)$. The latter algebra will be denoted by $\operatorname{alg}(PC, S)$, since it is clearly generated by the operators of multiplication by piecewise continuous functions and the Cauchy singular integral operator $S = 2P - I$.

For piecewise smooth curves Γ the following theorem goes back to [11], for so-called logarithmic Carleson curves Γ it was established in [2].

Theorem (Symbol Calculus for Singular Integral Operators). *Let Γ be a Carleson Jordan curve and let $1 < p < \infty$. Define the "bundle of skew spiralic horns" associated with $L^p(\Gamma)$ by*

$$\mathfrak{M} = \mathfrak{M}_{\Gamma,p} = \bigcup_{t \in \Gamma} \left(\{t\} \times \mathcal{S}(0, 1; \delta_t^-, \delta_t^+; 1/p, 1/p) \right).$$

Then

(a) the set $\mathcal{K}(L^p(\Gamma))$ of all compact operators on $L^p(\Gamma)$ is a subset of $\operatorname{alg}(PC, S)$ and the quotient algebra $\operatorname{alg}(PC, S)/\mathcal{K}(L^p(\Gamma))$ is inverse closed in $\mathcal{L}(L^p(\Gamma))/\mathcal{K}(L^p(\Gamma))$;

(b) for each point $(t, \mu) \in \mathfrak{M}$ the map

$$\sigma_{t,\mu} : \{aI : a \in PC(\Gamma)\} \cup \{S\} \to \mathbb{C}^{2 \times 2}$$

given by

$$\sigma_{t,\mu}(aI) = \begin{pmatrix} a(t-0)(1-\mu) + a(t+0)\mu & (a(t+0) - a(t-0))\sqrt{\mu(1-\mu)} \\ (a(t+0) - a(t-0))\sqrt{\mu(1-\mu)} & a(t-0)\mu + a(t+0)(1-\mu) \end{pmatrix}, \tag{77}$$

$$\sigma_{t,\mu}(S) = \begin{pmatrix} 1 & 0 \\ 0 & -1 \end{pmatrix} \tag{78}$$

extends to a Banach algebra homomorphism

$$\sigma_{t,\mu} : \operatorname{alg}(PC, S) \to \mathbf{C}^{2 \times 2}$$

with the property that $\sigma_{t,\mu}(K)$ is the zero matrix for every compact operator K;

(c) an operator $A \in \operatorname{alg}(PC, S)$ is Fredholm on $L^p(\Gamma)$ if and only if $\det \sigma_{t,\mu}(A) \neq 0$ for all $(t, \mu) \in \mathfrak{M}$.

Proof. Employing a local principle in conjunction with the "two projections theorem" of Finck, Roch, Silbermann [9] or an "extension theorem" a la Gohberg, Krupnik [13], this theorem can be derived from Theorem 9.2 by the method of [2]. ∎

We remark that in (77) we understand by $\sqrt{\mu(1-\mu)}$ any number whose square is $\mu(1-\mu)$. Note that on the sets of Figures 2 - 5 there is *no continuous* branch of $\sqrt{\mu(1-\mu)}$. Finally, in order to conform the above theorem to another version of the symbol calculus for singular integral operators, it should be noticed that the theorem remains literally true with (77) and (78) replaced by

$$\sigma_{t,\mu}(aI) = \begin{pmatrix} a(t+0) & 0 \\ 0 & a(t-0) \end{pmatrix} \quad \text{and} \quad \sigma_{t,\mu}(S) = \begin{pmatrix} 2\mu - 1 & 2\sqrt{\mu(1-\mu)} \\ 2\sqrt{\mu(1-\mu)} & 1 - 2\mu \end{pmatrix},$$

respectively (see [2]).

References

[1] C. Bennett and R. Sharpley: *Interpolation of Operators.* Academic Press, Boston 1988.

[2] A. Böttcher and Yu.I. Karlovich: *Toeplitz and singular integral operators on Carleson curves with logarithmic whirl points.* Integral Equations and Operator Theory **22** (1995), 127-161.

[3] D. Boyd: *Indices for the Orlicz spaces.* Pacific J. Math. **38** (1971), 315–323.

[4] A. Calderón: *Cauchy integrals on Lipschitz curves and related operators.* Proc. Nat. Acad. Sci. USA **74**, (1977), 1324-1327.

[5] G. David: *L'integrale de Cauchy sur le courbes rectifiables.* Prepublication Univ. Paris–Sud, Dept. Math. 82T05, 1982.

[6] G. David: *Opérateurs intégraux singuliers sur certaines courbes du plan complexe.* Ann. Sci. École Norm. Super. **17** (1984), 157–189.

[7] E.M. Dynkin: *Methods of the theory of singular integrals (Hilbert transform and Calderón-Zygmund theory).* Itogi Nauki Tekh., Ser. Sovr. Probl. Matem., Fundament. Napravl. **15** (1987), 197-292 [Russian].

[8] E.M. Dynkin and B.P. Osilenker: *Weighted norm estimates for singular integrals and their applications.* J. Sov. Math. **30** (1985), 2094–2154 [Russian original: Itogi Nauki Tekh., Ser. Mat. Anal. **21** (1983), 42–129].

[9] T. Finck, S. Roch, and B. Silbermann: *Two projections theorems and symbol calculus for operators with massive local spectra.* Math. Nachr. **162** (1993), 167–185.

[10] I. Gohberg: *On an application of the theory of normed rings to singular integral equations.* Uspekhi Matem. Nauk **7** (1952), 149–156 [Russian].

[11] I. Gohberg and N. Krupnik: *Singular integral operators with piecewise continuous coefficients and their symbols.* Izv. Akad. Nauk SSSR **35** (1971), 940–964 [Russian] (English transl. in Math. USSR Izv. **5** (1971), 955-979).

[12] I. Gohberg and N. Krupnik: *One-Dimensional Linear Singular Integral Equations, Vols. I and II.* Birkhäuser Verlag, Basel, Boston, Berlin 1992 [Russian original: Shtiintsa, Kishinev 1973].

[13] I. Gohberg and N. Krupnik: *Extension theorems for Fredholm and invertibility symbols.* Integral Equations and Operator Theory **16** (1993), 514–529.

[14] S.M. Grudsky: *Singular integral equations and the Riemann boundary value problem with infinite index in the space $L^p(\Gamma, w)$.* Izv. Akad. Nauk. SSSR **49** (1985), 55–80 [Russian].

[15] V.P. Havin: *Boundary properties of the Cauchy integral and of harmonic functions in domains with rectifiable boundary.* Matem. Sbornik **65** (1965), 499-517 [Russian].

[16] E. Hille and R.S. Phillips: *Functional Analysis and Semi-Groups.* Amer. Math. Soc. Coll. Publ., v. 31, revised edition, Providence, R.I, 1957.

[17] R. Hunt, B. Muckenhoupt, and R. Wheeden: *Weighted norm inequalities for the conjugate function and Hilbert transform.* Trans. Amer. Math. Soc. **176** (1973), 227–251.

[18] S.G. Krein, Yu.I. Petunin, and E.M. Semenov: *Interpolation of Linear Operators.* Transl. Math. Monogr. **54**, Amer. Math. Soc., Providence, R.I., 1982 [Russian original: Nauka, Moscow, 1978].

[19] V.A. Paatashvili and G.A. Khuskivadze: *On the boundedness of the Cauchy singular integral on Lebesgue spaces in the case of non-smooth contours.* Trudy Tbilisk. Mat. Inst. AN GSSR, 69 (1982), 93–107 [Russian].

[20] R.K. Seifullayev: *The Riemann boundary value problem on non-smooth open curves.* Matem. Sb. **112** (1980), 147–161 [Russian] (English transl. in Math. USSR Sb. **40** (1981)).

[21] I.B. Simonenko: *The Riemann boundary value problem with measurable coefficients.* Dokl. Akad. Nauk SSSR **135** (1960), 538–541 [Russian].

[22] I.B. Simonenko: *Some general questions of the theory of the Riemann boundary value problem*. Math. USSR Izv.**2** (1968), 1091–1099.

[23] I.B. Simonenko: *On the factorization and local factorization of measurable functions.t* Soviet Math. Dokl. **21** (1980), 271–274.

[24] I.B. Simonenko: *Stability of weight properties of functions with respect to the singular integral*. Matem. Zametki **33** (1983), 409–416 [Russian].

[25] I.M. Spitkovsky: *Singular integral operators with PC symbols on the spaces with general weights*. J. Funct. Anal. **105** (1992), 129–143.

[26] H. Widom: *Singular integral equations in L^p*. Trans. Amer. Math. Soc. **97** (1960), 131–160.

A. Böttcher
TU Chemnitz-Zwickau
Fakultät für Mathematik
09107 Chemnitz, Germany

Yu.I. Karlovich
Current address: Permanent address:
TU Chemnitz–Zwickau   Ukrainian Academy of Sciences
Fakultät für Mathematik   Marine Hydrophysical Institute
09107 Chemnitz, Germany   Hydroacoustic Department
                                           Preobrazhenskaya Street 3
                                           270100 Odessa, Ukraine

MSC 1991:   Primary 47B35
            Secondary 42A50, 45E05, 47A68

Operator Theory
Advances and Applications, Vol. 90
© 1996 Birkhäuser Verlag Basel/Switzerland

WEIGHTED UNIFORM CONVERGENCE OF THE QUADRATURE METHOD FOR CAUCHY SINGULAR INTEGRAL EQUATIONS

M. R. Capobianco, P. Junghanns, U. Luther, G. Mastroianni

Collocation and quadrature methods for Cauchy singular integral equations on an interval with variable coefficients are studied. Convergence rates are proved in weighted uniform and uniform norms.

1 INTRODUCTION

The present paper is mainly devoted to the investigation of the quadrature method for linear Cauchy singular integral equations over an interval (see equations (2.1) and (2.2)) in order to prove error estimates in weighted uniform norms. Recently there was dedicated more and more attention to the uniform convergence of polynomial approximation methods for such types of equations. For example, the papers [4] and [3] studied the collocation method. In [5] a perturbed collocation method is investigated in the case of constant coefficients of the singular integral operator, where perturbation means that the integrals occuring in the collocation equations are approximated with the help of a quadrature rule of order $2n + 1$, where $n - 1$ is the degree of the polynomial approximating the solution of the original equation. Here we study the "pure" quadrature method, where the order of the quadrature rule is equal to n.

At first, uniform convergence results for polynomial approximation methods for Cauchy singular integral equations were obtained by using the convergence rate, which could be proved in weighted L^2-norms (see, for example, [15, Corollary 9.33 and Theorem 9.34]). The paper [2] gives an approach to uniform error estimates by studying the regularized original and approximate equations in the Banach space of continuous functions and in discrete spaces with uniform norms, respectively, and using the concept of regular convergent operator sequences. In order to prove (quasi) optimal convergence rates in a wide range of cases it seems to be more natural to study the regularized equations in weighted spaces of continuous functions, done at the first time for the collocation method and the case of constant coefficients by one of the authors in [3]. The main aim of the present paper is to generalize the results of [3] to the quadrature method and to the case of variable coefficients.

The paper is organized as follows. In Section 2 we agree upon some notations and give some auxilliary material. Section 3 studies the quadrature method under a rather general assumption on the choise of the collocation points. In Section 4 special choises of collocation points are investigated, and in Section 5 we present some numerical results.

2   NOTATIONS AND PRELIMINARY RESULTS

We are interested in the numerical solution of linear Cauchy singular integral equations of the following types. Let $a, b : [-1, 1] \longrightarrow \mathbb{R}$, and $f, f_0 : [-1, 1] \longrightarrow \mathbb{C}$ as well as $h_0 : [-1, 1]^2 \longrightarrow \mathbb{C}$ be given continuous functions. We are looking for a function $v : [-1, 1] \longrightarrow \mathbb{C}$ satisfying

$$a(x)v(x) + \frac{1}{\pi} \int_{-1}^{1} \frac{b(t)v(t)}{t - x} \, dt + \frac{1}{\pi} \int_{-1}^{1} h_0(x, t)v(t) \, dt = f(x) \tag{2.1}$$

or

$$a(x)v(x) + \frac{b(x)}{\pi} \int_{-1}^{1} \frac{v(t)}{t - x} \, dt + \frac{1}{\pi} \int_{-1}^{1} h_0(x, t)v(t) \, dt = f_0(x) \tag{2.2}$$

for $x \in (-1, 1)$. We follow [7, 8] (see also [2] and [15, Chapter 9]) and assume that there exists a function

$$c(x) = \tilde{c}(x) \prod_{j=0}^{N+1} |x - c_j|^{-\gamma_j},$$

where $-1 = c_{N+1} < \cdots < c_0 = 1$; $\gamma_j > -1$, $j = 0, 1, \ldots, N + 1$; $N \in \mathbb{N} := \{0, 1, 2, \ldots\}$; $\tilde{c}(x) > 0$, $x \in [-1, 1]$, such that $B(x) := b(x)c(x)$ is a polynomial. Furthermore, in all what follows we suppose that

(A1) $a, b, \tilde{c} \in \mathbf{C}^{0,\eta}$, $0 < \eta < 1$;

(A2) $\gamma_0 \pm \alpha > -1$, $\gamma_{N+1} \pm \beta > -1$;

(A3) $B(\tilde{x}) = 0$ and $\tilde{x} \in [-1, 1]$ imply $b(\tilde{x}) = 0$;

(A4) $r(x) := \sqrt{[a(x)]^2 + [b(x)]^2} > 0$ for all $x \in [-1, 1]$.

Here we denote by $\mathbf{C}^{p,\gamma} = \mathbf{C}^{p,\gamma}[-1, 1]$, $p \in \mathbb{N}$, $0 < \gamma \le 1$, the Banach space of all complex-valued functions over $[-1, 1]$, which are p times continuously differentiable and whose p-th derivatives are Hölder continuous with exponent γ. A norm in $\mathbf{C}^{p,\gamma}$ is given by

$$\|u\|_{p,\gamma} = \sum_{k=0}^{p} \|u^{(k)}\|_\infty + C_\gamma \left(u^{(p)} \right),$$

where

$$C_\gamma(u) = \sup\{|x - t|^{-\gamma}|u(x) - u(t)| : x, t \in [-1, 1], x \neq t\},$$

and $\|u\|_\infty = \sup\{|u(x)| : x \in [-1, 1]\}$ denotes the norm in the Banach space $\mathbf{C} = \mathbf{C}[-1, 1]$ of all complex-valued and continuous functions over $[-1, 1]$. The real numbers α and β are defined by

$$\alpha = \lambda + g(1) \in (-1, 1) \text{ and } \beta = \lambda_- - g(-1) \in (-1, 1),$$

where λ and λ_- are integers and $g : [-1, 1] \longrightarrow \mathbb{R}$ is a continuous function satisfying

$$a(x) - i\,b(x) = r(x)e^{i\pi g(x)} \text{ for } x \in [-1, 1].$$

For $x \in (-1, 1)$ we define the weight functions

$$\sigma_0(x) = \frac{(1-x)^\lambda (1+x)^{\lambda_-}}{r(x)} \exp \int_{-1}^{1} \frac{g(t)}{t-x} dt, \quad \mu_0(x) = \frac{(1-x)^{-\lambda}(1+x)^{-\lambda_-}}{r(x)} \exp \int_{-1}^{1} \frac{g(t)}{x-t} dt,$$

and

$$\sigma(x) = \frac{\sigma_0(x)}{c(x)}, \quad \mu(x) = \frac{\mu_0(x)}{c(x)}.$$

Then, $\sigma(x)$ and $\mu(x)$ are generalized Jacobi weights, which can be written in the form (comp. [15, Prop. 9.7] and [2])

$$\sigma(x) = (1-x)^{\gamma_0 + \alpha}(1+x)^{\gamma_{N+1} + \beta} \prod_{j=1}^{N} |x - c_j|^{\gamma_j} \frac{w(x)}{r(x)\tilde{c}(x)},$$

$$\mu(x) = (1-x)^{\gamma_0 - \alpha}(1+x)^{\gamma_{N+1} - \beta} \prod_{j=1}^{N} |x - c_j|^{\gamma_j} \frac{1}{r(x)\tilde{c}(x)w(x)},$$

where $w \in C^{0,\eta}$ and $w(x) > 0$ for $x \in [-1, 1]$. Now, in case of Eq. (2.1) we shall seek the solution in the form $v = \sigma_0 u$ and define $h(x, t) = h_0(x, t)c(t)$, while in case of Eq. (2.2) we set $v = \sigma u$. Moreover, in case of Eq. (2.2) we define $h(x, t) = c(x)h_0(x, t)$ and $f(x) = c(x)f_0(x)$. Then, instead of (2.1) and (2.2) we investigate the equivalent equations

$$a(x)\sigma_0(x)u(x) + \frac{1}{\pi} \int_{-1}^{1} \frac{B(t)u(t)\sigma(t)}{t-x} dt + \frac{1}{\pi} \int_{-1}^{1} h(x, t)u(t)\sigma(t) dt = f(x) \tag{2.3}$$

and

$$a(x)\sigma_0(x)u(x) + \frac{B(x)}{\pi} \int_{-1}^{1} \frac{u(t)\sigma(t)}{t-x} dt + \frac{1}{\pi} \int_{-1}^{1} h(x, t)u(t)\sigma(t) dt = f(x), \tag{2.4}$$

respectively. We will write these equations in short form

$$A_0 u + K u = f \tag{2.5}$$

and

$$A u + K u = f, \tag{2.6}$$

where $A_0 = a\sigma_0 I + SB\sigma I = a\sigma_0 I + Sb\sigma_0 I$, $A = a\sigma_0 I + BS\sigma I$, and

$$(Sv)(x) = \frac{1}{\pi} \int_{-1}^{1} \frac{v(t)}{t-x} dt, \quad (Ku)(x) = \frac{1}{\pi} \int_{-1}^{1} h(x, t)u(t)\sigma(t) dt.$$

I denotes the identity operator. In the following we state some results concerning the mapping properties of the operators A_0 and A considered in the pair of Hilbert spaces L_σ^2 and L_μ^2, where $L_\omega^2 = L_\omega^2(-1, 1)$ denotes the Hilbert space of all quadratic integrable functions with respect to the weight ω, endowed with the scalar product

$$\langle u, v \rangle_\omega = \frac{1}{\pi} \int_{-1}^{1} u(t)\overline{v(t)}\omega(t) dt$$

and the norm $\|u\|_\omega = < u, u >_\omega^{1/2}$. By $\hat{A}_0$ and $\hat{A}$ we denote the operators

$$\hat{A}_0 = a\mu_0 I - SB\mu I = a\mu_0 I - Sb\mu_0 I \text{ and } \hat{A} = a\mu_0 I - BS\mu I,$$

respectively.

LEMMA 2.1 ([15], **Theorem 9.9, Remark 9.10**) *Let* $p(x) = x^n + \dots$ *be a monic polynomial of degree* n. *Then* $(A_0 p)(x) = (-1)^\lambda x^{n-\kappa} + \dots$ *is a polynomial of degree* $n - \kappa$, *where* $\kappa = -(\lambda + \lambda_-)$ *and* $(A_0 p)(x) \equiv 0$ *if* $n - \kappa < 0$. *Furthermore,* $(\hat{A}_0 p)(x) = (-1)^\lambda x^{n+\kappa} + \dots$, *where* $(\hat{A}_0 p)(x) \equiv 0$ *if* $n + \kappa < 0$. *Moreover,* $\hat{A}_0 A_0 p = p$ *if* $\kappa \leq 0$ *and* $A_0 \hat{A}_0 p = p$ *if* $\kappa \geq 0$.

PROPOSITION 2.2 ([15], **Theorem 9.17**) *The operator* $A : L^2_\sigma \longrightarrow L^2_\mu$ *is a Fredholm operator of index* κ. *Moreover,*

(i) $\dim \ker A = \max\{0, \kappa\}$ *and, in case* $\kappa > 0$, $\ker A = \mathrm{span}\,\{B(t), B(t)t, \dots, B(t)t^{\kappa-1}\}$,

(ii) $f \in \mathrm{im}\, A$ *if and only if* $\langle f, t^j \rangle_\mu = 0$ *for* $j = 0, 1, \dots, -\kappa - 1$,

(iii) $\hat{A}A = I$ *if* $\kappa \leq 0$ *and* $A\hat{A} = I$ *if* $\kappa \geq 0$,

(iv) $A^* = \hat{A}_0$ *and* $\hat{A}^* = A_0$.

COROLLARY 2.3 $A_0 : L^2_\sigma \longrightarrow L^2_\mu$ *is a Fredholm operator of index* κ *and*

(i) $\dim \ker A_0 = \max\{0, \kappa\}$ *and, in case of* $\kappa > 0$, $\ker A_0 = \mathrm{span}\,\{1, t, \dots, t^{\kappa-1}\}$,

(ii) $f \in \mathrm{im}\, A_0$ *if and only if* $\langle f, Bt^j \rangle_\mu = 0$ *for* $j = 0, 1, \dots, -\kappa - 1$,

(iii) $\hat{A}_0 A_0 = I$ *if* $\kappa \leq 0$ *and* $A_0 \hat{A}_0 = I$ *if* $\kappa \geq 0$,

(iv) $A_0^* = \hat{A}$ *and* $\hat{A}_0^* = A$.

PROOF. Since $A - A_0 = (BS - SB)\sigma I$ is a bounded operator of finite rank, A_0 is a Fredholm operator with the same index as A. Assertion (i) follows by the equivalence of $A_0 u = 0$ and $ABu = 0$ (in case $B \not\equiv 0$; the case $B \equiv 0$ is obvious). Since, in view of Proposition 2.2(ii), $\langle f, Bt^j \rangle_\mu = 0$ for $j = 0, 1, \dots, -\kappa - 1$ is equivalent to the existence of an $v \in L^2_\sigma$, such that $Av = Bf$, it follows, in case of $\kappa < 0$, $v = \hat{A}Bf = B\hat{A}_0 f$. This implies $u = \frac{1}{B}v \in L^2_\sigma$ and $A_0 u = f$. Assertion (iii) follows from Lemma 2.1 by continuity arguments. $\square$

Let $\alpha^+ = \max\{0, \alpha\}$, $\beta^+ = \max\{0, \beta\}$, and denote by $\mathbf{C}_{\alpha^+,\beta^+} = \mathbf{C}_{\alpha^+,\beta^+}[-1, 1]$ the Banach space of all continuous functions $u : (-1, 1) \longrightarrow \mathbb{C}$, for which $v^{\alpha^+,\beta^+} u \in \mathbf{C}$, where $v^{\gamma,\delta}(x) = (1-x)^\gamma (1+x)^\delta$. The norm in $\mathbf{C}_{\alpha^+,\beta^+}$ is defined by $\|u\|_{\infty,\alpha^+,\beta^+} = \|v^{\alpha^+,\beta^+} u\|_\infty$. Furthermore, let $\mathbf{C}^0_{\alpha^+,\beta^+}$ be the subspace of $\mathbf{C}_{\alpha^+,\beta^+}$ of functions $u \in \mathbf{C}_{\alpha^+,\beta^+}$ with $(v^{\alpha^+,\beta^+} u)(1) = 0$ if $\alpha^+ > 0$ and $(v^{\alpha^+,\beta^+} u)(-1) = 0$ if $\beta^+ > 0$. In the case of constant coefficients a and b there is proved in [3] (proof of Lemma 4.1) that $\hat{A} : C^{0,\gamma} \longrightarrow \mathbf{C}_{\alpha^+,\beta^+}$ is a linear bounded operator for all $0 < \gamma \leq 1$. In the following (see Prop. 2.10 (i)) we give a generalization of this result to the case of variable coefficients.

In all what follows, by c we denote a constant which can have different values at different places. Moreover, by $c \neq c(x, n, f, \dots)$ we will indicate, that c is independend from $x, n, f, \dots$

LEMMA 2.4 *For all* $\gamma > 0$ *and* $x \in (-1, 1)$ *we have*

$$\int_{-1}^1 |t - x|^{\gamma-1}(1-t)^{-\alpha}(1+t)^{-\beta}\, dt \leq c\,(1-x)^{-\alpha^+}(1+x)^{-\beta^+},$$

where $c \neq c(x)$.

PROOF. Write the integral as the sum of the integrals over the subintervals $[-1, \frac{x-1}{2}]$, $[\frac{x-1}{2}, \frac{x+1}{2}]$, and $[\frac{x+1}{2}, 1]$. $\square$

Let $\mathbb{P}_n$ denote the set of all algebraic polynomials of degree not greater than n. For a function $f : [-1,1] \longrightarrow \mathbb{C}$, we denote by $E_n(f)$ and $E_n^{\alpha^+,\beta^+}(f)$ the best uniform approximation and the best weighted uniform approximation of f by polynomials belonging to $\mathbb{P}_n$, respectively, that is

$$E_n(f) = \inf\{\|f - p\|_\infty : p \in \mathbb{P}_n\}, \quad E_n^{\alpha^+,\beta^+}(f) = \inf\{\|f - p\|_{\infty,\alpha^+,\beta^+} : p \in \mathbb{P}_n\}.$$

LEMMA 2.5 ([10], p.113, comp. also [16], IX.3.(9.31)) *There is a constant* $c \neq c(n,f)$ *such that, for each* $f \in C^{p,\gamma}$, *a sequence* $\{Q_n\}_{n=1}^\infty$ *of polynomials* $Q_n \in \mathbb{P}_n$ *exists which satisfy*

$$\|(f - Q_n)^{(i)}\|_\infty \leq c\frac{\|f\|_{p,\gamma}}{n^{p+\gamma-i}} \quad (n = 1,2,\ldots; \ i = 0,1,\ldots,p).$$

LEMMA 2.6 *There exists a constant* $c \neq c(f,n,t)$ *such that, for each continuously differentiable function* $f : [-1,1] \longrightarrow \mathbb{C}$, *the estimate*

$$v^{\alpha^+,\beta^+}(t) \int_{-1}^1 \left|\frac{f(x) - f(t)}{x - t}\right| \mu_0(x)dx \leq c\left(\|f\|_\infty \log n + \frac{\|f'\|_\infty}{n^2}\right)$$

is true for all $n = 2,3,\ldots$ *and* $t \in (-1,1)$.

PROOF. Writing the integral as the sum of the integrals over the subintervals $[-1, t - \frac{1+t}{2n^2}]$, $[t - \frac{1+t}{2n^2}, t + \frac{1-t}{2n^2}]$, $[t + \frac{1-t}{2n^2}, 1]$ and using the mean-value theorem, we deduce

$$\int_{-1}^1 \left|\frac{f(x) - f(t)}{x - t}\right| \mu_0(x)dx \leq c\|f\|_\infty \left(\int_{-1}^{t-\frac{1+t}{2n^2}} + \int_{t+\frac{1-t}{2n^2}}^1\right) \frac{v^{-\alpha^+,-\beta^+}(x)}{|x - t|}dx$$

$$+ c\|f'\|_\infty \int_{t-\frac{1+t}{2n^2}}^{t+\frac{1-t}{2n^2}} v^{-\alpha^+,-\beta^+}(x)dx.$$

With the substitution $1 + x = (1+t)(1 - \frac{1}{2n^2})y$ we get

$$\int_{-1}^{t-\frac{1+t}{2n^2}} \frac{v^{-\alpha^+,-\beta^+}(x)}{t - x}dx \leq (1-t)^{-\alpha^+} \int_{-1}^{t-\frac{1+t}{2n^2}} \frac{(1+x)^{-\beta^+}}{t - x}dx$$

$$= v^{-\alpha^+,-\beta^+}(t)\left(1 - \frac{1}{2n^2}\right)^{1-\beta^+} \int_0^1 \frac{y^{-\beta^+}}{1 - (1 - \frac{1}{2n^2})y}dy.$$

Taking into account, that $y^{-\beta^+} \leq y^{-\beta^+}(1 - (1 - \frac{1}{2n^2})y) + (1 - \frac{1}{2n^2})$, it follows

$$\int_{-1}^{t-\frac{1+t}{2n^2}} \frac{v^{-\alpha^+,-\beta^+}(x)}{t - x}dx \leq cv^{-\alpha^+,-\beta^+}(t) \int_0^1 \left(y^{-\beta^+} + \frac{1 - \frac{1}{2n^2}}{1 - (1 - \frac{1}{2n^2})y}\right)dy$$

$$\leq cv^{-\alpha^+,-\beta^+}(t) \log n. \tag{2.7}$$

Analogously we obtain

$$\int_{t+\frac{1-t}{2n^2}}^1 \frac{v^{-\alpha^+,-\beta^+}(x)}{x - t}dx \leq cv^{-\alpha^+,-\beta^+}(t) \log n. \tag{2.8}$$

Since

$$v^{-\alpha^+,-\beta^+}(x) \leq cv^{-\alpha^+,-\beta^+}(t) \quad \text{for} \quad x \in [t - \frac{1+t}{2n^2}, t + \frac{1-t}{2n^2}], \tag{2.9}$$

we have

$$\int_{t-\frac{1+t}{2n^2}}^{t+\frac{1-t}{2n^2}} v^{-\alpha^+,-\beta^+}(x)dx \leq c\frac{v^{-\alpha^+,-\beta^+}(t)}{n^2},$$

and the assertion of the lemma follows.   □

COROLLARY 2.7 *For each polynomial* $P \in \mathbb{P}_n$ *($n \geq 2$) we have*

$$\|\hat{A}_0 P\|_{\infty, \alpha+, \beta+} \, , \, \|\hat{A} P\|_{\infty, \alpha+, \beta+} \leq c\|P\|_\infty \log n,$$

where $c \neq c(n, P)$.

PROOF. From

$$(\hat{A}_0 P)(t) = P(t)(\hat{A}_0 1)(t) - \frac{1}{\pi} \int_{-1}^1 \frac{P(x) - P(t)}{x - t} b(x) \mu_0(x) dx,$$

Lemma 2.1 and Lemma 2.6 it follows

$$\|\hat{A}_0 P\|_{\infty, \alpha+, \beta+} \leq c \left(\|P\|_\infty \log n + \frac{\|P'\|_\infty}{n^2} \right).$$

Using Markov's inequality $\|P'\|_\infty \leq cn^2\|P\|_\infty$ we obtain $\|\hat{A}_0 P\|_{\infty, \alpha+, \beta+} \leq c\|P\|_\infty \log n$. Since $\hat{A} P = \hat{A}_0 P + (SB\mu I - BS\mu I)P$, we also get the assertion for $\hat{A}$.   □

LEMMA 2.8 *There exists a constant* $c \neq c(n, f)$ *such that, for each* $f \in C^{p,\gamma}$ *and the polynomial* $P_n = P_n(f) \in \mathbb{P}_n$ *of the best uniform approximation of* f, *the estimations*

$$\|\hat{A}_0(f - P_n)\|_{\infty, \alpha+, \beta+} \, , \, \|\hat{A}(f - P_n)\|_{\infty, \alpha+, \beta+} \leq c\frac{\log n}{n^{p+\gamma}}\|f\|_{p,\gamma}$$

and

$$E_n^{\alpha+, \beta+}(\hat{A}_0 f), \, E_n^{\alpha+, \beta+}(\hat{A} f) \leq c\frac{\log n}{n^{p+\gamma}}\|f\|_{p,\gamma}$$

are true for all $n \geq 2$.

PROOF. From

$$(\hat{A}_0(f - P_n))(t) = (f - P_n)(t)(\hat{A}_0 1)(t) - \frac{1}{\pi} \int_{-1}^1 \frac{(f - P_n)(x) - (f - P_n)(t)}{x - t} b(x) \mu_0(x) dx,$$

Lemma 2.1 and Lemma 2.5 we obtain

$$|(\hat{A}_0(f - P_n))(t)| \leq c\frac{\|f\|_{p,\gamma}}{n^{p+\gamma}} + c \int_{-1}^1 \left| \frac{(f - P_n)(x) - (f - P_n)(t)}{x - t} \right| \mu_0(x) dx.$$

In the case $p > 0$ we use Markov's inequality and get

$$\|(f - P_n)'\|_\infty \leq \|(f - Q_n)'\|_\infty + cn^2\|Q_n - P_n\|_\infty \leq cn^2\frac{\|f\|_{p,\gamma}}{n^{p+\gamma}},$$

where Q_n is the polynomial from Lemma 2.5. Now, the estimation $\|\hat{A}_0(f - P_n)\|_{\infty, \alpha+, \beta+} \leq c\frac{\log n}{n^{p+\gamma}}\|f\|_{p,\gamma}$ follows, for $p > 0$, from Lemma 2.6 and Lemma 2.5. In the case $p = 0$ we apply the inequalities (2.7) and (2.8) and deduce

$$\left(\int_{-1}^{t - \frac{1+t}{2n^2}} + \int_{t + \frac{1-t}{2n^2}}^1 \right) \left| \frac{(f - P_n)(x) - (f - P_n)(t)}{x - t} \right| \mu_0(x) dx \leq c\frac{\log n}{n^\gamma}\|f\|_{0,\gamma} v^{-\alpha+, -\beta+}(t),$$

where we took Lemma 2.5 into account. With the help of the estimation $\|P_n'\|_\infty \le cn^2\omega(f;1/n)$ (see [16], p.284) we get

$$|(f-P_n)(x)-(f-P_n)(t)| \le \|f\|_{0,\gamma}|x-t|^\gamma + \|P_n'\|_\infty|x-t| \le c(|x-t|^\gamma + n^{2-\gamma}|x-t|)\|f\|_{0,\gamma}.$$

This inequality together with (2.9) leads to

$$\int_{t-\frac{1+t}{2n^2}}^{t+\frac{1-t}{2n^2}} \left| \frac{(f-P_n)(x)-(f-P_n)(t)}{x-t} \right| \mu_0(x)dx \le c\frac{\|f\|_{0,\gamma}}{n^\gamma} v^{-\alpha+,-\beta+}(t).$$

Consequently $\|\hat{A}_0(f-P_n)\|_{\infty,\alpha+,\beta+} \le c\frac{\log n}{n^{p+\gamma}}\|f\|_{p,\gamma}$ holds true also in the case $p=0$. The corresponding assertion for $\hat{A}$ follows from $\hat{A}(f-P_n) = \hat{A}_0(f-P_n)+(SB\mu I - BS\mu I)(f-P_n)$ and Lemma 2.5. Since $\hat{A}_0P_n \in \mathbb{P}_{n+\kappa}$ (see Lemma 2.1) and $\hat{A}P_n = \hat{A}_0P_n + (SB\mu I - BS\mu I)P_n \in \mathbb{P}_{\max\{n+\kappa,\deg B-1\}}$ we can conclude $E_n^{\alpha+,\beta+}(\hat{A}_0f)$, $E_n^{\alpha+,\beta+}(\hat{A}f) \le c\frac{\log n}{n^{p+\gamma}}\|f\|_{p,\gamma}$. □

COROLLARY 2.9 *There is a constant* $c \ne c(n,f)$, *such that*

$$\|\hat{A}_0f\|_{\infty,\alpha+,\beta+}, \ \|\hat{A}f\|_{\infty,\alpha+,\beta+} \le c\left(\frac{\|f\|_{p,\gamma}}{n^{p+\gamma}} + \|f\|_\infty\right)\log n$$

for all $f \in C^{p,\gamma}$ *and* $n = 2,3,\ldots$.

PROOF. Since $\|\hat{A}_0f\|_{\infty,\alpha+,\beta+} \le \|\hat{A}_0(f-P_n)\|_{\infty,\alpha+,\beta+} + \|\hat{A}_0P_n\|_{\infty,\alpha+,\beta+}$ (analogously for $\hat{A}$), the assertion follows from Corollary 2.7 and Lemma 2.8. □

If X and Y are Banach spaces we denote by $\mathcal{L}(X,Y)$ the Banach space of all linear bounded operators from X into Y. In all what follows $h \in C_x^{p,\gamma}$ means that $h_t \in C^{p,\gamma}$ uniformly with respect to $t \in [-1,1]$, where $h_t(x) = h(x,t)$. Analogously we define $C_t^{p,\gamma}$.

PROPOSITION 2.10 *Let* $0 < \gamma \le 1$ *and* $h \in C_x^{p,\gamma}$. *Then*

(i) $\hat{A}_0, \hat{A} \in \mathcal{L}(C^{0,\gamma}, C_{\alpha+,\beta+}^0)$,

(ii) $K \in \mathcal{L}(C_{\alpha+,\beta+}, C^{p,\gamma})$,

(iii) $\hat{A}_0K, \hat{A}K \in \mathcal{K}(C_{\alpha+,\beta+}^0)$,

where $\mathcal{K}(X) = \mathcal{K}(X,X)$ *denotes the ideal of linear compact operators from* X *into* X.

PROOF. The first assertion is a consequence of Lemma 2.8 and Corollary 2.9. Assertion (ii) follows from the well known fact that $K \in \mathcal{L}(L_\sigma^2, C^{p,\gamma})$ and from the continuous embedding $C_{\alpha+,\beta+} \subset L_\sigma^2$. Since, for $0 < \delta < \gamma$, the embedding $C^{0,\gamma} \subset C^{0,\delta}$ is compact and since, in view of (i), $\hat{A}_0, \hat{A} \in \mathcal{L}(C^{0,\delta}, C_{\alpha+,\beta+}^0)$, assertion (iii) is obvious. □

3 THE QUADRATURE METHOD

Consider the Gaussian quadrature rule

$$\frac{1}{\pi}\int_{-1}^{1} u(t)\sigma(t)\,dt \approx Q_n^\sigma(u) := \sum_{k=1}^{n} \sigma_{nk} u(t_{nk})$$

and assume that $\{X_n\}_{n=1}^\infty$ is a sequence of partitions of the interval $[-1,1]$, $X_n = \{x_{n1},\dots,x_{nm_n}\}$ with $-1 \le x_{n1} < x_{n2} < \dots < x_{nm_n} \le 1$, satisfying

$$0 < c_1 \le \frac{m_n}{n} \le c_2 \quad (n = 1,2,\dots) \quad \text{and} \quad \lim_{n\to\infty}\frac{\log n}{n^{p+\gamma}}\|L^{X_n}\| = 0, \tag{3.1}$$

where $\|L^{X_n}\|$ denotes the Lebesgue constant of the respective Lagrange interpolation operator $L_n = L^{X_n}$, i. e. $(L_n f)(x_{nj}) = f(x_{nj})$ for $j = 1,\dots,m_n$ and $L_n f \in \mathbb{P}_{m_n-1}$ as well as

$$\|L_n\| = \sup\{\|L_n f\|_\infty : f \in C, \|f\|_\infty = 1\}.$$

Let us consider the case $\kappa = 0$. We look for $u_n \in \mathbb{P}_{m_n-1}$ satisfying

$$((A_0 + K_n)u_n)(x_{nj}) = f(x_{nj}), \quad j = 1,\dots,m_n, \tag{3.2}$$

or

$$((A + K_n)u_n)(x_{nj}) = f(x_{nj}), \quad j = 1,\dots,m_n, \tag{3.3}$$

where, for $u \in C_{\alpha+,\beta+}$,

$$(K_n u)(x) = \sum_{k=1}^{n} \sigma_{nk} h(x, t_{nk}) u(t_{nk}).$$

In view of Lemma 2.1 and $\Lambda = \Lambda_0 + (BS - SB)\sigma I$ we have $A_0 u_n$, $A u_n \in \mathbb{P}_{m_n-1}$ for $n \ge n_0$. Thus, for $n \ge n_0$, Eq.s (3.2) and (3.3) are equivalent to

$$(I + \hat{A}_0 L_n K_n)u_n = \hat{A}_0 L_n f \tag{3.4}$$

and

$$(I + \hat{A} L_n K_n)u_n = \hat{A} L_n f, \tag{3.5}$$

respectively, where we took Proposition 2.2 (iii) and Corollary 2.3 (iii) into account. Analogously the original equations (2.5) and (2.6) are equivalent to

$$(I + \hat{A}_0 K)u = \hat{A}_0 f \tag{3.6}$$

and

$$(I + \hat{A} K)u = \hat{A} f, \tag{3.7}$$

respectively. We remark that a solution $u_n \in C_{\alpha+,\beta+}$ of (3.4) or (3.5) belongs to $\mathbb{P}_{m_n-1}$ for all sufficiently large n. If we assume that $f \in C^{p,\gamma}$ and $h \in C_x^{p,\gamma}$, we have, in view of Proposition 2.10 (i) and $K \in \mathcal{L}(L_\sigma^2, C^{0,\gamma})$, that each solution $u \in L_\sigma^2$ of (3.6) or (3.7) as well as the right hand sides $\hat{A}_0 f$ or $\hat{A} f$ belong to $C_{\alpha+,\beta+}^0$. Consequently, it is possible to consider the operators on the left hand sides of (3.6) and (3.7) as operators in $C_{\alpha+,\beta+}^0$. In all what follows A can stand for A_0 or A.

PROPOSITION 3.1 (comp. also [3], Lemma 4.2) *If $f \in C^{p,\gamma}$ then, for $n \geq 2$,*

$$\|\hat{\mathcal{A}}(f - L_n f)\|_{\infty,\alpha+,\beta+} \leq c\|f\|_{p,\gamma}\frac{\log n}{n^{p+\gamma}}\|L_n\|,$$

where $c \neq c(n, f)$.

PROOF. Let $P_n \in \mathbb{P}_{m_n-1}$ be the polynomial of the best uniform approximation of f. Then, in view of Lemma 2.8 and Corollary 2.7,

$$\|\hat{\mathcal{A}}(f - L_n f)\|_{\infty,\alpha+,\beta+} \leq \|\hat{\mathcal{A}}(f - P_n)\|_{\infty,\alpha+,\beta+} + \|\hat{\mathcal{A}}L_n(P_n - f)\|_{\infty,\alpha+,\beta+} \leq c\|f\|_{p,\gamma}\frac{\log n}{n^{p+\gamma}}\|L_n\|,$$

where we took Lemma 2.5 into account.   □

LEMMA 3.2 ([9], § III.1, Lemma 1.5) *Assume $G \in C^\infty(-1,1)$ and $G^{(2r)}(t) \geq 0$ for all $t \in (-1,1)$ and all $r \in \mathbb{N}$. Then*

$$\sum_{k=1}^{n} \sigma_{nk} G(t_{nk}) \leq \frac{1}{\pi} \int_{-1}^{1} G(t)\sigma(t)\, dt.$$

COROLLARY 3.3 *There exists a constant c such that, for all $n \in \mathbb{N}$,*

$$\sum_{k=1}^{n}(1 - t_{nk})^{-\alpha+}(1 + t_{nk})^{-\beta+} \sigma_{nk} \leq c. \tag{3.8}$$

Furthermore, for each $u \in \mathbf{C}_{\alpha+,\beta+}$,

$$\left|Q_n^\sigma(u) - \frac{1}{\pi}\int_{-1}^{1} u(t)\sigma(t)\, dt\right| \leq c\, E_{2n-1}^{\alpha+,\beta+}(u).$$

PROOF. The functions $G(t) = (1 - t)^{-\alpha+}$ and $G(t) = (1 + t)^{-\beta+}$ fulfil the assumptions of Lemma 3.2. Consequently,

$$\sum_{k=1}^{n}(1 - t_{nk})^{-\alpha+}(1 + t_{nk})^{-\beta+}\sigma_{nk} \leq \sum_{t_{nk}\leq 0}(1 + t_{nk})^{-\beta+}\sigma_{nk} + \sum_{t_{nk}>0}(1 - t_{nk})^{-\alpha+}\sigma_{nk} \leq c.$$

Now, let $u \in \mathbf{C}_{\alpha+,\beta+}$ and $P \in \mathbb{P}_{2n-1}$. Then

$$\left|Q_n^\sigma(u) - \frac{1}{\pi}\int_{-1}^{1} u(t)\sigma(t)\, dt\right| \leq |Q_n^\sigma(u - P)| + \left|\frac{1}{\pi}\int_{-1}^{1}[P(t) - u(t)]\sigma(t)\, dt\right|$$

$$\leq \|u - P\|_{\infty,\alpha+,\beta+}\left[\sum_{k=1}^{n}(1 - t_{nk})^{-\alpha+}(1 + t_{nk})^{-\beta+}\sigma_{nk} + \frac{1}{\pi}\int_{-1}^{1}(1 - t)^{-\alpha+}(1 + t)^{-\beta+}\sigma(t)\, dt\right],$$

and the corollary is proved.   □

Let X be a Banach space and $A_n, A \in \mathcal{L}(X) = \mathcal{L}(X, X)$. We say that A_n converges compactly to A (if n tends to infinity) if $A_n \longrightarrow A$ strongly and if $u_n \in X$, $\|u_n\| \leq const$ implies that $\{A_n u_n\}_{n=1}^{\infty}$ is compact.

PROPOSITION 3.4 *Suppose that $h \in C_x^{p,\gamma} \cap C_t^{q,\delta}$. Then*

(i) the operators $K_n \in \mathcal{L}(\mathbf{C}_{\alpha+,\beta+}, C^{p,\gamma})$ are uniformly bounded,

(ii) $\|\hat{\mathcal{A}}(L_n K_n - K_n)\|_{C^0_{\alpha+,\beta+} \to C^0_{\alpha+,\beta+}} \leq c \frac{\log n}{n^{p+\gamma}} \|L_n\|$, $n \geq 2$,

(iii) for all $n \geq 2$ and $u \in \mathbf{C}_{\alpha+,\beta+}$,

$$\|\hat{\mathcal{A}} L_n K_n u - \hat{\mathcal{A}} K u\|_{\infty,\alpha+,\beta+} \leq c \left[\frac{\log n}{n^{p+\gamma}} \|L_n\| + \frac{\log n}{n^{q+\delta}} \right] \|u\|_{\infty,\alpha+,\beta+} + c\, E_n^{\alpha+,\beta+}(u),$$

where $c \neq c(n, u)$,

(iv) $\hat{\mathcal{A}} L_n K_n \longrightarrow \hat{\mathcal{A}} K$ compactly in $\mathbf{C}^0_{\alpha+,\beta+}$.

PROOF. The first assertion follows from

$$(K_n u)^{(r)}(x) = \sum_{k=1}^{n} \sigma_{nk} \frac{\partial^r}{\partial x^r} h(x, t_{nk}) u(t_{nk})$$

and (3.8). Taking Proposition 3.1 into account we conclude, for $u \in \mathbf{C}_{\alpha+,\beta+}$,

$$\|\hat{\mathcal{A}}(L_n K_n - K_n)u\|_{\infty,\alpha+,\beta+} \leq c\|K_n u\|_{p,\gamma} \frac{\log n}{n^{p+\gamma}} \|L_n\| \leq c\|u\|_{\infty,\alpha+,\beta+} \frac{\log n}{n^{p+\gamma}} \|L_n\|,$$

which gives (ii). Using (ii), for $u \in \mathbf{C}_{\alpha+,\beta+}$, we may estimate

$$\begin{aligned}
\|\hat{\mathcal{A}} L_n K_n u - \hat{\mathcal{A}} K u\|_{\infty,\alpha+,\beta+} &\leq \|\hat{\mathcal{A}}(L_n K_n - K_n)u\|_{\infty,\alpha+,\beta+} + \|\hat{\mathcal{A}}(K_n - K)u\|_{\infty,\alpha+,\beta+} \\
&\leq c\frac{\log n}{n^{p+\gamma}} \|L_n\|\|u\|_{\infty,\alpha+,\beta+} + \|\hat{\mathcal{A}}(K_n - K)u\|_{\infty,\alpha+,\beta+}.
\end{aligned} \tag{3.9}$$

For $x \in [-1, 1]$, let $P_{n-1}^{h,x} \in \mathbb{P}_{n-1}$ with $\|h_x - P_{n-1}^{h,x}\|_\infty = E_{n-1}(h_x)$. Moreover, define $P_n \in \mathbb{P}_n$ by $\|P_n - u\|_{\infty,\alpha+,\beta+} = E_n^{\alpha+,\beta+}(u)$. Then

$$(K_n - K)u = K_n(u - P_n) + (K_n - K)P_n - K(u - P_n)$$

and, in view of (i) and Proposition 2.10(ii),

$$\|K_n(u - P_n)\|_{0,\gamma}, \|K(u - P_n)\|_{0,\gamma} \leq c\|u - P_n\|_{\infty,\alpha+,\beta+} = c\, E_n^{\alpha+,\beta+}(u). \tag{3.10}$$

Furthermore, by Corollary 3.3 and Lemma 2.5, we have

$$\begin{aligned}
|((K_n - K)P_n)(x)| &\leq c\, E_{2n-1}^{\alpha+,\beta+}(h_x P_n) \\
&\leq c\|h_x P_n - P_{n-1}^{h,x} P_n\|_{\infty,\alpha+,\beta+} \leq c\|P_n\|_{\infty,\alpha+,\beta+} E_{n-1}(h_x) \\
&\leq \frac{c}{n^{q+\delta}} \|u\|_{\infty,\alpha+,\beta+}.
\end{aligned}$$

Since, with respect to Proposition 2.10(ii) and (i),

$$\|(K_n - K)P_n\|_{p,\gamma} \leq c\|P_n\|_{\infty,\alpha+,\beta+} \leq c\|u\|_{\infty,\alpha+,\beta+},$$

we can conclude, $\|\hat{\mathcal{A}}(K_n - K)P_n\|_{\infty,\alpha+,\beta+} \leq c\left(\frac{\log n}{n^{p+\gamma}} + \frac{\log n}{n^{q+\delta}}\right) \|u\|_{\infty,\alpha+,\beta+}$ with the help of Corollary 2.9. Consequently, by Proposition 2.10(i) and (3.10),

$$\begin{aligned}
\|\hat{\mathcal{A}}(K_n - K)u\|_{\infty,\alpha+,\beta+} &\leq \|\hat{\mathcal{A}}[K_n(u - P_n) - K(u - P_n)]\|_{\infty,\alpha+,\beta+} \\
&\qquad + \|\hat{\mathcal{A}}(K_n - K)P_n\|_{\infty,\alpha+,\beta+} \\
&\leq c\, E_n^{\alpha+,\beta+}(u) + c\left(\frac{\log n}{n^{p+\gamma}} + \frac{\log n}{n^{q+\delta}}\right) \|u\|_{\infty,\alpha+,\beta+}.
\end{aligned}$$

Together with (3.9) this leads to (iii). It remains to prove (iv). First of all, (iii) implies $\hat{A}L_n K_n \longrightarrow \hat{A}K$ strongly in $\mathbf{C}^0_{\alpha+,\beta+}$ since the set of polynomials is dense in $\mathbf{C}^0_{\alpha+,\beta+}$. (Remark that this do not hold true for the space $\mathbf{C}_{\alpha+,\beta+}$!) Moreover, suppose $u_n \in \mathbf{C}^0_{\alpha+,\beta+}$ and $\|u_n\|_{\infty,\alpha+,\beta+} \leq c$ for all $n = 1,2,\ldots$ From (i) we have $\|K_n u_n\|_{0,\gamma} \leq c$, which shows that $\{K_n u_n\}_{n=1}^\infty$ is compact in $C^{0,\tilde{\gamma}}$ for $0 < \tilde{\gamma} < \gamma$. Consequently, for each infinite subset $\mathbb{N}' \subset \mathbb{N}$ there exist an infinite subset $\mathbb{N}'' \subset \mathbb{N}'$ and an element $v \in \mathbf{C}^{0,\tilde{\gamma}}$ such that

$$\|K_n u_n - v\|_{0,\tilde{\gamma}} \longrightarrow 0 \text{ for } n \longrightarrow \infty, \, n \in \mathbb{N}''. \tag{3.11}$$

Taking into account (ii) and Proposition 2.10 (i) we obtain

$$\|\hat{A}L_n K_n u_n - \hat{A}v\|_{\infty,\alpha+,\beta+} \leq \|\hat{A}(L_n K_n - K_n)u_n\|_{\infty,\alpha+,\beta+} + \|\hat{A}(K_n u_n - v)\|_{\infty,\alpha+,\beta+}$$

$$\leq c\frac{\log n}{n^{p+\gamma}}\|L_n\|\,\|u_n\|_{\infty,\alpha+,\beta+} + c\|K_n u_n - v\|_{0,\tilde{\gamma}},$$

which, in view of (3.1) and (3.11), gives $\hat{A}L_n K_n u_n \longrightarrow u = \hat{A}v$ $(n \longrightarrow \infty, \, n \in \mathbb{N}'')$ in $\mathbf{C}^0_{\alpha+,\beta+}$. □

LEMMA 3.5 (comp. [2], Theorem 3.4 or [17],§3, Prop. (3)) *If $T_n \in \mathcal{K}(X)$ converges compactly to $T \in \mathcal{L}(X)$ and if $\dim \ker(I+T) = 0$ then, for all sufficiently large n, the inverse operators $(I + T_n)^{-1} \in \mathcal{L}(X)$ exist and $\sup\{\|(I+T_n)^{-1}\|_{X \to X} : n \geq n_0\} < \infty$.*

PROOF. If we assume that there exist a infinite subset $\mathbb{N}' \subset \mathbb{N}$ and a sequence $\{u_n\}_{n \in \mathbb{N}'}$ such that $\|u_n\| = 1$ and $\lim_{n \to \infty, n \in \mathbb{N}'} \|(I+T_n)u_n\| = 0$, then $\{T_n u_n\}_{n \in \mathbb{N}'}$ is compact. Consequently, there exist $\mathbb{N}'' \subset \mathbb{N}'$ and $u \in X$ with $T_n u_n \longrightarrow -u$ $(n \in \mathbb{N}'')$. It follows, for $n \longrightarrow \infty$ and $n \in \mathbb{N}''$,

$$u_n = (I + T_n)u_n - T_n u_n \longrightarrow u$$

and

$$T_n u_n = T_n(u_n - u) + (T_n u - Tu) + Tu \longrightarrow Tu,$$

which implies $(I+T)u = 0$. This contradicts $\dim \ker(I + T) = 0$. Thus, there is a constant $c_0 > 0$ with $\|(I + T_n)u\| \geq c_0\|u\|$ for all $u \in X$ and $n \geq n_0$. This implies $\dim \ker(I + T_n) = 0$ and, since the operators T_n are assumed to be compact, the existence of $(I + T_n)^{-1}$, where

$$\|(I + T_n)^{-1}\| \leq \frac{1}{c_0}, \quad n \geq n_0.$$

□

THEOREM 3.6 *Assume $\kappa = 0$, $f \in C^{p,\gamma}$, $h \in C^{p,\gamma}_x \cap C^{q,\delta}_t$, and that the original equations (2.3) or (2.4) possess only the trivial solution if $f \equiv 0$. Then, for all sufficiently large n, the respective equations (3.2) or (3.3) are uniquely solvable and the solutions u_n^* converge in the norm of $\mathbf{C}_{\alpha+,\beta+}$ to the unique solution u^* of (2.3) or (2.4), where*

$$\|u_n^* - u^*\|_{\infty,\alpha+,\beta+} \leq c\left[\frac{\log n}{n^{p+\gamma}}\|L_n\| + \frac{\log n}{n^{q+\delta}}\right]\|f\|_{p,\gamma}, \tag{3.12}$$

with a constant $c \neq c(n,f)$.

PROOF. Taking into account Proposition 2.10 (iii), dim im $K_n < \infty$, Proposition 3.4(iv), and Lemma 3.5 it remains to prove the estimate (3.12). For this end write

$$
\begin{aligned}
u^* - u_n^* &= (I + \hat{A}L_n K_n)^{-1}[(I + \hat{A}L_n K_n)u^* - \hat{A}L_n f] \\
&= (I + \hat{A}L_n K_n)^{-1}[\hat{A}(L_n K_n - K)u^* + \hat{A}(f - L_n f)].
\end{aligned}
$$

In view of Lemma 3.5, Proposition 3.4 (iii), and Proposition 3.1 it follows

$$
\|u_n^* - u^*\|_{\infty,\alpha+,\beta+} \leq c \left[\frac{\log n}{n^{p+\gamma}} \|L_n\| + \frac{\log n}{n^{q+\delta}} \right] \|u^*\|_{\infty,\alpha+,\beta+} + c\, E_n^{\alpha+,\beta+}(u^*)
$$

$$
+ c\|f\|_{p,\gamma} \frac{\log n}{n^{p+\gamma}} \|L_n\|,
$$

which implies together with Proposition 2.10 and $u^* = (I + \hat{A}K)^{-1}\hat{A}f$ the estimate

$$
\|u_n^* - u^*\|_{\infty,\alpha+,\beta+} \leq c \left[\frac{\log n}{n^{p+\gamma}} \|L_n\| + \frac{\log n}{n^{q+\delta}} \right] \|f\|_{p,\gamma} + c E_n^{\alpha+,\beta+}(u^*).
$$

As a consequence of $u^* = \hat{A}(f - Ku^*)$, $f - Ku^* \in C^{p,\gamma}$ (in view of Proposition 2.10 (ii)), Lemma 2.8, and $\|f - Ku^*\|_{p,\gamma} \leq c\|f\|_{p,\gamma}$, we have

$$
E_n^{\alpha+,\beta+}(u^*) \leq c\frac{\log n}{n^{p+\gamma}}\|f\|_{p,\gamma},
$$

which leads to (3.12).   □

REMARK 3.7 *If, instead of the quadrature method (3.2) or (3.3), we consider the collocation method*

$$
((\hat{A} + K)u_n)(x_{nj}) = f(x_{nj}), \quad j = 1, \dots, m_n,   \tag{3.13}
$$

which can also be written in the form

$$
(I + \hat{A}L_n K)u_n = \hat{A}L_n f,   \tag{3.14}
$$

then the following result holds true. Assume that $\kappa = 0$, $f \in C^{p,\gamma}$, $h \in C_x^{p,\gamma}$, and that the original equations (2.3) or (2.4) possess only the trivial solution if $f \equiv 0$. Then, for all sufficiently large n, the respective equation (3.14) is uniquely solvable and the solutions u_n^* converge in the norm of $C_{\alpha+,\beta+}$ to the unique solution u^* of (2.3) or (2.4), where

$$
\|u_n^* - u^*\|_{\infty,\alpha+,\beta+} \leq c\|f\|_{p,\gamma} \frac{\log n}{n^{p+\gamma}} \|L_n\|.   \tag{3.15}
$$

PROOF. From Proposition 2.10(ii) and Proposition 3.1 it follows

$$
\lim_{n\to\infty} \|\hat{A}L_n K - \hat{A}K\|_{C_{\alpha+,\beta+} \to C_{\alpha+,\beta+}} = 0,
$$

which implies, that, for all sufficiently large n, the invers operators $(I + \hat{A}L_n K)^{-1} \in \mathcal{L}(C_{\alpha+,\beta+})$ exist and that $\sup\{\|(I + \hat{A}L_n K)^{-1}\|_{C_{\alpha+,\beta+} \to C_{\alpha+,\beta+}} : n \geq n_0\} \leq c$. Consequently,

$$
\|u_n^* - u^*\|_{\infty,\alpha+,\beta+} \leq c(\|\hat{A}(L_n f - f)\|_{\infty,\alpha+,\beta+} + \|\hat{A}(L_n Ku^* - Ku^*)\|_{\infty,\alpha+,\beta+}),
$$

which leads to (3.15) taking into account Proposition 3.1.   □

REMARK 3.8 *If* $u_n \in \mathbb{P}_{mn}$, $u \in C_{\alpha+,\beta+}$, *and*

$$\|u_n - u\|_{\infty,\alpha+,\beta+} \leq c \frac{\log^k n}{n^\rho}$$

for some $k = 0, 1, 2, \ldots$ *and some* $\rho > 2 \max\{\alpha+, \beta+\}$, *then* $u \in C$ *and*

$$\|u_n - u\|_\infty \leq c \frac{\log^k n}{n^{\rho - 2 \max\{\alpha+,\beta+\}}} .$$

PROOF. See the proof of [3, Theorem 3.2]. $\qquad\qquad\qquad\qquad\qquad\qquad\qquad$ $\square$

COROLLARY 3.9 *If* $\|L_n\| \leq c \log n$ *and* $p+\gamma$, $q+\delta > 2 \max\{\alpha+, \beta+\}$, *then we obtain from Remark 3.8, Theorem 3.6, and Remark 3.7 the error estimate*

$$\|u_n^* - u^*\|_\infty \leq c \frac{\log^2 n}{n^{\min\{p+\gamma,q+\delta\} - 2 \max\{\alpha+,\beta+\}}} \|f\|_{p,\gamma} \tag{3.16}$$

for the quadrature method and

$$\|u_n^* - u^*\|_\infty \leq c \frac{\log^2 n}{n^{p+\gamma - 2 \max\{\alpha+,\beta+\}}} \|f\|_{p,\gamma} \tag{3.17}$$

for the collocation method under the assumptions of Theorem 3.6 and Remark 3.7, respectively.

REMARK 3.10 *If* $\kappa > 0$ *then we consider the equations (2.3) and (2.4) together with the additional conditions*

$$\int_{-1}^{1} u(t) B(t) t^i \sigma(t) \, dt = \int_{-1}^{1} u(t) b(t) t^i \sigma_0(t) \, dt = 0 , \quad i = 0, 1, \ldots, \kappa - 1 , \tag{3.18}$$

and

$$\int_{-1}^{1} u(t) t^i \sigma(t) \, dt = 0 , \quad i = 0, 1, \ldots, \kappa - 1 , \tag{3.19}$$

respectively. Denote

$$\mathbf{L}_{\sigma,B,\kappa}^2 = \{u \in \mathbf{L}_\sigma^2 : u \text{ satisfies } (3.18)\}$$

and

$$\mathbf{L}_{\sigma,\kappa}^2 = \{u \in \mathbf{L}_\sigma^2 : u \text{ satisfies } (3.19)\} .$$

Then the operators $A_0 \in \mathcal{L}(\mathbf{L}_{\sigma,B,\kappa}^2, \mathbf{L}_\mu^2)$ *and* $A \in \mathcal{L}(\mathbf{L}_{\sigma,\kappa}^2, \mathbf{L}_\mu^2)$ *are invertible, where* $A_0^{-1} = \hat{A}_0 \in \mathcal{L}(\mathbf{L}_\mu^2, \mathbf{L}_{\sigma,B,\kappa}^2)$ *and* $A^{-1} = \hat{A} \in \mathcal{L}(\mathbf{L}_\mu^2, \mathbf{L}_{\sigma,\kappa}^2)$.

PROOF. (See also [2, Lemma 4.4].) For each $v \in \mathbf{L}_\sigma^2$ we have, in view of Proposition 2.2(iii), $A(\hat{A}\hat{A}v - v) = 0$, i.e. $\hat{A}\hat{A}v - v \in \ker A$. Let $u \in \mathbf{L}_{\sigma,B,\kappa}^2 = \{v \in \mathbf{L}_\sigma^2 :< v, f >_\sigma = 0 \text{ for all } f \in \ker A\}$. Then, for all $v \in \mathbf{L}_\sigma^2$, $< \hat{A}_0 A_0 u, v >_\sigma =< u, \hat{A}\hat{A}v >_\sigma =< u, v >_\sigma$, where we took into consideration Corollary 2.3(iv). Consequently, $\hat{A}_0 A_0 u = u$ for all $u \in \mathbf{L}_{\sigma,B,\kappa}^2$. Remark that $g \in \mathbf{L}_\mu^2$ and $f \in \ker A$ imply $< \hat{A}_0 g, f >_\sigma =< g, Af >_\mu = 0$, i.e. $\hat{A}_0(\mathbf{L}_\mu^2) \subset \mathbf{L}_{\sigma,B,\kappa}^2$. Together with Corollary 2.3 this completes the proof for the operator A_0 . For the operator A the proof is analogous. $\qquad$ $\square$

COROLLARY 3.11 *We see, that if $u \in L^2_{\sigma,B,\kappa}$ is a solution of (2.5) then u is a solution of (3.6). Conversely, if $u \in L^2_\sigma$ is a solution of (3.6) then $u = \hat{A}_0(f - Ku) \in L^2_{\sigma,B,\kappa}$. Consequently, in case $\kappa > 0$ equation (2.3) together with (3.18) is equivalent to (3.6). If, for $n \geq n_0$, we assume that $m_n \leq 2(n - \kappa) - \deg B + 1$, then the same holds true for*

$$((A_0 + K_n)u_n)(x_{nj}) = f(x_{nj}), \quad j = 1, \ldots, m_n, \quad u_n \in \mathbb{P}_{m_n+\kappa-1},$$

together with the additional conditions

$$\sum_{k=1}^n \sigma_{nk} u_n(t_{nk}) B(t_{nk}) t_{nk}^i = 0, \quad i = 0, 1, \ldots, \kappa - 1,$$

and

$$(I + \hat{A}_0 L_n K_n)u_n = \hat{A}_0 L_n f,$$

and all sufficiently large n. Thus, one can see that for the quadrature method or the collocation method the assertions of Theorem 3.6 and Remark 3.7 remain valid in case $\kappa > 0$. The same is true for equations (2.4) and (3.19) together with the respective quadrature or collocation method.

REMARK 3.12 *For the numerical evaluation of $A_0 u_n$ and $A u_n$, $u_n \in \mathbb{P}_{m_n-1}$, in the approximate equations one can use [15, Theorem 9.14]. If $p_m^\rho(x)$ denotes the orthonormal polynomial of degree m with positive leading coefficient with respect to the weight function $\rho(x)$ then, for $m \geq \kappa + \deg B$,*

$$A p_m^\sigma = (-1)^\lambda p_{m-\kappa}^\mu. \tag{3.20}$$

This implies, for $m \geq \max\{\kappa + \deg B, \deg B\}$,

$$A_0 p_m^\sigma = (-1)^\lambda p_{m-\kappa}^\mu. \tag{3.21}$$

For $u \in \mathbb{P}_{2n}$ and $x \in (-1,1)$, $x \neq t_{nk}$ $(k = 1, \ldots, n)$, we have

$$\begin{aligned}
(Au)(x) &= u(x)(A1)(x) + \frac{B(x)}{\pi} \int_{-1}^1 \frac{u(t) - u(x)}{t - x} \sigma(t)\, dt \\
&= u(x)(A1)(x) + B(x) \sum_{k=1}^n \sigma_{nk} \frac{u(t_{nk}) - u(x)}{t_{nk} - x}.
\end{aligned}$$

In particular, for $u = p_n^\sigma$, $n \geq \kappa + \deg B$, we obtain from (3.20)

$$(-1)^\lambda p_{n-\kappa}^\mu(x) = p_n^\sigma(x) \left[(A1)(x) - B(x) \sum_{k=1}^n \frac{\sigma_{nk}}{t_{nk} - x} \right],$$

which leads to

$$(Au)(x) = (-1)^\lambda u(x) \frac{p_{n-\kappa}^\mu(x)}{p_n^\sigma(x)} + B(x) \sum_{k=1}^n \sigma_{nk} \frac{u(t_{nk})}{t_{nk} - x}. \tag{3.22}$$

Analogously, for $u \in \mathbb{P}_{2n-\deg B}$, $n \geq \max\{\kappa + \deg B, \deg B\}$, and $x \in (-1,1)$, $x \neq t_{nk}$ $(k = 1, \ldots, n)$, from (3.21) follows

$$(A_0 u)(x) = (-1)^\lambda u(x) \frac{p_{n-\kappa}^\mu(x)}{p_n^\sigma(x)} + \sum_{k=1}^n \sigma_{nk} \frac{B(t_{nk}) u(t_{nk})}{t_{nk} - x}. \tag{3.23}$$

4   SPECIAL INTERPOLATION PROCESSES

If $a = a(m)$ and $b = b(m)$ are two expressions depending on $m \in \mathbb{N}$ then we write $a \sim b$ uniformly in $m \in \mathbb{N}$ if and only if there are two positive constants c_1 and c_2 such that $c_1|b(m)| \leq |a(m)| \leq c_2|b(m)|$ for all $m \in \mathbb{N}$. Let $x_k = x_{mk}^{\alpha,\beta}$, $k = 1,\ldots,m$, be the zeros of the m-th orthonormal polynomial $p_m = p_m^{\alpha,\beta}$ with positive leading coefficient with respect to the Jacobi weight $v^{\alpha,\beta}(x) = (1 - x)^\alpha(1 + x)^\beta$, where $\alpha, \beta > -1$. Given two nonnegative integers r and s, we introduce the additional points $y_j = y_{mj}$, $j = 1,\ldots,s$, and $z_i = z_{mi}$, $i = 1,\ldots,r$, such that

$$-1 \leq y_1 < \ldots < y_s < x_1 < \ldots < x_m < z_1 \ldots < z_r \leq 1, \tag{4.1}$$

$$x_1 - y_s \sim m^{-2}, \quad z_1 - x_m \sim m^{-2}, \tag{4.2}$$

$$y_j - y_{j-1} \sim m^{-2}, j = 2,\ldots,s, \quad z_i - z_{i-1} \sim m^{-2}, i = 2,\ldots,r, \tag{4.3}$$

uniformly in $m \in \mathbb{N}$. For a given continuous function f we denote by $L_{m,r,s}(v^{\alpha,\beta}; f)$ the Lagrange interpolation polynomial of degree less than $m + r + s$ interpolating f at the points $y_1,\ldots,y_s,x_1,\ldots,x_m,z_1\ldots,z_r$. Our aim is to get estimations for the interpolation error $f - L_{m,r,s}(v^{\alpha,\beta}; f)$ in the weighted uniform norm $\|\cdot\|_{\infty,\varrho,\sigma}$, where ϱ and σ are nonnegative constants. Additionally to the conditions (4.1)-(4.3) we provide

in case $\varrho > 0 : 1 - z_r \sim m^{-2}$ uniformly in $m \in \mathbb{N}$, \hfill (4.4)

in case $\sigma > 0 : 1 + y_1 \sim m^{-2}$ uniformly in $m \in \mathbb{N}$. \hfill (4.5)

Remembering $1 - x_m \sim m^{-2}$ and $1 + x_1 \sim m^{-2}$ (which follows from [14, Theorem 9.22]) we see, that the conditions (4.1)-(4.5) can easily be fulfilled. Defining the polynomials

$$A_0(x) = 1, \quad A_s(x) = \prod_{j=1}^{s}(x - y_j), \ s > 0,$$

$$B_0(x) = 1, \quad B_r(x) = \prod_{i=1}^{r}(x - z_i), \ r > 0,$$

we can write

$$L_{m,r,s}(v^{\alpha,\beta}; f; x) = A_s(x)B_r(x)L_m\left(v^{\alpha,\beta}; \frac{f}{A_sB_r}; x\right) + \\ + A_s(x)p_m(x)L_r\left(\frac{f}{A_sp_m}; x\right) + B_r(x)p_m(x)L_s\left(\frac{f}{B_rp_m}; x\right), \tag{4.6}$$

where $L_m(v^{\alpha,\beta}; g) \in \mathbb{P}_{m-1}$ interpolates g at the points $x_1,\ldots,x_m$, $L_r(g) \in \mathbb{P}_{r-1}$ interpolates g at the points $z_1,\ldots,z_r$ and $L_s(g) \in \mathbb{P}_{s-1}$ interpolates g at the points $y_1,\ldots,y_s$. The following theorem is a generalization of [12, Remark 3.3].

THEOREM 4.1 *Let the conditions (4.1)-(4.3), in case $\varrho > 0$ condition (4.4), and in case $\sigma > 0$ condition (4.5) be satisfied and let $f \in C_{\rho,\sigma}$. If*

$$\frac{\alpha}{2} + \frac{1}{4} - \varrho \leq r \leq \frac{\alpha}{2} + \frac{5}{4} - \varrho, \quad \frac{\beta}{2} + \frac{1}{4} - \sigma \leq s \leq \frac{\beta}{2} + \frac{5}{4} - \sigma, \tag{4.7}$$

then

$$\|f - L_{m,r,s}(v^{\alpha,\beta}; f)\|_{\infty,\varrho,\sigma} \leq c\, E_{m+r+s-1}^{\varrho,\sigma}(f)\log m$$

where $c \neq c(m, f)$.

PROOF. For every polynomial $Q \in \mathbb{P}_{m+r+s-1}$ we have

$$f - L_{m,r,s}(v^{\alpha,\beta}; f) = f - Q - L_{m,r,s}(v^{\alpha,\beta}; f - Q).$$

Thus we obtain

$$\|f - L_{m,r,s}(v^{\alpha,\beta}; f)\|_{\infty,\varrho,\sigma} \leq \|f - Q\|_{\infty,\varrho,\sigma} + \|L_{m,r,s}(v^{\alpha,\beta}; f - Q)\|_{\infty,\varrho,\sigma}. \qquad (4.8)$$

Since

$$\|L_{m,r,s}(v^{\alpha,\beta}; f - Q)\|_{\infty,\varrho,\sigma} \leq c \sup_{|x| \leq 1-m^{-2}} |L_{m,r,s}(v^{\alpha,\beta}; f - Q; x)v^{\varrho,\sigma}(x)|$$

(see [6, Theorem 8.4.8]) we only have to estimate $L_{m,r,s}(v^{\alpha,\beta}; f - Q; x)$ for $|x| \leq 1-m^{-2}$. In view of (4.6) we get

$$|L_{m,r,s}(v^{\alpha,\beta}; f - Q; x)| \leq \left| A_s(x) B_r(x) L_m \left(v^{\alpha,\beta}; \frac{f-Q}{A_s B_r}; x \right) \right|$$

$$+ \left| A_s(x) p_m(x) L_r \left(\frac{f-Q}{A_s p_m}; x \right) \right| + \left| B_r(x) p_m(x) L_s \left(\frac{f-Q}{B_r p_m}; x \right) \right|. \qquad (4.9)$$

Using the estimations $|p_m(x)| \leq c(1-x)^{-\alpha/2-1/4}(1+x)^{-\beta/2-1/4}$, $|x| \leq 1-m^{-2}$ (cf. [1, Theorem 1.1]), and $|A_s(x)| \leq c(1+x)^s$, $|x| \leq 1-m^{-2}$ (since $1+x_1 \sim m^{-2}$) we obtain, for $|x| \leq 1-m^{-2}$,

$$\left| A_s(x) p_m(x) L_r \left(\frac{f-Q}{A_s p_m}; x \right) \right| \leq c(1-x)^{-\alpha/2-1/4}(1+x)^{s-\beta/2-1/4} \left| L_r \left(\frac{f-Q}{A_s p_m}; x \right) \right|.$$

We have

$$\left| L_r \left(\frac{f-Q}{A_s p_m}; x \right) \right| = \left| \sum_{k=1}^r \frac{f(z_k) - Q(z_k)}{A_s(z_k) p_m(z_k)} \prod_{i=1, i \neq k}^r \frac{x - z_i}{z_k - z_i} \right|$$

$$\leq \|f - Q\|_{\infty,\varrho,\sigma} \sum_{k=1}^r \left| \frac{1}{A_s(z_k) p_m(z_k) v^{\varrho,\sigma}(z_k)} \prod_{i=1, i \neq k}^r \frac{x - z_i}{z_k - z_i} \right|.$$

By $1 + x_1 \sim m^{-2}$ and $1 - x_m \sim m^{-2}$ we get $A_s(z_k) \geq c$. Since $z_k \in [x_m + cm^{-2}, 1]$ (cf. (4.2)) the estimation $|p_m(z_k)| \geq cm^{\alpha+1/2}$ is valid (cf. [14, Theorem 9.33 and Theorem 6.3.28] and use $1 - x_m \sim m^{-2}$). In view of conditions (4.3) and (4.4) we have $|v^{\varrho,\sigma}(z_k)| \geq cm^{-2\varrho}$ and $|z_k - z_i| \geq cm^{-2}$. Furthermore we get $|x - z_i| \leq c(1 - x)$, $|x| \leq 1 - m^{-2}$. Consequently, we obtain, for $|x| \leq 1 - m^{-2}$,

$$\left| A_s(x) p_m(x) L_r \left(\frac{f-Q}{A_s p_m}; x \right) \right| \leq c \frac{(1-x)^{r-\alpha/2-5/4}(1+x)^{s-\beta/2-1/4}}{m^{\alpha+1/2-2\varrho-2r+2}} \|f - Q\|_{\infty,\varrho,\sigma}.$$

From (4.7) we get $\alpha + 1/2 - 2\varrho - 2r + 2 = 2(\alpha/2 + 1/4 - \varrho - r + 1) \geq 0$. Thus $m^{-(\alpha+1/2-2\varrho-2r+2)} \leq (1-x)^{\alpha/2+1/4-\varrho-r+1}$ since $m^{-2} \leq 1 - x$. Together with $s + \sigma - \beta/2 - 1/4 \geq 0$ (cf. (4.7)) we obtain

$$\left| A_s(x) p_m(x) L_r \left(\frac{f-Q}{A_s p_m}; x \right) \right| \leq c \frac{\|f - Q\|_{\infty,\varrho,\sigma}}{v^{\varrho,\sigma}(x)}, \quad |x| \leq 1 - m^{-2}.$$

In a similar way we deduce

$$\left| B_r(x) p_m(x) L_s \left(\frac{f-Q}{B_r p_m}; x \right) \right| \leq c \frac{\|f - Q\|_{\infty,\varrho,\sigma}}{v^{\varrho,\sigma}(x)}, \quad |x| \leq 1 - m^{-2}.$$

Now we consider the first term on the right hand side of (4.9). Denoting by x_d the zero of p_m closest to $x \in [-1+m^{-2}, 1-m^{-2}]$, we can write

$$\left| A_s(x) B_r(x) L_m \left(v^{\alpha,\beta}; \frac{f-Q}{A_s B_r}; x \right) \right| \le |\Gamma_1(x)| + |\Gamma_2(x)|$$

where

$$\Gamma_1(x) = A_s(x) B_r(x) l_{md}(x) \frac{f(x_d) - Q(x_d)}{A_s(x_d) B_r(x_d)} ,$$

$$\Gamma_2(x) = A_s(x) B_r(x) \sum_{k=1, k \ne d}^{m} \frac{f(x_k) - Q(x_k)}{A_s(x_k) B_r(x_k)} l_{mk}(x)$$

with

$$l_{mk}(x) = \prod_{j=1, j \ne k}^{m} \frac{x - x_j}{x_k - x_j} .$$

In view of $1 + x_1 \sim m^{-2}$ and $1 - x_m \sim m^{-2}$ we deduce

$$|A_s(x) B_r(x)| \le c (1-x)^r (1+x)^s , \quad |x| \le 1 - \frac{c}{m^2}$$

and, since, for example,

$$x_k - y_j = x_1 - y_j + x_k - x_1 \overset{(4.2)}{\ge} \frac{c}{m^2} + x_k - x_1 \ge c(1 + x_k),$$

we also have

$$|A_s(x_k) B_r(x_k)| \ge c (1-x_k)^r (1+x_k)^s .$$

Since $|f(x_k) - Q(x_k)| \le \|f - Q\|_{\infty,\varrho,\sigma} / v^{\varrho,\sigma}(x_k)$ it follows, for $|x| \le 1 - m^{-2}$,

$$|\Gamma_1(x)| \le c \left(\frac{1-x}{1-x_d} \right)^{r+\varrho} \left(\frac{1+x}{1+x_d} \right)^{s+\sigma} |l_{md}(x)| \frac{\|f - Q\|_{\infty,\varrho,\sigma}}{v^{\varrho,\sigma}(x)} ,$$

and

$$|\Gamma_2(x)| \le c(1-x)^r (1+x)^s \sum_{k=1, k \ne d}^{m} \frac{|l_{m,k}(x)|}{(1-x_k)^{r+\varrho}(1+x_k)^{s+\sigma}} \|f - Q\|_{\infty,\varrho,\sigma} .$$

We have $1 - x \le 1 - x_d + |x_d - x|$, $1 + x \le 1 + x_d + |x - x_d|$, and

$$
\begin{aligned}
|x_d - x| &= |\cos\theta_d - \cos\theta| = 2 \sin \frac{\theta + \theta_d}{2} \sin \frac{|\theta - \theta_d|}{2} \\
&\le |\theta - \theta_d| \sin \frac{\theta + \theta_d}{2} = |\theta - \theta_d| \left(\sin\theta_d \cos \frac{\theta - \theta_d}{2} + \cos\theta_d \sin \frac{\theta - \theta_d}{2} \right) \\
&\le |\theta - \theta_d| \left(\sin\theta_d + \frac{|\theta - \theta_d|}{2} \right) = |\theta - \theta_d| \left(\sqrt{1 - x_d^2} + \frac{|\theta - \theta_d|}{2} \right) ,
\end{aligned}
$$

where $x_d = \cos\theta_d$ and $x = \cos\theta$ with $\theta_d, \theta \in [0, \pi]$. Since $|\theta - \theta_d| \le cm^{-1}$ (cf. [14, Theorem 9.22]) $m^{-1} \le c\sqrt{1 - x_d}$, and $m^{-1} \le c\sqrt{1 + x_d}$ we get

$$|x - x_d| \le c(1 - x_d), \quad |x - x_d| \le c(1 + x_d) .$$

Consequently, $1 - x \le c(1 - x_d)$, $1 + x \le c(1 + x_d)$. Furthermore we have $|l_{md}(x)| \sim 1$ (cf. [14, Proof of Theorem 9.33]). Thus, we obtain

$$|\Gamma_1(x)| \le c \frac{\|f - Q\|_{\infty,\varrho,\sigma}}{v^{\varrho,\sigma}(x)}, \quad |x| \le 1 - m^{-2}.$$

Using the estimations (cf. [12, p. 45] and [1, Theorem 1.1])

$$|l_{mk}(x)| \le c|p_m(x)|\frac{(1 - x_k)^{\alpha/2+3/4}(1 + x_k)^{\beta/2+3/4}}{m|x - x_k|},$$

$$|p_m(x)| \le c(1 - x)^{-\alpha/2-1/4}(1 + x)^{-\beta/2-1/4}, \quad |x| \le 1 - m^{-2}$$

we get, for $|x| \le 1 - m^{-2}$,

$$|\Gamma_2(x)| \le c\, v^{r-\alpha/2-1/4,s-\beta/2-1/4}(x) \sum_{k=1,k\neq d}^{m} \frac{v^{\alpha/2+3/4-r-\varrho,\beta/2+3/4-s-\sigma}(x_k)}{m|x - x_k|}\|f - Q\|_{\infty,\varrho,\sigma}.$$

Using (4.7) we have $-1/2 \le \alpha/2 + 3/4 - r - \varrho, \beta/2 + 3/4 - s - \sigma \le 1/2$. Thus we can apply [12, Lemma 4.1] to estimate the sum and get, for $|x| \le 1 - m^{-2}$,

$$\sum_{k=1,k\neq d}^{m} \frac{v^{\alpha/2+3/4-r-\varrho,\beta/2+3/4-s-\sigma}(x_k)}{m|x - x_k|} \le c(1 - x)^{\alpha/2+1/4-r-\varrho}(1 + x)^{\beta/2+1/4-s-\sigma} \log m.$$

Consequently,

$$\left| A_s(x) B_r(x) L_m \left(v^{\alpha,\beta}; \frac{f - Q}{A_s B_r}; x \right) \right| \le c \frac{\|f - Q\|_{\infty,\varrho,\sigma}}{v^{\varrho,\sigma}(x)} \log m, \quad |x| \le 1 - m^{-2}. \tag{4.10}$$

From (4.8)-(4.10) we deduce, for all $Q \in \mathbb{P}_{m+r+s-1}$,

$$\|f - L_{m,r,s}(v^{\alpha,\beta}; f)\|_{\infty,\varrho,\sigma} \le c\|f - Q\|_{\infty,\varrho,\sigma} \log m.$$

Taking the infimum on Q the assertion of the theorem is completely proved.   □

The next corollary follows immediately from Theorem 4.1.

COROLLARY 4.2 (comp. [12], Remark 3.3) *Let the conditions (4.1)-(4.3) be satisfied. If*

$$\frac{\alpha}{2} + \frac{1}{4} \le r \le \frac{\alpha}{2} + \frac{5}{4}, \quad \frac{\beta}{2} + \frac{1}{4} \le s \le \frac{\beta}{2} + \frac{5}{4},$$

then $\|L_{m,r,s}(v^{\alpha,\beta})\| \le c \log m$, *where* $\|L_{m,r,s}(v^{\alpha,\beta})\| := \sup\{\|L_{m,r,s}(v^{\alpha,\beta}; f)\|_{\infty} : f \in \mathbf{C}, \|f\|_{\infty} = 1\}$. *Thus, the interpolation operators* $L^{X_n} = L_{n,r,s}(v^{\alpha,\beta})$ *satisfy (3.1).*

Let us consider the case of constant coefficients $a(x) = a \in \mathbb{R}$ and $b(x) = b \in \mathbb{R} \setminus 0$ with $a^2 + b^2 = 1$. Then $\sigma(x) = (1 - x)^\alpha(1 + x)^\beta$ and $\mu(x) = (1 - x)^{-\alpha}(1 + x)^{-\beta}$ are classical Jacobi weights. Assume $\kappa \ge 0$, i.e. $\kappa = 0$ or $\kappa = 1$. Let r and s be nonnegative integers with

$$-\frac{\alpha}{2} + \frac{1}{4} \le r \le -\frac{\alpha}{2} + \frac{5}{4}, \quad -\frac{\beta}{2} + \frac{1}{4} \le s \le -\frac{\beta}{2} + \frac{5}{4}, \tag{4.11}$$

and choose $L_n = L_{n-\kappa,r,s}(v^{-\alpha,-\beta})$, and consider the quadrature method (3.3) (with $m_n = n - \kappa + r + s$). From Theorem 3.6 and Corollary 4.2 we obtain

$$\|u_n^* - u^*\|_{\infty,\alpha+,\beta+} \leq \left[c\frac{\log^2 n}{n^{p+\gamma}} + \frac{\log n}{n^{q+\delta}}\right]\|f\|_{p,\gamma}, \tag{4.12}$$

if $f = f_0 \in C^{p,\gamma}$ and $h = h_0 \in C_x^{p,\gamma} \cap C_t^{q,\delta}$.

To compare the results obtained here with some previous results of [2] for the "classical" quadrature method we consider the case $p = q$, $\gamma = \delta$. Then, from (3.16) it follows, with $\mu^* = \max\{\alpha+,\beta+\}$,

$$\|u_n^* - u^*\|_\infty \leq c(f)n^{-p-\gamma+2\mu^*}\log^2 n \leq c(\varepsilon,f)\,n^{-p-\gamma+2\mu^*+\varepsilon}, \tag{4.13}$$

while in [2, Theorem 6.4] there was obtained the estimate

$$\|u_n^* - u^*\|_\infty \leq c(\varepsilon,f)n^{-p-\gamma+2\max\{a^*,\mu^*\}+\varepsilon}, \tag{4.14}$$

where

$$a^* = \min\left\{\mu^* + \sigma^{**}, \mu^* + \mu^{**}, \frac{1}{4} + \sigma^{**}\right\},$$

$$\sigma^{**} = \frac{1}{2}\max\left\{0, \frac{1}{2} + \alpha, \frac{1}{2} + \beta\right\}, \quad \mu^{**} = \frac{1}{2}\max\left\{0, \frac{1}{2} - \alpha, \frac{1}{2} - \beta\right\}.$$

In case $\kappa = 1$ (i.e. $\alpha+ = \beta+ = 0$, $\alpha + \beta = -1$) (4.14) becomes

$$\|u_n^* - u^*\|_\infty \leq c(\varepsilon,f)\,n^{-p-\gamma+\max\{0,\frac{1}{2}+\alpha,\frac{1}{2}+\beta\}+\varepsilon}, \tag{4.15}$$

and the estimates (4.13) (with ε) and (4.15) coincide only for $\alpha = \beta = -\frac{1}{2}$ (i.e. $a = 0$, $b = 1$).

Since $|\alpha| < 1$ and $|\beta| < 1$ the nonnegative integers r and s in (4.11) can only be 0 or 1. In the case of constant coefficients a and b we know that $x_{n-\kappa,j} = x_{n-\kappa,j}^{-\alpha,-\beta} \neq t_{nk}$ for all $k = 1,\ldots,n$ and all $j = 1,\ldots,n - \kappa$ (see [15]). Thus, applying (3.22) we obtain, for $u_n \in \mathbb{P}_{n+r+s-1}$,

$$(Au_n)(x_{n-\kappa,j}) = b\sum_{k=1}^n \sigma_{nk}\frac{u_n(t_{nk})}{t_{nk} - x_{n-\kappa,j}}, \quad j = 1,\ldots,n-\kappa.$$

The polynomial u_n is defined by its values $u_n(t_{nk})$, $k = 1,\ldots,n$, $u_n(1)$ if $r = 1$, and $u_n(-1)$ if $s = 1$. Thus, if we consider the quadrature method (3.5) for $L_n = L_{n-\kappa,r,s}(v^{-\alpha,-\beta})$, then we see that the values $u_n(t_{nk})$, $k = 1,\ldots,n$, are the solutions of the "classical" (without additional collocation points) quadrature method

$$\sum_{k=1}^n \sigma_{nk}\left[\frac{b}{t_{nk} - x_{n-\kappa,j}} + h(x_{n-\kappa,j}, t_{nk})\right]u_n(t_{nk}) = f(x_{n-\kappa,j}), \quad j = 1,\ldots,n-\kappa,$$

and, if $\kappa = 1$,

$$\sum_{k=1}^n \sigma_{nk}u_n(t_{nk}) = 0.$$

If we choose $y_1 = -1$ (if $s = 1$) and $z_1 = 1$ (if $r = 1$), then the remaining values $u_n(-1)$ (if $s = 1$) and $u_n(1)$ (if $r = 1$) are determined by

$$(-1)^\lambda \frac{p_{n-\kappa}^\mu(\pm 1)}{p_n^\sigma(\pm 1)}u_n(\pm 1) + \sum_{k=1}^n \sigma_{nk}\left[\frac{b}{t_{nk} - (\pm 1)} + h(\pm 1, t_{nk})\right]u_n(t_{nk}) = f(\pm 1).$$

Concerning the zero distribution of p_n^σ and $p_{n-\kappa}^\mu$ in case of variable coefficients and formulas for $(Au_n)(x_{n-\kappa,j})$ if $x_{n-\kappa,j}$ coincides with one of the nodes of the quadrature rule, we refer the reader to [15].

In the remaining part of this section we show how it is possible to obtain analogous results to Proposition 3.1 under the condition that f' belongs to a suitable weighted space of continuous functions. Let $-1 < \alpha, \beta < 1$, α^+ and β^+ be defined as in Section 2, and $\alpha^- = \max\{0, -\alpha\}$, $\beta^- = \max\{0, -\beta\}$.

LEMMA 4.3 Let $0 < c_1 \le c_2$, $-1 + \frac{c_1}{n^2} \le x_n \le -1 + \frac{c_2}{n^2}$, and $\beta \ne 0$, $f_n \in C_{\alpha^-,\beta^-}$, $f'_n \in C_{\frac{1}{2}+\alpha^-,\frac{1}{2}+\beta^-}$ with $f_n(x_n) = 0$. Then, for $-1 < t \le 0$, the function

$$\Phi_n(t) = v^{\alpha^+,\beta^+}(t) \int_{-1}^{1} v^{-\alpha,-\beta}(x)\frac{f_n(x)}{x-t}\,dx$$

satisfies the estimate

$$|\Phi_n(t)| \le c\left(\|f_n\|_{\infty,\alpha^-,\beta^-}\log n + \frac{1}{n}\|f'_n\|_{\infty,\frac{1}{2}+\alpha^-,\frac{1}{2}+\beta^-}\right),\qquad(4.16)$$

where $c \ne c(n, f_n, t)$.

PROOF. We write

$$\Phi_n(t) = v^{\alpha^+,\beta^+}(t)\left[\int_{-1}^{2t+1} + \int_{2t+1}^{1}\right]v^{-\alpha,-\beta}(x)\frac{f_n(x)}{x-t}\,dx =: G(t) + H(t).$$

First of all we estimate $|H(t)|$ for $-1 < t \le -1 + \frac{c_1}{2n^2}$ and write

$$H(t) = v^{\alpha^+,\beta^+}(t)\left[\int_{2t+1}^{x_n} + \int_{x_n}^{1}\right]v^{-\alpha,-\beta}(x)\frac{f_n(x)}{x-t}\,dx =: H_1(t) + H_2(t)$$

taking into consideration $2t+1 \le x_n$. It follows

$$|H_1(t)| \le c(1+t)^{\beta^+}\int_{2t+1}^{x_n}\frac{(1+x)^{-\beta}}{x-t}\int_{x}^{x_n}|f'_n(\tau)|\,d\tau\,dx$$

$$\le c\|f'_n\|_{\infty,\frac{1}{2}+\alpha^-,\frac{1}{2}+\beta^-}(1+t)^{\beta^+}\int_{2t+1}^{x_n}\frac{(1+x)^{-\beta}}{x-t}\int_{x}^{x_n}\frac{d\tau}{(1+\tau)^{\frac{1}{2}+\beta^-}}$$

$$\le c\|f'_n\|_{\infty,\frac{1}{2}+\alpha^-,\frac{1}{2}+\beta^-}(1+t)^{\beta^+}\int_{2t+1}^{x_n}\frac{(1+x)^{-\beta}}{x-t}\,*$$

$$*\left\{\begin{array}{ll}\log\frac{1+x_n}{1+x} &, \ \beta^- = \frac{1}{2}\\ \left|(1+x_n)^{\frac{1}{2}-\beta^-} - (1+x)^{\frac{1}{2}-\beta^-}\right| &, \ \beta^- \ne \frac{1}{2}\end{array}\right\}dx$$

In case $\beta^- = \frac{1}{2}$ we obtain

$$\int_{2t+1}^{x_n}\frac{\sqrt{1+x}}{x-t}\log\frac{1+x_n}{1+x}\,dx = \sqrt{1+t}\int_{2}^{\frac{1+x_n}{1+t}}\frac{\sqrt{\tau}}{\tau-1}\log\frac{1+x_n}{(1+t)\tau}\,d\tau$$

$$\le 2\sqrt{1+t}\int_{2}^{\frac{1+x_n}{1+t}}\tau^{-\frac{1}{2}}\log\frac{1+x_n}{(1+t)\tau}\,d\tau$$

$$= 4\sqrt{1+t}\left[\sqrt{2}\log\frac{2(1+t)}{1+x_n} + 2\left(\sqrt{\frac{1+x_n}{1+t}}-\sqrt{2}\right)\right]$$

$$\le c\sqrt{1+x_n} \le \frac{c}{n}.$$

Now, for $\beta^- \neq \frac{1}{2}$, consider

$$I := (1+t)^{\beta^+} \int_{2t+1}^{x_n} \frac{(1+x)^{-\beta}}{x-t} \left| (1+x_n)^{\frac{1}{2}-\beta^-} - (1+x)^{\frac{1}{2}-\beta^-} \right| dx .$$

If $0 < \beta^- < \frac{1}{2}$ then

$$I \leq (1+x_n)^{\frac{1}{2}-\beta^-} \int_{2t+1}^{x_n} \frac{(1+x)^{\beta^-}}{x-t} dx = (1+x_n)^{\frac{1}{2}-\beta^-}(1+t)^{\beta^-} \int_2^{\frac{1+x_n}{1+t}} \frac{\tau^{\beta^-}}{\tau-1} d\tau$$

$$\leq 2(1+x_n)^{\frac{1}{2}-\beta^-}(1+t)^{\beta^-} \int_2^{\frac{1+x_n}{1+t}} \tau^{\beta^--1} d\tau \leq c\sqrt{1+x_n} ,$$

while in case $\frac{1}{2} < \beta^- < 1$

$$I \leq \int_{2t+1}^{x_n} \frac{\sqrt{1+x}}{x-t} dx \leq c\sqrt{1+x_n} .$$

At last, for $0 < \beta^+ < 1$, we have

$$I \leq \sqrt{1+x_n}(1+t)^{\beta^+} \int_{2t+1}^{x_n} \frac{(1+x)^{-\beta^+}}{x-t} dx \leq 2\sqrt{1+x_n} \int_2^{\frac{1+x_n}{1+t}} \tau^{-\beta^+-1} d\tau \leq \frac{2^{1-\beta^+}}{\beta^+}\sqrt{1+x_n} .$$

Thus,

$$|H_1(t)| \leq \frac{c}{n} \|f_n'\|_{\infty,\frac{1}{2}+\alpha^-,\frac{1}{2}+\beta^-} , \quad -1 < t < -1 + \frac{c_1}{2n^2} .$$

Since, for $x_n \leq x$ and $-1 < t \leq -1 + \frac{c_1}{2n^2}$, we have $x - t \geq x + 1 - \frac{c_1}{2n^2} \geq \frac{x+1}{2}$, it follows

$$|H_2(t)| \leq c\,(1+t)^{\beta^+} \int_{x_n}^1 (1-x)^{-\alpha}(1+x)^{-\beta-1}|f_n(x)| dx$$

$$\leq c\,\|f_n\|_{\infty,\alpha^-,\beta^-}(1+t)^{\beta^+} \int_{x_n}^1 (1-x)^{-\alpha^+}(1+x)^{-\beta^+-1} dx$$

$$\leq c\,\|f_n\|_{\infty,\alpha^-,\beta^-}(1+t)^{\beta^+} \left[\int_{x_n}^0 (1+x)^{-\beta^+-1} dx + \int_0^1 (1-x)^{-\alpha^+} dx \right]$$

$$\leq c\,\|f_n\|_{\infty,\alpha^-,\beta^-}(\log n + 1) .$$

Thus, for $-1 < t \leq -1 + \frac{c_1}{2n^2}$,

$$|H(t)| \leq c\left(\|f_n\|_{\infty,\alpha^-,\beta^-} \log n + \frac{1}{n}\|f_n'\|_{\infty,\frac{1}{2}+\alpha^-,\frac{1}{2}+\beta^-} \right) . \tag{4.17}$$

Now, let $-1 + \frac{c_1}{2n^2} < t \leq 0$. In this case we have

$$|H(t)| \leq c\,\|f_n\|_{\infty,\alpha^-,\beta^-} \int_{2t+1}^1 \frac{(1-x)^{-\alpha^+}}{x-t} dx .$$

Substituting $x = 1 + 2t(1-\tau)$ we obtain

$$\int_{2t+1}^1 \frac{(1-x)^{-\alpha^+}}{x-t} dx = (-2t)^{1-\alpha^+} \int_0^1 \frac{(1-\tau)^{-\alpha^+} d\tau}{1+t-2t\tau}$$

$$\leq \begin{cases} (-2t)^{-\alpha^+} \int_0^1 \frac{(1-\tau)^{-\alpha^+}}{\tau - \frac{1+t}{2t}} d\tau , & t \leq -\frac{1}{2} \\ 2(-2t)^{1-\alpha^+} \int_0^1 (1-\tau)^{-\alpha^+} d\tau , & -\frac{1}{2} < t \leq 0 \end{cases}$$

$$\leq c \left\{ \begin{array}{ll} \int_0^{\frac{1}{2}} \frac{d\tau}{\tau - \frac{1+t}{2t}} + \int_{\frac{1}{2}}^1 (1-\tau)^{-\alpha^+} d\tau \ , & t \leq -\frac{1}{2} \\ 1 & , \ -\frac{1}{2} < t \leq 0 \end{array} \right\}$$

$$\leq c(\log n + 1).$$

Thus, for $-1 + \frac{c_1}{2n^2} < t \leq 0$,

$$|H(t)| \leq c \|f_n\|_{\infty,\alpha-,\beta-} \log n .$$
(4.18)

To estimate $G(t)$ we remark that $\int_{-1}^{2t+1} \frac{dx}{x-t} = 0$ and write

$$G(t) = v^{\alpha^+,\beta^+}(t) \int_{-1}^{2t+1} \frac{v^{-\alpha,-\beta}(x) - v^{-\alpha,-\beta}(t)}{x - t} f_n(x)\, dx$$

$$+ v^{\alpha^-,\beta^-}(t) \int_{-1}^{2t+1} \frac{f_n(x) - f_n(t)}{x - t}\, dx =: G_1(t) + G_2(t).$$

$G_1(t)$ can be written in the form

$$G_1(t) = v^{\alpha^+,\beta^+}(t) \int_{-1}^{2t+1} \frac{v^{\alpha^-,\beta^-}(x)[v^{-\alpha^+,-\beta^+}(x)) - v^{-\alpha^+,-\beta^+}(t)]}{x - t} f_n(x)\, dx$$

$$+ \int_{-1}^{2t+1} \frac{v^{\alpha^-,\beta^-}(x) - v^{\alpha^-,\beta^-}(t)}{x - t} f_n(x)\, dx ,$$

such that

$$|G_1(t)| \leq c \|f_n\|_{\infty,\alpha-,\beta-} \left[v^{\alpha^+,\beta^+}(t) \int_{-1}^{2t+1} \left| \frac{v^{-\alpha^+,-\beta^+}(x) - v^{-\alpha^+,-\beta^+}(t)}{x - t} \right| dx \right.$$

$$\left. + \int_{-1}^{2t+1} \left| \frac{v^{\alpha^-,\beta^-}(x) - v^{\alpha^-,\beta^-}(t)}{(x - t)v^{\alpha^-,\beta^-}(x)} \right| dx \right] .$$

We have

$$v^{\alpha^+,\beta^+}(t) \int_{-1}^{2t+1} \left| \frac{v^{-\alpha^+,-\beta^+}(x) - v^{-\alpha^+,-\beta^+}(t)}{x - t} \right| dx \leq$$

$$v^{\alpha^+,\beta^+}(t) \int_{-1}^{2t+1} \frac{|(1+x)^{-\beta^+} - (1+t)^{-\beta^+}|}{|x - t|(1-x)^{\alpha^+}}\, dx + (1-t)^{\alpha^+} \int_{-1}^{2t+1} \frac{|(1-x)^{-\alpha^+} - (1-t)^{-\alpha^+}|}{|x - t|}\, dx$$

and

$$\int_{-1}^{2t+1} \left| \frac{v^{\alpha^-,\beta^-}(x) - v^{\alpha^-,\beta^-}(t)}{(x - t)v^{\alpha^-,\beta^-}(x)} \right| dx$$

$$\leq \int_{-1}^{2t+1} \frac{|(1+x)^{\beta^-} - (1+t)^{\beta^-}|}{|x - t|(1+x)^{\beta^-}}\, dx + (1+t)^{\beta^-} \int_{-1}^{2t+1} \frac{|(1-x)^{\alpha^-} - (1-t)^{\alpha^-}|}{|x - t|(1-x)^{\alpha^-}(1+x)^{\beta^-}}\, dx$$

as well as, for $0 < \gamma < 1, 0 \leq \delta < 1, -1 < t \leq 0$,

$$\int_{-1}^{2t+1} \frac{|(1+x)^{-\gamma} - (1+t)^{-\gamma}|}{|x - t|(1-x)^\delta}\, dx$$

$$= (1+t)^{-\gamma} \int_0^2 \left| \frac{\tau^{-\gamma}-1}{\tau-1} \right| \frac{d\tau}{[2-(1+t)\tau]^\delta} \leq (1+t)^{-\gamma} \int_0^2 \left| \frac{\tau^{-\gamma}-1}{\tau-1} \right| \frac{d\tau}{(2-\tau)^\delta},$$

$$\int_{-1}^{2t+1} \frac{|(1-x)^{-\gamma}-(1-t)^{-\gamma}|}{|x-t|} dx$$

$$= (1-t)^{-\gamma} \int_{\frac{-2t}{1-t}}^{\frac{2}{1-t}} \left| \frac{\tau^{-\gamma}-1}{\tau-1} \right| d\tau \leq \int_0^2 \left| \frac{\tau^{-\gamma}-1}{\tau-1} \right| d\tau,$$

$$\int_{-1}^{2t+1} \frac{|(1+x)^\gamma-(1+t)^\gamma|}{|x-t|(1+x)^\gamma} dx = \int_0^2 \left| \frac{\tau^\gamma-1}{\tau-1} \right| \frac{d\tau}{\tau^\gamma},$$

$$\int_{-1}^{2t+1} \frac{|(1-x)^\gamma-(1-t)^\gamma|}{|x-t|(1-x)^\gamma(1+x)^\delta} dx \leq c \int_{-1}^{2t+1} \frac{|x-t|^{\gamma-1}dx}{(1-x)^\gamma(1+x)^\delta}$$

$$= c(1+t)^{\gamma-\delta} \int_0^2 \frac{|\tau-1|^{\gamma-1}d\tau}{[2-(1+t)\tau]^\gamma\tau^\delta} \leq c(1+t)^{\gamma-\delta} \int_0^2 \frac{|\tau-1|^{\gamma-\delta}d\tau}{(2-\tau)^\gamma\tau^\delta}.$$

Thus,

$$|G_1(t)| \leq c\|f_n\|_{\infty,\alpha-,\beta-}. \tag{4.19}$$

For the estimation of $G_2(t)$, at first we consider the case $-1 < t \leq -1+\frac{c_1}{2n^2}$. We conclude

$$|G_2(t)| \leq \int_{-1}^{2t+1} \frac{dx}{|x-t|} \cdot \left| \int_t^x |f_n'(\tau)|d\tau \right|$$

$$\leq c(1+t)^{\beta-} \|f_n'\|_{\infty,\frac{1}{2}+\alpha-,\frac{1}{2}+\beta-} \int_{-1}^{2t+1} \frac{dx}{|x-t|} \left| \int_t^x (1+\tau)^{-\frac{1}{2}-\beta-} d\tau \right|$$

$$\leq c\sqrt{1+t}\|f_n'\|_{\infty,\frac{1}{2}+\alpha-,\frac{1}{2}+\beta-} \begin{cases} \int_0^2 \left| \frac{\tau^{\frac{1}{2}-\beta-}-1}{1-\tau} \right| d\tau &, \quad \beta- \neq \frac{1}{2} \\ \int_0^2 \frac{|\log\tau|}{|1-\tau|} d\tau &, \quad \beta- = \frac{1}{2} \end{cases}$$

$$\leq \frac{c}{n}\|f_n'\|_{\infty,\frac{1}{2}+\alpha-,\frac{1}{2}+\beta-}$$

For $-1+\frac{c_1}{2n^2} < t \leq 0$, we estimate

$$|G_2(t)| \leq v^{\alpha-,\beta-}(t) \left[\int_{-1}^{t-\frac{1+t}{2n^2}} + \int_{t-\frac{1+t}{2n^2}}^{t+\frac{1-t}{2n^2}} + \int_{t+\frac{1-t}{2n^2}}^1 \right] \left| \frac{f_n(x)-f_n(t)}{x-t} \right| dx$$

$$=: G_{21}(t) + G_{22}(t) + G_{23}(t)$$

and

$$G_{21}(t) + G_{23}(t) \leq \left\{ v^{\alpha-,\beta-}(t) \left[\int_{-1}^{t-\frac{1+t}{2n^2}} + \int_{t+\frac{1-t}{2n^2}}^1 \right] \frac{(1-x)^{-\alpha-}(1+x)^{-\beta-}}{|x-t|} dx \right.$$

$$\left. + \left[\int_{-1}^{t-\frac{1+t}{2n^2}} + \int_{t+\frac{1-t}{2n^2}}^1 \right] \frac{dx}{|x-t|} \right\} \|f_n\|_{\infty,\alpha-,\beta-}$$

$$\leq c\|f_n\|_{\infty,\alpha-,\beta-}\log n$$

in view of (2.7) and (2.8) (with $\alpha-,\beta-$ instead of $\alpha+,\beta+$). At least we consider $G_{22}(t)$. Using the mean value theorem we obtain

$$G_{22}(t) \leq v^{\alpha-,\beta-}(t) \int_{t-\frac{1+t}{2n^2}}^{t+\frac{1-t}{2n^2}} \frac{dx}{\sqrt{1-\xi^2}(1-\xi)^{\alpha-}(1+\xi)^{\beta-}} \|f_n'\|_{\infty,\frac{1}{2}+\alpha-,\frac{1}{2}+\beta-},$$

where $\xi = \xi(x,t)$ lies between x and t. It follows

$$G_{22}(t) \leq \frac{c}{n^2\sqrt{1-t^2}}\,\|f_n'\|_{\infty,\frac{1}{2}+\alpha-,\frac{1}{2}+\beta-} \leq \frac{c}{n}\,\|f_n'\|_{\infty,\frac{1}{2}+\alpha-,\frac{1}{2}+\beta-}\,,$$

since $\frac{c_1}{2n^7} \leq 1+t$. So we can summarize

$$G_2(t) \leq c \left(\|f_n\|_{\infty,\alpha-,\beta-}\,\log n + \frac{1}{n}\,\|f_n'\|_{\infty,\frac{1}{2}+\alpha-,\frac{1}{2}+\beta-} \right).$$

Together with (4.17)-(4.19) this leads to the assertion. $\square$

REMARK 4.4 *The estimate (4.16) remains true if we consider instead of $\Phi_n(t)$ the functions*

$$\Psi_n^0(t) = v^{\alpha+,\beta+}(t) \int_{-1}^1 b(x)\mu_0(x)\frac{f_n(x)}{x-t}\,dx$$

or

$$\Psi_n(t) = v^{\alpha+,\beta+}(t)B(t) \int_{-1}^1 \mu(x)\frac{f_n(x)}{x-t}\,dx\,,$$

where $\mu_0(x)$ and $\mu(x)$ are defined in Section 2 and $\beta \neq 0$ is assumed.

PROOF. Since $\mu_0(x)b(x) = v^{-\alpha,-\beta}(x)v(x)$, where $v \in C^{0,\eta}$, we have

$$\Psi_n^0(t) = v(t)\Phi_n(t) + v^{\alpha+,\beta+}(t) \int_{-1}^1 \frac{v(x)-v(t)}{x-t}v^{-\alpha+,-\beta+}(x)v^{\alpha-,\beta-}(x)f_n(x)\,dx$$

and

$$\int_{-1}^1 \left|\frac{v(x)-v(t)}{x-t}\right| v^{-\alpha+,-\beta+}(x)\,dx \leq c\,v^{-\alpha+,-\beta+}(t)$$

in view of Lemma 2.4. Analogously we can consider

$$\Psi_n(t) = \Psi_n^0(t) + v^{\alpha+,\beta+}(t) \int_{-1}^1 \frac{B(t)-B(x)}{x-t}\mu(x)f_n(x)\,dx\,,$$

where

$$\int_{-1}^1 \left|\frac{B(t)-B(x)}{x-t}\right|\mu(x)v^{-\alpha-,-\beta-}(x)\,dx \leq c\,.$$

$\square$

COROLLARY 4.5 *Let r and s be nonnegative integers satisfying*

$$\frac{\gamma}{2} + \frac{1}{4} \leq r + \alpha^- \leq \frac{\gamma}{2} + \frac{5}{4}\,, \qquad \frac{\delta}{2} + \frac{1}{4} \leq r + \beta^- \leq \frac{\delta}{2} + \frac{5}{4}\,,$$

where γ and δ are some real numbers lying in $(-1,1)$. If $L_n = L_{m,r,s}(v^{\gamma,\delta})$, $n = m+r+s$, is the interpolation operator defined in (4.6) and satisfying (4.1)-(4.5) for γ, δ, α^-, and β^- instead of α, β, ρ, and σ, respectively, and if $f' \in C_{\frac{1}{2}+\alpha-,\frac{1}{2}+\beta-}$, then

$$\|\hat{A}(f-L_nf)\|_{\infty,\alpha+,\beta+} \leq c\left[\|f-L_nf\|_{\infty,\alpha-,\beta-}\,\log n + \frac{1}{n}\,\|(f-L_nf)'\|_{\infty,\frac{1}{2}+\alpha-,\frac{1}{2}+\beta-}\right]\,, \qquad (4.20)$$

provided that $\alpha \neq 0$ and $\beta \neq 0$.

REMARK 4.6 *If the assumptions of Corollary 4.5 are fulfilled then*

$$\|\widehat{\mathcal{A}}(f - L_n f)\|_{\infty,\alpha+,\beta+} \leq \frac{c}{n} E_{n-2}^{\frac{1}{2}+\alpha^-,\frac{1}{2}+\beta^-}(f') \log n. \tag{4.21}$$

PROOF. With $\varphi(x) = \sqrt{1-x^2}$ and $w(x) = v^{\alpha^-,\beta^-}(x)$ we recall [6], (E), p. 91: For some constant $C > 0$ we have

$$\mathcal{K}_\varphi(C/2, f, t)_w \sim \Omega_\varphi(f, t)_w, \tag{4.22}$$

where

$$\Omega_\varphi(f, t)_w = \sup_{0 \leq h \leq t} \sup \left\{ w(x) |(\Delta_{h\varphi} f)(x)| : -1 + 2h^2 \leq x \leq 1 - 2h^2 \right\},$$

$$(\Delta_h f)(x) = f(x + h/2) - f(x - h/2),$$

and

$$\mathcal{K}_\varphi(C/2, f, t)_w = \sup_{0 \leq h \leq t} \inf \Big\{ \sup_{-1+Ch^2 \leq x \leq 1-Ch^2} w(x)|f(x) - g(x)|$$

$$+ h \sup_{-1+Ch^2 \leq x \leq 1-Ch^2} w(x)\varphi(x)|g'(x)| : g \in A.C.[-1 + Ch^2, 1 - Ch^2] \Big\}.$$

Choosing $g = f$ in the definition of $\mathcal{K}_\varphi$ we obtain

$$\mathcal{K}_\varphi(C/2, f, t) \leq c\, t\, \|w\varphi f'\|_\infty.$$

Together with [6], (8.2.1) we conclude

$$E_n^{\alpha^-,\beta^-}(f) \sim \int_0^{1/n} \Omega_\varphi(f, \tau) \frac{d\tau}{\tau} \leq \frac{c}{n} \|w\varphi f'\|_\infty. \tag{4.23}$$

Now, if q is a polynomial of degree $n - 2$ and $Q' = q$, then by relation (4.23)

$$E_{n-1}^{\alpha^-,\beta^-}(f) = E_{n-1}^{\alpha^-,\beta^-}(f - Q) \leq \frac{c}{n} \|w\varphi(f' - q)\|_\infty.$$

Taking the infimum we obtain

$$E_{n-1}^{\alpha^-,\beta^-}(f) \leq \frac{c}{n} E_{n-1}^{\frac{1}{2}+\alpha^-,\frac{1}{2}+\beta^-}(f'). \tag{4.24}$$

In view of [11], relation (2) we have

$$\|(f - L_n f)'\|_{\infty,\frac{1}{2}+\alpha^-,\frac{1}{2}+\beta^-} \leq c\, E_{n-2}^{\frac{1}{2}+\alpha^-,\frac{1}{2}+\beta^-}(f') = n\, \|f - L_n f\|_{\infty,\alpha^-,\beta^-} \tag{4.25}$$

and so, by relation (4.20), Theorem 4.1, and relation (4.24),

$$\|\widehat{\mathcal{A}}(f - L_n f)\|_{\infty,\alpha+,\beta+} \leq c \left[\|f - L_n f\|_{\infty,\alpha^-,\beta^-} \log n + \frac{1}{n} E_{n-2}^{\frac{1}{2}+\alpha^-,\frac{1}{2}+\beta^-}(f') \right]$$

$$\leq c \left[E_{n-1}^{\alpha^-,\beta^-}(f) \log^2 n + \frac{1}{n} E_{n-2}^{\frac{1}{2}+\alpha^-,\frac{1}{2}+\beta^-}(f') \right] \leq \frac{c}{n} E_{n-2}^{\frac{1}{2}+\alpha^-,\frac{1}{2}+\beta^-}(f') \log^2 n.$$

$$\square$$

5   Numerical examples

Example 1. Let $a(x) \equiv 0$ and $b(x) \equiv 1$. Then, for $g = -\frac{1}{2}$, $\lambda = \lambda_- = 0$, we get $\alpha = -\frac{1}{2}$, $\beta = \frac{1}{2}$ and

$$\sigma(x) = \sqrt{\frac{1+x}{1-x}}, \quad \mu(x) = \sqrt{\frac{1-x}{1+x}}.$$

For these weight functions, the zeros t_{nk} of p_n^σ and x_{nk} of p_n^μ are given by

$$t_{nk} = \cos\frac{(2k-1)\pi}{2n+1}, \quad x_{nk} = \cos\frac{2k\pi}{2n+1}, \quad k = 1,\ldots,n.$$

Furthermore we have

$$\sigma_{nk} = \frac{1+t_{nk}}{n+\frac{1}{2}}, \quad k = 1,\ldots,n, \quad \text{and} \quad \frac{p_n^\mu(\pm 1)}{p_n^\sigma(\pm 1)} = (2n+1)^{\pm 1}.$$

One can easily check, that $u^*(t) = (1-t)^{\frac{3}{2}}(1+t)^{\frac{1}{2}}$ is the solution of $(A+K)u = f$ for

$$f(x) = \frac{1}{\pi}\left((1-x^2)\log\frac{1-x}{1+x} - \frac{x^4}{6} + x^2 - 2x + \frac{1}{2}\right), \quad h(x,t) = |x-t|.$$

If we choose $L_n = L_{n,r,s}(v^{\frac{1}{2},-\frac{1}{2}})$ with $r = s = 1$ or $r = 1$, $s = 0$, we get from (4.12)

$$\|u_n^* - u^*\|_{\infty,0,\frac{1}{2}} \leq c\frac{\log^2 n}{n}$$

for the approximate solutions $u_n^* \in \mathbb{P}_{n+r+s-1}$ obtained by the quadrature method. The following table shows the numerical results for the case $r = s = 1$ and $y_1 = -1$, $z_1 = 1$.

n	$\|u_n^* - u^*\|_{\infty,0,\frac{1}{2}}$
10	0.0134
20	0.0039
30	0.0018
40	0.0010
50	0.0007
60	0.0005
70	0.0004
80	0.0003

All these values fulfil the estimate $\|u_n^* - u^*\|_{\infty,0,\frac{1}{2}} \leq 2n^{-2}$, which is better than the theoretical result.

Example 2. We take a, b, σ, μ as in Example 1. An easy calculation shows, that we get $(Au^*)(x) = \sqrt{1-x^2}$ for

$$u^*(t) = (\widehat{A}Au^*)(t) = \frac{1}{\pi}\left(2 - (1-t)\log\frac{1-t}{1+t}\right).$$

We set $h(x,t) = \sqrt{|x|(1-t^2)}$. Since the integral $\int_{-1}^1 (1-t^2)\log\frac{1-t}{1+t}dt$ vanishes (the integrand is odd), we deduce $(Ku^*)(x) = \frac{4}{\pi^2}\sqrt{|x|}$. Consequently, u^* is the solution of $(A+K)u = f$ with $f(x) = \sqrt{1-x^2} + \frac{4}{\pi^2}\sqrt{|x|}$. Now we choose $L_n = L_{n,1,0}(v^{\frac{1}{2},-\frac{1}{2}})$ with $z_1 = 1$ and get the following numerical results.

n	$\|u_n^* - u^*\|_{\infty,0,\frac{1}{4}}$
10	0.0620
20	0.0336
30	0.0230
40	0.0175
50	0.0141
60	0.0118
70	0.0098
80	0.0085

For all these values the estimation $\|u_n^* - u^*\|_{\infty,0,\frac{1}{4}} \le 0.75\, n^{-1}$ holds true, while we only have the theoretical result (comp. (4.12))

$$\|u_n^* - u^*\|_{\infty,0,\frac{1}{2}} \le c \frac{\log^2 n}{n^{\frac{1}{2}}}.$$

Example 3. Let $a(x) = \sqrt{1-x^2}$ and $b(x) = -x$ (i.e. $c(x) \equiv 1$). Then we have $r(x) \equiv 1$ and $g(x) = \frac{1}{\pi}\arcsin x$. For $\lambda = \lambda_- = 0$, we get

$$\alpha = \beta = \frac{1}{2}, \quad \sigma(x) = \exp\left(\frac{1}{\pi}\int_{-1}^{1}\frac{\arcsin t}{t-x}dt\right) = 2\sqrt{1-x^2}$$

taking into account (see [13, II.§6, Lemma 6.1])

$$\left(\int_{-1}^{1}\frac{\arcsin t}{t-x}dt\right)' = (\pi\log 2\sqrt{1-x^2})' + \int_{-1}^{1}\frac{1}{t-x}\frac{dt}{\sqrt{1-t^2}},$$

$$\int_{-1}^{1}\frac{1}{t-x}\frac{dt}{\sqrt{1-t^2}} = 0, \quad \text{and} \quad \int_{-1}^{1}\frac{\arcsin t}{t}dt = \pi\log 2.$$

Moreover,

$$\mu(x) = \frac{1}{\sigma(x)} = \frac{1}{2\sqrt{1-x^2}}.$$

For these weight functions, the zeros t_{nk} of p_n^σ and x_{nk} of p_n^μ are given by

$$t_{nk} = \cos\frac{k\pi}{n+1}, \quad x_{nk} = \cos\frac{(2k-1)\pi}{2n}, \quad k = 1,\ldots,n.$$

Furthermore we have

$$\sigma_{n,k} = 2\frac{1-t_{n,k}^2}{n+1}, \quad k = 1,\ldots,n.$$

One can easily check, that $u^*(t) = \sqrt{1-t^2}$ is the solution of $(A+K)u = f$ for

$$f(x) = 2(1-x^2)^{\frac{3}{2}} + \frac{4x^2}{\pi} - \frac{2x(1-x^2)}{\pi}\log\frac{1-x}{1+x}, \quad h(x,t) = t\sqrt{|x|(1-t^2)}.$$

(We remark that $Ku^* = 0$, since the integrand is odd.) With $L_n = L_n(v^{-\frac{1}{2},-\frac{1}{2}})$ we get (see Corollary 4.2 and Theorem 3.6)

$$\|u_n^* - u^*\|_{\infty,\frac{1}{2},\frac{1}{2}} \le c\frac{\log^2 n}{n^{\frac{1}{2}}}.$$

The following results show, that $\|u_n^* - u^*\|_\infty$ also seems to converge to zero, although Corollary 3.9 is not applicable.

n	$\|u_n^* - u^*\|_{\infty,\frac{1}{2},\frac{1}{2}}$	$\|u_n^* - u^*\|_\infty$
10	0.0069	0.1504
20	0.0018	0.0772
30	0.0008	0.0521
40	0.0005	0.0393
50	0.0003	0.0316
60	0.0002	0.0264
70	0.0002	0.0227
80	0.0001	0.0199

For these values of n the estimate $\|u_n^* - u^*\|_{\infty,\frac{1}{2},\frac{1}{2}} \leq 1/n^2$ holds true, which is again better than the theoretical result.

References

[1] V. M. Badkov, Convergence in the mean and almost everywhere of Fourier series in polynomials orthogonal on an interval, Math. USSR-Sb., **24** (1974), 223-256.

[2] D. Berthold and P. Junghanns, New error bounds for the quadrature method for the solution of Cauchy singular integral equations, SIAM J. Numer. Anal., **30** (1993), 1351-1372.

[3] M. R. Capobianco, The stability and the convergence of a collocation method for a class of Cauchy singular integral equations, Math. Nachr., **162** (1993), 45-58.

[4] J. A. Cuminato, Uniform convergence of a collocation method for the numerical solution of Cauchy-type singular integral equations: a generalization, IMA J. Numer. Anal., **12** (1992), 31-45.

[5] J. A. Cuminato, On the uniform convergence of a perturbed collocation method for a class of Cauchy singular integral equations, Applied Numer. Math., **16** (1995), 417-438.

[6] Z. Ditzian and V. Totik, Moduli of Smoothness, Springer-Verlag, 1987.

[7] D. Elliott, Orthogonal polynomials associated with singular integral equations having a Cauchy kernel, SIAM J. Math. Anal., **13** (1982), 1041-1052.

[8] D. Elliott, The classical collocation method for singular integral equations, SIAM J. Numer. Anal., **19** (1982), 816-832.

[9] G. Freud, Orthogonale Polynome, VEB Deutscher Verlag der Wissenschaften, Berlin, 1969.

[10] I. E. Gopengauz, On a theorem of A. F. Timan on the approximation of functions by polynomials on an bounded interval (in Russian), Math. Zametki, **1** (1967), 163-172, also in: Math. Notes, **1** (1967), 110-116.

[11] N. X. Ky, On simultaneous approximation by polynomials with weight, Colloquia Mathematica Societatis Janos Bolyai, 49. Alfred Haar Memorial Conference, Budapest, 1985.

[12] G. Mastroianni, Uniform convergence of derivatives of Lagrange interpolation, J. Comp. Appl. Math., **43** (1992), 37-51.

[13] S. G. Michlin and S. Prössdorf, Singular Integral Operators, Akademie-Verlag, Berlin, 1986.

[14] P. Nevai, Orthogonal Polynomials, Mem. Amer. Math. Soc. **213**, Amer. Math. Soc., Providence, RI, 1979.

[15] S. Prössdorf and B. Silbermann, Numerical Analysis for Integral and Related Operator Equations, Akademie Verlag, Berlin, 1991.

[16] J. Szabados and P. Vértesi, Interpolation of Functions, World Scientific Publishing Co. Pte. Ltd., Singapore, 1990.

[17] G. M. Vainikko, Functional Analysis of Discretization Methods, Teubner, Leipzig, 1976 (in German).

Maria Rosaria Capobianco, Giuseppe Mastroianni
Istituto per Applicazioni della Mathematica, C.N.R.
Via Pietro Castellino 111
80131 Napoli, Italy

Peter Junghanns, Uwe Luther
Technische Universität Chemnitz-Zwickau
Fakultät Mathematik
D–09107 Chemnitz, Germany

AMS-Subject-Classification: 41A05, 45E05, 45L05, 65R20

Operator Theory
Advances and Applications, Vol. 90
© 1996 Birkhäuser Verlag Basel/Switzerland

Symbol calculus for singular integrals with operator-valued PQC-coefficients

Torsten Ehrhardt   Steffen Roch*

Bernd Silbermann

The symbol calculus for singular integral operators with piecewise continuous coefficients created by Gohberg and Krupnik in the early seventies as well as its generalization to piecewise quasicontinuous coefficients due to Sarason and Silbermann are derived for operators with operator-valued coefficients. The approach is based upon localization techniques in combination with a general two-projections-theorem.

1 Introduction

One of the main achievements in the theory of one-dimensional singular integral operators with piecewise continuous coefficients is the celebrated symbol calculus by Gohberg and Krupnik ([8], [9], and see also [13], Chapter V, §8). This calculus associates homomorphically with each singular integral operator a certain two-by-two matrix function living on the cylinder $\mathbb{T} \times [0, 1]$ and having the property that the operator is Fredholm if and only if the function is invertible. This function is usually called the *symbol* of the operator. A similar symbol calculus can be derived for singular integral operators with piecewise quasicontinuous coefficients (see [15], [16] for basic properties of single singular integral and Toeplitz operators and [17] – [19] for algebras generated by them). In this setting one has to replace the cylinder $\mathbb{T} \times [0, 1]$ by the set $M(QC) \times [0, 1]$ where $M(QC)$ refers to the maximal ideal space of the algebra QC of the quasicontinuous functions (for an introduction to QC we recommend [6], Chapter 9, and [15], [16]), and the symbol becomes also more involved (it degenerates to a 1×1 matrix-valued function at some points of $M(QC) \times [0, 1]$). A current standard approach to these results is via localization theorems (see the local principle by Allan and Douglas in Section 2.3 below) in combination with two-projections-theorems in order to describe the local algebras ([12], [14], [10]).

It is the goal of the present paper to generalize these results to the context of singular integral operators with operator-valued coefficients. Our approach is again based upon

---
*Research supported by a DFG Heisenberg grant

a coupling of localization techniques and 2-projections-theorems. Fortunately, there is a version of the 2-projections-theorem which exactly fits to our goal: In [5] there is studied the algebra generated by two projections and by an arbitrary coefficient algebra. On the other hand, the needs of the local principle require certain special generalizations of the notions "quasicontinuous function" and "compact operator" to the operator-valued setting. The "correct" definition of what an operator-valued quasicontinuous function is will be discussed in Section 2 below. Concerning the other generalization, let us remark here that the offered symbol calculus is with respect to a larger ideal than that of the compact operators. This ideal – consisting of so-called Q_n-compact operators – plays an important role in the theory of the finite section method for singular integral operators with operator-valued coefficients (see [2], [3], [7]). So our symbol calculus will not only provide us with necessary conditions for the Fredholmness of the operators under consideration (which are even sufficient in the finite-dimensional case), but it is also closely related to the finite section method for these operators. These connections will be pointed out in a forthcoming paper ([4]).

2   Operator-valued quasicontinuous functions

2.1   Operator-valued measurable functions

Let H be a Hilbert space with inner product $\langle .,. \rangle$ and $L(H)$ the C^*-algebra of all bounded linear operators on H. Further let $\mathbf{T}$ denote the unit circle with Lebesgue measure m. A function f on $\mathbf{T}$ with values in $L(H)$ is called *measurable* if, for each $\varepsilon > 0$, there is a compact $K \subseteq \mathbf{T}$ such that $m(\mathbf{T} \setminus K) \leq \varepsilon$ and the restriction of f onto K is continuous. An *essentially bounded* function is a measurable function $f : \mathbf{T} \to L(H)$ for which there exists an $r > 0$ with

$$m(\{t \in \mathbf{T} : \|f(t)\|_{L(H)} > r\}) = 0.$$

The set $L^\infty_{L(H)}$ of all essentially bounded functions on $\mathbf{T}$ with values in $L(H)$ is an algebra (with pointwise operations), and

$$\|f\|_{L^\infty_{L(H)}} := \inf\{r \in \mathbb{R} : \|f(t)\|_{L(H)} \leq r \quad \text{almost everywhere on } \mathbf{T}\}$$

defines a semi-norm on $L^\infty_{L(H)}$. Moreover, $\|f\|_{L^\infty_{L(H)}} = 0$ if and only if $f(t) = 0$ a.e., thus, factoring out the functions which are zero a.e., one gets a normed algebra which is even a C^*-algebra (with pointwise involution) as one can easily show.

The *essential range* ess range f of a measurable function $f : \mathbf{T} \to L(H)$ consists of all operators $a \in L(H)$ for which

$$m(\{t \in \mathbf{T} : \|f(t) - a\|_{L(H)} < \varepsilon\}) > 0 \quad \text{for all} \quad \varepsilon > 0.$$

It is always a closed subset of $L(H)$.

Proposition 1 *If $f \in L^\infty_{L(H)}$ then*

$$m(\{t \in \mathbf{T} : f(x) \notin \text{ess range} f\}) = 0, \tag{1}$$

$$\|f\|_{L^\infty_{L(H)}} = \sup_{a \in \text{ess range} f} \|a\|_{L(H)}. \tag{2}$$

For a proof we need the following almost obvious lemma.

Lemma 1 *Let $r \in \mathbb{R}^+$ and $f \in L^\infty_{L(H)}$. Then $\|f\|_{L^\infty_{L(H)}} \leq r$ if and only if $\|f(t)\|_{L(H)} \leq r$ almost everywhere.*

Proof of Proposition 1. Given $\varepsilon > 0$ choose a compact $K \subseteq T$ such that $m(T \setminus K) < \varepsilon$ and $f|_K$ is continuous. Then

$$m(\{t \in T : f(t) \notin \text{ess range } f\}) \leq m(\{t \in K : f(t) \notin \text{ess range } f\}) + \varepsilon.$$

The set $U := \{t \in K : f(t) \notin \text{ess range } f\}$ is open in K (due to the closedness of ess range f and the continuity of $f|_K$). Let K_0 be any compact subset of K such that $K_0 \subseteq U$ and $m(U \setminus K_0) \leq \varepsilon$. Thus,

$$m(\{t \in T : f(t) \notin \text{ess range } f\}) \leq 2\varepsilon + m(K_0). \tag{3}$$

We claim that $m(K_0) = 0$. Indeed, since $K_0 \subseteq U$, one can find for each $t \in K_0$ a $\delta(t)$ such that

$$U_{\delta(t)} := \{s \in K_0 : \|f(t) - f(s)\|_{L(H)} < \delta(t)\}$$

has measure zero. The sets $\{U_{\delta(t)}\}_{t \in K_0}$ form an open covering of K_0 from which one can pick a finite subcovering. Thus, $m(K_0) = 0$, and since (3) holds for every $\varepsilon > 0$, we get (1).

For the second assertion, let $a \in \text{ess range } f$. Then $m(\{t \in T : \|f(t) - a\|_{L(H)} < \varepsilon\}) > 0$ for all $\varepsilon > 0$ and, hence, $m(\{t \in T : \|a\|_{L(H)} - \varepsilon < \|f(t)\|_{L(H)}\}) > 0$. The lemma implies that $\|a\|_{L(H)} - \varepsilon < \|f\|_{L^\infty_{L(H)}}$ for all $\varepsilon > 0$. Letting ε go to zero yields $\|a\|_{L(H)} \leq \|f\|_{L^\infty_{L(H)}}$.

For the reverse inequality, suppose that $\sup_{a \in \text{ess range} f} \|a\|_{L(H)} < \|f\|_{L^\infty_{L(H)}}$. Then, again by the lemma,

$$m(\{t \in T : \sup_{a \in \text{ess range} f} \|a\|_{L(H)} < \|f(t)\|_{L(H)}\}) > 0,$$

hence, $m(\{t \in T : f(t) \notin \text{ess range } f\}) > 0$ which contradicts (1). ∎

It is evident from the definition that the functions f in $L^\infty_{L(H)}$ are *weakly measurable* in the sense that $\langle x, fy \rangle(t) := \langle x, f(t)y \rangle$ is a customary scalar-valued L^∞-function for all $x, y \in H$. Moreover, one has the following result, where $B(H)$ refers to the unit ball $\{x \in H : \|x\| \leq 1\}$ in H.

Proposition 2 *For all $f \in L^\infty_{L(H)}$, $\|f\|_{L^\infty_{L(H)}} = \sup_{x, y \in B(H)} \|\langle x, fy \rangle\|_{L^\infty}$.*

Proof. Clearly, $|\langle x, f(t)y \rangle| \leq \|f(t)\|_{L(H)}$, and $\|f(t)\|_{L(H)} \leq \|f\|_{L^\infty_{L(H)}}$ almost everywhere. Thus, $|\langle x, f(t)y \rangle| \leq \|f\|_{L^\infty_{L(H)}}$ a.e., and $\|\langle x, fy \rangle\|_{L^\infty} \leq \|f\|_{L^\infty_{L(H)}}$ for all $x, y \in B(H)$.

Suppose that $\sup_{x, y \in B(H)} \|\langle x, fy \rangle\|_{L^\infty} < \|f\|_{L^\infty_{L(H)}}$. Then, by (2), there is an operator $a \in \text{ess range } f$ such that $\sup_{x, y \in B(H)} \|\langle x, fy \rangle\|_{L^\infty} < \|a\|_{L(H)}$. Since

$$\{t \in T : \|f(t) - a\|_{L(H)} < \varepsilon\} \subseteq \{t \in T : |\langle x, f(t)y \rangle - \langle x, ay \rangle| < \varepsilon\},$$

we conclude that $\langle x, ay \rangle \in \text{ess range } \langle x, fy \rangle$ for all $x, y \in B(H)$. Thus, again by (2), $\|\langle x, fy \rangle\|_{L^\infty} \geq |\langle x, ay \rangle|$ and, consequently,

$$\sup_{x, y \in B(H)} \|\langle x, fy \rangle\|_{L^\infty} \geq \sup_{x, y \in B(H)} |\langle x, ay \rangle| = \|a\|_{L(H)},$$

which is a contradiction. ∎

Besides the space $L^\infty_{L(H)}$ we introduce the following function spaces:

- $C_{L(H)}$ is the algebra of all continuous functions on **T** with values in $L(H)$,

- $H^\infty_{L(H)}$ is the subspace of $L^\infty_{L(H)}$ consisting of all functions f with vanishing negative Fourier coefficients, i.e. $\int_0^1 e^{-2\pi i n t}\langle x, f(e^{2\pi i t})y\rangle dt = 0$ for all $x, y \in B(H)$ and $n < 0$, and

- $\overline{H^\infty_{L(H)}}$ is the subspace of $L^\infty_{L(H)}$ consisting of all functions with vanishing positive Fourier coefficients.

Let us further agree upon omitting the subscript $L(H)$ in case $H = \mathbb{C}$.

2.2   Three definitions of quasicontinuous functions

The set QC of the scalar-valued quasicontinuous functions is defined as the intersection $(C + H^\infty) \cap (C + \overline{H^\infty})$ which in fact is a (commutative) C^*-algebra with respect to the supremum norm. The maximal ideal space of this algebra (which is of a rather exotic nature) will be denoted by $M(QC)$. Thus, by the Gelfand-Naimark theorem, QC is isometrically isomorphic to the algebra $C(M(QC))$ of all continuous functions on $M(QC)$. For a better acquaintance with the algebra QC we recommend [1], [6], and [15].

For operator-valued functions there seem to at least three natural ways to introduce quasicontinuity. Picking up the definition of QC it is convenient to define the set $QC_{L(H)}$ of the operator-valued quasicontinuous functions as $(C_{L(H)} + H^\infty_{L(H)}) \cap (C_{L(H)} + \overline{H^\infty_{L(H)}})$. This set is actually a C^*-subalgebra of $L^\infty_{L(H)}$ (see [3]). Besides this algebra we consider the smallest closed subalgebra $QC^s_{L(H)}$ of $L^\infty_{L(H)}$ which contains the constant functions $t \mapsto a$ for all $a \in L(H)$ as well as all functions $t \mapsto f(t)e$ where e is the identity operator and f runs through QC. Finally, we let $QC^w_{L(H)}$ denote the set of all functions f in $L^\infty_{L(H)}$ such that $\langle x, fy\rangle$ is (scalar-valued) quasicontinuous for all $x, y \in L(H)$.

Theorem 1 $QC^s_{L(H)} \subseteq QC_{L(H)} \subseteq QC^w_{L(H)}$, and in case H is infinite-dimensional, all inclusions are proper.

It is evident that the inclusions hold. The proof of the properness will be given in subsection 2.5 below. In case $H = \mathbb{C}^n$ it is easy to see that three sets of quasicontinuous functions coincide. In what follows we speak about functions in $QC^s_{L(H)}$ and $QC^w_{L(H)}$ as *strongly* and *weakly* quasicontinuous, respectively. Further, we shall always think of the algebras $L(H)$ and QC as being embedded into $L^\infty_{L(H)}$ via identifying the operators $a \in L(H)$ with the constant functions $t \mapsto a$ and the scalar-valued quasicontinuous functions f with the functions $t \mapsto f(t)e$.

2.3   Our choice of the definition

Our choice of the "correct" definition of quasicontinuity is mainly determined by the needs of the local principle by Allan and Douglas (see, e.g., [1], Theorem 1.34).

Local Principle. *Let* **B** *be a unital C^*-algebra and* **C** *be a* *-subalgebra in the center of* **B** *(which means that each element of* **C** *commutes with each element of* **B**). *For each maximal ideal ξ of* **C** *let I_ξ denote the smallest closed two-sided ideal of* **B** *which contains ξ. Then*

I_ξ is a proper ideal of $\mathbf{B}$, and an element b of $\mathbf{B}$ is invertible in $\mathbf{B}$ if and only if the cosets $b + I_\xi$ are invertible in the quotient algebras $\mathbf{B}/I_\xi$ for all ξ.

We want to apply this principle to an algebra $\mathbf{B}$ of operators which have operator-valued quasicontinuous functions as their coefficients, and the algebra QC of scalar-valued quasi-continuous functions (which is considered as being well-known here) should play the role of the central subalgebra. Thereby we have to suppose that the operator-valued QC-functions *behave locally* (at the points of $M(QC)$) as simple as possible, i.e. *as constant operator-valued functions*. Well, and here is the exact definition of what is meant:

Let B be a C^*-algebra between $QC^s_{L(H)}$ and $QC^w_{L(H)}$, and let $I_{B,\xi}$ denote the ideal of B associated with the maximal ideal ξ of QC via the local principle, i.e.

$$I_{B,\xi} = \operatorname{clos}_B \left\{ \sum_{i=1}^n b_i f_i \quad \text{with} \quad b_i \in B, \ f_i \in \xi \right\}.$$

The algebra B is called *locally trivial at $\xi \in M(QC)$* if, for each b in B, there is a constant function $a \in L(H)$ such that $b - a \in I_{B,\xi}$. Clearly, the algebra $QC^s_{L(H)}$ is locally trivial at each point of $M(QC)$ (see also the paragraph preceding the proof of Theorem 2 below). The following theorem states that it is the only algebra with this property.

Theorem 2 *Let B be a C^*-algebra between $QC^s_{L(H)}$ and $QC^w_{L(H)}$. If B is locally trivial at all points $\xi \in M(QC)$ then $B = QC^s_{L(H)}$.*

The proof is splitted into several steps. We start with introducing a linear mapping which associates with each function in $QC^w_{L(H)}$ a bounded function on $M(QC)$.

Let $f \in QC^w_{L(H)}$ and $\xi \in M(QC)$. The mapping $H \times H \to \mathbb{C}$, $(x, y) \mapsto \xi(\langle x, fy\rangle)$ is a continuous bilinear form on H, hence there exists a uniquely determined operator $\Phi_\xi(f) \in L(H)$ such that $\xi(\langle x, fy\rangle) = \langle x, \Phi_\xi(f)y\rangle$ for all $x, y \in H$. Here are some elementary properties of the mapping $\Phi_\xi : QC^w_{L(H)} \to L(H)$.

Proposition 3 *The mapping Φ_ξ is linear, symmetric, and continuous with norm 1. Furthermore,*

$$\Phi_\xi(a) = a \quad \text{for all} \quad a \in L(H), \tag{4}$$

$$\Phi_\xi(f) = \xi(f) \quad \text{for all} \quad f \in QC, \tag{5}$$

$$\Phi_\xi(fg) = \xi(f)\Phi_\xi(g) \quad \text{for all} \quad f \in QC, g \in QC^w_{L(H)}, \tag{6}$$

$$\Phi_\xi(fg) = \Phi_\xi(f)\Phi_\xi(g) \quad \text{for all} \quad f, g \in QC^s_{L(H)}. \tag{7}$$

Proof. The linearity, symmetry and continuity of Φ_ξ are obvious and, for all $f \in QC^w_{L(H)}$, one has $\|\Phi_\xi(f)\|_{L(H)} \le \|f\|_{L^\infty_{L(H)}}$. That the norm is even equal to 1 is a consequence of (4) which, on its hand, is also obvious.

For (5) observe that $\langle x, fy\rangle(t) = \langle x, f(t)y\rangle = f(t)\langle x, y\rangle$ and, hence, $\xi(\langle x, fy\rangle) = \xi(f\langle x, y\rangle) = \xi(f)\langle x, y\rangle$. Similarly one gets for (6), $\langle x, fgy\rangle = f\langle x, gy\rangle$ and, consequently,

$$\xi(\langle x, fgy\rangle) = \xi(f\langle x, gy\rangle) = \xi(f)\xi(\langle x, gy\rangle) = \xi(f)\langle x, \Phi_\xi(g)y\rangle.$$

Finally, the multiplicativity of Φ_ξ over $QC^s_{L(H)}$ follows from (4), (5), (6) and from the fact that each strongly quasicontinuous function can be approximated as closely as desired by linear combinations of functions $f_i a_i$ with $f_i \in QC$ and $a_i \in L(H)$ and from $fa = af$ for all $f \in QC$ and $a \in L(H)$.   ■

Corollary 1 *Let B be a C^*-algebra between $QC^s_{L(H)}$ and $QC^w_{L(H)}$ which is locally trivial at $\xi \in M(QC)$. Then, for all $b \in B$, there is at most one constant function a with $b - a \in I_{B,\xi}$, and this function coincides with $\Phi_\xi(b)$.*

Proof. By formula (6), the ideal $I_{B,\xi}$ belongs to the kernel of Φ_ξ. Hence, if $b - a \in I_{B,\xi}$ with $a \in L(H)$ then $\Phi_\xi(b) = \Phi_\xi(a)$ whence, by (4), follows $\Phi_\xi(b) = a$.   ■

Corollary 2 *Let B be as in Corollary 1. Then the restriction of Φ_ξ onto B is multiplicative.*

Proof. Let $b, c \in B$. Then

$$bc - \Phi_\xi(b)\Phi_\xi(c) = (b - \Phi_\xi(b))c + \Phi_\xi(b)(c - \Phi_\xi(c)).$$

The functions c and $\Phi_\xi(b)$ are in B, and the functions $b - \Phi_\xi(b)$ and $c - \Phi_\xi(c)$ belong to the ideal $I_{B,\xi}$ of B. Hence, $bc - \Phi_\xi(b)\Phi_\xi(c) \in I_{B,\xi}$, and from Corollary 1 one concludes that $\Phi_\xi(b)\Phi_\xi(c) = \Phi_\xi(bc)$.   ■

Let $B_{L(H)}(M(QC))$ stand for the set of all bounded functions on $M(QC)$ with values in $L(H)$. Provided with pointwise operations, pointwise involution, and with supremum norm, this set becomes a C^*-algebra. The mapping $M(QC) \to L(H)$, $\xi \mapsto \Phi_\xi(f)$ associates with each function $f \in QC^w_{L(H)}$ a function in $B_{L(H)}(M(QC))$ which is denoted by $\Phi(f)$. The following proposition summarizes some properties of the mapping $\Phi : QC^w_{L(H)} \to B_{L(H)}(M(QC))$, $f \mapsto \Phi(f)$.

Proposition 4 *The mapping Φ is a linear and symmetric isometry, and*

$$\Phi(a) = a \quad \text{for all } a \in L(H), \tag{8}$$

$$\Phi(f) = \hat{f} \quad \text{with } \hat{f} \text{ the Gelfand transform of } f \in QC, \tag{9}$$

$$\Phi(fg) = \hat{f}\Phi(g) \quad \text{for all } f \in QC, \, g \in QC^w_{L(H)}, \tag{10}$$

$$\Phi(fg) = \Phi(f)\Phi(g) \quad \text{for all } f, g \in QC^s_{L(H)}. \tag{11}$$

Here, as above, we identify the operators $a \in L(H)$ with the constant functions $\xi \mapsto a$ in $B_{L(H)}(M(QC))$ and the scalar-valued functions $\hat{f} : M(QC) \to \mathbb{C}$ with the functions $\xi \mapsto \hat{f}(\xi)e$ in $B_{L(H)}(M(QC))$.

Proof. By Proposition 2, we have for each $f \in QC^w_{L(H)}$,

$$
\begin{aligned}
\|f\|_{L^\infty_{L(H)}} &= \sup_{x,y \in B(H)} \|\langle x, fy\rangle\|_{L^\infty} \\
&= \sup_{x,y \in B(H)} \sup_{\xi \in M(QC)} |\xi(\langle x, fy\rangle)| \quad \text{(the Gelfand-Naimark theorem)} \\
&= \sup_{\xi \in M(QC)} \sup_{x,y \in B(H)} |\langle x, \Phi_\xi(f)y\rangle| \quad \text{(the definition of } \Phi_\xi) \\
&= \sup_{\xi \in M(QC)} \|\Phi_\xi(f)\|_{L(H)} = \sup_{\xi \in M(QC)} \|\Phi(f)(\xi)\|_{L(H)} \\
&= \|\Phi(f)\|_{B_{L(H)}(M(QC))}.
\end{aligned}
$$

The other assertions are either obvious or immediate consequences of (4) – (7).   ■

Proposition 5 *Let B be a C^*-algebra between $QC^s_{L(H)}$ and $QC^w_{L(H)}$. The algebra B is locally trivial at $\xi \in M(QC)$ if and only if the functions $\Phi(b)$ are continuous at ξ for all $b \in B$.*

Proof. Let B be locally trivial at ξ and $b \in B$. Then, by Corollary 1, $b - \Phi_\xi(b) \in I_{B,\xi}$ and, by (8), $\Phi(b) - \Phi_\xi(b) \in \Phi(I_{B,\xi})$. The constant function $\Phi_\xi(b)$ is continuous on all of $M(QC)$, and we claim that the functions in $\Phi(I_{B,\xi})$ are continuous at ξ. Indeed, a function $h \in I_{B,\xi}$ can be approximated as closely as desired by functions $\sum_i b_i f_i$ with $b_i \in B$ and $f_i \in \xi$. Hence, and by Proposition 4, $\Phi(h)$ can be approximated by $\sum_i \Phi(b_i f_i) = \sum_i \hat{f}_i \Phi(b_i)$. The functions $\hat{f}_i \Phi(b_i)$ are continuous at ξ since $\Phi(b_i)$ is bounded and $\hat{f}_i$ is continuous on $M(QC)$ and $\hat{f}_i(\xi) = 0$. Consequently, $\Phi(h)$ is continuous at ξ, and this yields the continuity of $\Phi(b)$ at ξ, too.

For the reverse implication, let $\Phi(b)$ be continuous at ξ. We claim that $b - \Phi_\xi(b) \in I_{B,\xi}$. Given $\varepsilon > 0$ one can find a neighborhood U_ε of ξ such that $\|\Phi_\eta(b) - \Phi_\xi(b)\| < \varepsilon$ for all $\eta \in U_\varepsilon$. Choose a function $f_\varepsilon \in QC$ which takes values in $[0,1]$ only, and which is equal to 1 at ξ and vanishes outside U_ε (which is possible by the Gelfand-Naimark theorem), and consider the function

$$a_\varepsilon = (b - \Phi_\xi(b)) f_\varepsilon = (b - \Phi_\xi(b)) + (b - \Phi_\xi(b))(f_\varepsilon - 1). \tag{12}$$

The function $b - \Phi_\xi(b)$ is in B, and $f_\varepsilon - 1$ is in $\xi \in M(QC)$, hence, the second item of (12) is in $I_{B,\xi}$. Further,

$$
\begin{aligned}
\|a_\varepsilon\|_{L^\infty_{L(H)}} &= \|a_\varepsilon\|_{B_{L(H)}(M(QC))} && \text{(by Proposition 4)} \\
&= \|\hat{f}_\varepsilon(\Phi(b) - \Phi_\xi(b))\|_{B_{L(H)}(M(QC))} && \text{(by (8), (10) and (12))} \\
&= \sup_{\eta \in M(QC)} |\hat{f}_\varepsilon(\eta)| \|\Phi_\eta(b) - \Phi_\xi(b)\|_{L(H)}.
\end{aligned}
$$

Hence, $\|a_\varepsilon\|_{L^\infty_{L(H)}} < \varepsilon$. It follows from (12) by an approximation argument that $b - \Phi_\xi(b) \in I_{B,\xi}$. Consequently, B is locally trivial at ξ. ∎

For a more direct proof without invoking the functions Φ and Φ_ξ see the proof of Proposition 8 below.

Let $C_{L(H)}(M(QC))$ be the *-subalgebra of $B_{L(H)}(M(QC))$ consisting of all continuous operator-valued functions on $M(QC)$. It is an immediate consequence of Proposition 4 that $\Phi(f) \in C_{L(H)}(M(QC))$ for all $f \in QC^s_{L(H)}$. Hence, by Proposition 5, $QC^s_{L(H)}$ is locally trivial at all points of $M(QC)$.

Proof of Theorem 2. If B is locally trivial at each point of $M(QC)$ then, by the preceding proposition, $\Phi(b)$ is a continuous function on $M(QC)$ for all $b \in B$ and, hence, Φ is a linear (evident), multiplicative (by Corollary 2) *-isometry (by Proposition 4) from B onto a *-subalgebra of $C_{L(H)}(M(QC))$.

We claim that the restriction of Φ onto $QC^s_{L(H)}$ is even an isometrical isomorphism onto all of $C_{L(H)}(M(QC))$. Indeed, each function $f \in C_{L(H)}(M(QC))$ can be approximated in the norm by functions $f_n = \sum_{i=1}^n a_i g_i$ where $a_i \in L(H)$ and $g_i \in C(M(QC))$ (this can be verified, e.g., by invoking the local inclusion theorem of [11], Theorem 1.14(b); see also the proof of Proposition 8 below). Let $\check{g}_i$ be QC-functions with $\hat{\check{g}}_i = g_i$ (these functions exist by the Gelfand-Naimark theorem), and set $\check{f}_n := \sum_{i=1}^n a_i \check{g}_i \in QC^s_{L(H)}$. Then $\Phi(\check{f}_n) = f_n$, and the isometry of Φ entails that the sequence $(\check{f}_n)$ converges to a certain function $\check{f} \in QC^s_{L(H)}$ such that $\Phi(\check{f}) = f$. Hence, Φ maps $QC^s_{L(H)}$ onto $C_{L(H)}(M(QC))$.

Consequently, $QC^s_{L(H)}$ and $C_{L(H)}(M(QC))$ are isometrically isomorphic via Φ, hence, B and $C_{L(H)}(M(QC))$ are also isometrically isomorphic via Φ (the mapping $\Phi : B \to C_{L(H)}(M(QC))$ is onto), and since $QC^s_{L(H)} \subseteq B$ we finally obtain $QC^s_{L(H)} = B$. ∎

Thus, the most favourite candidate among the definitions of quasicontinuity is that of the *strongly* quasicontinuous functions.

2.4 Properties of strongly quasicontinuous functions

Theorem 3 *The following assertions are equivalent for $f \in QC^w_{L(H)}$:*
(a) f belongs to $QC^s_{L(H)}$.
(b) The range of $\Phi(f)$ is compact in $L(H)$.
(c) The essential range of f is compact in $L(H)$.
(d) $\{\langle x, fy \rangle : x, y \in B(H)\}$ is precompact in QC.
(e) $\Phi(f)$ belongs to $C_{L(H)}(M(QC))$.

Proof. $(a) \Rightarrow (b)$: If $f \in QC^s_{L(H)}$ then, as we have seen in the proof of Theorem 2, $\Phi(f)$ is a continuous function on the compact set $M(QC)$, hence the range of $\Phi(f)$ is compact.

$(b) \Rightarrow (c)$: We are going to show that ess range $f \subseteq$ range $\Phi(f)$ for all $f \in QC^w_{L(H)}$. Indeed, let $a \in$ ess range f. Then, for all $\varepsilon > 0$,

$$m(\{t \in \mathbf{T} : \|f(t) - a\|_{L(H)} < \varepsilon\}) > 0,$$

and hence, the function $\|c\|$ defined by $\|c\|(t) = \|c(t)\|_{L(H)}$ with $c = f - a$ (which is an element of L^∞ by the definition of $L^\infty_{L(H)}$) cannot be invertible in L^∞. So there must be a non-trivial multiplicative functional $r \in M(L^\infty)$ such that $r(\|c\|) = 0$.

Let $x, y \in B(H)$. Then

$$|r(\langle x, cy \rangle)| = r(|\langle x, cy \rangle|) = r(|\langle x, cy \rangle| - \|c\|) \le 0$$

and consequently, $r(\langle x, cy \rangle) = 0$ for all choices of x and y. Let $\xi \in M(QC)$ denote the restriction of r onto QC. Then $\xi(\langle x, cy \rangle) = 0$ for all $x, y \in B(H)$ (recall that $\langle x, cy \rangle$ is a QC-function), hence $\Phi_\xi(c) = 0$. Then $\Phi_\xi(f) = a$, i.e., a is in the range of $\Phi(f)$.

This gives our claim and proves the compactness of ess range f since essential ranges are always closed.

$(c) \Rightarrow (d)$: Given $\varepsilon > 0$ choose a finite ε-net $\{a_1, \ldots, a_n\} \subseteq L(H)$ for ess range f as well as finite ε-nets $\{c_{i1}, \ldots, c_{in_i}\} \subseteq \mathbb{C}$ for $\{\langle x, a_i y \rangle : x, y \in B(H)\}$ with $i = 1, \ldots, n$. Further set

$$N := \{t \in \mathbf{T} : \|f(t) - a_i\|_{L(H)} \ge \varepsilon \quad \text{for all} \quad i = 1, \ldots, n\},$$

and choose, for each $i = 1, \ldots, n$, a measurable subset T_i of $\{t \in \mathbf{T} : \|f(t) - a_i\|_{L(H)} < \varepsilon\}$ such a way that $\cup_{i=1}^n T_i \cup N = \mathbf{T}$ and that the T_i are pairwise disjoint. Since N is a subset of $\{t \in \mathbf{T} : f(t) \notin$ ess range $f\}$, it has measure zero by Proposition 1.

For each n-tuple $(\alpha_1, \ldots, \alpha_n)$ of integers $1 \le \alpha_i \le n_i$, consider the functions

$$f_{(\alpha_1, \ldots, \alpha_n)} := \sum_{i=1}^n c_{i\alpha_i} \chi_{T_i} \quad (\in L^\infty)$$

where χ_{T_i} stands for the characteristic function of T_i. The finitely many functions $f_{(\alpha_1, \ldots, \alpha_n)}$ form a (2ε)-net for $\{\langle x, fy \rangle : x, y \in B(H)\}$ in L^∞. Indeed, if $x, y \in B(H)$ are fixed then

$$
\begin{aligned}
\|\langle x, fy \rangle - f_{(\alpha_1, \ldots, \alpha_n)}\|_{L^\infty} &\leq \max_{1 \leq i \leq n} \sup_{t \in T_i} |\langle x, f(t)y \rangle - f_{(\alpha_1, \ldots, \alpha_n)}(t)| \\
&= \max_{1 \leq i \leq n} \sup_{t \in T_i} |\langle x, f(t)y \rangle - c_{i\alpha_i}| \\
&\leq \max_{1 \leq i \leq n} \sup_{t \in T_i} (|\langle x, a_i y \rangle - c_{i\alpha_i}| + \|f(t) - a_i\|_{L(H)}) \\
&\leq \max_{1 \leq i \leq n} \sup_{t \in T_i} |\langle x, a_i y \rangle - c_{i\alpha_i}| + \varepsilon
\end{aligned}
$$

by our choice of T_i. Now pick, for each $i = 1, \ldots, n$, an α_i in $\{1, \ldots, n\}$ such that $|\langle x, a_i y \rangle - c_{i\alpha_i}| < \varepsilon$. Then, $\|\langle x, fy \rangle - f_{(\alpha_1, \ldots, \alpha_n)}\|_{L^\infty} \leq 2\varepsilon$ and, thus, the set $\{\langle x, fy \rangle : x, y \in B(H)\}$ is precompact in L^∞ and, consequently, also in QC.

$(d) \Rightarrow (e)$: Let $\{f_1, \ldots, f_n\} \subset QC$ be a finite ε-net for the subset $\{\langle x, fy \rangle : x, y \in B(H)\}$ of QC. Further, given $\xi \in M(QC)$ consider its neighbourhood $U_\varepsilon := \{\eta \in M(QC) : |\eta(f_i) - \xi(f_i)| < \varepsilon$ for all $i = 1, \ldots, n\}$.

Let $\eta \in U_\varepsilon$ and x, y be fixed vectors in $B(H)$. Then

$$
|\langle x, (\Phi_\eta(f) - \Phi_\xi(f))y \rangle| = |\eta(\langle x, fy \rangle) - \xi(\langle x, fy \rangle)|
$$

$$
\leq |\eta(\langle x, fy \rangle - f_i)| + |\xi(\langle x, fy \rangle - f_i)| + |\eta(f_i) - \xi(f_i)|.
$$

We can choose a functions f_i such that the first two items become less than ε, and the third term is automatically less that ε due to the choice of η. Since this holds for all $x, y \in B(H)$, we conclude that $\|\Phi_\eta(f) - \Phi_\xi(f)\|_{L(H)} \leq 3\varepsilon$ for all $\eta \in U_\varepsilon$, hence, $\Phi(f)$ is continuous at ξ.

$(e) \Rightarrow (a)$: This implication has been already established in the proof of Theorem 2. ∎

2.5   Proof of Theorem 1

The properness of the inclusion $QC^s_{L(H)} \subseteq QC_{L(H)}$ for infinite-dimensional Hilbert spaces will be shown by constructing a function in $QC_{L(H)}$ with non-compact essential range. Our construction is based on the following theorem, which is a main tool for the construction of (non-trivial) quasicontinuous functions.

Theorem 4 *Let* x_n, y_n, z_n *be real-valued continuous functions on the unit circle such that*

(i) $x_n + iy_n$ *is in* H^∞ *for all* $n \geq 1$

(ii) $\sup_{t \in \mathbb{T}} \sum_{n=1}^\infty |z_n(t)| < \infty$,

(iii) $\sum_{n=1}^\infty \sup_{|t-1| \geq \varepsilon} |z_n(t)| < \infty$ *for all* $\varepsilon > 0$,

(iv) $\sum_{n=1}^\infty \|y_n\|_C < \infty$,

(v) $\sum_{n=1}^\infty \|x_n - z_n\|_C < \infty$,

and let $a_n \in B(L(H))$. *Then the series* $f(t) = \sum_{n=1}^\infty z_n(t)a_n$ *converges pointwise on* $\mathbb{T}$ *as well as uniformly on* $\{t \in \mathbb{T} : |t - 1| \geq \varepsilon\}$ *for all* $\varepsilon > 0$, *and* f *is in* $QC_{L(H)}$.

Proof. The convergence of the series is a consequence of (ii) and (iii). Moreover, (ii) and (iii) yield that f is bounded on $\mathbf{T}$ and continuous on $\mathbf{T} \setminus \{1\}$. Write f as

$$f = \sum_{n=1}^{\infty} a_n(z_n - x_n) + \sum_{n=1}^{\infty} a_n(x_n + iy_n) - i \sum_{n=1}^{\infty} a_n y_n.$$

By (iv) and (v), the series $\sum_{n=1}^{\infty} a_n(z_n - x_n)$ and $\sum_{n=1}^{\infty} a_n y_n$ converge uniformly to functions in $C_{L(H)}$. Thus, the series $\sum_{n=1}^{\infty} a_n(x_n + iy_n)$ converges (pointwise on $\mathbf{T}$ and uniformly on compact subsets of $\mathbf{T} \setminus \{1\}$) to a certain function which is bounded on $\mathbf{T}$ and continuous on $\mathbf{T} \setminus \{1\}$ again and, hence, measurable and essentially bounded in sense of Section 2.1. We are going to show that this function even belongs to $H_{L(H)}^{\infty}$. Indeed, for all $\varepsilon > 0$, $k < 0$ and $N > 0$,

$$\left| \int_0^1 e^{-2\pi i k t} \left(\sum_{n=1}^{\infty} a_n \left(x_n(e^{2\pi i t}) + i y_n(e^{2\pi i t}) \right) \right) dt \right|$$

$$= \left| \int_0^1 e^{-2\pi i k t} \left(\sum_{n=N}^{\infty} a_n \left(x_n(e^{2\pi i t}) + i y_n(e^{2\pi i t}) \right) \right) dt \right| \qquad \text{(analyticity of } x_n + i y_n \text{)}$$

$$\leq \left| \left(\int_0^\varepsilon + \int_{1-\varepsilon}^1 \right) e^{-2\pi i k t} \left(\sum_{n=N}^{\infty} a_n \left(x_n(e^{2\pi i t}) + i y_n(e^{2\pi i t}) \right) \right) dt \right| +$$

$$+ \left| \int_\varepsilon^{1-\varepsilon} e^{-2\pi i k t} \left(\sum_{n=N}^{\infty} a_n \left(x_n(e^{2\pi i t}) + i y_n(e^{2\pi i t}) \right) \right) dt \right|. \qquad (13)$$

The first term in (13) is less than

$$2\varepsilon \sup_{t \in \mathbf{T}} \sum_{n=1}^{\infty} \left(|x_n(e^{2\pi i t})| + |y_n(e^{2\pi i t})| \right)$$

$$\leq 2\varepsilon \left(\sum_{n=1}^{\infty} \|y_n\|_C + \sum_{n=1}^{\infty} \|x_n - z_n\|_C + \sup_{t \in \mathbf{T}} \sum_{n=1}^{\infty} |z_n(e^{2\pi i t})| \right) \leq 2\varepsilon M$$

with M referring to the constant term in parentheses which is independent of ε. The second item in (13) can be estimated by

$$\sup_{\varepsilon \leq t \leq 1-\varepsilon} \sum_{n=N}^{\infty} \left(|x_n(e^{2\pi i t})| + |y_n(e^{2\pi i t})| \right)$$

$$\leq \sum_{n=N}^{\infty} \|y_n\|_C + \sum_{n=N}^{\infty} \|x_n - z_n\|_C + \sum_{n=N}^{\infty} \sup_{\varepsilon \leq t \leq 1-\varepsilon} |z_n(e^{2\pi i t})|.$$

If N is large enough, this expression becomes smaller than ε, hence, if $k < 0$, then the kth Fourier coefficient of $\sum_{n=1}^{\infty} a_n(x_n + i y_n)$ has absolute value less than $(2M + 1)\varepsilon$ for all $\varepsilon > 0$, therefore it vanishes. Thus, $\sum_{n=1}^{\infty} a_n(x_n + i y_n) \in H_{L(H)}^{\infty}$ and $f \in C_{L(H)} + H_{L(H)}^{\infty}$.

Analogously, the decomposition

$$f = \sum_{n=1}^{\infty} a_n(z_n - x_n) + \sum_{n=1}^{\infty} a_n(x_n - iy_n) + i\sum_{n=1}^{\infty} a_n y_n.$$

entails that $f \in C_{L(H)} + \overline{H^{\infty}_{L(H)}}$. Thus, $f \in QC_{L(H)}$. ∎

Obviously, the functions $x_n = y_n = z_n \equiv 0$ satisfy the assumptions of the preceding theorem. However, what we need in what follows are functions x_n, y_n and z_n satisfying not only (i) – (v) of Theorem 4, but also leading to a function f with a certain oscillating behaviour in a neighbourhood of the point 1 and with prescribed essential range. Functions with these properties will be constructed in the proof of the following proposition.

Proposition 6 *There exist non-trivial functions x_n, y_n, z_n with properties (i) – (v).*

Proof. Set $r_n = 1 - e^{-n^2}$ and consider the functions

$$F_n(z) = \frac{\ln(1 - r_n z)}{\ln(1 - r_n)} = -\frac{\ln(1 - r_n z)}{n^2},$$

where ln stands for the continuous branch of the logarithm which is defined on $\mathbb{C} \setminus (-\infty, 0]$ by $\ln z = \ln|z| + i\arg z$ with the argument of z being chosen in $(-\pi, \pi)$. The function F_n has its only pole at $1/r_n$, hence it is analytic on the open unit disc and bounded on $\mathbb{T}$, i.e. $F_n \in H^{\infty}$.

Set $X_n = \operatorname{Re} F_n$ and $Y_n = \operatorname{Im} F_n$. Then, cleary,

$$\|Y_n\|_C \leq -\frac{\pi}{2}\frac{1}{\ln(1-r_n)} = \frac{\pi}{2n^2}, \tag{14}$$

and

$$X_n(e^{i\theta}) = \frac{\ln|1 - r_n e^{i\theta}|}{\ln(1 - r_n)} = \frac{\ln(1 + r_n^2 - 2r_n\cos\theta)}{2\ln(1 - r_n)}.$$

Because of

$$(1 - r_n)^2 = 1 + r_n^2 - 2r_n \leq 1 + r_n^2 - 2r_n\cos\theta \leq 1 + r_n^2 + 2r_n = (1 + r_n)^2 \leq 2$$

we conclude that

$$\|X_n\|_C \leq \max\left\{-\frac{\ln 2}{2\ln(1-r_n)}, 1\right\} \leq 1. \tag{15}$$

If, moreover, $|\theta| \geq 1/n^2$, then

$$\sin^2\frac{1}{n^2} \leq (r_n - \cos\frac{1}{n^2})^2 + \sin^2\frac{1}{n^2} = 1 + r_n^2 - 2r_n\cos\frac{1}{n^2} \leq 1 + r_n^2 - 2r_n\cos\theta$$

and, consequently,

$$|X_n(e^{i\theta})| \leq -\frac{1}{\ln(1-r_n)}\max\left\{\frac{\ln 2}{2}, -\ln\sin\frac{1}{n^2}\right\} \leq \frac{C\ln(n+1)}{n^2} \tag{16}$$

with a certain constant C.

Let D_n be a continuous function on $\mathbf{T}$ with values in $[0, 1]$ such that

$$D_n(e^{i\theta}) = \begin{cases} 1 & \text{for } |\theta| \leq 1/n^2 \\ 0 & \text{for } |\theta| \geq 2/n^2, \end{cases}$$

and set $Z_n = D_n X_n$. Estimate (16) involves that

$$\|X_n - Z_n\|_C \leq \frac{C \ln(n+1)}{n^2}. \tag{17}$$

Now define x_n, y_n, z_n as rotations of X_n, Y_n, Z_n:

$$x_n(e^{i\theta}) = X_n(e^{i(\theta-6/n)}), \quad y_n(e^{i\theta}) = Y_n(e^{i(\theta-6/n)}), \quad z_n(e^{i\theta}) = Z_n(e^{i(\theta-6/n)}).$$

These functions satisfy (i) – (v). Indeed, (v) is a consequence of (17) and (iv) follows from (14). Further, the support of $z_n = z_n(e^{i\theta})$ is contained in $[6/n - 2/n^2, 6/n + 2/n^2]$, and the pairwise disjointness of these intervals together with the inequalities $\|z_n\| \leq \|x_n\| \leq 1$ (which hold by (15)) yields (ii). Moreover, for each $\varepsilon > 0$, there are only finitely many functions z_n the supports of which are contained in $\{|t - 1| \geq \varepsilon\}$. This gives (iii). Finally, (i) is satisfied since $x_n + iy_n$ is a rotation of the function F_n which is in H^∞. $\blacksquare$

Now one can easily see that the inclusion $QC^s_{L(H)} \subseteq QC_{L(H)}$ is proper for infinite-dimensional Hilbert spaces. For this goal, choose a non-precompact subset $\{a_1, a_2, \ldots\}$ of $B(L(H))$, and let x_n, y_n and z_n be as in the proof of Proposition 6. Then, by Theorem 4, $f = \sum a_n z_n$ is in $QC_{L(H)}$ and, since f is continuous on $\mathbf{T} \setminus \{1\}$ and since $f(e^{i6/n}) = a_n$, all operators a_n belong to the essential range of f. Thus, ess range f cannot be compact whereas all functions in $QC^s_{L(H)}$ have compact essential ranges by Theorem 3.

For a proof that the inclusion $QC_{L(H)} \subseteq QC^w_{L(H)}$ is also proper (in case H is infinite-dimensional) recall that $QC_{L(H)}$ is always an algebra (see [3]), whereas for $QC^w_{L(H)}$ the following holds.

Proposition 7 *If H is infinite-dimensional then $QC^w_{L(H)}$ is not an algebra.*

Proof. It is sufficient to prove the assertion for $H = l^2(\mathbf{Z}^+)$, the Hilbert space of all squared-summable sequences. For $n \geq 2$ define continuous and piecewise linear functions a_n on $\mathbf{R}$ by

$$a_n(t) = \begin{cases} 0 & \text{if } t < \frac{1}{n} \text{ or } t > \frac{1}{n-1} \\ 2n(n-1)t - 2(n-1) & \text{if } t \in \left[\frac{1}{n}, \frac{1}{2}\left(\frac{1}{n} + \frac{1}{n-1}\right)\right] \\ -2n(n-1)t + 2n & \text{if } t \in \left[\frac{1}{2}\left(\frac{1}{n} + \frac{1}{n-1}\right), \frac{1}{n-1}\right]. \end{cases}$$

Thus, $a_n(\frac{1}{n-1}) = a_n(\frac{1}{n}) = 0$ and $a_n(\frac{1}{2}(\frac{1}{n}+\frac{1}{n-1})) = 1$. Further, let τ be the conformal mapping which maps the open unit disk onto the open upper half plane, i.e.

$$\tau(w) = i\frac{1-w}{1+w} \quad \text{for } w \in \mathbf{C},$$

and set $b_n(s) = \sqrt{a_n(\tau(s))}$ for $s \in \mathbf{T}$.

Then define $f : \mathbf{T} \to L(l^2)$ to be the operator-valued function for which $f(s)$ is the matrix with $(1n)$th entry being equal to $b_n(s)$ ($n = 2, 3, \ldots$) and with all other entries being equal to zero. Similarly, let $g : \mathbf{T} \to L(l^2)$ be the function for which $g(s)$ is the matrix with $(n1)$th entry $b_n(s)$ (again, $n = 2, 3, \ldots$) and with all other entries being zero.

Let $(e_i)_{i=1}^{\infty}$ denote the standard basis of $l^2(\mathbb{Z}^+)$. Evidently, $\langle e_i, fe_j \rangle$ and $\langle e_i, ge_j \rangle$ are continuous functions for all i and j, thus the functions $\langle x, fy \rangle$ and $\langle x, gy \rangle$ are continuous for all finitely supported sequences $x, y \in l^2$, and since these sequences are dense in l^2, and since C is a Banach algebra, we obtain that $\langle x, fy \rangle$ and $\langle x, gy \rangle$ are continuous for all $x, y \in l^2$. This shows, in particular, that f and g are weakly quasicontinuous.

Now consider the product fg. Clearly, this is a function in $L^{\infty}_{L(H)}$ with the function $h = \sum_{i=2}^{\infty} b_i^2 = \sum_{i=2}^{\infty} a_i \circ \tau$ standing at the (11)st place , whereas all other entries vanish. We are going to show that h is not in QC, hence, fg cannot be weakly quasicontinuous.

Indeed, the function h is quasicontinuous if and only if the function $k := \sum_{i=2}^{\infty} a_i$ has vanishing mean oscillation (VMO) on $\mathbb{R}$ (see [6], Ch. IV, 1.3 for the correspondence between $QC(\mathbf{T})$ and $QC(\mathbb{R})$, Ch. VI, Sect. 5 for the definition of VMO, and Ch. IX, 2.3 for the equality $QC(\mathbb{R}) = VMO \cap L^{\infty}$). Let us compute the mean oscillation of k over the interval $I = [0, 1/n]$. The arithmetic mean k_I of k is

$$k_I = \frac{1}{|I|} \int_0^{1/n} k(t)\, dt = 1/2,$$

and its mean oscillation $\Delta_I k$ is equal to

$$\Delta_I k = \frac{1}{|I|} \int_0^{1/n} |k(t) - k_I|\, dt = 1/4.$$

Hence, $M_\varepsilon(k) := \sup_{|I| < \varepsilon} \Delta_I k \geq 1/4$ for all $\varepsilon > 0$, i.e., k has no vanishing mean oscillation, and h is not quasicontinuous. ∎

2.6   Piecewise quasicontinuous functions

A function $f : \mathbf{T} \to L(H)$ is said to be *piecewise continuous* if the one-sided limits $f(t \pm 0)$ in the operator norm of $L(H)$ exist at each point $t \in \mathbf{T}$. The set $PC_{L(H)}$ of all piecewise continuous functions forms a *-subalgebra of $L^{\infty}_{L(H)}$.

Proposition 8 *The algebra $PC_{L(H)}$ is the smallest closed subalgebra of $L^{\infty}_{L(H)}$ which contains the algebra PC of the scalar-valued piecewise continuous functions (in the sense that $f \in PC$ is identified with $t \mapsto f(t)I$) and the constant operator-valued functions (in the sense that $a \in L(H)$ is viewed of as $t \mapsto a$).*

Proof. The inclusions $PC \subseteq PC_{L(H)}$ and $L(H) \subseteq PC_{L(H)}$ are evident. For the reverse inclusions, let us denote for a moment the smallest closed subalgebra of $L^{\infty}_{L(H)}$ which contains both PC and $L(H)$ by $PC'_{L(H)}$. The functions $f \in C$ belong to the center of $PC_{L(H)}$, so this algebra can be localized over the maximal ideal space $\mathbf{T}$ of C via the local principle. It is easy to see that a function $f \in PC_{L(H)}$ behaves locally, at $t \in \mathbf{T}$, as the function

$$f(t - 0)\chi_t^- + f(t + 0)\chi_t^+$$

where $f(t \pm 0) \in L(H)$ are constant functions and where χ_t^+ and χ_t^- refer to the (scalar-valued and piecewise continuous) characteristic functions of the arcs $(t, -t) \subseteq \mathbf{T}$ and $(-t, t) \subseteq \mathbf{T}$, respectively. Thus, every function in $PC_{L(H)}$ behaves locally as a function in $PC'_{L(H)}$, and the local inclusion theorem (see [11], Theorem 1.14(b)) yields $PC_{L(H)} \subseteq PC'_{L(H)}$. ∎

Now let $PQC_{L(H)}$ stand for the smallest closed subalgebra of $L^\infty_{L(H)}$ which both contains $PC_{L(H)}$ and $QC^s_{L(H)}$. The functions in $PQC_{L(H)}$ are called *piecewise quasicontinuous*. It is evident from the definition of $QC^s_{L(H)}$ and from the preceding proposition that $PQC_{L(H)}$ can be also characterized as the smallest closed subalgebra of $L^\infty_{L(H)}$ which contains the algebras PC and QC of scalar-valued piecewise continuous and quasicontinuous functions as well as the constant operator-valued functions. Let us agree upon omitting the subscript $L(H)$ in $PQC_{L(H)}$ in case $H = \mathbb{C}$.

3   Symbol calculus

3.1   Algebras of singular integral operators

Let $l^2_H(\mathbb{Z})$ denote the Hilbert space of all sequences $(x_i)_{i \in \mathbb{Z}}$ of vectors $x_i \in H$ with scalar product

$$\langle (x_i), (y_i) \rangle = \sum_{i \in \mathbb{Z}} \langle x_i, y_i \rangle < \infty$$

where the $\langle .,. \rangle$ at the right hand side refers to the initial scalar product on H.

Each function $a \in L^\infty_{L(H)}$ induces a *Laurent operator* $T^0(a)$ on $l^2_H(\mathbb{Z})$: if a_n denotes the nth Fourier coefficient of a then $T^0(a)$ is the operator given by the infinite matrix $(a_{i-j})_{i,j \in \mathbb{Z}}$; this operator is bounded, and

$$\|T^0(a)\|_{L(l^2_H(\mathbb{Z}))} = \|a\|_{L^\infty_{L(H)}}.$$

Further we introduce the projection $P : l^2_H(\mathbb{Z}) \to l^2_H(\mathbb{Z})$, $(x_i) \mapsto (\dots, 0, 0, x_0, x_1, \dots)$. Clearly, the image of P can be identified with the space $l^2_H(\mathbb{Z}^+)$ of the one-sided sequences and the operator $PT^0(a)P$ acting on Im P with the *Toeplitz operator* $T(a) = (a_{i-j})_{i,j=0}^\infty$ acting on $l^2_H(\mathbb{Z}^+)$.

Let $S(PQC_{L(H)})$ stand for the smallest closed subalgebra of $L(l^2_H(\mathbb{Z}))$ containing all Laurent operators $T^0(a)$ with generating function $a \in PQC_{L(H)}$ as well as the projection P. The characterization of $PQC_{L(H)}$ given in Section 2.6 entails that $S(PQC_{L(H)})$ is the smallest closed subalgebra of $L(l^2_H(\mathbb{Z}))$ which contains both the algebra $S(PQC)$ of operators with scalar-valued generating functions and the algebra $L(H)$ of constant operator-valued functions. Further it is evident that $S(PQC_{L(H)})$ is the smallest closed subalgebra of $L(l^2_H(\mathbb{Z}))$ which contains all *singular integral operators* $T^0(a)P + T^0(b)Q$ where $Q = I - P$ and $a, b \in PQC_{L(H)}$.

Given $n \geq 1$, let $P_n \in L(l^2_H(\mathbb{Z}))$ denote the projection operator

$$P_n : (x_i) \to (\dots, 0, 0, x_{-n}, \dots, x_{n-1}, 0, 0, \dots).$$

These operators belong to $S(PQC_{L(H)})$ (and even to $S(C_{L(H)})$) which can be seen as follows. Let τ_k with $k \in \mathbb{Z}$ refer to the function $\tau_k(t) = t^k$ on $\mathbb{T}$. Then $U_k := T^0(\tau_k)$ is the two-sided shift operator $U_k : (x_i) \mapsto (x_{i-k})$ and, clearly, this operator is in $S(C_{L(H)})$. Now it is easy to check that $P_n = U_{-n}QU_n \cdot U_nPU_{-n}$ and, thus, $P_n \in S(C_{L(H)})$.

So it makes sense to consider the smallest closed two-sided ideal $\mathcal{K}_{L(H)}$ of $\mathcal{S}(PQC_{L(H)})$ which contains the projection P_1. The operators in $\mathcal{K}_{L(H)}$ are called Q_n-*compact* with $Q_n = I - P_n$. This notation was proposed in [3] for reasons described in the following proposition.

Proposition 9 *(a) An operator $K \in \mathcal{S}(PQC_{L(H)})$ belongs to $\mathcal{K}_{L(H)}$ if and only if $\|KQ_n\| \to 0$ and $\|Q_n K\| \to 0$ as $n \to \infty$.*
(b) In case H is finite-dimensional, the ideal $\mathcal{K}_{L(H)}$ coincides with the ideal of all compact operators on $l_H^2(\mathbb{Z})$.
(c) If K is a compact operator on $l_{\mathbb{C}}^2(\mathbb{Z})$ with matrix representation (k_{ij}) with respect to the standard basis, then the operator K' acting on $l_H^2(\mathbb{Z})$ via $K'(x_i) = (y_i)$ with $y_i = \sum k_{ij} x_j$ belongs to $\mathcal{K}_{L(H)}$.

Proof. Let $\mathcal{Q}$ denote the set of all operators $K \in \mathcal{S}(PQC_{L(H)})$ with $\|KQ_n\| \to 0$ and $\|Q_n K\| \to 0$. It can be shown exactly as in [3], Proposition 3.1, that $\mathcal{Q}$ is a closed two-sided ideal of $\mathcal{S}(PQC_{L(H)})$. Since, evidently, $P_1 \in \mathcal{Q}$, we conclude that $\mathcal{K}_{L(H)} \subseteq \mathcal{Q}$. Conversely, the operators P_n can be written as

$$P_n = \sum_{k=-n}^{n-1} U_k P P_1 U_{-k}$$

and, thus, they belong to $\mathcal{K}_{L(H)}$. Hence, all operators KP_n with $K \in \mathcal{S}(PQC_{L(H)})$ are in $\mathcal{K}_{L(H)}$ and if, in particular, $K \in \mathcal{Q}$, then $\|KP_n - K\| = \|KQ_n\| \to 0$ by the definition of $\mathcal{Q}$. So, K is in the closure of $\mathcal{K}_{L(H)}$, i.e. in $\mathcal{K}_{L(H)}$ itself. This verifies assertion (a). Assertion (b) is an immediate consequence of part (a) since, for finite-dimensional H, the operator P_1 is compact and so all operators in $\mathcal{K}_{L(H)}$ are compact and, conversely, if K is compact, then $K \in \mathcal{S}(PQC_{L(H)})$ (see Proposition 4.5 in [1]) and, hence, $K \in \mathcal{Q}$. Thus,

$$\mathcal{K}_{L(H)} \subseteq \text{compact operators} \subseteq \mathcal{Q}.$$

For assertion (c), let K be a compact operator on $l_{\mathbb{C}}^2(\mathbb{Z})$ with matrix representation (k_{ij}). One easily checks that, for $n > 0$,

$$P_n K P_n = \sum_{i,j=-n}^{n-1} k_{ij} U_i P_1 U_{-j}$$

(where the U_k and P_1 act on scalar-valued sequences). Thus,

$$K_n' := (P_n K P_n)' = \sum_{i,j=-n}^{n-1} k_{ij} U_i P_1 U_{-j},$$

but now with U_k and P_1 acting on operator-valued sequences. Hence, K_n' belongs to $\mathcal{S}(PQC_{L(H)})$ and even to $\mathcal{K}_{L(H)}$. Further, it is easy to see that, for each fixed n, the C^*-algebra generated by the identity and all operators $P_n K P_n$ is isomorphic to the C^*-algebra generated by the identity and all operators $(P_n K P_n)'$ with K compact. Since isomorphisms between C^*-algebras are isometries, we conclude that $\|P_n K P_n\| = \|(P_n K P_n)'\|$ for all compact operators K. Now observe that $\|P_n K P_n - K\| \to 0$, hence, $(P_n K P_n)_{n \geq 1}$ is a Cauchy sequence. The isometry gives that $((P_n K P_n)')_{n \geq 1}$ is a Cauchy sequence, too. Thus, the latter sequence converges, its limit is in $\mathcal{S}(PQC_{L(H)})$ and even in $\mathcal{K}_{L(H)}$ and, clearly, this limit equals K'.   ∎

Let us emphasize once more that, for infinite-dimensional Hilbert spaces H, the ideal of the compact operators on $l_H^2(\mathbb{Z})$ is properly contained in $\mathcal{K}_{L(H)}$ (for example, P_1 is in $\mathcal{K}_{L(H)}$, but this operator fails to be compact).

In what follows we let $\mathcal{S}^\pi(PQC_{L(H)})$ refer to the quotient algebra $\mathcal{S}(PQC_{L(H)})/\mathcal{K}_{L(H)}$. The desired symbol calculus for this algebra (or, in other words, criteria for the invertibility of operators in $\mathcal{S}(PQC_{L(H)})$ modulo $\mathcal{K}_{L(H)}$ and, hence, necessary conditions for the Fredholmness of operators in $\mathcal{S}(PQC_{L(H)})$)) will be derived by employing the local principle by Allan and Douglas. The applicability of this principle follows from the fact that, whenever $f \in QC$ is scalar-valued then the coset $T^0(f) + \mathcal{K}_{L(H)}$ belongs to the center of $\mathcal{S}^\pi(PQC_{L(H)})$. Indeed, the Laurent operator $T^0(f) \in L(l_H^2(\mathbb{Z}))$ commutes evidently with all Laurent operators $T^0(a)$ with $a \in PQC_{L(H)}$, and it commutes with P modulo $\mathcal{K}_{L(H)}$ since the operator $T^0(f) \in L(l_{\mathbb{C}}^2(\mathbb{Z}))$ commutes with $P \in L(l_{\mathbb{C}}^2(\mathbb{Z}))$ modulo the compact operators (see [15] and [1], Theorem 2.54, and take into account Proposition 9(c)).

So we can localize $\mathcal{S}^\pi(PQC_{L(H)})$ with respect to its central subalgebra $\mathcal{B} := \{T^0(f) + \mathcal{K}_{L(H)}, f \in QC\}$.

Proposition 10 *The algebras $\mathcal{B}$ and QC are isometrically isomorphic.*

Proof. It is sufficient to show that $T^0(QC) \cap \mathcal{K}_{L(H)} = \{0\}$. Then the third isomorphism theorem for C^*-algebras,

$$(\mathcal{B} \cong) \ (T^0(QC) + \mathcal{K}_{L(H)})/\mathcal{K}_{L(H)} \cong T^0(QC)/(T^0(QC) \cap \mathcal{K}_{L(H)}),$$

yields the assertion.

One easily verifies that $U_{-n}T^0(a)U_n = T^0(a)$ for all $a \in PQC_{L(H)}$ and $U_{-n}PU_n \to I$ strongly as $n \to \infty$. Thus, the strong limit s-$\lim_{n\to\infty} U_{-n}AU_n$ exists for each operator $A \in \mathcal{S}(PQC_{L(H)})$ and, moreover, it is always an element of $T^0(PQC_{L(H)})$. Let us denote this limit by $W(A)$. Clearly, the mapping $W : A \mapsto W(A)$ is a symmetric and continuous algebra homomorphism.

If $K \in T^0(QC) \cap \mathcal{K}_{L(H)}$ then, on the one hand, $W(K) = K$ since $K \in T^0(QC)$ whereas, on the other hand, $W(K) = 0$ since $K \in \mathcal{K}_{L(H)}$, the ideal $\mathcal{K}_{L(H)}$ is generated by P_1, and $W(P_1) = W(U_{-1}PU_2QU_{-1}) = 0$.   ∎

So the maximal ideal space of $\mathcal{B}$ is homeomorphic to $M(QC)$, and we can localize $\mathcal{S}^\pi(PQC_{L(H)})$ over this set. Before doing this, let us recall some facts about $M(QC)$. These results are due to Sarason and can be found in [15] and in [1], 3.24 - 3.36.

The points of the unit circle $\mathbf{T}$ can be identified with the maximal ideals of the algebra C via $t : f \mapsto f(t)$ for all $f \in C$ and $t \in \mathbf{T}$ (recall that there is a ono-to-one correspondence between the maximal ideals and the non-zero multiplicative functionals of a commutative Banach algebra). Let $t \in \mathbf{T}$. Then $M_t(QC)$, the *fiber of $M(QC)$ over t*, is the set of all $\xi \in M(QC)$ the restriction of which onto C coincides with the functional $f \mapsto f(t)$ of evaluation at t (i. e. the set of all $\xi \in M(QC)$ for which $\xi(f) = f(t)$ for all $f \in C$). The fibers $M_t(QC)$ are non-empty and compact. Let furthermore $M_t^+(QC)$ (resp. $M_t^-(QC)$) denote the set of all $\xi \in M_t(QC)$ such that $\xi(f) = 0$ whenever $f \in QC$ and $\limsup_{s\to t+0}|f(s)| = 0$ (resp. $\limsup_{s\to t-0}|f(s)| = 0$). Finally, set $M_t^0(QC) = M_t^+(QC) \cap M_t^-(QC)$. The sets

$M_t^+(QC) \setminus M_t^0(QC)$, $M_t^0(QC)$, and $M_t^-(QC) \setminus M_t^0(QC)$ are pairwise disjoint, each of these sets is a proper and non-empty subset of the fiber $M_t(QC)$, and

$$M_t(QC) = \left(M_t^+(QC) \setminus M_t^0(QC)\right) \cup M_t^0(QC) \cup \left(M_t^-(QC) \setminus M_t^0(QC)\right).$$

3.2   The symbol calculus

Localizing $S^\pi(PQC_{L(H)})$ over its central subalgebra $\mathcal{B}$ yields, at $\xi \in M(QC)$, local ideals

$$\mathcal{J}_\xi = \text{clos id } \{T^0(f) + \mathcal{K}_{L(H)}, \quad f \in QC, \, \xi(f) = 0\}$$

of $S^\pi(PQC_{L(H)})$, and local algebras $S_\xi^\pi(PQC_{L(H)}) := S^\pi(PQC_{L(H)})/\mathcal{J}_\xi$ with canonical homomorphisms $\Lambda_\xi : S(PQC_{L(H)}) \to S_\xi^\pi(PQC_{L(H)})$. By the local principle, an operator $A \in S(PQC_{L(H)})$ is invertible modulo $\mathcal{K}_{L(H)}$ if and only if all local cosets $\Lambda_\xi(A)$ are invertible. The following theorem gives a complete representation of these local algebras as algebras of operators.

Theorem 5 *Let* $t \in \mathbf{T}$ *and* $\xi \in M_t(QC)$.
(a) If $\xi \in M_t^\pm(QC) \setminus M_t^0(QC)$, *then the local algebra* $S_\xi^\pi(PQC_{L(H)})$ *is isometrically isomorphic to the* C^*-algebra $L(H) \times L(H)$, *and the isomorphism sends*

$$\begin{aligned}
\Lambda_\xi(T^0(a)) &\quad into \quad (\Phi_\xi(a), \Phi_\xi(a)) &\quad if \quad a \in QC_{L(H)}^s, \\
\Lambda_\xi(T^0(a)) &\quad into \quad (a(t \pm 0), a(t \pm 0)) &\quad if \quad a \in PC_{L(H)}, \\
\Lambda_\xi(P) &\quad into \quad (I, 0). &
\end{aligned}$$

(b) If $\xi \in M_t^0(QC)$, *then the local algebra* $S_\xi^\pi(PQC_{L(H)})$ *is isometrically isomorphic to a subalgebra of the* C^*-algebra of all continuous functions on the interval $[0, 1]$ with values in $L(H)_{2\times2}$, and the isomorphism sends

$$\Lambda_\xi(T^0(a)) \quad into \quad s \mapsto \begin{pmatrix} \Phi_\xi(a) & 0 \\ 0 & \Phi_\xi(a) \end{pmatrix} \quad if \quad a \in QC_{L(H)}^s,$$

$$\Lambda_\xi(T^0(a)) \quad into \quad s \mapsto$$

$$\begin{pmatrix} a(t+0)s + a(t-0)(1-s) & (a(t+0) - a(t-0))\sqrt{s(1-s)} \\ (a(t+0) - a(t-0))\sqrt{s(1-s)} & a(t+0)(1-s) + a(t-0)s \end{pmatrix} \quad if \, a \in PC_{L(H)},$$

$$\Lambda_\xi(P) \quad into \quad s \mapsto \begin{pmatrix} I & 0 \\ 0 & 0 \end{pmatrix}.$$

Here, $L(H) \times L(H)$ stands for the algebra of all ordered pairs (a, b) with $a, b \in L(H)$, provided with componentwise operations and with supremum norm, and $L(H)_{2\times2}$ is the algebra of all two-by-two matrices with entries in $L(H)$ and with usual matrix operations.

The remaining part of this section is devoted to the proof of this theorem.

Proof. We have already remarked that $S(PQC_{L(H)})$ is generated by all Laurent operators $T^0(a)$ with $a \in QC_{L(H)}^s$ and $a \in PC_{L(H)}$ and by the projection P. Further, since the local ideal $\mathcal{J}_\xi$ is larger than the ideal $I_{QC_{L(H)}^s,\xi}$ introduced in Section 2.3, we conclude that

$$\Lambda_\xi(T^0(a)) = \Lambda_\xi(\Phi_\xi(a)) \text{ for all } a \in QC_{L(H)}^s \text{ and } \xi \in M(QC), \tag{18}$$

and moreover, as we have observed in Section 2.6,

$$\Lambda_\xi(T^0(a)) = \Lambda_\xi(a(t+0)T^0(\chi_t^+) + a(t-0)T^0(\chi_t^-)) \tag{19}$$

whenever $a \in PC_{L(H)}$ and $\xi \in M_t(QC)$. (To be precise: It is evident that the operators $T^0(a)$ and $T^0(a(t+0)\chi_t^+ + a(t-0)\chi_t^-) = a(t+0)T^0(\chi_t^+) + a(t-0)T^0(\chi_t^-)$ coincide modulo the local ideal

$$I_t = \text{clos id}_{\mathcal{S}^\pi(PQC_{L(H)})}\{T^0(f) + \mathcal{K}_{L(H)}, \ f \in C(\mathbf{T}), \ f(t) = 0\}$$

resulting from Allan's local principle when localizing over $\mathbf{T} = M(C)$. But this ideal is, for $\xi \in M_t(QC)$, contained in the ideal

$$\mathcal{J}_\xi = \text{clos id}_{\mathcal{S}^\pi(PQC_{L(H)})}\{T^0(f) + \mathcal{K}_{L(H)}, \ f \in QC, \ \xi(f) = 0\}$$

and thus, (19) holds.)

The identities (18) and (19) show that the local algebras $\mathcal{S}_\xi^\pi(PQC_{L(H)})$ with $\xi \in M_t(QC)$ are generated by the cosets $\Lambda_\xi(T^0(\chi_t^+))$, $\Lambda_\xi(P)$ and $\Lambda_\xi(a)$ with $a \in L(H)$. Herein, the cosets $\Lambda_\xi(T^0(\chi_t^+))$ and $\Lambda_\xi(P)$ are self-adjoint idempotents, whereas the cosets $\Lambda_\xi(a)$ can be viewed as *coefficients*, namely they commute (by definition) both with $\Lambda_\xi(T^0(\chi_t^+))$ and $\Lambda_\xi(P)$.

Proposition 11 *The coefficient algebra $\Lambda_\xi(L(H))$ of the local algebra $\mathcal{S}_\xi^\pi(PQC_{L(H)})$ is isometrically isomorphic to $L(H)$.*

Proof. The image $L(H)^\pi$ of $L(H)$ in the quotient algebra $\mathcal{S}^\pi(PQC_{L(H)})$ is isometrically isomorphic to $L(H)$ itself. This can be seen by repeating the arguments in the proof of Proposition 10 (notice that $U_{-n}aU_n = a$ for each constant function a).

So what remains to check is whether $L(H)^\pi \cap \mathcal{J}_\xi = \{0\}$. Then the third isomorphism theorem entails that $\Lambda_\xi(L(H)) \cong L(H)$, and we are done.

Clearly, $L(H)^\pi \cap \mathcal{J}_\xi$ is a closed two-sided ideal of $L(H)^\pi$ which is isomorphic to a closed two-sided ideal of $L(H)$. The only closed two-sided ideals of the latter algebra are $L(H)$ itself, the ideal $K(H)$ of the compact operators on H, and the zero ideal $\{0\}$.

The case $L(H)^\pi \cap \mathcal{J}_\xi \cong L(H)$ would imply that the identity operator belongs to $\mathcal{J}_\xi$ which is impossible since the ideal $\mathcal{J}_\xi$ is proper in $\mathcal{S}^\pi(PQC_{L(H)})$ by the local principle. Suppose that $L(H)^\pi \cap \mathcal{J}_\xi \cong K(H)$. Fix a non-zero vector v in H and let p_v denote the orthogonal projection from H onto its (one-dimensional) subspace $\mathbb{C}v$. As usually, we identify the compact operator p_v with the constant function $t \mapsto p_v$ and obtain from our assumption that the coset $p_v + \mathcal{K}_{L(H)}$ is in $\mathcal{J}_\xi$. Thus, given $\varepsilon > 0$, there are operators $K \in \mathcal{K}_{L(H)}$ and $A_i, C \in \mathcal{S}(PQC_{L(H)})$ with $\|C\| < \varepsilon$ as well as functions $q_i \in QC$ with $\xi(q_i) = 0$ such that

$$p_v = \sum_{i=1}^n A_i T^0(q_i) + K + C.$$

Applying the homomorphism W introduced in the proof of Proposition 10 to both sides of this equality yields

$$p_v = \sum_{i=1}^n W(A_i)T^0(q_i) + W(C) \tag{20}$$

where now $W(A_i) = T^0(a_i)$ with a certain function $a_i \in PQC_{L(H)}$ and $\|W(C)\| < \varepsilon$. Multiplying (20) from both sides with p_v (which is both a projection and a coefficient and, hence, commutes with all operators $T^0(q)$ with $q \in QC$) gives

$$p_v = \sum_{i=1}^{n} p_v T^0(a_i) p_v \cdot p_v T^0(q_i) p_v + p_v W(C) p_v.$$

This equality can be viewed as holding on $l^2_{\mathbb{C}v}(\mathbb{Z}) \cong l^2(\mathbb{Z})$ rather than on $l^2_H(\mathbb{Z})$, hence,

$$I = \sum_{i=1}^{n} T^0(\hat{a}_i) T^0(q_i) + \hat{C}$$

where now $\hat{a}_i \in PQC$, $\hat{C} \in L(l^2(\mathbb{Z}))$ with $\|\hat{C}\| < \varepsilon$, and I is the identity operator on $l^2(\mathbb{Z})$. This shows that the identity operator belongs to the local ideal of the algebra of scalar-valued PQC-functions generated by the QC-functions q with $\xi(q) = 0$, and this result again contradicts the properness of local ideals stated by the local principle. Thus, $L(H)^\pi \cap \mathcal{J}_\xi = \{0\}$. ∎

Let now $\xi \in M_t^+(QC) \setminus M_t^0(QC)$. Then $\Lambda_\xi(T^0(\chi_t^-)) = 0$ (see [15] or [1], Proposition 4.85) and, hence, the projection $\Lambda_\xi(T^0(\chi_t^+))$ is nothing but the identity element in the local algebra $S_\xi^\pi(PQC_{L(H)})$. On the other hand, $\Lambda_\xi(P)$ is a non-trivial projection in this algebra. Indeed, suppose for example, the point 1 is not in the spectrum of $\Lambda_\xi(P)$. Then $\Lambda_\xi(I - P)$ is invertible, i.e. there exist operators $B, C \in S(PQC_{L(H)})$ such that $C + \mathcal{K}_{L(H)} \in \mathcal{J}_\xi$ and $B(I - P) = I + C$. Applying the homomorphism W to both sides of this equality gives $0 = I + W(C)$ where $W(C)$ is a Laurent operator, say $W(C) = T^0(c)$, with c being a $PQC_{L(H)}$-function which belongs to the local ideal associated with $\xi \in M(QC)$. This again contradicts the local principle. Thus, the point 1 belongs to the spectrum of $\Lambda_\xi(P)$. Analogously (replace W by $\tilde{W} : A \mapsto$ s-lim$_{n\to\infty} U_n A U_{-n}$) one can check that the point 0 is contained in this spectrum, too, i.e. $\Lambda_\xi(P)$ is an non-trivial projection. This observation together with the characterization of $\Lambda_\xi(L(H))$ in Proposition 11 yields assertion (a) of the theorem in case of the "+"-sign, and the proof for the "−"-sign is analogous.

Let now $\xi \in M_t^0(QC)$. Then both $p := \Lambda_\xi(P)$ and $q := \Lambda_\xi(T^0(\chi_t^+))$ are non-trivial projections in $S_\xi^\pi(PQC_{L(H)})$ and, moreover,

$$\text{spectrum of } pqp = [0, 1]. \tag{21}$$

Indeed, since p and q are self-adjoint projections, the spectrum of pqp is contained in $[0, 1]$. On the other hand, repeating arguments from the proof of Proposition 11 and taking into account that p_v commutes with p and q and that $p_v \mathcal{K}_{L(H)} p_v$ resp. $p_v \mathcal{J}_\xi p_v$ are contained in $\mathcal{K}_{\mathbb{C}}$ resp. $\mathcal{J}_\xi$ for the scalar case, one can show that, whenever $pqp - ve$ with $e := \Lambda_\xi(I)$ is invertible in the operator-valued setting then it is also invertible in the scalar case. But the spectrum of pqp in the scalar case is actually $[0, 1]$ (see [15] or [1], Proposition 4.85), and this gives the reverse inclusion.

What results is that we arrived in the world described by the two-projections theorem with arbitrary coefficient algebra as considered in [5]. In order to specify the results of [5] to our context we need some more notations. One can show (see, e.g., [14]) that the spectrum of $b := pqp + (e - p)(e - q)(e - p)$ is $[0, 1]$ again, and that b belongs to the center of alg $(e, p, q) = S_\xi^\pi(PQC_{L(H)})$. So we can localize this algebra once more, now with respect to the maximal ideal space if its central subalgebra (simply) generated by b which is homeomorphic to the interval $[0, 1]$. Given $s \in [0, 1]$, let us denote by $\mathcal{J}_{\xi,s}$ the associated local ideal in $S_\xi^\pi(PQC_{L(H)})$, by $S_{\xi,s}^\pi(PQC_{L(H)})$ the local algebra $S_\xi^\pi(PQC_{L(H)})/\mathcal{J}_{\xi,s}$, and by $\Lambda_{\xi,s}$ the canonical homomorphism from $\mathcal{S}(PQC_{L(H)})$ onto $S_{\xi,s}^\pi(PQC_{L(H)})$.

Further we let $\mathcal{J}_{\xi,0,0}$ and $\mathcal{J}_{\xi,0,1}$ stand for the smallest closed two-sided ideals of the local algebra $S_{\xi,0}^\pi(PQC_{L(H)})$ which contain the cosets $(e - \frac{p+q}{2}) + \mathcal{J}_{\xi,0}$ and $\frac{p+q}{2} + \mathcal{J}_{\xi,0}$, respectively, and we denote by $\Lambda_{\xi,0,0}$ and $\Lambda_{\xi,0,1}$ the canonical homomorphisms from $\mathcal{S}(PQC_{L(H)})$ onto $S_{\xi,0}^\pi(PQC_{L(H)})/\mathcal{J}_{\xi,0,0}$ and $S_{\xi,0}^\pi(PQC_{L(H)})/\mathcal{J}_{\xi,0,1}$. Analogously, one introduces the homomorphisms $\Lambda_{\xi,1,0}$ and $\Lambda_{\xi,1,1}$ associated with the other end-point of the interval. Then the two-projections theorem of [5] can be restated as follows.

Proposition 12 *There is a homomorphism Ψ from $S_\xi^\pi(PQC_{L(H)})$ into the algebra of all functions on $[0, 1]$ which take at $s \in (0, 1)$ a value in $(\Lambda_{\xi,s}(L(H)))_{2\times2}$ and at $s \in \{0, 1\}$ a value in $\Lambda_{\xi,s,0}(L(H)) \times \Lambda_{\xi,s,1}(L(H))$. This homomorphism sends $a \in \Lambda_\xi(L(H)) \cong L(H)$, p and q into the functions*

$$s \mapsto \begin{cases} \begin{pmatrix} 1 & 0 \\ 0 & 1 \end{pmatrix} \Lambda_{\xi,s}(a) & \text{if } s \in (0, 1) \\ (\Lambda_{\xi,s,0}(a), \Lambda_{\xi,s,1}(a)) & \text{if } s \in \{0, 1\}, \end{cases}$$

$$s \mapsto \begin{cases} \begin{pmatrix} 1 & 0 \\ 0 & 0 \end{pmatrix} \Lambda_{\xi,s}(e) & \text{if } s \in (0, 1) \\ (\Lambda_{\xi,s,0}(e), 0) & \text{if } s \in \{0, 1\}, \end{cases}$$

$$s \mapsto \begin{cases} \begin{pmatrix} s & \sqrt{s(1-s)} \\ \sqrt{s(1-s)} & 1-s \end{pmatrix} \Lambda_{\xi,s}(a) & \text{if } s \in (0, 1) \\ (\Lambda_{\xi,s,0}(e), 0) & \text{if } s = 0 \\ (0, \Lambda_{\xi,s,1}(e)) & \text{if } s = 1, \end{cases}$$

respectively. A coset $\Lambda_\xi(A) \in S_\xi^\pi(PQC_{L(H)})$ is invertible if and only if the function $\Psi(\Lambda_\xi(A))$ is invertible at each point $s \in [0, 1]$.

In order to get assertion (b) of Theorem 5, it remains to show that all local algebras $\Lambda_{\xi,s}(L(H))$, $\Lambda_{\xi,s,0}(L(H))$ and $\Lambda_{\xi,s,1}(L(H))$ are isometrically isomorphic to $L(H)$ itself. In any case, the intersection of $\Lambda_\xi(L(H))$ with a local ideal can be either isomorphic to $L(H)$, to $K(H)$, or to $\{0\}$. The first possibility would imply that the local ideal coincides with the whole algebra $\Lambda_\xi(L(H))$ which contradicts the local principle. If the intersection would be $K(H)$ for at least one point $s \in [0, 1]$, then the symbol function $\Psi(\Lambda_\xi(k))$ would be zero at s for all compact operators k, and in particular for the operator p_v introduced in the proof of Proposition 11. But then, as the arguments employed in this proof show, the symbol function of the scalar identity operator would be zero at s which is evidently false. Thus, the intersections are $\{0\}$ in any case, and this finishes the proof of Theorem 5. ∎

References

[1] A. BÖTTCHER, B. SILBERMANN, Analysis of Toeplitz operators. – Akademie Verlag, Berlin, 1989, and Springer Verlag, Berlin, 1990.

[2] A. BÖTTCHER, B. SILBERMANN, Operator-valued Szegö-Widom limit theorems. – In: Toeplitz operators and related topics: The Harold Widom anniversary Volume. – Operator Theory 71, Birkhäuser Verlag, Basel, Boston, Berlin, 1994, 33 – 53.

[3] A. BÖTTCHER, B. SILBERMANN, Infinite Toeplitz and Hankel matrices with operator-valued entries. – SIAM Journal on Mathematical Analysis (to appear).

[4] T. EHRHARDT, S. ROCH, B. SILBERMANN, Finite section method for singular integrals with operator-valued PQC-coefficients. – this volume.

[5] T. FINCK, S. ROCH, B. SILBERMANN, Two projection theorems and symbol calculus for operators with massive local spectra. – Math. Nachr. 162(1993), 167 – 185.

[6] J. B. GARNETT, Bounded analytic functions. – Academic Press, New York, 1981.

[7] I. GOHBERG, M. A. KAASHOEK, Projection method for Block Toeplitz operators with operator-valued symbols. – In: Toeplitz operators and related topics: The Harold Widom anniversary Volume. – Operator Theory 71, Birkhäuser Verlag, Basel, Boston, Berlin, 1994, 79 – 104.

[8] I. GOHBERG, N. KRUPNIK, On the algebra generated by one-dimensional singular integral operators with piecewise continuous coefficients. – Funkts. Analiz i ego prilozh. 4, 3(1970), 26 – 36 (Russian).

[9] I. GOHBERG, N. KRUPNIK, Singular integral operators with piecewise continuous coefficients and their symbols. – Izv. AN SSSR 35, 4(1971), 940 – 964 (Russian).

[10] I. GOHBERG, N. KRUPNIK, Extension theorems for invertibility symbols in Banach algebras. – IEOT 15(1992), 991 - 1010.

[11] R. HAGEN, S. ROCH, B. SILBERMANN, Spectral Theory of Approximation Methods for Convolution Equations. – Operator Theory 74, Birkhäuser Verlag, Basel, Boston, Berlin, 1995.

[12] P. R. HALMOS, Two subspaces. – Trans. Amer. Math. Soc. 144(1969), 381 – 389.

[13] S. G. MICHLIN, S. PRÖSSDORF, Singuläre Integraloperatoren. – Akademie-Verlag, Berlin 1980 (Extended English translation: Singular integral operators.)

[14] S. ROCH, B. SILBERMANN, Algebras generated by idempotents and the symbol calculus for singular integral operators. – IEOT 11(1988), 385 – 419.

[15] D. SARASON, Toeplitz operators with piecewise quasicontinuous symbols. – Indiana Univ. Math. J. 26:5 (1977), 817- 838.

[16] D. SARASON, Function theory on the unit circle. – Virginia Polytechnic Institute and State Univ., Blacksburg, 1978.

[17] B. SILBERMANN, Asymptotics for Toeplitz operators with piecewise quasicontinuous symbols and related questions. – Math. Nachr. 125 (1986), 179-190.

[18] B. SILBERMANN, Local objects in the theory of Toeplitz operators, IEOT **9**(1986), 706 – 738.

[19] B. SILBERMANN, The C^*-algebra generated by Toeplitz and Hankel operators with piecewise quasicontinuous symbols. – IEOT **10**(1987), 730-738.

[20] E. M. STEIN, Singular integrals and differentiability properties of functions. – Princeton Univ. Press, Princeton, New Jersey, 1970.

Torsten Ehrhardt and Bernd Silbermann
Technische Universität Chemnitz-Zwickau
Fakultät für Mathematik
D 09107 Chemnitz
Germany
ehrhardt@mathematik.tu-chemnitz.de
silbermn@mathematik.tu-chemnitz.de

Steffen Roch
Universität Leipzig
Mathematisches Institut
Augustusplatz 10 – 11
D 04109 Leipzig
Germany
roch@miaix550.mathematik.uni-leipzig.de

MSC 1991:   Primary   47B35
            Secondary 47A35, 46E40, 45E10, 45F15, 46L05

Operator Theory
Advances and Applications, Vol. 90
© 1996 Birkhäuser Verlag Basel/Switzerland

Finite Section Method for singular integrals with operator-valued PQC-coefficients

Torsten Ehrhardt   Steffen Roch*

Bernd Silbermann

Let S be the smallest closed subalgebra of operators on $L_H^2(\mathbf{Z})$ which contains all one-dimensional singular integral operators with operator-valued piecewise quasi-continuous coefficients. We consider the stability problem for operator sequences belonging to the algebra which is generated by the sequences of finite sections of operators in S. The investigations are strongly based on C*-algebra methods, and we prove that a sequence belonging to that algebra is stable if and only if a certain well-defined collection of operators consists only of invertible ones.

1. Introduction

Let X be a Hilbert space, and let A be a linear bounded operator on X. The approximate solution of the equation

$$Ax = y \quad (x, y \in X) \tag{1}$$

is frequently carried out by the following scheme: One chooses a sequence of projections $R_n \in \mathcal{L}(X)$ which converge strongly to the identity I as $n \to \infty$, as well as an appropriate sequence of operators $A_n \in \mathcal{L}(\text{Im } R_n)$. Instead of (1), one considers the sequence of equations

$$A_n x_n = R_n y \quad (x_n \in \text{Im } R_n, \, y \in X). \tag{2}$$

The approximation method (2) is said to be *applicable* to the operator A if there is an n_0 such that (2) is uniquely solvable for all $n \geq n_0$ and all $y \in X$, and if the sequence $(x_n)_{n \geq n_0}$

---

*Research supported by a DFG Heisenberg grant

converges in the norm to a solution x of (1). The investigation of this question is heavily related the stability of the sequence (A_n). We recall that a sequence (A_n) of operators is said to be *stable* if there is an n_0 such that A_n is invertible in $\mathcal{L}(\text{Im } R_n)$ for all $n \geq n_0$ and

$$\sup_{n \geq n_0} \|A_n^{-1}\|_{\mathcal{L}(\text{Im } R_n)} < \infty. \tag{3}$$

The following result is well-known (see [9], Section 1.3.1).

Proposition 1.1 *Let $A_n \to A$ strongly. Then the method (2) applies to the operator A if and only if A is invertible and the sequence (A_n) is stable.*

Let $\mathcal{S}(PQC_{\mathcal{L}(H)})$ be the smallest closed algebra which contains all singular integral operator on $\mathbb{T}$ with operator-valued piecewise quasicontinuous coefficients. This algebra contains also all Toeplitz operators with piecewise quasicontinuous coefficients. Further, let $\mathcal{F}(PQC_{\mathcal{L}(H)})$ be the smallest closed algebra which contains all sequences of finite sections of operators in $\mathcal{S}(PQC_{\mathcal{L}(H)})$. The precise definition of these algebras will be given in Section 5. Since one can show that each sequence in $\mathcal{F}(PQC_{\mathcal{L}(H)})$ converges strongly to an operator $A \in \mathcal{S}(PQC_{\mathcal{L}(H)})$, the sequences in $\mathcal{F}(PQC_{\mathcal{L}(H)})$ can be regarded as approximation sequences for operators in $\mathcal{S}(PQC_{\mathcal{L}(H)})$, and the problem which remains to consider is the stability of these sequences.

It is the aim of the present paper to establish stability criteria for all sequences (A_n) which belong to $\mathcal{F}(PQC_{\mathcal{L}(H)})$. We prove that such a sequence is stable if and only if a certain collection of operators, which are assigned to the sequence (A_n) via *-homomorphisms, consists only of invertible operators. This result extends significantly former results from [19], [20] and [15]. Some historical remarks can be found in [9] and also in [15].

2. Preliminaries

In this section we fix some notation and mention basic results, which we need subsequently.

Let $\mathbb{C}$ (resp. $\mathbb{R}$, $\mathbb{R}^+$) denote the set of all complex (resp. real, non-negative real) numbers. Let $\mathbb{T} = \{z \in \mathbb{C} : |z| = 1\}$ be the unit circle, and let $\mathbb{Z}$ (resp. $\mathbb{Z}^+$) refer to the set of integers (resp. non-negative integers). Finally, set $\mathbb{Z}_n = \{-n, -n+1, \ldots, n-1\}$.

Throughout the paper let H be a Hilbert space. By $l_H^2(\mathbb{Z})$ we denote the Hilbert space of all two-sided sequences $x = (x_n)_{n \in \mathbb{Z}}$ with $x_n \in H$ for which

$$\|x\|_{l_H^2(\mathbb{Z})} := \left(\sum_{n \in \mathbb{Z}} \|x_n\|_H^2 \right)^{1/2} < \infty. \tag{4}$$

Replacing $\mathbb{Z}$ by $\mathbb{Z}^+$ or $\mathbb{Z}_n$, we can similarily define the Hilbert spaces $l_H^2(\mathbb{Z}^+)$ and $l_H^2(\mathbb{Z}_n)$. Further, for $M = \mathbb{R}$ (resp. $M = \mathbb{R}^+$, $M = [-1,1]$, $M = [0,1]$) we denote by $L_H^2(M)$ the Hilbert space of all H-valued Lebesgue measurable functions f on M for which

$$\|f\|_{L_H^2(M)} := \left(\int_M \|f(x)\|_H^2 \, dx \right)^{1/2} < \infty. \tag{5}$$

These Hilbert spaces are equipped with a scalar product $\langle \cdot, \cdot \rangle$ and with algebraic operations as customary. In the case where $H = \mathbb{C}$ we omit the index H.

For a Banach space X, let $\mathcal{L}(X)$ refer to the Banach algebra of all linear bounded operators on X, and let $\mathcal{K}(X)$ refer to the ideal of all compact operators. In particular, $\mathcal{L}(H)$ is a C*-algebra, whose unit element we denote by e.

Let $L^\infty_{\mathcal{L}(H)}$ stand for the C*-algebra of all Lebesgue measurable and essentially bounded functions a on $\mathbb{T}$ with values in $\mathcal{L}(H)$ (cf. [7], Section 2.1). The norm in $L^\infty_{\mathcal{L}(H)}$ is defined by

$$\|a\|_{L^\infty_{\mathcal{L}(H)}} \;:=\; \operatorname{ess\,sup}_{t\in\mathbb{T}} \|a(t)\|_{\mathcal{L}(H)}. \tag{6}$$

Again, if $H = \mathbb{C}$, we omit the index $\mathcal{L}(H)$. The C*-algebras $\mathcal{L}(H)$ and L^∞ are *-isomorphic to certain *-subalgebras of $L^\infty_{\mathcal{L}(H)}$. In fact, we can assign $a \in \mathcal{L}(H)$ to the constant function $t \mapsto a$ on $\mathbb{T}$ and $f \in L^\infty$ to the "scalar" function $t \mapsto f(t)e$, $t \in \mathbb{T}$. We will use this identification subsequently without noting.

The *Laurent operator* $L(a)$ of a function $a \in L^\infty_{\mathcal{L}(H)}$ is defined by

$$L(a) \;:\; (x_n)_{n\in\mathbb{Z}} \;\mapsto\; \left(\sum_{k\in\mathbb{Z}} a_{n-k} x_k \right)_{n\in\mathbb{Z}}, \tag{7}$$

where a_n is the n-th Fourier coefficients of a :

$$a_n \;=\; \frac{1}{2\pi} \int_0^{2\pi} a(e^{i\phi}) e^{-in\phi}\, d\phi. \tag{8}$$

In [11], it is shown that $L(a)$ is a linear bounded operator on $l^2_H(\mathbb{Z})$. We note that the reverse is also true: Suppose that $(a_n)_{n\in\mathbb{Z}}$ is a sequence with $a_n \in \mathcal{L}(H)$. Then, if the operator defined by (7) is bounded on $l^2_H(\mathbb{Z})$, there is a (uniquely determined) function $a \in L^\infty_{\mathcal{L}(H)}$, the Fourier coefficients of which are a_n. Moreover, $\|L(a)\|_{\mathcal{L}(l^2_H(\mathbb{Z}))} = \|a\|_{L^\infty_{\mathcal{L}(H)}}$, $L(ab) = L(a)L(b)$ and $L(a^*) = L(a)^*$. Hence, $L^\infty_{\mathcal{L}(H)}$ is *-isomorphic to the *-subalgebra of all Laurent operators in $\mathcal{L}(l^2_H(\mathbb{Z}))$. Again, we will use this identification without noting, and for brevity we will occasionally write a instead of $L(a)$.

Further, we consider the following linear bounded operators on $l^2_H(\mathbb{Z})$:

$$P \;:\; (x_n)_{n\in\mathbb{Z}} \;\mapsto\; (y_n)_{n\in\mathbb{Z}} \quad \text{with} \quad y_n = \begin{cases} x_n & \text{if} \quad n \geq 0 \\ 0 & \text{if} \quad n < 0, \end{cases} \tag{9}$$

$$J \;:\; (x_n)_{n\in\mathbb{Z}} \;\mapsto\; (x_{-1-n})_{n\in\mathbb{Z}}. \tag{10}$$

We denote the identity mapping on $l^2_H(\mathbb{Z})$ by I, and we set $Q := I - P$. For these operators the relations $P^* = P = P^2$, $J = J^*$, $J^2 = I$ and $JPJ = Q$ hold. The *singular integral operator* on $\mathbb{T}$ is given by $S_\mathbb{T} = P - Q = 2P - I$. Instead of $S_\mathbb{T}$ we will consider P and Q.

For $a \in L^\infty_{\mathcal{L}(H)}$, the *Toeplitz operator* is $T(a) = PL(a)P$, and the *Hankel operator* is $H(a) = PL(a)JP$. Identifying the image of the projection P with $l^2_H(\mathbb{Z}^+)$, these operators can be considered as acting on $l^2_H(\mathbb{Z}^+)$. For $a \in L^\infty_{\mathcal{L}(H)}$, let $\tilde{a} \in L^\infty_{\mathcal{L}(H)}$ denote the function $\tilde{a}(t) = a(1/t)$, $t \in \mathbb{T}$. Then $JL(a)J = L(\tilde{a})$ and therefore

$$T(ab) \;=\; T(a)T(b) + H(a)H(\tilde{b}). \tag{11}$$

For $n \in \mathbb{Z}^+$, we introduce the following linear bounded operators:

$$P_n : \; (x_k)_{k\in\mathbb{Z}} \;\mapsto\; (y_k)_{k\in\mathbb{Z}}, \qquad y_k = \begin{cases} x_k & \text{if } -n \le k < n \\ 0 & \text{if } k < -n \text{ or } k \ge n \end{cases} \tag{12}$$

$$W_n : \; (x_k)_{k\in\mathbb{Z}} \;\mapsto\; (y_k)_{k\in\mathbb{Z}}, \qquad y_k = \begin{cases} x_{n-1-k} & \text{if } 0 \le k < n \\ x_{-n-1-k} & \text{if } -n \le k < 0 \\ 0 & \text{if } k < -n \text{ or } k \ge n \end{cases} \tag{13}$$

$$V_n : \; (x_k)_{k\in\mathbb{Z}} \;\mapsto\; (y_k)_{k\in\mathbb{Z}}, \qquad y_k = \begin{cases} 0 & \text{if } -n \le k < n \\ x_{k-n} & \text{if } k \ge n \\ x_{k+n} & \text{if } k < -n \end{cases} \tag{14}$$

$$V_{-n} : \; (x_k)_{k\in\mathbb{Z}} \;\mapsto\; (y_k)_{k\in\mathbb{Z}}, \qquad y_k = \begin{cases} x_{k+n} & \text{if } k \ge 0 \\ x_{k-n} & \text{if } k < 0 \end{cases} \tag{15}$$

$$U_n : \; (x_k)_{k\in\mathbb{Z}} \;\mapsto\; (x_{k-n})_{k\in\mathbb{Z}} \tag{16}$$

$$U_{-n} : \; (x_k)_{k\in\mathbb{Z}} \;\mapsto\; (x_{k+n})_{k\in\mathbb{Z}} \tag{17}$$

Further, set $Q_n := I - P_n$. Then we have $P_n^* = P_n = P_n^2 = W_n^2$, $W_n^* = W_n = W_n P_n = P_n W_n$, $V_n V_{-n} = Q_n$, $V_{-n} V_n = I$, $V_n^* = V_{-n}$, $U_n = L(t^n)$, and $U_{-n} = L(t^{-n})$. The remarkable identity

$$P_n T(ab) P_n \;=\; P_n T(a) P_n T(b) P_n + P_n H(a) H(\tilde{b}) P_n + W_n H(\tilde{a}) H(b) W_n \tag{18}$$

goes back to Widom [21]. It can be derived from (11) and from

$$P_n T(a) Q_n T(b) P_n \;=\; W_n H(\tilde{a}) H(b) W_n. \tag{19}$$

We introduce the following *-subalgebras of $L^\infty_{\mathcal{L}(H)}$. Let $PC_{\mathcal{L}(H)}$ denote the set of all $\mathcal{L}(H)$–valued piecewise continuous functions, i. e., functions $p \in L^\infty_{\mathcal{L}(H)}$ for which the one-sided limits $\lim\limits_{t\to\tau\pm0} p(t)$ exist at all points $\tau \in \mathbb{T}$. Then we have (see [7], Proposition 8)

$$PC_{\mathcal{L}(H)} \;=\; \operatorname{clos}_{L^\infty_{\mathcal{L}(H)}} \left\{ \sum_i a_i p_i \;:\; a_i \in \mathcal{L}(H),\; p_i \in PC \right\}. \tag{20}$$

By $C_{\mathcal{L}(H)}(\mathbb{T})$ we denote the set of all $\mathcal{L}(H)$–valued continuous functions on $\mathbb{T}$. $H^\infty_{\mathcal{L}(H)}$ (resp. $\overline{H^\infty_{\mathcal{L}(H)}}$) refers to the set of all functions $a \in L^\infty_{\mathcal{L}(H)}$ whose Fourier coefficients a_n vanish for $n < 0$ (resp. $n > 0$).

It was already discussed in [7], Section 2, that there exist several possibilities of defining quasicontinuous functions in the $\mathcal{L}(H)$–valued case. We quote the following two:

$$QC_{\mathcal{L}(H)} \ := \ \left(C_{\mathcal{L}(H)}(\mathbb{T}) + H^\infty_{\mathcal{L}(H)} \right) \cap \left(C_{\mathcal{L}(H)}(\mathbb{T}) + \overline{H^\infty_{\mathcal{L}(H)}} \right) \tag{21}$$

$$QC^s_{\mathcal{L}(H)} \ := \ \mathrm{clos}_{L^\infty_{\mathcal{L}(H)}} \left\{ \sum_i a_i q_i \ : \ a_i \in \mathcal{L}(H), \ q_i \in QC \right\} \tag{22}$$

Both $QC^s_{\mathcal{L}(H)}$ and $QC_{\mathcal{L}(H)}$ are *-subalgebras of $L^\infty_{\mathcal{L}(H)}$, and $QC^s_{\mathcal{L}(H)} \subseteq QC_{\mathcal{L}(H)}$, where the inclusion is proper if and only if $\dim H = \infty$. In Section 10 we will discuss $\mathcal{L}(H)$–valued quasicontinuous functions once more. Finally, let $PQC_{\mathcal{L}(H)}$ be the smallest closed sub-algebra of $L^\infty_{\mathcal{L}(H)}$ including both $PC_{\mathcal{L}(H)}$ and $QC^s_{\mathcal{L}(H)}$. We call such functions *piecewise quasicontinuous functions*.

The space of all maximal ideals of the commutative C*-algebra QC is denoted by $M(QC)$. Furthermore we introduce $M_\tau(QC)$, $M^0_\tau(QC)$, $M^+_\tau(QC)$ and $M^-_\tau(QC)$ (see [7], Section 3.1, [1] or [12]).

Also the following notions and results are adopted from [7], Section 2. By a construc-tion, which is described there, to each $q \in QC_{\mathcal{L}(H)}$ and $\xi \in M(QC)$, we can assign an element $\Phi_\xi(q) \in \mathcal{L}(H)$. A closed algebra $\mathfrak{B}$ with $QC^s_{\mathcal{L}(H)} \subseteq \mathfrak{B} \subseteq QC_{\mathcal{L}(H)}$ is called *locally trivial* at $\xi \in M(QC)$ if for each $q \in \mathfrak{B}$ there is an $a \in \mathcal{L}(H)$ such that $q - a \in I_{\xi,\mathcal{L}(H)}$, where $I_{\xi,\mathcal{L}(H)}$ is the closed ideal

$$I_{\xi,\mathcal{L}(H)} \ := \ \mathrm{clos\,id}_{\mathfrak{B}} \left\{ f \ : \ f \in QC, \ \xi(f) = 0 \right\}. \tag{23}$$

In this case, we have $\Phi_\xi(q) = a$ (which means in particular that a is uniquely determined). By [7], Theorem 2, $QC^s_{\mathcal{L}(H)}$ is locally trivial at all points $\xi \in M(QC)$, and there is no larger closed algebra with this property.

3. A generalization of compactness and of strong convergence

To investigate stability, we pursue a method that was presented for the scalar case, $H = \mathbb{C}$, in [1], Chapter 7. A detailed examination yields that this method makes use of the following trivial facts: Firstly, a Hankel operator of a continuous function is compact. Secondly, the set of all compact operators is an ideal in $\mathcal{L}(l^2(\mathbb{Z}))$. Finally, if $A_n \to A$ (resp. $A^*_n \to A^*$) strongly and if K is compact, then $A_n K \rightrightarrows AK$ (resp. $KA_n \rightrightarrows KA$). Hence, in the scalar case, this approach relies essentially on compactness and strong convergence.

However, if $\dim H = \infty$, a Hankel operator of a continuous function fails in general to be compact. A simple counterexample is $H(te) = PP_1$. For that reason, we will consider Q_n–compact operators instead of compact ones. This notion was proposed in [3] and is justified by the circumstance that Hankel operators of continuous functions are Q_n–compact (see Proposition 3.5). Furthermore, since the set $\mathcal{K}$ of Q_n–compact operators is no longer an ideal in $\mathcal{L}(l^2_H(\mathbb{Z}))$ if $\dim H = \infty$, we must replace $\mathcal{L}(l^2_H(\mathbb{Z}))$ by another algebra $\mathcal{A}$, in which $\mathcal{K}$ is yet an ideal. Finally, if $\dim H = \infty$, also the above relationship concerning strong and norm convergence fails for $K \in \mathcal{K}$, and hence we have to introduce a different notion of "strong convergence". To start with, we define $\mathcal{K}$ and $\mathcal{A}$.

Let $\mathcal{K}$ stand for the set of all operators $K \in \mathcal{L}(l_H^2(\mathbb{Z}))$ for which

$$\|KQ_n\|_{\mathcal{L}(l_H^2(\mathbb{Z}))} \to 0 \quad \text{and} \quad \|Q_n K\|_{\mathcal{L}(l_H^2(\mathbb{Z}))} \to 0$$

as $n \to \infty$. Further, let $\mathcal{A}$ denote the set of all operators $A \in \mathcal{L}(l_H^2(\mathbb{Z}))$ for which both $AK \in \mathcal{K}$ and $KA \in \mathcal{K}$ whenever $K \in \mathcal{K}$.

Proposition 3.1

(a) $\mathcal{A}$ is a *-subalgebra of $\mathcal{L}(l_H^2(\mathbb{Z}))$, and $\mathcal{K}$ is a *-ideal of $\mathcal{A}$.

(b) $\|A\|_\mathcal{A} = \displaystyle\sup_{K \in \mathcal{K} \backslash \{0\}} \frac{\|AK\|_\mathcal{A}}{\|K\|_\mathcal{A}}$ for all $A \in \mathcal{A}$.

Proof. Assertion (a) can be shown straightforwardly. (b): Since $Q_n^* = Q_n \to 0$ strongly on $l_H^2(\mathbb{Z})$, we have $\mathcal{K}(l_H^2(\mathbb{Z})) \subseteq \mathcal{K}$. For $x \in l_H^2(\mathbb{Z})$, the operator K_x defined by

$$K_x y = x \cdot \langle x, y \rangle_{l_H^2(\mathbb{Z})} \qquad (y \in l_H^2(\mathbb{Z}))$$

is compact, and $\|K_x\|_\mathcal{A} = \|x\|_{l_H^2(\mathbb{Z})}^2$ and $\|AK_x\|_\mathcal{A} = \|Ax\|_{l_H^2(\mathbb{Z})}\|x\|_{l_H^2(\mathbb{Z})}$. Thus

$$\sup_{K \in \mathcal{K} \backslash \{0\}} \frac{\|AK\|_\mathcal{A}}{\|K\|_\mathcal{A}} \geq \sup_{x \in l_H^2(\mathbb{Z}) \backslash \{0\}} \frac{\|AK_x\|_\mathcal{A}}{\|K_x\|_\mathcal{A}} = \sup_{x \in l_H^2(\mathbb{Z}) \backslash \{0\}} \frac{\|Ax\|_{l_H^2(\mathbb{Z})}}{\|x\|_{l_H^2(\mathbb{Z})}} = \|A\|_\mathcal{A}.$$

The reverse inequality is evident. ∎

Remark 3.2 If $\dim H < \infty$, then $\mathcal{K} = \mathcal{K}(l_H^2(\mathbb{Z}))$ and $\mathcal{A} = \mathcal{L}(l_H^2(\mathbb{Z}))$. In contrast, if $\dim H = \infty$, then $\mathcal{K} \supset \mathcal{K}(l_H^2(\mathbb{Z}))$ and $\mathcal{A} \subset \mathcal{L}(l_H^2(\mathbb{Z}))$, and the inclusions are proper.

Example 3.3 We have $I, P, Q, J \in \mathcal{A}$. Moreover, $L(a) \in \mathcal{A}$ for all $a \in L^\infty_{\mathcal{L}(H)}$.

Proof. (See also [3].) That $I, P, Q, J \in \mathcal{A}$ is readily shown by using the fact that these operators commute with each Q_n. To prove $L(a) \in \mathcal{A}$, we have to show that $\|Q_n L(a)K\| \to 0$ and $\|KL(a)Q_n\| \to 0$ for $K \in \mathcal{K}$. Given any $\varepsilon > 0$, we choose an n_0 such that $\|Q_{n_0}K\| < \varepsilon$. Then we have

$$\|Q_n L(a)K\| \leq \varepsilon \|L(a)\| + \|Q_n L(a)P_{n_0}\| \|K\|.$$

We estimate $Q_n L(a)P_{n_0}$, considering its matrix representation, by

$$\|Q_n L(a)P_{n_0}\|_{\mathcal{L}(l_H^2(\mathbb{Z}))}^2 \leq 2n_0 \sum_{|k| > n - n_0} \|a_k\|_{\mathcal{L}(H)}^2.$$

Since $L^\infty_{\mathcal{L}(H)} \subset L^2_{\mathcal{L}(H)}$, we have $\sum \|a_k\|_{\mathcal{L}(H)}^2 \leq \|a\|_{L^\infty_{\mathcal{L}(H)}}^2 < \infty$. Hence $\|Q_n L(a)P_{n_0}\| \to 0$ as $n \to \infty$, and it follows that $\|Q_n L(a)K\| \to 0$. That $\|KL(a)Q_n\| \to 0$ is shown similarly. ∎

Example 3.4 $P_n A P_n \in \mathcal{K}$ for all $A \in \mathcal{L}(l_H^2(\mathbb{Z}))$. In particular, $P_n \in \mathcal{K}$ and $W_n \in \mathcal{K}$.

Proof. The assertion is evident since $Q_m P_n = P_n Q_m = 0$ for all $m \geq n$. ∎

The following proposition describes the relationship between Hankel operators of quasi-continuous functions and the above introduced concept of Q_n–compactness. It is the immediate consequence of the $\mathcal{L}(H)$–valued version of the well-known Hartman Theorem (see [3], Proposition 3.2, and [8]). Recall that the Hankel operators are given by $H(f) = PfQJ$ and $H(\tilde{f}) = JQfP$.

Proposition 3.5 Let $f \in L^\infty_{\mathcal{L}(H)}$. Then $f \in QC_{\mathcal{L}(H)}$ if and only if $PfQ \in \mathcal{K}$ and $QfP \in \mathcal{K}$.

Now we introduce a new concept of "strong convergence" for operators contained in $\mathcal{A}$. Let $(A_n)_{n=1}^\infty$ be a sequence of operators $A_n \in \mathcal{A}$. We say that A_n converges $\mathcal{K}$-strongly to an operator A if, for all $K \in \mathcal{K}$, both

$$\|K(A_n - A)\|_{\mathcal{L}(l_H^2(\mathbb{Z}))} \to 0 \quad \text{and} \quad \|(A_n - A)K\|_{\mathcal{L}(l_H^2(\mathbb{Z}))} \to 0$$

as $n \to \infty$. In this case we write "$A_n \to A$ $\mathcal{K}$-strongly" or "$A = \text{s-}\lim_{n\to\infty} A_n$".

Remark 3.6 We briefly mention how $\mathcal{K}$-strong convergence is related to customary strong convergence. If $\dim H < \infty$, then $A = \text{s-}\lim A_n$ if and only if both $A_n \to A$ and $A_n^* \to A^*$ strongly. If $\dim H = \infty$, then $A = \text{s-}\lim A_n$ implies that $A_n \to A$ and $A_n^* \to A^*$ strongly; whereas the reverse direction fails in general to be true.

Proposition 3.7

(a) If $A_n \to A$ $\mathcal{K}$-strongly, then $A \in \mathcal{A}$ and

$$\|A\|_{\mathcal{A}} \leq \liminf_{n\to\infty} \|A_n\|_{\mathcal{A}} \leq \sup_{n\geq 1} \|A_n\|_{\mathcal{A}} < \infty.$$

(b) If $A_n \to A$ and $B_n \to B$ $\mathcal{K}$-strongly, then $A_n B_n \to AB$ and $A_n + B_n \to A + B$ $\mathcal{K}$-strongly.

(c) If $A_n \to A$ $\mathcal{K}$-strongly and $\lambda \in \mathbb{C}$, then $A_n^* \to A^*$ and $\lambda A_n \to \lambda A$ $\mathcal{K}$-strongly.

Proof. (a): Let $A_n \to A$ $\mathcal{K}$-strongly, and let $K \in \mathcal{K}$. Then we have $A_n K \in \mathcal{K}$ and $KA_n \in \mathcal{K}$ since, by definition, $A_n \in \mathcal{A}$. Further, $A_n K \rightrightarrows AK$ and $KA_n \rightrightarrows KA$. Hence, since $\mathcal{K}$ is closed, $AK \in \mathcal{K}$ and $KA \in \mathcal{K}$. So $A \in \mathcal{A}$.

To show the inequality, we consider $\mathcal{K}$ as a Banach space with norm $\|\cdot\|_{\mathcal{A}}$. Then, by Proposition 3.1, $\mathcal{A}$ can be regarded as a subalgebra of $\mathcal{L}(\mathcal{K})$, and the operator norm $\|\cdot\|_{\mathcal{L}(\mathcal{K})}$ is equal to the original norm $\|\cdot\|_{\mathcal{A}}$ of $\mathcal{A}$. Now, $A_n \to A$ $\mathcal{K}$-strongly implies by definition that the operators $A_n \in \mathcal{L}(\mathcal{K})$ converge strongly (in the usual sense with respect to the Banach space $\mathcal{K}$) to $A \in \mathcal{L}(\mathcal{K})$. From the Banach-Steinhaus Theorem we obtain

$$\|A\|_{\mathcal{L}(\mathcal{K})} \leq \liminf_{n\to\infty} \|A_n\|_{\mathcal{L}(\mathcal{K})} \leq \sup_{n\geq 1} \|A_n\|_{\mathcal{L}(\mathcal{K})} < \infty,$$

which proves the assertion by the above mentioned norm coincidence.

Assertions (b) and (c) can be shown straightforwardly by using (a) and Proposition 3.1. ∎

Example 3.8 $Q_n \to 0$, $P_n \to I$ and, if $L \in \mathcal{K}$, $W_n L W_n \to 0$ $\mathcal{K}$-strongly.

Proof. The assertion for P_n and Q_n is evident. Concerning W_n, we must prove that $\|KW_nL\| \to 0$ for all $K, L \in \mathcal{K}$. Given $\varepsilon > 0$, there is an n_0 such that $\|LQ_{n_0}\| < \varepsilon$ and $\|Q_{n_0}K\| < \varepsilon$. Then

$$\begin{aligned} \|LW_nK\| &\leq \|LQ_{n_0}W_nK\| + \|LP_{n_0}W_nQ_{n_0}K\| + \|LP_{n_0}W_nP_{n_0}K\| \\ &\leq \varepsilon(\|L\| + \|K\|) + \|LP_{n_0}W_nP_{n_0}K\|. \end{aligned}$$

Since $P_{n_0}W_nP_{n_0} = 0$ for $n \geq 2n_0$, the assertion follows immediately. ∎

4. Algebraization of stability and Lifting Theorem

In this section we formulate the problem of stability in an algebraic language. More precisely, we construct a C*-algebra $\mathcal{F}$ such that a sequence of operators is stable if and only if a certain element of $\mathcal{F}$, to which the sequence is assigned, is invertible. The reason for this approach is to tackle the corresponding invertibility problem by means of Banach and C*-algebra techniques. We note that all algebras which occur subsequently are C*-algebras, although at many points it would suffice to make use only of their properties as Banach algebras. Nevertheless, we will be concerned with C*-algebras throughout the paper since it simplifies some argumentations. In the second part of this section we establish the "Lifting Theorem". It puts down the invertibility problem in $\mathcal{F}$ to another one, which is the starting point of further considerations.

Let $\mathcal{F}$ stand for the set of all sequences $(A_n)_{n=1}^\infty$ of operators $A_n \in \mathcal{L}(l_H^2(\mathbb{Z}_n))$ for which

$$\|(A_n)\|_{\mathcal{F}} := \sup_{n \geq 1} \|A_n\|_{\mathcal{L}(l_H^2(\mathbb{Z}_n))} < \infty. \tag{24}$$

Provided with the norm of (24) and algebraic operations

$$\begin{aligned} \lambda(A_n) &:= (\lambda A_n), & (A_n) + (B_n) &:= (A_n + B_n), \\ (A_n)(B_n) &:= (A_n B_n), & (A_n)^* &:= (A_n^*), \end{aligned} \tag{25}$$

$\mathcal{F}$ becomes a C*-algebra with the unit element (P_n).

Let $\mathcal{N}$ be the set of all sequences $(C_n) \in \mathcal{F}$ for which $\|C_n\|_{\mathcal{L}(l_H^2(\mathbb{Z}_n))} \to 0$ as $n \to \infty$. Obviously, $\mathcal{N}$ is a *-ideal of $\mathcal{F}$, and hence the quotient algebra $\mathcal{F}/\mathcal{N}$ is a C*-algebra.

Proposition 4.1 *Let* $(A_n) \in \mathcal{F}$. *Then* (A_n) *is stable if and only if* $(A_n) + \mathcal{N}$ *is invertible in* $\mathcal{F}/\mathcal{N}$.

Proof. See [9], Section 1.3.2.   ∎

Let $\mathcal{F}_0$ be the set of all sequences $(A_n) \in \mathcal{F}$ for which the $\mathcal{K}$-strong limits

$$P(A_n) \;\; := \;\; \text{s-}\lim_{n\to\infty} P_n A_n P_n, \tag{26}$$

$$W(A_n) \;\; := \;\; \text{s-}\lim_{n\to\infty} W_n A_n W_n \tag{27}$$

exist. By $\mathcal{J}$ we denote the set

$$\mathcal{J} = \left\{ (P_n K P_n + W_n L W_n + C_n)_{n=1}^{\infty} \; : \; K, L \in \mathcal{K}, \, (C_n) \in \mathcal{N} \right\}. \tag{28}$$

Proposition 4.2

(a) $\mathcal{F}_0$ is a *-subalgebra of $\mathcal{F}$.

(b) P and W are *-homomorphisms from $\mathcal{F}_0$ into $\mathcal{A}$.

(c) $\mathcal{J}$ is a *-ideal of $\mathcal{F}_0$.

Proof. (a),(b): By Proposition 3.7(bc), $\mathcal{F}_0$ is an algebra with involution, and P and W are linear, multiplicative, and symmetric. From Example 3.4, we have $P_n A_n P_n \in \mathcal{A}$ and $W_n A_n W_n \in \mathcal{A}$. Hence, using Proposition 3.7(a), we obtain $P(A_n) \in \mathcal{A}$ and $W(A_n) \in \mathcal{A}$. Further,

$$\|P(A_n)\|_{\mathcal{A}} \leq \|(A_n)\|_{\mathcal{F}} \quad \text{and} \quad \|W(A_n)\|_{\mathcal{A}} \leq \|(A_n)\|_{\mathcal{F}}.$$

This yields the continuity of P and W. A standard argument shows the closedness of $\mathcal{F}_0$.

(c): Let $(A_n) = (P_n K P_n + W_n L W_n + C_n) \in \mathcal{J}$. From Example 3.8, the limits (26) and (27) exist, and $P(A_n) = K$ and $W(A_n) = L$. Hence $\mathcal{J} \subseteq \mathcal{F}_0$. Moreover, we have $\|K\|_{\mathcal{A}} \leq \|(A_n)\|_{\mathcal{F}}$ and $\|L\|_{\mathcal{A}} \leq \|(A_n)\|_{\mathcal{F}}$. A standard argument shows the closedness of $\mathcal{J}$ in $\mathcal{F}_0$. Next, let $(B_n) \in \mathcal{F}_0$. Then we have

$$\begin{aligned}
B_n A_n \;\; &= \;\; B_n(P_n K P_n + W_n L W_n + C_n) \\
&= \;\; P_n(P_n B_n P_n - P(B_n))K P_n + W_n(W_n B_n W_n - W(B_n))L W_n \\
&\quad + B_n C_n + P_n P(B_n) K P_n + W_n W(B_n) L W_n \\
&= \;\; C_n' + P_n K' P_n + W_n L' W_n.
\end{aligned}$$

We note that $(P_n B_n P_n - P(B_n))K$ and $(W_n B_n W_n - W(B_n))L$ converge in the norm to zero and that $P(B_n)K$ and $W(B_n)L$ belong to $\mathcal{K}$. Hence $(B_n)(A_n) \in \mathcal{J}$ whenever $(B_n) \in \mathcal{F}_0$ and $(A_n) \in \mathcal{J}$. It can be shown analogously that $(A_n)(B_n) \in \mathcal{J}$ in this case. Linearity and symmetry of $\mathcal{J}$ are evident. Therefore, $\mathcal{J}$ is a *-ideal of $\mathcal{F}_0$.   ∎

From the preceding proposition we draw the conclusion that it is possible to form the quotient algebra $\mathcal{F}_0/\mathcal{J}$, which is a C*-algebra. Further, $\mathcal{N}$ is a *-ideal of $\mathcal{F}_0$, and the quotient algebra $\mathcal{F}_0/\mathcal{N}$ is a *-subalgebra of $\mathcal{F}/\mathcal{N}$ and therefore inverse closed.

Theorem 4.3 (Lifting Theorem)

Let $(A_n) \in \mathcal{F}_0$. Then the following assertions are equivalent :

(a) The sequence (A_n) is stable.

(b) $(A_n) + \mathcal{N}$ is invertible in $\mathcal{F}_0/\mathcal{N}$.

(c) Both $\mathcal{P}(A_n)$ and $\mathcal{W}(A_n)$ are invertible in $\mathcal{A}$, and $(A_n) + \mathcal{J}$ is invertible in $\mathcal{F}_0/\mathcal{J}$.

Proof. (a) $\Leftrightarrow$ (b): Observe Proposition 4.1 and that $\mathcal{F}_0/\mathcal{N}$ is inverse closed in $\mathcal{F}/\mathcal{N}$.
(b) $\Rightarrow$ (c): See Proposition 4.2, and notice that $\mathcal{N} \subseteq \mathcal{J}$, $\mathcal{N} \subseteq \ker \mathcal{P}$ and $\mathcal{N} \subseteq \ker \mathcal{W}$.
(c) $\Rightarrow$ (b): Suppose that $(B_n) + \mathcal{J}$ is the left inverse of $(A_n) + \mathcal{J}$. Then

$$B_n A_n = P_n + P_n K P_n + W_n L W_n + C_n,$$

where $K, L \in \mathcal{K}$ and $(C_n) \in \mathcal{N}$. Define a sequence (B'_n) by

$$B'_n = B_n - P_n K \mathcal{P}(A_n)^{-1} P_n - W_n L \mathcal{W}(A_n)^{-1} W_n.$$

We obtain that $B'_n A_n$ is equal to

$$P_n + P_n K P_n + W_n L W_n + C_n - P_n K \mathcal{P}(A_n)^{-1} P_n A_n - W_n L \mathcal{W}(A_n)^{-1} W_n A_n$$
$$= P_n + P_n C_n^{(1)} P_n + W_n C_n^{(2)} W_n + C_n,$$

where both sequences

$$\begin{aligned}
C_n^{(1)} &= K \mathcal{P}(A_n)^{-1} \left(\mathcal{P}(A_n) - P_n A_n P_n \right), \\
C_n^{(2)} &= L \mathcal{W}(A_n)^{-1} \left(\mathcal{W}(A_n) - W_n A_n W_n \right)
\end{aligned}$$

converge in the norm to zero. Hence $(B'_n) + \mathcal{N}$ is the left inverse of $(A_n) + \mathcal{N}$ in $\mathcal{F}_0/\mathcal{N}$. The right invertibility can be shown analogously. ∎

5. The algebras $\mathcal{S}(PQC_{\mathcal{L}(H)})$ and $\mathcal{F}(PQC_{\mathcal{L}(H)})$

Although $\mathcal{F}_0$ is already a much smaller algebra than $\mathcal{F}$, it yet contains too many sequences, and we are not able to describe the structure of $\mathcal{F}_0/\mathcal{J}$ in a suitable way. So we restrict our considerations to a smaller algebra $\mathcal{F}(PQC_{\mathcal{L}(H)})$, which contains all sequences we are interested in. The stability criteria is established for sequences of this algebra (Section 10).

Let $\mathcal{S}(PQC_{\mathcal{L}(H)})$ be the smallest closed subalgebra of $\mathcal{L}(l_H^2(\mathbb{Z}))$ which contains all Laurent operators $L(f)$ with $f \in PQC_{\mathcal{L}(H)}$ and the operators P and Q. By $\mathcal{F}(PQC_{\mathcal{L}(H)})$ we denote the smallest closed subalgebra of $\mathcal{F}$ which includes the ideal $\mathcal{J}$ and which contains all sequences $(P_n A P_n)$ with $A \in \mathcal{S}(PQC_{\mathcal{L}(H)})$. We will now examine these algebras.

Proposition 5.1 $S(PQC_{\mathcal{L}(H)})$ is a *-subalgebra of $\mathcal{A}$, and $\mathcal{K}$ is a *-ideal of $S(PQC_{\mathcal{L}(H)})$.

Proof. The assertion follows from Example 3.3, from Proposition 3.1(a), and from $\mathcal{K} \subseteq S(PQC_{\mathcal{L}(H)})$, which we prove now. Let $K_{ij,a}$ refer to the operator which has in its matrix representation at the (i,j)–position the entry $a \in \mathcal{L}(H)$ and elsewhere zero. Since

$$K_{ij,a} \;=\; U_i(P - U_1 PU_{-1})U_{-j}\,a, \tag{29}$$

we have $K_{ij,a} \in S(PQC_{\mathcal{L}(H)})$. Further, by the definition of $\mathcal{K}$, each $K \in \mathcal{K}$ can be approximated as closely as desired by operators $P_n K P_n$, which, in turn, are finite sums of operators $K_{ij,a}$. Hence $K \in S(PQC_{\mathcal{L}(H)})$.   ∎

Proposition 5.2 *For all* $A \in S(PQC_{\mathcal{L}(H)})$ *the* $\mathcal{K}$-*strong limit*

$$\mathcal{U}(A) \;:=\; s\text{-}\lim_{n\to\infty} \begin{pmatrix} U_{-n}PAPU_n & U_{-n}PAQU_{-n} \\ U_nQAPU_n & U_nQAQU_{-n} \end{pmatrix} \tag{30}$$

exists. Furthermore, the mapping $\mathcal{U}$ *is a* *-homomorphism from* $S(PQC_{\mathcal{L}(H)})$ *onto the* C^*-*algebra* $PQC_{\mathcal{L}(H)} \oplus PQC_{\mathcal{L}(H)} \subseteq \mathcal{A}^{2\times 2}$. *We have* $\mathcal{K} \subseteq \ker \mathcal{U}$, *and* $\mathcal{U}$ *acts on the generating elements of* $S(PQC_{\mathcal{L}(H)})$ *as follows :*

$$\mathcal{U}(P) \;=\; \begin{pmatrix} I & 0 \\ 0 & 0 \end{pmatrix} \tag{31}$$

$$\mathcal{U}(f) \;=\; \begin{pmatrix} f & 0 \\ 0 & f \end{pmatrix} \qquad (f \in PQC_{\mathcal{L}(H)}) \tag{32}$$

Proof. By Example 3.8, $U_nPU_{-n} = PQ_n \to 0$, $U_{-n}QU_n = QQ_n \to 0$, $U_{-n}PU_n = I - QQ_n \to I$ and $U_nQU_{-n} = I - PQ_n \to I$. This proves (31) and that $\mathcal{U}(I)$ is the unit element. Further, we have

$$
\begin{aligned}
U_{-n}PfPU_n &= U_{-n}PU_n\, f\, U_{-n}PU_n & \to & \quad f, \\
U_nQfQU_{-n} &= U_nQU_{-n}\, f\, U_nQU_{-n} & \to & \quad f, \\
U_{-n}PfQU_{-n} &= U_{-2n}\,(U_nPU_{-n}\, f\, U_nQU_{-n}) \\
&= (U_{-n}PU_n\, f\, U_{-n}QU_n)\,U_{-2n}, \\
U_nQfPU_n &= U_{2n}\,(U_{-n}QU_n\, f\, U_{-n}PU_n) \\
&= (U_nQU_{-n}\, f\, U_nPU_{-n})\,U_{2n}.
\end{aligned}
$$

We verify easily that the expressions in the braces tend to zero $\mathcal{K}$-strongly. Hence, by definition, $U_{-n}PfQU_{-n} \to 0$ and $U_nQfPU_n \to 0$, and we get (32). Therefore, the limit (30) exists for the generating elements of $S(PQC_{\mathcal{L}(H)})$. Since

$$(PU_n\,,\,QU_{-n})\begin{pmatrix} U_{-n}P \\ U_nQ \end{pmatrix} \;=\; I,$$

$$(PU_n\,,\,QU_{-n})^* \;=\; \begin{pmatrix} U_{-n}P \\ U_nQ \end{pmatrix},$$

we immediately conclude from Proposition 3.7 that (30) exists for all $A \in \mathcal{S}(PQC_{\mathcal{L}(H)})$ and that $\mathcal{U}$ is a *-homomorphism. By (31) and (32), $\mathcal{U}$ maps onto $PQC_{\mathcal{L}(H)} \oplus PQC_{\mathcal{L}(H)}$. Further, $\mathcal{U}(K_{ij,a}) = 0$ (cf. (29)). Using the same approximation argument as in the proof of Proposition 5.1, we obtain that $\mathcal{K} \subseteq \ker \mathcal{U}$. ∎

Corollary 5.3 *For all $A \in \mathcal{S}(PQC_{\mathcal{L}(H)})$ the $\mathcal{K}$-strong limit*

$$\mathcal{R}(A) \quad := \quad s\text{-}\lim_{n \to \infty} \begin{pmatrix} W_n A W_n & W_n A V_n \\ V_{-n} A W_n & V_{-n} A V_n \end{pmatrix} \tag{33}$$

exists, and it can be calculated by

$$\mathcal{R}(A) \quad = \quad \begin{pmatrix} PJ & QJ \\ P & Q \end{pmatrix} \mathcal{U}(A) \begin{pmatrix} JP & P \\ JQ & Q \end{pmatrix}. \tag{34}$$

Proof. (See also [18] for the scalar case.) By using

$$\begin{pmatrix} W_n \\ V_{-n} \end{pmatrix} = \begin{pmatrix} PJ & QJ \\ P & Q \end{pmatrix} \begin{pmatrix} U_{-n}P \\ U_n Q \end{pmatrix},$$

$$(W_n, V_n) = (PU_n, QU_{-n}) \begin{pmatrix} JP & P \\ JQ & Q \end{pmatrix},$$

the assertion follows immediately from the preceding proposition. ∎

Corollary 5.4

(a) $\mathcal{F}(PQC_{\mathcal{L}(H)})$ is a *-subalgebra of $\mathcal{F}_0$, and $\mathcal{J}$ is *-ideal of $\mathcal{F}(PQC_{\mathcal{L}(H)})$.

(b) On the generating elements of $\mathcal{F}(PQC_{\mathcal{L}(H)})$, $\mathcal{P}$ and $\mathcal{W}$ act as follows :

$$\mathcal{P}(P_n A P_n) \quad = \quad A \tag{35}$$

$$\mathcal{W}(P_n A P_n) \quad = \quad (PJ, QJ) \mathcal{U}(A) \begin{pmatrix} JP \\ JQ \end{pmatrix} \tag{36}$$

for $A \in \mathcal{S}(PQC_{\mathcal{L}(H)})$, and

$$\mathcal{P}(A_n) \quad = \quad K, \qquad\qquad \mathcal{W}(A_n) \quad = \quad L \tag{37}$$

for $(A_n) = (P_n K P_n + W_n L W_n + C_n) \in \mathcal{J}$.

(c) $\mathcal{P}$ is a *-homomorphism from $\mathcal{F}(PQC_{\mathcal{L}(H)})$ onto $\mathcal{S}(PQC_{\mathcal{L}(H)})$, and $\mathcal{W}$ is a *-homomorphism from $\mathcal{F}(PQC_{\mathcal{L}(H)})$ into $\mathcal{S}(PQC_{\mathcal{L}(H)})$.

Proof. (a),(b): We show that $\mathcal{F}(PQC_{\mathcal{L}(H)}) \subseteq \mathcal{F}_0$. This means, we have to ensure that the limits $\mathcal{P}$ and $\mathcal{W}$ exist for all elements of $\mathcal{F}(PQC_{\mathcal{L}(H)})$. By Prop. 3.7, it suffices to prove the existence only for the generating elements. Indeed, (36) follows from the preceding corollary, whereas (35) and (37) from Example 3.8. Finally, take into account Prop. 4.2 (c).

(c): The assertion follows from Proposition 4.2(b) by using (35)–(37). As for $\mathcal{W}$, observe that the mapping $f \mapsto JfJ$ sends $f \in PQC_{\mathcal{L}(H)}$ to $\tilde{f} \in PQC_{\mathcal{L}(H)}$. ∎

Remark 5.5 That $\mathcal{F}(PQC_{\mathcal{L}(H)})$ includes the ideal $\mathcal{J}$ was explicitly required by definition. One might ask whether this is actually necessary. To this end, let $\mathcal{F}'(PQC_{\mathcal{L}(H)})$ be the smallest closed algebra which contains all sequences $(P_n A P_n)$ with $A \in S(PQC_{\mathcal{L}(H)})$. Then one can prove that $\mathcal{F}'(PQC_{\mathcal{L}(H)}) \cap \mathcal{J}$ contains exactly the sequences $(P_n K P_n + W_n L W_n + C_n)$ for which $K, L \in \mathcal{K}$, $(C_n) \in \mathcal{N}$ and $PLQ = QLP = 0$. Hence $\mathcal{F}'(PQC_{\mathcal{L}(H)})$ does not include the whole set $\mathcal{J}$, and thus $\mathcal{F}(PQC_{\mathcal{L}(H)})$ is larger than $\mathcal{F}'(PQC_{\mathcal{L}(H)})$. Nevertheless, it is possible to deal with $\mathcal{F}'(PQC_{\mathcal{L}(H)})$ instead of $\mathcal{F}(PQC_{\mathcal{L}(H)})$. This approach would merely necessitate some minor modifications. A further question, which might arise with regard to Corollary 5.4(c), is what is the image of $\mathcal{F}(PQC_{\mathcal{L}(H)})$ under the homomorphism $\mathcal{W}$. Here one obtains without essential effort that the image is the set of operators $A \in S(PQC_{\mathcal{L}(H)})$ for which $PAQ, QAP \in \mathcal{K}$. In comparison, the image of $\mathcal{F}'(PQC_{\mathcal{L}(H)})$ under $\mathcal{W}$ is the set of operators $A \in S(PQC_{\mathcal{L}(H)})$ for which $PAQ = QAP = 0$.

From Theorem 4.3, Proposition 5.1 and Corollary 5.4(ac), we obtain immediately:

Corollary 5.6 *Let $(A_n) \in \mathcal{F}(PQC_{\mathcal{L}(H)})$. Then the following two assertions are equivalent:*

(a) The sequence (A_n) is stable.

(b) $\mathcal{P}(A_n)$ and $\mathcal{W}(A_n)$ are invertible in $S(PQC_{\mathcal{L}(H)})$, and $(A_n) + \mathcal{J}$ is invertible in $\mathcal{F}(PQC_{\mathcal{L}(H)})/\mathcal{J}$.

6. Localization

Corollary 5.6 raises the question when an element $(A_n) + \mathcal{J}$ is invertible. To answer this question, we will investigate the C*-algebra $\mathcal{F}(PQC_{\mathcal{L}(H)})/\mathcal{J}$. Fortunately, this algebra possessess a sufficiently large center, and therefore the "Local Principle" by Allan/Douglas ([5], Proposition 4.5) can effectively be employed. A proof is in [1], Section 1.34.

Theorem 6.1 (Local Principle by Allan / Douglas)
Let $\mathcal{B}$ be a C*-algebra and let $\mathcal{C}$ be a *-subalgebra contained in the center of $\mathcal{B}$. (That means, each element of $\mathcal{C}$ commutes with each element of $\mathcal{B}$.) For each maximal ideal ξ of $\mathcal{C}$ let $\mathcal{J}_\xi$ denote the smallest closed two-sided ideal of $\mathcal{B}$ which contains ξ. Then an element b of $\mathcal{B}$ is invertible in $\mathcal{B}$ if and only if the cosets $b + \mathcal{J}_\xi$ are invertible in $\mathcal{B}/\mathcal{J}_\xi$ for all ξ.

We need some preliminary lemmas.

Lemma 6.2 *Let* $A \in \mathcal{S}(PQC_{\mathcal{L}(H)})$. *Then*

(a) $Af - fA \in \mathcal{K}$ *for all* $f \in QC$,

(b) $(P_n AQ_n fP_n) \in \mathcal{J}$ *and* $(P_n fQ_n AP_n) \in \mathcal{J}$ *for all* $f \in QC_{\mathcal{L}(H)}$.

Proof. (a): By Proposition 5.1, it suffices to show the assertion for the generating elements of $\mathcal{S}(PQC_{\mathcal{L}(H)})$. For $A = g \in PQC_{\mathcal{L}(H)}$, we have $gf - fg = 0$ since f is scalar. For $A = P$, we have $Pf - fP = QfP - PfQ \in \mathcal{K}$, by Propositon 3.5.

(b): We show that $(P_n AQ_n fP_n) \in \mathcal{J}$. To this end, consider the identity

$$
\begin{aligned}
V_{-n}(PfP &+ QfQ)W_n \\
&= (PU_{-n}P + QU_nQ)(PfP + QfQ)(PU_nJP + QU_{-n}JQ)P_n \\
&= (PU_{-n}PfPU_nJP + QU_nQfQU_{-n}JQ)P_n \\
&= (PfJP + QfJQ)P_n.
\end{aligned} \tag{38}
$$

It follows that

$$
\begin{aligned}
P_n AQ_n fP_n &= P_n AV_n V_{-n}(PfP + QfQ)P_n + P_n AQ_n(PfQ + QfP)P_n \\
&= W_n(W_n AV_n)(PfQ + QfP)JW_n + P_n AQ_n(PfQ + QfP)P_n.
\end{aligned}
$$

Further, $PfQ + QfP \in \mathcal{K}$, by Proposition 3.5, and $Q_n \to 0$ and $W_n AV_n \to \tilde{A}$ $\mathcal{K}$-strongly, by Corollary 5.3, where $\tilde{A} \in \mathcal{A}$ is a certain operator. Therefore,

$$
P_n AQ_n fP_n = W_n LW_n + C_n,
$$

where $(C_n) \in \mathcal{N}$ and $L = \tilde{A}(PfQ + QfP)J \in \mathcal{K}$. Hence $(P_n AQ_n fP_n) \in \mathcal{J}$. It can be shown analogously that $(P_n fQ_n AP_n) \in \mathcal{J}$. ∎

Lemma 6.3 *The set* $\mathcal{C} = \{(P_n fP_n) + \mathcal{J} : f \in QC\}$ *is a *-subalgebra contained in the center of* $\mathcal{F}(PQC_{\mathcal{L}(H)})/\mathcal{J}$. *Further,* $\mathcal{C}$ *is *-isomorphic to* QC.

Proof. We first show that $\mathcal{C}$ is contained in the center. By Corollary 5.4(a), it suffices to prove that $(P_n fP_n)$ commutes modulo $\mathcal{J}$ with $(P_n AP_n)$ for all $A \in \mathcal{S}(PQC_{\mathcal{L}(H)})$ and $f \in QC$. Indeed,

$$
P_n AP_n fP_n - P_n fP_n AP_n = P_n(Af - fA)P_n + P_n fQ_n AP_n - P_n AQ_n fP_n,
$$

whence the assertion follows from the preceding lemma.

To show the remaining, let Φ be the mapping $\Phi : f \mapsto (P_n fP_n) + \mathcal{J}$. Obviously, Φ maps QC onto $\mathcal{C}$. Moreover, Φ is a *-homomorphism. (The multiplicativity follows from what we have just proved.) We claim that the kernel of Φ is trivial. Indeed, assume that

$f \in QC$ and $(P_n f P_n) \in \mathcal{J}$. Then, by Corollary 5.4(b), we have $\mathcal{P}(P_n f P_n) = f = K \in \mathcal{K}$. From Proposition 5.2 we obtain $0 = \mathcal{U}(K) = \mathcal{U}(f) = f \oplus f$. Hence $f = 0$. Therefore, Φ is a *-isomorphism. So $\mathcal{C}$ is a *-subalgebra of $\mathcal{F}(PQC_{\mathcal{L}(H)})/\mathcal{J}$, which is *-isomorphic to QC. ■

For $\xi \in M(QC)$, we denote by $\mathcal{J}_{\xi,\mathcal{L}(H)}$ the smallest closed ideal of $\mathcal{F}(PQC_{\mathcal{L}(H)})$ including $\mathcal{J}$ and containing all sequences $(P_n f P_n)$ with $f \in QC$ and $\xi(f) = 0$, i. e.,

$$\mathcal{J}_{\xi,\mathcal{L}(H)} := \operatorname{clos id}_{\mathcal{F}(PQC_{\mathcal{L}(H)})}\left(\mathcal{J} \cup \left\{(P_n f P_n) : f \in QC, \xi(f) = 0\right\}\right). \tag{39}$$

The quotient algebra $\mathcal{F}(PQC_{\mathcal{L}(H)})/\mathcal{J}_{\xi,\mathcal{L}(H)}$ is denoted by $\mathcal{F}_\xi(PQC_{\mathcal{L}(H)})$.

Corollary 6.4 *Let $(A_n) \in \mathcal{F}(PQC_{\mathcal{L}(H)})$. Then $(A_n) + \mathcal{J}$ is invertible in $\mathcal{F}(PQC_{\mathcal{L}(H)})/\mathcal{J}$ if and only if $(A_n) + \mathcal{J}_{\xi,\mathcal{L}(H)}$ is invertible in $\mathcal{F}_\xi(PQC_{\mathcal{L}(H)})$ for all $\xi \in M(QC)$.*

Proof. On account of Lemma 6.3, we can employ Theorem 6.1 in the setting $\mathcal{B} = \mathcal{F}(PQC_{\mathcal{L}(H)})/\mathcal{J}$ and $\mathcal{C} = \{(P_n f P_n) + \mathcal{J} : f \in QC\}$. Moreover, since $\mathcal{C} \cong QC$, the space of all maximal ideals of $\mathcal{C}$ is homeomorphic to that of QC, and, for $\xi \in M(QC)$,

$$J_\xi = \operatorname{clos id}_{\mathcal{B}}\left\{(P_n f P_n) + \mathcal{J} : f \in QC, \xi(f) = 0\right\}.$$

We obtain : $(A_n) + \mathcal{J}$ is invertible in $\mathcal{F}(PQC_{\mathcal{L}(H)})/\mathcal{J}$ if and only if $((A_n) + \mathcal{J}) + J_\xi$ is invertible in $(\mathcal{F}(PQC_{\mathcal{L}(H)})/\mathcal{J})/J_\xi$ for all ξ. Finally, the (correctly defined) mapping

$$\Phi : (A_n) + \mathcal{J}_{\xi,\mathcal{L}(H)} \mapsto ((A_n) + \mathcal{J}) + J_\xi$$

is a *-isomorphism from $\mathcal{F}_\xi(PQC_{\mathcal{L}(H)})$ onto $(\mathcal{F}(PQC_{\mathcal{L}(H)})/\mathcal{J})/J_\xi$. ■

The algebras $\mathcal{F}_\xi(PQC_{\mathcal{L}(H)})$, which arise from "localization", are called *local algebras* (at ξ). In the sequel, we are concerned with its description, i. e., we show that they are *-isomorphic to certain other (more explicitly given) C*-algebras.

For $\xi \in M(QC)$, let $\mathcal{I}_{\xi,\mathcal{L}(H)}$ denote the smallest closed ideal of $\mathcal{S}(PQC_{\mathcal{L}(H)})$ including $\mathcal{K}$ and containing all operators $f \in QC$ with $\xi(f) = 0$, i. e.,

$$\mathcal{I}_{\xi,\mathcal{L}(H)} := \operatorname{clos id}_{\mathcal{S}(PQC_{\mathcal{L}(H)})}\left(\mathcal{K} \cup \left\{f : f \in QC, \xi(f) = 0\right\}\right). \tag{40}$$

The quotient algebra $\mathcal{S}(PQC_{\mathcal{L}(H)})/\mathcal{I}_{\xi,\mathcal{L}(H)}$ is denoted by $\mathcal{S}_\xi(PQC_{\mathcal{L}(H)})$.

Lemma 6.5 $(P_n B P_n) \in \mathcal{J}_{\xi,\mathcal{L}(H)}$ *whenever $B \in \mathcal{I}_{\xi,\mathcal{L}(H)}$.*

Proof. Using Lemma 6.2(a), we see that B can be approximated as closely as desired by operators of the form $\sum_i A_i f_i + K$ where $K \in \mathcal{K}$, $A_i \in \mathcal{S}(PQC_{\mathcal{L}(H)})$, $f_i \in QC$ and $\xi(f_i) = 0$. Again from Lemma 6.2, we obtain by simple computations that

$$\left(P_n(\sum_i A_i f_i + K)P_n\right) = \sum_i (P_n A_i P_n)(P_n f_i P_n) \mod \mathcal{J}.$$

This expression is contained in $\mathcal{J}_{\xi,\mathcal{L}(H)}$. Hence $(P_n B P_n) \in \mathcal{J}_{\xi,\mathcal{L}(H)}$. ∎

In order to investigate the local algebras $\mathcal{F}_\xi(PQC_{\mathcal{L}(H)})$, we have to distinguish two cases. The first case is the one where $\xi \in M_\tau(QC) \backslash M_\tau^0(QC)$, $\tau \in \mathbb{T}$, i. e., where ξ belongs either to $M_\tau(QC) \backslash M_\tau^+(QC)$ or to $M_\tau(QC) \backslash M_\tau^-(QC)$. The following theorem settles this situation. It turns out that in this case it is, with regard to the final result, redundant to identify $\mathcal{F}_\xi(PQC_{\mathcal{L}(H)})$. For that reason, we are actually not interested in the local algebras (but see Remark 6.7 below). The second case, where $\xi \in M_\tau^0(QC)$, is more complicated. The next three sections are devoted to the description of the local algebras in this situation.

Theorem 6.6 *Let $\tau \in \mathbb{T}$ and $\xi \in M_\tau(QC) \backslash M_\tau^0(QC)$. If $(A_n) \in \mathcal{F}(PQC_{\mathcal{L}(H)})$ and $P(A_n)$ is invertible in $\mathcal{S}(PQC_{\mathcal{L}(H)})$, then $(A_n) + \mathcal{J}_{\xi,\mathcal{L}(H)}$ is invertible in $\mathcal{F}_\xi(PQC_{\mathcal{L}(H)})$.*

Proof. We introduce two *-homomorphisms which act between $\mathcal{F}_\xi(PQC_{\mathcal{L}(H)})$ and $\mathcal{S}_\xi(PQC_{\mathcal{L}(H)})$. (They are actually *-isomorphisms.) Let $\mathcal{P}'$ and Φ' be the mappings

$$\mathcal{P}': \quad \mathcal{F}_\xi(PQC_{\mathcal{L}(H)}) \to \mathcal{S}_\xi(PQC_{\mathcal{L}(H)}), \quad (A_n) + \mathcal{J}_{\xi,\mathcal{L}(H)} \mapsto P(A_n) + \mathcal{I}_{\xi,\mathcal{L}(H)},$$

$$\Phi': \quad \mathcal{S}_\xi(PQC_{\mathcal{L}(H)}) \to \mathcal{F}_\xi(PQC_{\mathcal{L}(H)}), \quad A + \mathcal{I}_{\xi,\mathcal{L}(H)} \mapsto (P_n A P_n) + \mathcal{J}_{\xi,\mathcal{L}(H)}.$$

By Corollary 5.4(c), $\mathcal{P}$ is a *-homomorphism from $\mathcal{F}(PQC_{\mathcal{L}(H)})$ into $\mathcal{S}(PQC_{\mathcal{L}(H)})$. Since $\mathcal{P}$ sends the generating elements of $\mathcal{J}_{\xi,\mathcal{L}(H)}$ in those of $\mathcal{I}_{\xi,\mathcal{L}(H)}$, $\mathcal{P}$ maps the *-ideal $\mathcal{J}_{\xi,\mathcal{L}(H)}$ into the *-ideal $\mathcal{I}_{\xi,\mathcal{L}(H)}$. Hence, the mapping $\mathcal{P}'$ is correctly defined and a *-homomorphism. By the preceding lemma, Φ' is a well-defined linear and symmetric mapping from $\mathcal{S}_\xi(PQC_{\mathcal{L}(H)})$ into $\mathcal{F}_\xi(PQC_{\mathcal{L}(H)})$.

We are going to show that Φ' is multiplicative. For this, we assert that for each function $p \in PC_{\mathcal{L}(H)}$ there is an $a \in \mathcal{L}(H)$ such that $p - a \in \mathcal{I}_{\xi,\mathcal{L}(H)}$. In fact, suppose that $\xi \in M_\tau(QC) \backslash M_\tau^+(QC)$. Then there is an $f \in QC$ with $\xi(f) \neq 0$ and $\limsup_{t \to \tau+0} |f(t)| = 0$. We choose $a \in \mathcal{L}(H)$ such that $\limsup_{t \to \tau-0} \|p(t) - a\| = 0$. Thus $f(p - a) \in PQC_{\mathcal{L}(H)}$ is continuous at τ and vanishes there. Since $\xi \in M_\tau(QC)$, we conclude by an approximation argument that $f(p - a) \in \mathcal{I}_{\xi,\mathcal{L}(H)}$. On the other hand, we have $f - \xi(f)e \in \mathcal{I}_{\xi,\mathcal{L}(H)}$ and therefore $(f - \xi(f)e)(p - a) \in \mathcal{I}_{\xi,\mathcal{L}(H)}$. Since $\xi(f) \neq 0$, we obtain $p - a \in \mathcal{I}_{\xi,\mathcal{L}(H)}$. The case where $\xi \in M_\tau(QC) \backslash M_\tau^-(QC)$ can be treated analogously.

Let $\mathcal{S}(QC_{\mathcal{L}(H)}^s)$ be the smallest closed algebra containing all operators $f \in QC_{\mathcal{L}(H)}^s$ and the operators P and Q. The assertion proved in the preceding paragraph yields

$$\mathcal{S}(PQC_{\mathcal{L}(H)}) \;=\; \text{clos} \Big\{ A + B \;:\; A \in \mathcal{S}(QC_{\mathcal{L}(H)}^s), \; B \in \mathcal{I}_{\xi,\mathcal{L}(H)} \Big\}.$$

Since P commutes modulo $\mathcal{K}$ with each function $q \in QC_{\mathcal{L}(H)}^s$ (see Proposition 3.5),

$$\mathcal{S}_\xi(PQC_{\mathcal{L}(H)}) \;=\; \text{clos} \Big\{ (q_1 P + q_2 Q) + \mathcal{I}_{\xi,\mathcal{L}(H)} \;:\; q_1, q_2 \in QC_{\mathcal{L}(H)}^s \Big\}.$$

Using Lemma 6.2(b) and again Proposition 3.5, we obtain by simple computations

$$(P_n(q_1 P + q_2 Q)P_n)(P_n(q_3 P + q_4 Q)P_n) \;=\; (P_n(q_1 P + q_2 Q)(q_3 P + q_4 Q)P_n) \mod \mathcal{J}.$$

Hence Φ' is multiplicative, and thus we have proved that Φ' is a *-homomorphism.

Next, we claim that the mapping $\Phi' \circ \mathcal{P}' \;:\; \mathcal{F}_\xi(PQC_{\mathcal{L}(H)}) \to \mathcal{F}_\xi(PQC_{\mathcal{L}(H)})$ is the identity mapping. It suffices to show that the generating elements of $\mathcal{F}_\xi(PQC_{\mathcal{L}(H)})$ are sended to themselves. Indeed, by (35), we have $\mathcal{P}' : (P_n A P_n) + \mathcal{J}_{\xi,\mathcal{L}(H)} \mapsto A + \mathcal{I}_{\xi,\mathcal{L}(H)}$, and this shows the assertion.

To complete the proof, we argue as follows. Let $(A_n) \in \mathcal{F}_\xi(PQC_{\mathcal{L}(H)})$, and assume that $\mathcal{P}(A_n)$ is invertible in $\mathcal{S}(PQC_{\mathcal{L}(H)})$. Then $\mathcal{P}(A_n) + \mathcal{I}_{\xi,\mathcal{L}(H)}$ is invertible in $\mathcal{S}_\xi(PQC_{\mathcal{L}(H)})$. On the other hand, by definition, $\mathcal{P}(A_n) + \mathcal{I}_{\xi,\mathcal{L}(H)} = \mathcal{P}'((A_n) + \mathcal{J}_{\xi,\mathcal{L}(H)})$. Hence

$$\cdot\,(\Phi' \circ \mathcal{P}')\,((A_n) + \mathcal{J}_{\xi,\mathcal{L}(H)}) \;=\; (A_n) + \mathcal{J}_{\xi,\mathcal{L}(H)}$$

is invertible in $\mathcal{F}_\xi(PQC_{\mathcal{L}(H)})$.   ∎

Remark 6.7 Continuing the considerations of the preceding proof, one can prove that, in the case where $\xi \in M_\tau(QC)\backslash M_\tau^0(QC)$, the local algebra $\mathcal{F}_\xi(PQC_{\mathcal{L}(H)})$ is *-isomorphic to $\mathcal{S}_\xi(PQC_{\mathcal{L}(H)})$. Further, by [7], Theorem 5(a), $\mathcal{S}_\xi(PQC_{\mathcal{L}(H)})$ is *-isomorphic to $\mathcal{L}(H)\oplus\mathcal{L}(H)$.

7. Identification of local algebras

In the sequel we analyse the local algebras $\mathcal{F}_\xi(PQC_{\mathcal{L}(H)})$ in the case where $\xi \in M_\tau^0(QC)$.

Let $\mathcal{S}(PC_{\mathcal{L}(H)})$ be the smallest closed subalgebra of $\mathcal{L}(l_H^2(\mathbb{Z}))$ which contains all Laurent operators $L(f)$ with $f \in PC_{\mathcal{L}(H)}$ and the operators P and Q. Further, let $\mathcal{F}(PC_{\mathcal{L}(H)})$ stand for the smallest closed subalgebra of $\mathcal{F}$ which includes the ideal $\mathcal{J}$ and which contains all sequences $(P_n A P_n)$ with $A \in \mathcal{S}(PC_{\mathcal{L}(H)})$.

Then $\mathcal{S}(PC_{\mathcal{L}(H)})$ is a *-subalgebra of $\mathcal{S}(PQC_{\mathcal{L}(H)})$, which includes the *-ideal $\mathcal{K}$ (cf. the proof of Propositon 5.1), and $\mathcal{F}(PC_{\mathcal{L}(H)})$ is a *-subalgebra of $\mathcal{F}(PQC_{\mathcal{L}(H)})$. Finally, similar to $\mathcal{I}_{\xi,\mathcal{L}(H)}$ and $\mathcal{J}_{\xi,\mathcal{L}(H)}$, we define the *-ideals

$$\mathcal{I}_{1,\mathcal{L}(H)} := \operatorname{clos\,id}_{\mathcal{S}(PC_{\mathcal{L}(H)})}\big(\mathcal{K} \cup \{f \,:\, f \in C(\mathbb{T}),\, f(1) = 0\}\big), \tag{41}$$

$$\mathcal{J}_{1,\mathcal{L}(H)} := \operatorname{clos\,id}_{\mathcal{F}(PC_{\mathcal{L}(H)})}\big(\mathcal{J} \cup \{(P_n f P_n) \,:\, f \in C(\mathbb{T}),\, f(1) = 0\}\big). \tag{42}$$

We abbreviate the quotient algebras $\mathcal{S}(PC_{\mathcal{L}(H)})/\mathcal{I}_{1,\mathcal{L}(H)}$ and $\mathcal{F}(PC_{\mathcal{L}(H)})/\mathcal{J}_{1,\mathcal{L}(H)}$ by $\mathcal{S}_1(PC_{\mathcal{L}(H)})$ and $\mathcal{F}_1(PC_{\mathcal{L}(H)})$, respectively. For $\tau \in \mathbb{T}$, we introduce the operators Y_τ by

$$Y_\tau \,:\, (x_n)_{n=-\infty}^\infty \mapsto (\tau^{-n} x_n)_{n=-\infty}^\infty, \qquad l_H^2(\mathbb{Z}) \to l_H^2(\mathbb{Z}). \tag{43}$$

Obviously, $Y_\tau^{-1} = Y_{\tau^{-1}} = Y_\tau^*$.

Proposition 7.1 *Let* $\tau \in \mathbb{T}$ *and* $\xi \in M_\tau^0(QC)$. *Then the mappings*

$$\Phi_{\tau,S} : \quad S_1(PC_{\mathcal{L}(H)}) \to S_\xi(PQC_{\mathcal{L}(H)}), \qquad A + \mathcal{I}_{1,\mathcal{L}(H)} \mapsto Y_\tau A Y_\tau^{-1} + \mathcal{I}_{\xi,\mathcal{L}(H)}, \quad (44)$$

$$\Phi_{\tau,\mathcal{F}} : \quad \mathcal{F}_1(PC_{\mathcal{L}(H)}) \to \mathcal{F}_\xi(PQC_{\mathcal{L}(H)}), \quad (A_n) + \mathcal{J}_{1,\mathcal{L}(H)} \mapsto (Y_\tau A_n Y_\tau^{-1}) + \mathcal{J}_{\xi,\mathcal{L}(H)} \quad (45)$$

are *-homomorphisms, which map onto the corresponding algebras.

Proof. We have $(Y_\tau P_n Y_\tau^{-1}) = (P_n)$, $Y_\tau P Y_\tau^{-1} = P$, and $Y_\tau L(a) Y_\tau^{-1} = L(a_\tau)$, where a_τ refers to the function $a_\tau(t) = a(t/\tau)$, $t \in \mathbb{T}$. From this, we obtain a remarkable property of the algebras $S(\cdot)$ and $\mathcal{F}(\cdot)$, which we name *rotation invariance* (Section 2.5.1 of [9]) :

$$A \in S(PC_{\mathcal{L}(H)}) \quad \Longleftrightarrow \quad Y_\tau A Y_\tau^{-1} \in S(PC_{\mathcal{L}(H)}), \qquad (46)$$

$$(A_n) \in \mathcal{F}(PC_{\mathcal{L}(H)}) \quad \Longleftrightarrow \quad (Y_\tau A_n Y_\tau^{-1}) \in \mathcal{F}(PC_{\mathcal{L}(H)}). \qquad (47)$$

(The same is valid for $S(PQC_{\mathcal{L}(H)})$ and $\mathcal{F}(PQC_{\mathcal{L}(H)})$.) Therefore, the mappings

$$\Phi'_{\tau,S} : \quad S(PC_{\mathcal{L}(H)}) \mapsto S(PQC_{\mathcal{L}(H)}), \qquad A \mapsto Y_\tau A Y_\tau^{-1}$$

$$\Phi'_{\tau,\mathcal{F}} : \quad \mathcal{F}(PC_{\mathcal{L}(H)}) \mapsto \mathcal{F}(PQC_{\mathcal{L}(H)}), \qquad (A_n) \mapsto (Y_\tau A_n Y_\tau^{-1})$$

are *-homomorphisms. Since $f \in C(\mathbb{T})$ and $f(1) = 0$ implies $f_\tau \in QC$ and $\xi(f_\tau) = 0$, we conclude that $\Phi'_{\tau,S}$ maps $\mathcal{I}_{1,\mathcal{L}(H)}$ into $\mathcal{I}_{\xi,\mathcal{L}(H)}$ and that $\Phi'_{\tau,\mathcal{F}}$ maps $\mathcal{J}_{1,\mathcal{L}(H)}$ into $\mathcal{J}_{\xi,\mathcal{L}(H)}$. The mappings $\Phi_{\tau,S}$ and $\Phi_{\tau,\mathcal{F}}$ are therefore correctly defined and *-homomorphisms.

It remains to show that they are "onto". Let $\mathcal{I}_{\xi,\mathcal{L}(H)}$ refer to (23) with $B = QC_{\mathcal{L}(H)}^s$. Since $QC_{\mathcal{L}(H)}^s \subseteq S(PQC_{\mathcal{L}(H)})$, we have $\mathcal{I}_{\xi,\mathcal{L}(H)} \subseteq \mathcal{I}_{\xi,\mathcal{L}(H)}$. Hence, because $QC_{\mathcal{L}(H)}^s$ is locally trivial at ξ, for each function $q \in QC_{\mathcal{L}(H)}^s$, there is an $a \in \mathcal{L}(H)$ such that $q - a \in \mathcal{I}_{\xi,\mathcal{L}(H)}$. We therefore obtain

$$S(PQC_{\mathcal{L}(H)}) = \text{clos}\left\{ A + B \ : \ A \in S(PC_{\mathcal{L}(H)}),\ B \in \mathcal{I}_{\xi,\mathcal{L}(H)} \right\}.$$

Consequently, $\Phi_{\tau,S}$ is "onto". Using Lemma 6.5, we obtain furthermore

$$\mathcal{F}(PQC_{\mathcal{L}(H)}) = \text{clos}\left\{ (A_n) + (B_n) \ : \ (A_n) \in \mathcal{F}(PC_{\mathcal{L}(H)}),\ (B_n) \in \mathcal{J}_{\xi,\mathcal{L}(H)} \right\}.$$

Hence also $\Phi_{\tau,\mathcal{F}}$ is "onto", too. ∎

It is now time to explain how we will proceed. We have already mentioned that we want to "identify" the local algebra $\mathcal{F}_\xi(PQC_{\mathcal{L}(H)})$. That means, we will show that $\mathcal{F}_\xi(PQC_{\mathcal{L}(H)})$ is *-isomorphic to a C*-algebra $\Xi_{\mathcal{L}(H)}$ of operators on $L_H^2([-1,1])$. In order to describe the *-isomorphism, we will also be concerned with $S_\xi(PQC_{\mathcal{L}(H)})$ and with a C*-algebra $\Sigma_{\mathcal{L}(H)}$ of operators on $L_H^2(\mathbb{R})$. This aim is no easy task, and in the course of our considerations we will prove the following scheme (as far as it is necessary for our purposes):

$$
\begin{array}{ccccc}
\mathcal{F}_\xi(PQC_{\mathcal{L}(H)}) & \overset{\Phi_{\tau,\mathcal{F}}}{\rightleftarrows} & \mathcal{F}_1(PC_{\mathcal{L}(H)}) & \overset{\tilde{\mathcal{E}}_{\mathcal{F},\mathcal{L}(H)}}{\rightleftarrows} & \Xi_{\mathcal{L}(H)} \\
\downarrow & & \downarrow & & \downarrow \\
S_\xi(PQC_{\mathcal{L}(H)}) & \underset{\Phi_{\tau,S}}{\rightleftarrows} & S_1(PC_{\mathcal{L}(H)}) & \underset{\tilde{\mathcal{E}}_{S,\mathcal{L}(H)}}{\rightleftarrows} & \Sigma_{\mathcal{L}(H)}
\end{array}
\qquad (48)
$$

In this scheme, a double arrow indicates a *-isomorphism, and a vertical single arrow means a *-homomorphism "onto". In the preceding proposition, we have already established $\Phi_{\tau,S}$ and $\Phi_{\tau,\mathcal{F}}$, and the mappings $\tilde{\mathcal{E}}_{S,\mathcal{L}(H)}$ and $\tilde{\mathcal{E}}_{\mathcal{F},\mathcal{L}(H)}$ will be introduced below[1].

For a subset M of the real axis, we denote by χ_M the characteristic function of M. Obviously, χ_M can also be regarded as a multiplication operator. The context will clarify what is meant. Let E_n and E_{-n} stand for the linear bounded operators defined by $(n \geq 1)$

$$E_n : \quad l_H^2(\mathbb{Z}) \to L_H^2(\mathbb{R}), \qquad (x_i)_{i=-\infty}^{\infty} \mapsto \sqrt{n} \sum_{i=-\infty}^{\infty} x_i \chi_{[\frac{i}{n},\frac{i+1}{n}]}, \tag{49}$$

$$E_{-n} : \quad L_H^2(\mathbb{R}) \to l_H^2(\mathbb{Z}), \qquad f \mapsto \left(\sqrt{n} \int_{-\infty}^{\infty} f(x) \chi_{[\frac{i}{n},\frac{i+1}{n}]}(x) \, dx \right)_{i=-\infty}^{\infty} \tag{50}$$

Obviously, $E_{-n}^* = E_n$ and $E_{-n} E_n = I$.

Let $S_{\mathbb{R}}$ be the *singular integral operator on the real axis*, i. e., (if $H = \mathbb{C}$)

$$(S_{\mathbb{R}} f)(x) \quad = \quad \frac{1}{\pi i} \int_{-\infty}^{\infty} \frac{f(y)}{y - x} \, dy \, , \qquad x \in \mathbb{R}. \tag{51}$$

For $H \neq \mathbb{C}$, $S_{\mathbb{R}}$ can be defined in an appropriate manner (e. g., by identifying with $S_{\mathbb{R}} \otimes e$, see [17], IV.7.12). In any case, $S_{\mathbb{R}}$ is bounded on $L_H^2(\mathbb{R})$ and $S_{\mathbb{R}}^* = S_{\mathbb{R}}$.

Let $\Sigma_{\mathcal{L}(H)}$ denote the smallest closed subalgebra of $\mathcal{L}(L_H^2(\mathbb{R}))$ which contains the multiplication operator $\chi_{[0,\infty)}$, the singular integral operator $S_{\mathbb{R}}$, and all constants $a \in \mathcal{L}(H)$, considered as operators $f \in L_H^2(\mathbb{R}) \mapsto af \in L_H^2(\mathbb{R})$, i. e.,

$$\Sigma_{\mathcal{L}(H)} \quad := \quad \text{alg} \left\{ \chi_{[0,\infty)}, \, S_{\mathbb{R}}, \, a \in \mathcal{L}(H) \right\}. \tag{52}$$

Furthermore, let $\Xi_{\mathcal{L}(H)}$ be the smallest closed subalgebra of $\mathcal{L}(L_H^2([-1,1]))$ which contains all operators $\chi_{[-1,1]} A \chi_{[-1,1]}$ with $A \in \Sigma_{\mathcal{L}(H)}$, i. e.,

$$\Xi_{\mathcal{L}(H)} \quad := \quad \text{alg} \left\{ \chi_{[-1,1]} A \chi_{[-1,1]} \, : \, A \in \Sigma_{\mathcal{L}(H)} \right\}. \tag{53}$$

Evidently, $\Sigma_{\mathcal{L}(H)}$ and $\Xi_{\mathcal{L}(H)}$ are C*-algebras.

The next definitions are restricted to the case $H = \mathbb{C}$. Let σ denote the function

$$\sigma(e^{2\pi i \phi}) \quad = \quad -\frac{\sin^2(\pi \phi)}{\pi^2} \sum_{m \in \mathbb{Z}} \frac{\text{sgn}(m + \frac{1}{2})}{(\phi + m)^2} \qquad \phi \in (0, 1). \tag{54}$$

This function σ is continuous on $\mathbb{T} \setminus \{1\}$ and possess a jump discontinuity at 1. The one-sided limits are $\sigma(1 + 0) = -1$ and $\sigma(1 - 0) = 1$. Moreover, for all $n \geq 1$, we have

$$L(\sigma) \quad = \quad E_{-n} S_{\mathbb{R}} E_n. \tag{55}$$

---

[1]The "vertical" *-homomorphisms are of minor importance. Only for completeness, we will mention them: The "left" *-homomorphism is just Φ' (as defined in the proof of Theorem 6.6), and has the continuous cross-section $\mathcal{P}'$. The "middle" *-homomorphism is defined similarly. The "right" *-homomorphism is given by means of a mapping Φ_Z, which is defined in the next section.

The operator $L_n := E_n E_{-n}$ is an ortho-projection on $L^2(\mathbb{R})$ which converges strongly on $L^2(\mathbb{R})$ to the identity operator I as $n \to \infty$ (see e. g. Section 2.2.1 and 2.2.3 of [9]).

For a sequence B_n of operators and an operator B in $\mathcal{L}(L^2(\mathbb{R}))$ (resp. $\mathcal{L}(L^2([-1,1])))$, we write, in accordance with the notation introduced in Section 3 (cf. also Remark 3.6), $B = \text{s-}\lim_{n \to \infty} B_n$ if both B_n converges strongly to B and B_n^* converges strongly to B^*.

Proposition 7.2

(a) *There exist two *-homomorphisms $\mathcal{E}_{S,\mathcal{L}(H)}$ and $\mathcal{E}_{\mathcal{F},\mathcal{L}(H)}$*

$$\mathcal{E}_{S,\mathcal{L}(H)} : \quad S(PC_{\mathcal{L}(H)}) \to \Sigma_{\mathcal{L}(H)} \quad with \quad \mathcal{I}_{1,\mathcal{L}(H)} \subseteq \ker \mathcal{E}_{S,\mathcal{L}(H)}, \quad (56)$$

$$\mathcal{E}_{\mathcal{F},\mathcal{L}(H)} : \quad \mathcal{F}(PC_{\mathcal{L}(H)}) \to \Xi_{\mathcal{L}(H)} \quad with \quad \mathcal{J}_{1,\mathcal{L}(H)} \subseteq \ker \mathcal{E}_{\mathcal{F},\mathcal{L}(H)}, \quad (57)$$

which map onto the corresponding algebras, and for which

$$\mathcal{E}_{S,\mathcal{L}(H)}(P) = \chi_{[0,\infty)}, \quad (58)$$

$$\mathcal{E}_{S,\mathcal{L}(H)}(a) = a(1+0)\frac{(I-S_{\mathbb{R}})}{2} + a(1-0)\frac{(I+S_{\mathbb{R}})}{2} \quad (a \in PC_{\mathcal{L}(H)}), \quad (59)$$

$$\mathcal{E}_{\mathcal{F},\mathcal{L}(H)}(P_n A P_n) = \chi_{[-1,1]}\,\mathcal{E}_{S,\mathcal{L}(H)}(A)\,\chi_{[-1,1]} \quad (A \in S(PC_{\mathcal{L}(H)})). \quad (60)$$

(b) *Let $H = \mathbb{C}$. For $A \in S(PC)$ and $(A_n) \in \mathcal{F}(PC)$, the limits*

$$\mathcal{E}_S(A) := \text{s-}\lim_{n \to \infty} E_n A E_{-n}, \quad (61)$$

$$\mathcal{E}_{\mathcal{F}}(A_n) := \text{s-}\lim_{n \to \infty} E_n A_n E_{-n} \quad (62)$$

exist. Moreover, the mappings $\mathcal{E}_S$ and $\mathcal{E}_{\mathcal{F}}$ comply with the conditions of (a).

Proof. (b): We first show the existence of the limits for the generating elements of $S(PC)$ and $\mathcal{F}(PC)$. Since $L_n = L_n^* \to I$ strongly, $\mathcal{E}_S(I) = I$. Further, $E_n P E_{-n} = L_n \chi_{[0,\infty)}$ gives (58). Assertion (59) follows essentially from (55), namely $E_n L(\sigma) E_{-n} = L_n S_{\mathbb{R}} L_n$, and since

$$a = a(1+0)\frac{1-\sigma}{2} + a(1-0)\frac{1+\sigma}{2} + d, \quad (63)$$

where d is a function on $\mathbb{T}$ which is continuous at 1 and vanishes there. For details see [9], Section 2.3.3, Proposition 2.10(a). There it is also shown that $\mathcal{K} \subseteq \ker \mathcal{E}_S$.

Formula (60) follows from $E_n P_n = \chi_{[-1,1]} E_n$ and $P_n E_{-n} = E_{-n} \chi_{[-1,1]}$. Define an operator $\tilde{w}$ on $L^2(\mathbb{R})$ by

$$(\tilde{w}f)(t) = \begin{cases} f(1-t) & \text{if} \quad t \in [0,1] \\ f(-1-t) & \text{if} \quad t \in [-1,0) \\ 0 & \text{if} \quad |t| > 1. \end{cases}$$

Then $E_n W_n = \tilde{w} E_n$ and $W_n E_{-n} = E_{-n} \tilde{w}$, and for $K, L \in \mathcal{K}$ and $(C_n) \in \mathcal{N}$ we obtain

$$\mathcal{E}_{\mathcal{F}} \left(P_n K P_n + W_n L W_n + C_n \right) = \chi_{[-1,1]} \, \mathcal{E}_{\mathcal{S}}(K) \, \chi_{[-1,1]} + \tilde{w} \, \mathcal{E}_{\mathcal{S}}(L) \, \tilde{w}.$$

This expression vanishes since $\mathcal{K} \subseteq \ker \mathcal{E}_{\mathcal{S}}$. Hence $\mathcal{J} \subseteq \ker \mathcal{E}_{\mathcal{F}}$.

Since $E_{-n} E_n = I$ and $E_{-n}^* = E_n$, the existence of the limits $\mathcal{E}_{\mathcal{S}}$ and $\mathcal{E}_{\mathcal{F}}$ for the generating elements of $\mathcal{S}(PC)$ and $\mathcal{F}(PC)$ implies the existence of the limits for all elements of these algebras. Further, $\mathcal{E}_{\mathcal{S}}$ and $\mathcal{E}_{\mathcal{F}}$ are obviously *-homomorphisms "onto" Σ and Ξ, respectively. Finally, from (59) and (60), $\mathcal{I}_1 \subseteq \ker \mathcal{E}_{\mathcal{S}}$ and $\mathcal{J}_1 \subseteq \ker \mathcal{E}_{\mathcal{F}}$.

(a): We obtain the result from (b) by considering tensor products of represented C*-algebras (cf. [10]). In fact, $\mathcal{S}(PC)$ is represented on $l^2(\mathbb{Z})$, and Σ is represented on $L^2(\mathbb{R})$. Hence the *-homomorphism $\mathcal{E}_{\mathcal{S}} \otimes \mathrm{id}$ maps $\mathcal{S}(PC) \otimes \mathcal{L}(H)$ onto $\Sigma \otimes \mathcal{L}(H)$, and $\ker(\mathcal{E}_{\mathcal{S}} \otimes \mathrm{id}) \supseteq \mathcal{I}_1 \otimes \mathcal{L}(H)$. Now we identify the Hilbert space tensor product $l^2(\mathbb{Z}) \otimes H$ with $l_H^2(\mathbb{Z})$, and $L^2(\mathbb{R}) \otimes H$ with $L_H^2(\mathbb{R})$. Then $\mathcal{E}_{\mathcal{S}} \otimes \mathrm{id}$ is the desired *-homomorphism $\mathcal{E}_{\mathcal{S},\mathcal{L}(H)}$. By definition, $\Sigma \otimes \mathcal{L}(H) \cong \Sigma_{\mathcal{L}(H)}$. Further $\mathcal{K} \cong \mathcal{K}(l^2(\mathbb{Z})) \otimes \mathcal{L}(H)$, or, equivalently, $\mathcal{K} = \mathrm{clos}\{\sum_i K_i a_i : K_i \in \mathcal{K}(l^2(\mathbb{Z})), a_i \in \mathcal{L}(H)\}$, which can be shown, e. g., by using the same argument as in the proof of Proposition 5.1. From this and from (20) it follows

$$\mathcal{S}(PC_{\mathcal{L}(H)}) = \mathrm{clos}\left\{ \sum_i A_i a_i : A_i \in \mathcal{S}(PC), a_i \in \mathcal{L}(H) \right\},$$

$$\mathcal{I}_{1,\mathcal{L}(H)} = \mathrm{clos}\left\{ \sum_i A_i a_i : A_i \in \mathcal{I}_1, a_i \in \mathcal{L}(H) \right\}.$$

Hence $\mathcal{S}(PC_{\mathcal{L}(H)}) \cong \mathcal{S}(PC) \otimes \mathcal{L}(H)$ and $\mathcal{I}_{1,\mathcal{L}(H)} \cong \mathcal{I}_1 \otimes \mathcal{L}(H)$. Using again (20), we obtain (59), whereas (58) is obvious.

As to $\mathcal{E}_{\mathcal{F},\mathcal{L}(H)}$, the argumentation is similar. We remark the main steps. $\mathcal{F}(PC)$ is represented on $\sum_{n=1}^{\infty} \oplus l^2(\mathbb{Z}_n)$, and Ξ on $L^2([-1,1])$. We identify $(\sum_{n=1}^{\infty} \oplus l^2(\mathbb{Z}_n)) \otimes H$ with $\sum_{n=1}^{\infty} \oplus l_H^2(\mathbb{Z}_n)$, and $L^2([-1,1]) \otimes H$ with $L_H^2([-1,1])$, and we consider $\mathcal{E}_{\mathcal{F},\mathcal{L}(H)} \cong \mathcal{E}_{\mathcal{F}} \otimes \mathrm{id}$. It can be shown by an approximation argument that $\mathcal{N}^{(\mathcal{L}(H))} \cong \mathcal{N}^{(\mathcal{L}(\mathbb{C}))} \otimes \mathcal{L}(H)$. Using this and what we have derived above, we get (60) and

$$\mathcal{F}(PC_{\mathcal{L}(H)}) = \mathrm{clos}\left\{ \sum_i (A_n^{(i)} a_i) : (A_n^{(i)}) \in \mathcal{F}(PC), a_i \in \mathcal{L}(H) \right\},$$

$$\mathcal{J}_{1,\mathcal{L}(H)} = \mathrm{clos}\left\{ \sum_i (A_n^{(i)} a_i) : (A_n^{(i)}) \in \mathcal{J}_1, a_i \in \mathcal{L}(H) \right\},$$

whence follows that $\mathcal{F}(PC_{\mathcal{L}(H)}) \cong \mathcal{F}(PC) \otimes \mathcal{L}(H)$ and $\mathcal{J}_{1,\mathcal{L}(H)} \cong \mathcal{J}_1 \otimes \mathcal{L}(H)$.   ∎

From the preceding proposition we conclude that the (unambiguously defined) mappings

$$\tilde{\mathcal{E}}_{\mathcal{S},\mathcal{L}(H)} : \mathcal{S}_1(PC_{\mathcal{L}(H)}) \to \Sigma_{\mathcal{L}(H)} \qquad A + \mathcal{I}_1 \mapsto \tilde{\mathcal{E}}_{\mathcal{S},\mathcal{L}(H)}(A) \qquad (64)$$

$$\tilde{\mathcal{E}}_{\mathcal{F},\mathcal{L}(H)} : \mathcal{F}_1(PC_{\mathcal{L}(H)}) \to \Xi_{\mathcal{L}(H)} \qquad (A_n) + \mathcal{J}_1 \mapsto \tilde{\mathcal{E}}_{\mathcal{F},\mathcal{L}(H)}(A_n) \qquad (65)$$

are *-homomorphisms "onto". These mappings are even those which occur in scheme (48).

It remains to show that the kernels of $\tilde{\mathcal{E}}_{\mathcal{S},\mathcal{L}(H)}$, $\tilde{\mathcal{E}}_{\mathcal{F},\mathcal{L}(H)}$, $\Phi_{\tau,\mathcal{S}}$ and $\Phi_{\tau,\mathcal{F}}$ are trivial. For this, we proceed as follows: First we analyse the algebras $\Sigma_{\mathcal{L}(H)}$ and $\Xi_{\mathcal{L}(H)}$ in the case $H = \mathbb{C}$. (As agreed before, we omit the indices H and $\mathcal{L}(H)$ in this case.) Then we construct the inverse mappings $\tilde{\mathcal{E}}_{\mathcal{S}}^{-1}$ and $\tilde{\mathcal{E}}_{\mathcal{F}}^{-1}$, and we obtain the $\mathcal{L}(H)$–valued case afterwards by resorting to a result on tensor products of C*-algebras. Finally, we treat $\Phi_{\tau,\mathcal{S}}$ and $\Phi_{\tau,\mathcal{F}}$.

8. The algebras Σ and Ξ

In order to analyse the algebras Σ and Ξ, we introduce two algebras Σ_1 and Ξ_1, and two ideals of them, Σ_1^0 and Ξ_1^0. Then Σ and Ξ are expressed in terms of these algebras and ideals. Further, each element of the latter algebras and ideals is the sum of certain operators, which are given explicitly. We note that the results for Σ, Σ_1 and Σ_1^0 are already known, even in a more general situation than treated here (see [9], Section 2.1, [4] and [16]). However, on account of the C*-algebra case, the considerations simplify essentially here.

We denote by η the mapping $f \mapsto (f_1, f_2)^T$ with $f_1(x) = f(x)$ and $f_2(x) = f(-x)$, $\forall x \in \mathbb{R}^+$. Evidently, η is an isometry from $L^2(\mathbb{R})$ onto $L^2(\mathbb{R}^+) \oplus L^2(\mathbb{R}^+)$, as well as from $L^2([-1,1])$ onto $L^2([0,1]) \oplus L^2([0,1])$. Therefore, the *-isomorphism Φ_η defined by

$$\Phi_\eta \;:\; A \mapsto \eta A \eta^{-1} \tag{66}$$

maps $\mathcal{L}(L^2(\mathbb{R}))$ onto $\mathcal{L}(L^2(\mathbb{R}^+))^{2\times2}$ and $\mathcal{L}(L^2([-1,1]))$ onto $\mathcal{L}(L^2([0,1]))^{2\times2}$. Now we introduce the following operators on $L^2(\mathbb{R}^+)$. Let $\tilde{P}$ denote the operator $\chi_{[0,1]}$ of multiplication, let $S = S_{\mathbb{R}^+}$ stand for the *singular integral operator on the semi-axis*, and let N be the *Hankel operator*:

$$(Sf)(x) \;=\; \frac{1}{\pi i} \int_0^\infty \frac{f(y)}{y - x}\, dy\,, \qquad x \in \mathbb{R}^+, \tag{67}$$

$$(Nf)(x) \;=\; \frac{1}{\pi i} \int_0^\infty \frac{f(y)}{y + x}\, dy\,, \qquad x \in \mathbb{R}^+. \tag{68}$$

Then we have (I refers to the identity operator on $L^2(\mathbb{R}^+)$)

$$\Phi_\eta(\chi_{[0,\infty)}) = \begin{pmatrix} I & 0 \\ 0 & 0 \end{pmatrix}, \qquad \Phi_\eta(S_{\mathbb{R}}) = \begin{pmatrix} S & -N \\ N & -S \end{pmatrix},$$

$$\Phi_\eta(\chi_{[-1,1]}) = \begin{pmatrix} \tilde{P} & 0 \\ 0 & \tilde{P} \end{pmatrix}. \tag{69}$$

We define the C*-algebras Σ_2^0 and Ξ_2^0 by

$$\Sigma_2^0 \;:=\; \mathrm{alg}_{\,\mathcal{L}(L^2(\mathbb{R}^+))^{2\times2}} \left\{ \begin{pmatrix} S & -N \\ N & -S \end{pmatrix}, \begin{pmatrix} I & 0 \\ 0 & 0 \end{pmatrix} \right\}, \tag{70}$$

$$\Xi_2^0 \;:=\; \mathrm{alg}_{\,\mathcal{L}(L^2([0,1]))^{2\times2}} \left\{ \begin{pmatrix} \tilde{P} & 0 \\ 0 & \tilde{P} \end{pmatrix} A \begin{pmatrix} \tilde{P} & 0 \\ 0 & \tilde{P} \end{pmatrix} \;:\; A \in \Sigma_2^0 \right\}. \tag{71}$$

Corollary 8.1 *The *-isomorphism Φ_n maps Σ onto Σ_2^0 and Ξ onto Ξ_2^0.*

Therefore, we will examine Σ_2^0 and Ξ_2^0 instead of Σ and Ξ. Recall that the Mellin transform M and its inverse M^{-1} are given by

$$(M f)(z) = \int_0^\infty x^{-iz-\frac{1}{2}} f(x)\, dx, \quad z \in \mathbb{R}, \qquad M : L^2(\mathbb{R}^+) \to L^2(\mathbb{R}),$$

$$(M^{-1} f)(x) = \frac{1}{2\pi} \int_{-\infty}^\infty x^{iz-\frac{1}{2}} f(z)\, dz, \quad x \in \mathbb{R}^+, \qquad M^{-1} : L^2(\mathbb{R}) \to L^2(\mathbb{R}^+).$$

For a multiplication operator $b \in L^\infty(\mathbb{R})$, let $M^0(b)$ denote the *Mellin convolution operator*

$$M^0(b) := M^{-1} b M : \quad L^2(\mathbb{R}^+) \to L^2(\mathbb{R}^+). \tag{72}$$

Evidently, $\|M^0(b)\| = \|b\|$, $M^0(b)^* = M^0(b^*)$ and $M^0(b_1 b_2) = M^0(b_1)\, M^0(b_2)$. Thus the mapping $b \mapsto M^0(b) \in \mathcal{L}(L^2(\mathbb{R}^+))$ is a *-isomorphism. In [9], Section 2.1.2, it is shown that S and N are certain Mellin convolution operators :

$$S = M^0(s), \quad s(z) = \tanh(\pi z) \qquad (z \in \mathbb{R}), \tag{73}$$
$$N = M^0(n), \quad n(z) = -i\,(\cosh(\pi z))^{-1} \quad (z \in \mathbb{R}). \tag{74}$$

Hence, we have (since $s^2 - n^2 = 1$)

$$SN = NS \qquad \text{and} \qquad S^2 - N^2 = I. \tag{75}$$

Let $PC_\infty(\mathbb{R})$ denote the set of all continuous functions f on $\mathbb{R}$ for which the limits $\lim_{x\to\infty} f(x)$ and $\lim_{x\to-\infty} f(x)$ exist and are finite, and let $C_\infty^0(\mathbb{R})$ be the set of all continuous functions f on $\mathbb{R}$ for which $\lim_{x\to\pm\infty} f(x) = 0$. Obviously, $PC_\infty(\mathbb{R})$ is a C*-algebra with the *-ideal $C_\infty^0(\mathbb{R})$. Further, $PC_\infty(\mathbb{R})$ is the smallest closed subalgebra of $L^\infty(\mathbb{R})$ which contains the function s, and $C_\infty^0(\mathbb{R})$ is the smallest closed ideal of $PC_\infty(\mathbb{R})$ which contains the function n.

In order to describe Σ_2^0, we introduce

$$\Sigma_1 := \left\{ \alpha I + \beta S + M^0(b) : \alpha, \beta \in \mathbb{C}, b \in C_\infty^0(\mathbb{R}) \right\}, \tag{76}$$

$$\Sigma_1^0 := \left\{ M^0(b) : b \in C_\infty^0(\mathbb{R}) \right\}. \tag{77}$$

Proposition 8.2

(a) Σ_1 *is a C*-algebra, and Σ_1^0 is a *-ideal of Σ_1.*

(b) $\Sigma_2^0 = \left\{ \begin{pmatrix} A & B \\ C & D \end{pmatrix} : A, D \in \Sigma_1, B, C \in \Sigma_1^0 \right\}.$

Proof. (a): The mapping $b \mapsto M^0(b)$ is a *-isomorphism, which maps $PC_\infty(\mathbb{R})$ onto $\Sigma_1 = \{ M^0(a) : a \in PC_\infty(\mathbb{R}) \}$ and $C^0_\infty(\mathbb{R})$ onto Σ^0_1.

(b): Let $\mathfrak{X}$ be the set on the right hand side. By (a), $\mathfrak{X}$ is a C*-algebra. Further, $\mathfrak{X}$ contains the generating elements of Σ^0_2. Hence $\Sigma^0_2 \subseteq \mathfrak{X}$. From what we have mentioned above, we conclude that Σ_1 is the smallest closed subalgebra of $\mathcal{L}(L^2(\mathbb{R}^+))$ which contains S and that Σ^0_1 is the smallest closed ideal of Σ_1 which contains N. By simple considerations, we now obtain $\mathfrak{X} \subseteq \Sigma^0_2$. ∎

We pass on to consider Ξ^0_2. This situation is more involved. Let

$$\Xi_1 := \text{alg}_{\mathcal{L}(L^2([-1,1]))} \{ \tilde{P} A \tilde{P} : A \in \Xi_1 \}, \tag{78}$$

$$\Xi^0_1 := \text{clos id}_{\Xi_1} \{ \tilde{P} A \tilde{P} : A \in \Xi^0_1 \}. \tag{79}$$

From (71) and Proposition 8.2 we obtain immediately :

Corollary 8.3 $\qquad \Xi^0_2 = \left\{ \begin{pmatrix} A & B \\ C & D \end{pmatrix} : A, D \in \Xi_1, B, C \in \Xi^0_1 \right\}.$

In the remaining part of this section, we will describe Ξ_1 and Ξ^0_1. For this purpose, we define the following operators on $L^2(\mathbb{R}^+)$:

$$(\tilde{W}f)(x) = \begin{cases} f(1-x) & \text{if} \quad x \in [0,1] \\ 0 & \text{if} \quad x > 1 \end{cases} \tag{80}$$

$$(\tilde{V}_+f)(x) = \begin{cases} f(x-1) & \text{if} \quad x \geq 1 \\ 0 & \text{if} \quad x \in [0,1) \end{cases} \tag{81}$$

$$(\tilde{V}_-f)(x) = f(1+x), \quad x \in \mathbb{R}^+ \tag{82}$$

Further, we set $\tilde{Q} := I - \tilde{P}$, where I denotes the identity mapping on $L^2(\mathbb{R}^+)$. Then we have the relations $I = \tilde{V}_-\tilde{V}_+$, $\tilde{Q} = \tilde{V}_+\tilde{V}_-$, $\tilde{W}^2 = \tilde{P}$, $\tilde{W}\tilde{P} = \tilde{P}\tilde{W} = \tilde{W}$, and

$$\begin{aligned} \tilde{W}S\tilde{W} &= -\tilde{P}S\tilde{P}, & \tilde{W}S\tilde{V}_+ &= \tilde{P}N, \\ \tilde{V}_-S\tilde{W} &= -N\tilde{P}, & \tilde{V}_-S\tilde{V}_+ &= S. \end{aligned} \tag{83}$$

From this and from (75) we obtain in particular

$$\tilde{P}S\tilde{Q}S\tilde{P} = -\tilde{W}N^2\tilde{W}, \qquad \tilde{P}S\tilde{P}S\tilde{P} = \tilde{P} + \tilde{P}N^2\tilde{P} + \tilde{W}N^2\tilde{W}. \tag{84}$$

Further, for $a, b \in L^\infty(\mathbb{R})$ we have

$$\tilde{P}M^0(a)\tilde{P}M^0(b)\tilde{P} = \tilde{P}M^0(ab)\tilde{P} - \tilde{P}M^0(a)\tilde{Q}M^0(b)\tilde{P}, \tag{85}$$

$$\tilde{W}M^0(a)\tilde{P}M^0(b)\tilde{W} = \tilde{W}M^0(ab)\tilde{W} - \tilde{W}M^0(a)\tilde{Q}M^0(b)\tilde{W}. \tag{86}$$

Proposition 8.4

(a) *The operators $\tilde{P}M^0(b)\tilde{Q}$ and $\tilde{Q}M^0(b)\tilde{P}$ are compact for all $b \in C^0_\infty(\mathbb{R})$.*

(b) *All compact operators $K \in \mathcal{K}(L^2([0,1]))$ are contained in the smallest closed algebra (without unit element) which contains all operators $\tilde{P}M^0(b)\tilde{P}$ with $b \in C^0_\infty(\mathbb{R})$.*

Proof. We can write $M = F \circ T$, where F is the Fourier transform and where T is the isometry $L^2(\mathbb{R}^+) \to L^2(\mathbb{R})$ defined by $(Tf)(z) = 2\pi e^{\pi z} f(e^{2\pi z})$, $z \in \mathbb{R}$. Since $M^0(b) = T^{-1}M_\mathbb{R}(b)T$ (where $M_\mathbb{R}(b) = F^{-1}bF$ is the Fourier convolution operator), $\tilde{P} = T^{-1}\chi_{(-\infty,0]}T$ and $\tilde{Q} = T^{-1}\chi_{[0,\infty)}T$, we are left to verify the following assertions :

(a): The operators $\chi_{(-\infty,0]}M_\mathbb{R}(b)\chi_{[0,\infty)}$ and $\chi_{[0,\infty)}M_\mathbb{R}(b)\chi_{(-\infty,0]}$ are compact for all $b \in C^0_\infty(\mathbb{R})$. But this follows from the consideration in Section 9.1–9.8 of [1].

(b): All compact operators $K' \in \mathcal{K}(L^2(\mathbb{R}^-))$ are contained in the smallest closed algebra which contains all operators $\chi_{(-\infty,0]}M_\mathbb{R}(b)\chi_{(-\infty,0]}$ with $b \in C^0_\infty(\mathbb{R})$. This assertion can be proved in the same way as (i) in Section 3.1.1 of [9] (see also Section 3.9/E 3.1). ∎

Let Z_n be the isometry on $L^2(\mathbb{R}^+)$ defined by $(Z_n f)(x) = \sqrt{n}f(nx)$, $x \in \mathbb{R}^+$. We consider the strong limit (if it exists)

$$\Phi_Z(A) := \text{s-}\lim_{n\to\infty} Z_n^{-1}AZ_n \tag{87}$$

for certain operators $A \in \mathcal{L}(L^2(\mathbb{R}^+))^2$.

Proposition 8.5

(a) $\Phi_Z(\tilde{P}M^0(a)\tilde{P}) = M^0(a)$ *for all $a \in L^\infty(\mathbb{R})$.*

(b) $\Phi_Z(K) = 0$ *if K is compact.*

(c) $\Phi_Z(\tilde{W}M^0(b)\tilde{W}) = 0$ *if $b \in L^\infty(\mathbb{R})$ and $\lim_{x\to\pm\infty} b(x) = 0$.*

Proof. (a): We can write $MZ_n = t_n Z_n$, where $t_n \in L^\infty(\mathbb{R})$ refers to the multiplication operator $t_n(z) = n^{iz}$, $z \in \mathbb{R}$. Hence $Z_n^{-1}M^0(a)Z_n = M^{-1}t_n^{-1}at_n M = M^0(a)$. Now observe that $Z_n^{-1}\tilde{P}Z_n = \chi_{[0,n]} \to I$ strongly.

(b): It is not hard to verify that $Z_n \to 0$ weakly on $L^2(\mathbb{R}^+)$.

(c): Since the functions b under consideration can be approximated by functions cn^2, it suffices to show that $\Phi_Z(\tilde{W}M^0(cn^2)\tilde{W}) = 0$ for $c \in L^\infty(\mathbb{R})$. Indeed, we have (see (86))

$$\tilde{W}M^0(cn^2)\tilde{W} = \tilde{W}M^0(c)\tilde{Q}M^0(n^2)\tilde{W} + (\tilde{W}M^0(c)\tilde{W})(\tilde{W}N^2\tilde{W}). \tag{88}$$

Since the first term on the right hand side is compact (Proposition 8.4(a)), we can apply (b) to it. That $\Phi_Z(\tilde{W}N^2\tilde{W}) = 0$ follows from (a) and (84). This settles the issue. ∎

---

[2]Recall scheme (48). The mapping Φ_Z proves to be a *-homomorphism from Ξ_1 onto Σ_1 with continuous cross-section $A \in \Sigma_1 \mapsto \tilde{P}A\tilde{P} \in \Xi_1$. The "left vertical" homomorphism in (48) can be defined similarly (observe the operator-valued case and that Ξ instead of Ξ_1 is considered).

Theorem 8.6 $\Xi_1^0 \;=\; \left\{\, \tilde{P}M^0(b)\tilde{P} + K \;:\; b \in C_\infty^0(\mathbb{R}),\, K \in \mathcal{K}(L^2([0,1])) \,\right\}.$

Proof. Let $\mathfrak{I}$ denote the set on the right hand side. It follows from Proposition 8.4(b) and from the definitions (77) and (79) of Σ_1^0 and Ξ_1^0 that $\mathfrak{I} \subseteq \Xi_1^0$.

We are going to show that $\mathfrak{I}$ is closed. Let Φ be the mapping $\Phi_M \circ \Phi_Z$, where Φ_M stands for the *-isomorphism $M^0(b) \mapsto b$. From Proposition 8.5, we obtain $\Phi(\tilde{P}M^0(b)\tilde{P} + K) = b$. Hence, by simple computations,

$$\frac{1}{2}\max\{\,\|b\|,\,\|K\|\,\} \;\leq\; \|\tilde{P}M^0(b)\tilde{P} + K\| \;\leq\; 2\max\{\,\|b\|,\,\|K\|\,\}.$$

An approximation argument shows the closedness of $\mathfrak{I}$.

Now we show that $\mathfrak{I}$ is a two-sided ideal of Ξ_1. Since Ξ_1 is generated by elements $\tilde{P}M^0(a)\tilde{P}$ with $a \in PC_\infty(\mathbb{R})$, it suffices to show that $\tilde{P}M^0(a)\tilde{P}B \in \mathfrak{I}$ and $B\tilde{P}M^0(a)\tilde{P} \in \mathfrak{I}$ whenever $a \in PC_\infty(\mathbb{R})$ and $B \in \mathfrak{I}$. Indeed, for $B = \tilde{P}M^0(b)\tilde{P} + K$, we have $\tilde{P}M^0(a)\tilde{P}B = \tilde{P}M^0(ab)\tilde{P} - \tilde{P}M^0(a)\tilde{Q}M^0(b)\tilde{P} + K'$ (see (85)), which is in $\mathfrak{I}$ since $ab \in C_\infty^0(\mathbb{R})$ and since the last two terms are compact. We obtain similarily that $B\tilde{P}M^0(a)\tilde{P} \in \mathfrak{I}$. Consequently, $\mathfrak{I}$ is a closed two-sided ideal of Ξ_1, and, since $\mathfrak{I}$ contains all operators $\tilde{P}M^0(b)\tilde{P}$ with $b \in C_\infty^0(\mathbb{R})$, we conclude that $\mathfrak{I} \supseteq \Xi_1^0$. ∎

Theorem 8.7

$$\Xi_1 \;=\; \left\{\, \alpha\tilde{P} + \beta\tilde{P}S\tilde{P} + \tilde{P}M^0(b)\tilde{P} + \tilde{W}M^0(c)\tilde{W} + K \,\right\}$$
$$\text{where}\quad \alpha,\beta \in \mathbb{C},\; b,c \in C_\infty^0(\mathbb{R}),\; K \in \mathcal{K}(L^2([0,1])).$$

Proof. Denote the set on the right hand side by $\mathfrak{X}$. We first show that $\mathfrak{X} \subseteq \Xi_1$. With regard to the definition of Ξ_1 and Σ_1 and the preceding theorem, it remains to show that $\tilde{W}M^0(c)\tilde{W} \in \Xi_1$ for all $c \in C_\infty^0(\mathbb{R})$. Since $\tilde{W}N^2\tilde{W} \in \Xi_1$, by (84), and since the first term on the right hand side of (88) is compact (hence in Ξ_1), we conclude : If $\tilde{W}M^0(c)\tilde{W} \in \Xi_1$, then $\tilde{W}M^0(n^2c)\tilde{W} \in \Xi_1$. Since $\tilde{W}M^0(1)\tilde{W} = \tilde{P} \in \Xi_1$ and $\tilde{W}M^0(s)\tilde{W} \in \Xi_1$, by (83), and since linear combinations of the functions sn^{2k} and n^{2k} are dense in $C_\infty^0(\mathbb{R})$, we obtain the assertion $\mathfrak{X} \subseteq \Xi_1$.

We are going to prove that $\mathfrak{X}$ is closed. Let Φ be same mapping as in the proof of the preceding theorem, and consider $A = \alpha\tilde{P} + \beta\tilde{P}S\tilde{P} + \tilde{P}M^0(b)\tilde{P} + \tilde{W}M^0(c)\tilde{W} + K$. Again it follows from Proposition 8.5 (take also (83) into account) that

$$\Phi(A) = \alpha + \beta s + b, \qquad\qquad \Phi(\tilde{W}A\tilde{W}) = \alpha - \beta s + c.$$

Hence, by simple computations, we obtain

$$\frac{1}{5}\,\|A\| \;\leq\; \max\{\,|\alpha|,\,|\beta|,\,\|b\|,\,\|c\|,\,\|K\|\,\} \;\leq\; 4\,\|A\|,$$

and an approximation argument shows the closedness of $\mathfrak{X}$.

In order to show that $\mathfrak{X}$ is an algebra, we prove that, if both A and B are equal to one of the operators $\tilde{P}$, $\tilde{P}S\tilde{P}$, $\tilde{P}M^0(b)\tilde{P}$, $\tilde{W}M^0(c)\tilde{W}$ or K, where $b,c \in C_\infty^0(\mathbb{R})$ and $K \in \mathcal{K}(L^2([-1,1]))$, then $AB \in \mathfrak{X}$. Observing $\tilde{P}S\tilde{P} = \tilde{P}M^0(s)\tilde{P} = -\tilde{W}M^0(s)\tilde{W}$, we must treat the following cases (except for changing A and B; the respecting cases are similar) :

(A) $A = \tilde{P}$ and $B \in \mathfrak{X}$ arbitrary,

(B) $A \in \mathcal{K}(L^2([-1,1]))$ and $B \in \mathfrak{X}$ arbitrary,

(C) $A = \tilde{P} M^0(a) \tilde{P}$, $a \in PC_\infty(\mathbb{R})$, and $B = \tilde{P} M^0(b) \tilde{P}$, $b \in C_\infty^0(\mathbb{R})$,

(D) $A = \tilde{W} M^0(a) \tilde{W}$, $a \in PC_\infty(\mathbb{R})$, and $B = \tilde{W} M^0(b) \tilde{W}$, $b \in C_\infty^0(\mathbb{R})$,

(E) $A = \tilde{P} M^0(b) \tilde{P}$, $b \in C_\infty^0(\mathbb{R})$, and $B = \tilde{W} M^0(c) \tilde{W}$, $c \in C_\infty^0(\mathbb{R})$,

(F) $A = \tilde{P} S \tilde{P}$ and $B = \tilde{P} S \tilde{P}$.

The assertion is evident for (A) and (B). For (C) and (D), it follows from (85) and (86) by observing that the first terms on the right hand side involve a Mellin convolution operator with $ab \in C_\infty^0(\mathbb{R})$ and the second terms are compact, by Proposition 8.4(a).

Case (F) follows from (84). Concerning Case (E), we claim that AB is compact. Since the functions $n^2 c$ ($c \in C_\infty^0(\mathbb{R})$) are dense in $C_\infty^0(\mathbb{R})$, we can substitute c by $n^2 c$ without loss of generality, i. e., we consider $B = \tilde{W} M^0(n^2 c) \tilde{W}$, and A as above. Then we obtain that

$$B = B_1 B_2 + K_1, \quad \text{where} \quad B_1 = \tilde{W} N^2 \tilde{W}, \quad B_2 = \tilde{W} M^0(c) \tilde{W},$$
$$K_1 = \tilde{W} N^2 \tilde{Q} M^0(c) \tilde{W}, \tag{89}$$

and K_1 is compact. Further, by simple computations and from (73) and (84),

$$K_2 := AB_1 = -\tilde{P} M^0(b) \tilde{P} S \tilde{Q} S \tilde{P}$$
$$= \tilde{P} M^0(b) \tilde{Q} S \tilde{Q} S \tilde{P} - \tilde{P} M^0(bs) \tilde{Q} S \tilde{P}, \tag{90}$$

where K_2 is compact (observe that $b, bs \in C_\infty^0(\mathbb{R})$). Therefore, since we can write

$$AB = AB_1 B_2 + AK_1 = K_2 B_2 + AK_1, \tag{91}$$

AB is compact. Thus we have proved that $\mathfrak{X}$ is a closed algebra, which obviously contains the generating elements of Ξ_1. Consequently, $\mathfrak{X} \supseteq \Xi_1$. ∎

The following result follows from Corollary 8.1 and 8.3, Theorem 8.6 and 8.7.

Corollary 8.8 $\mathcal{K}(L^2([-1,1])) \subsetneq \Xi$.

9. Identification of local algebras continued

In the first part of this section, we will construct the inverses $\tilde{\mathcal{E}}_{S,\mathcal{L}(H)}^{-1}$ and $\tilde{\mathcal{E}}_{F,\mathcal{L}(H)}^{-1}$ (see scheme (48)). We start with the scalar case. The mappings E_n and E_{-n} defined in (49) and (50) can also be considered as acting on

$$E_n : l^2(\mathbb{Z}^+) \to L^2(\mathbb{R}^+), \qquad E_{-n} : L^2(\mathbb{R}^+) \to l^2(\mathbb{Z}^+).$$

For $b \in L^\infty(\mathbb{R})$, the operator $G(b) \in \mathcal{L}(l^2(\mathbb{Z}^+))$ given by

$$G(b) := E_{-1} M^0(b) E_1 \tag{92}$$

is called *discretized Mellin convolution operator*. The following facts are known.

Proposition 9.1

(a) $G(b) = E_{-n} M^0(b) E_n$ for all $b \in L^\infty(\mathbb{R})$ and all $n \geq 1$.

(b) $G(s) = T(\sigma)$ and $G(1) = T(1) = P$.

(c) $G(b_1) G(b_2) - G(b_1 b_2)$ is compact for all $b_1, b_2 \in C^0_\infty(\mathbb{R})$.

(d) $G(b) T(\sigma) - G(bs)$ and $T(\sigma) G(b) - G(bs)$ are compact for all $b \in C^0_\infty(\mathbb{R})$.

(e) $G(n) + H(\sigma)$ and $G(n) - H(\tilde{\sigma})$ are compact.

Proof. (a): We can write $E_n = Z_n E_1$ and $E_{-n} = E_{-1} Z_n^{-1}$, where Z_n is the isometry on $L^2(\mathbb{R}^+)$ defined in the preceding section. Then $E_{-n} M^0(b) E_n = E_{-1} Z_n^{-1} M^0(b) Z_n E_1 = E_{-1} M^0(b) E_1$, as it has already been stated in the proof of Proposition 8.5(a).

(b): We have $E_{-1} S E_1 = E_{-1} \chi_{[0,\infty)} S_{\mathbb{R}} \chi_{[0,\infty)} E_1 = P E_{-1} S_{\mathbb{R}} E_1 P = P L(\sigma) P$, by (55).

(c)–(e): See [9], Sect. 2.4.2. It is stated that $G(n) - H(f)$ is compact, where f is the function $f(e^{i\phi}) = 1 - \phi/\pi$, $\phi \in (0, 2\pi)$. The continuity of $f + \sigma$ and $f - \tilde{\sigma}$ involves (c). ∎

With regard to (76), we define the linear bounded mapping $\mathcal{E}^0_{\mathcal{S}}$ by

$$\mathcal{E}^0_{\mathcal{S}} : \Sigma_1 \rightarrow \mathcal{L}(l^2(\mathbb{Z}^+)), \qquad M^0(b) \mapsto G(b). \qquad (93)$$

Then we have $\mathcal{E}^0_{\mathcal{S}}(A) = E_{-n} A E_n$ for all $A \in \Sigma_1$ and all $n \geq 1$. Further, let $\mathcal{F}_+$ stand for the set of all sequences $(A_n) \in \mathcal{F}$ with $(P A_n P) = (A_n)$, and let $\mathcal{E}^0_{\mathcal{F}}$ be the linear bounded mapping

$$\mathcal{E}^0_{\mathcal{F}} : \Xi_1 \rightarrow \mathcal{F}_+, \qquad A \mapsto (E_{-n} A E_n). \qquad (94)$$

Lemma 9.2

(a) $\mathcal{E}^0_{\mathcal{S}}(AB) - \mathcal{E}^0_{\mathcal{S}}(A) \mathcal{E}^0_{\mathcal{S}}(B)$ is compact if $A \in \Sigma_1$ and $B \in \Sigma^0_1$ or if $A \in \Sigma^0_1$ and $B \in \Sigma_1$.

(b) $\mathcal{E}^0_{\mathcal{S}}(AB) - \mathcal{E}^0_{\mathcal{S}}(A) \mathcal{E}^0_{\mathcal{S}}(B) \in \mathcal{I}_1$ if $A \in \Sigma_1$ and $B \in \Sigma_1$.

Proof. The assertion follows immediately from (76),(77) and Proposition 9.1(b-e). The only non-trivial case is $A = B = S$. Then $AB = I + N^2$, by (75), and hence

$$\begin{aligned}
\mathcal{E}^0_{\mathcal{S}}(AB) - \mathcal{E}^0_{\mathcal{S}}(A) \mathcal{E}^0_{\mathcal{S}}(B) &= T(1) + G(n^2) - T(\sigma) T(\sigma) \\
&= T(1) - H(\sigma) H(\tilde{\sigma}) - T(\sigma) T(\sigma) + \text{compact} \\
&= T(1 - \sigma^2) + \text{compact},
\end{aligned}$$

by (11). Since $1 - \sigma^2$ is continuous on $\mathbb{T}$ and vanishes at 1, this expression is in $\mathcal{I}_1$. ∎

Lemma 9.3

(a) $E_{-n}KE_n \to 0$ strongly if K is compact, and $W_n G(b)W_n \to 0$ strongly if $b \in C^0_\infty(\mathbb{R})$.

(b) $\mathcal{E}^0_{\mathcal{F}}(AB) - \mathcal{E}^0_{\mathcal{F}}(A)\mathcal{E}^0_{\mathcal{F}}(B) \in \mathcal{J}$ if $A \in \Xi_1$ and $B \in \Xi^0_1$ or if $A \in \Xi^0_1$ and $B \in \Xi_1$.

(c) $\mathcal{E}^0_{\mathcal{F}}(AB) - \mathcal{E}^0_{\mathcal{F}}(A)\mathcal{E}^0_{\mathcal{F}}(B) \in \mathcal{J}_1$ if $A \in \Xi_1$ and $B \in \Xi_1$.

Proof. By Theorem 8.6 and 8.7, each element of Ξ_1 (resp. Ξ^0_1) can be written as the sum of certain elements, on which $\mathcal{E}^0_{\mathcal{F}}$ acts as follows :

$$\mathcal{E}^0_{\mathcal{F}}(\tilde{P}) = (P_n T(1)P_n) \tag{95}$$
$$\mathcal{E}^0_{\mathcal{F}}(\tilde{P}S\tilde{P}) = (P_n T(\sigma)P_n) \tag{96}$$
$$\mathcal{E}^0_{\mathcal{F}}(\tilde{P}M^0(b)\tilde{P}) = (P_n G(b)P_n) \tag{97}$$
$$\mathcal{E}^0_{\mathcal{F}}(\tilde{W}M^0(c)\tilde{W}) = (W_n G(c)W_n) \tag{98}$$
$$\mathcal{E}^0_{\mathcal{F}}(K) = (E_{-n}KE_n) \tag{99}$$

Further, since $\tilde{Q} = \tilde{Q}(I - L_n) + E_n Q_n E_{-n}$ and $L_n = L^*_n \to I$ strongly, both sequences

$$(E_{-n}\tilde{P}M^0(a)\tilde{Q}M^0(b)\tilde{P}E_n) - (P_n G(a)Q_n G(b)P_n) \tag{100}$$
$$(E_{-n}\tilde{W}M^0(a)\tilde{Q}M^0(b)\tilde{W}E_n) - (W_n G(a)Q_n G(b)W_n), \tag{101}$$

are contained in $\mathcal{N}$ if $a \in C^0_\infty(\mathbb{R})$ or if $b \in C^0_\infty(\mathbb{R})$ (see Proposition 8.4(a)).

(a): It is readily seen that E_n converges weakly to zero. Hence $E_{-n}KE_n \to 0$ strongly. Further, from (19) we obtain that $W_n H(\sigma)H(\tilde{\sigma})W_n \to 0$. Therefore, by Proposition 9.1(de) and Example 3.8, we have $W_n G(n^2)W_n \to 0$. We show that $W_n G(bn^2)W_n \to 0$, which is sufficient since the functions bn^2 are dense in $C^0_\infty(\mathbb{R})$. Indeed, we can write

$$(W_n G(bn^2)W_n) = (W_n K W_n) + (W_n G(b)W_n)(W_n G(n^2)W_n)$$
$$+ (W_n G(b)Q_n G(n^2)W_n),$$

with $K = G(bn^2) - G(b)G(n^2)$ compact. Hence $W_n K W_n \to 0$ strongly. Further, by (101),

$$(W_n G(b)Q_n G(n^2)W_n) = (C_n) + (E_{-n}K'E_n),$$

where $(C_n) \in \mathcal{N}$ and $K' = \tilde{W}M^0(b)\tilde{Q}N^2\tilde{W}$ is compact.

(bc): As in the second part of the proof of Theorem 8.7, we distinguish between several cases (A)–(F), for which we prove the assertion. Case (A) is obvious.

Case (B): Let A be compact and B be arbitrary. Then $\mathcal{E}^0_{\mathcal{F}}(AB) - \mathcal{E}^0_{\mathcal{F}}(A)\mathcal{E}^0_{\mathcal{F}}(B)$ equals

$$(E_{-n}ABE_n) - (E_{-n}AE_n E_{-n}BE_n) = (E_{-n}A(I - L_n)BE_n)$$

This expression is in $\mathcal{N} \subseteq \mathcal{J}$ since $L_n = L^*_n \to I$ strongly.

Case (C): Let $A = \tilde{P}M^0(a)\tilde{P}$ and $B = \tilde{P}M^0(b)\tilde{P}$ with $a \in PC_\infty(\mathbb{R})$ and $b \in C_\infty^0(\mathbb{R})$. Then, by (85),

$$\mathcal{E}_\mathcal{F}^0(AB) = (P_n G(ab)P_n) - (E_{-n}\tilde{P}M^0(a)\tilde{Q}M^0(b)\tilde{P}E_n),$$
$$\mathcal{E}_\mathcal{F}^0(A)\mathcal{E}_\mathcal{F}^0(B) = (P_n G(a)G(b)P_n) - (P_n G(a)Q_n G(b)P_n).$$

From (100) and Proposition 9.1(cd) we deduce

$$\mathcal{E}_\mathcal{F}^0(AB) - \mathcal{E}_\mathcal{F}^0(A)\mathcal{E}_\mathcal{F}^0(B) = (P_n K P_n) + (C_n),$$

where $K = G(ab) - G(a)G(b)$ is compact and $(C_n) \in \mathcal{N}$.

Case (D): This case is similar to Case (C). Using (86) and (101), we obtain

$$\mathcal{E}_\mathcal{F}^0(AB) - \mathcal{E}_\mathcal{F}^0(A)\mathcal{E}_\mathcal{F}^0(B) = (W_n K W_n) + (C_n).$$

Case (E): Again, without loss of generality, we can assume that $A = \tilde{P}M^0(b)\tilde{P}$ and $B = \tilde{W}M^0(cn^2)\tilde{W}$. Taking into account the formulas (89)–(91), we obtain

$$\mathcal{E}_\mathcal{F}^0(AB) - \mathcal{E}_\mathcal{F}^0(A)\mathcal{E}_\mathcal{F}^0(B) =$$
$$\left[\mathcal{E}_\mathcal{F}^0(AK_1) - \mathcal{E}_\mathcal{F}^0(A)\mathcal{E}_\mathcal{F}^0(K_1)\right] + \left[\mathcal{E}_\mathcal{F}^0(K_2 B_2) - \mathcal{E}_\mathcal{F}^0(K_2)\mathcal{E}_\mathcal{F}^0(B_2)\right] +$$
$$\left[\mathcal{E}_\mathcal{F}^0(AB_1) - \mathcal{E}_\mathcal{F}^0(A)\mathcal{E}_\mathcal{F}^0(B_1)\right]\mathcal{E}_\mathcal{F}^0(B_2) + \mathcal{E}_\mathcal{F}^0(A)\left[\mathcal{E}_\mathcal{F}^0(B_1)\mathcal{E}_\mathcal{F}^0(B_2) - \mathcal{E}_\mathcal{F}^0(B_1 B_2)\right].$$

The first two terms on the right hand side are in $\mathcal{N}$ as it has already been proved in Case (B). The fourth term is equal to (see Case (D))

$$(P_n G(b)P_n)\left[(W_n L W_n) + (C_n)\right] = (W_n(W_n G(b)W_n)L W_n) + (C'_n).$$

Hence, by (a), this expression is in $\mathcal{N}$. To handle the third term, we will prove that $\mathcal{E}_\mathcal{F}^0(AB_1) - \mathcal{E}_\mathcal{F}^0(A)\mathcal{E}_\mathcal{F}^0(B_1) \in \mathcal{N}$. From (90) we obtain that

$$\mathcal{E}_\mathcal{F}^0(AB_1) = (E_{-n}\tilde{P}M^0(b)\tilde{Q}S\tilde{Q}S\tilde{P}E_n) - (E_{-n}\tilde{P}M^0(bs)\tilde{Q}S\tilde{P}E_n).$$

Substituting in this expression the $\tilde{Q}$'s by $\tilde{Q} = \tilde{Q}(I - L_n) + E_n Q_n E_{-n}$ and observing that $\tilde{P}M^0(b)\tilde{Q}$ and $\tilde{P}M^0(bs)\tilde{Q}$ are compact and $L_n = L_n^* \to I$ strongly, we conclude that

$$\mathcal{E}_\mathcal{F}^0(AB_1) = (P_n G(b)Q_n T(\sigma)Q_n T(\sigma)P_n) - (P_n G(bs)Q_n T(\sigma)P_n) + (C_n),$$
$$\mathcal{E}_\mathcal{F}^0(A)\mathcal{E}_\mathcal{F}^0(B_1) = (P_n G(b)P_n W_n G(n^2)W_n),$$

where $(C_n) \in \mathcal{N}$. Therefore, $\mathcal{E}_\mathcal{F}^0(AB_1) - \mathcal{E}_\mathcal{F}^0(A)\mathcal{E}_\mathcal{F}^0(B_1)$ is equal to

$$(P_n(G(b)T(\sigma) - G(bs))Q_n T(\sigma)P_n) - (W_n(W_n G(b)W_n)(G(n^2) + H(\tilde{\sigma})H(\sigma))W_n)$$
$$+ (P_n G(b)(W_n H(\tilde{\sigma})H(\sigma)W_n - P_n T(\sigma)Q_n T(\sigma)P_n)) + (C_n).$$

The first term is in $\mathcal{N}$ (cf. Proposition 9.1(d)), the third term vanishes (cf. (19)), and the second term is again in $\mathcal{N}$. (See (a) and use the fact that $G(n^2) + H(\tilde{\sigma})H(\sigma)$ is compact.)

Case (F): Let $A = B = \check{P}S\check{P}$. From (84) we obtain

$$\begin{aligned}
\mathcal{E}^0_{\mathcal{F}}(AB) &= (P_n + P_nG(n^2)P_n + W_nG(n^2)W_n) \\
\mathcal{E}^0_{\mathcal{F}}(A)\mathcal{E}^0_{\mathcal{F}}(B) &= (P_nT(\sigma)P_nT(\sigma)P_n)
\end{aligned}$$

Hence, by (18) and Proposition 9.1(ce), $\mathcal{E}^0_{\mathcal{F}}(AB) - \mathcal{E}^0_{\mathcal{F}}(A)\mathcal{E}^0_{\mathcal{F}}(B)$ is equal to

$$(P_nT(1 - \sigma^2)P_n + P_nKP_n + W_nLW_n),$$

where $K = G(n^2) + H(\sigma)H(\tilde{\sigma})$ and $L = G(n^2) + H(\tilde{\sigma})H(\sigma)$ are compact, and $1 - \sigma^2$ is continuous on $\mathbb{T}$ and vanishes at 1. ∎

Lemma 9.4

(a) $(P_nBP_n) \in \mathcal{J}_1$ whenever $B \in \mathcal{I}_1$.

(b) If $A \in \mathcal{I}_1$, then $JAJ \in \mathcal{I}_1$, and, if $A \in \mathcal{K}$, then $JA \in \mathcal{K}$ and $AJ \in \mathcal{K}$.

(c) If $(A_n) \in \mathcal{J}_1$, then $(JA_nJ) \in \mathcal{J}_1$, and, if $(A_n) \in \mathcal{J}$, then $(JA_n) \in \mathcal{J}$ and $(A_nJ) \in \mathcal{J}$.

Proof. Assertion (a) can be proved similarily as Lemma 6.5. Assertion (bc) follows easily from the relations $JL(a)J = L(\tilde{a})$, $JPJ = Q$, $J^2 = I$, $JP_nJ = P_n$, $JW_nJ = W_n$ and the fact that both $\mathcal{K}$ and $\mathcal{N}$ are ideals. ∎

By $\mathcal{E}''_{\mathcal{S}} : \Sigma^0_2 \to \mathcal{L}(l^2(\mathbb{Z}^+))^{2\times2}$ and $\mathcal{E}''_{\mathcal{F}} : \Xi^0_2 \to \mathcal{F}^{2\times2}_+$ we denote the mappings

$$\mathcal{E}''_{\mathcal{S}} : \begin{pmatrix} A & B \\ C & D \end{pmatrix} \mapsto \begin{pmatrix} \mathcal{E}^0_{\mathcal{S}}(A) & \mathcal{E}^0_{\mathcal{S}}(B) \\ \mathcal{E}^0_{\mathcal{S}}(C) & \mathcal{E}^0_{\mathcal{S}}(D) \end{pmatrix}, \quad \mathcal{E}''_{\mathcal{F}} : \begin{pmatrix} A & B \\ C & D \end{pmatrix} \mapsto \begin{pmatrix} \mathcal{E}^0_{\mathcal{F}}(A) & \mathcal{E}^0_{\mathcal{F}}(B) \\ \mathcal{E}^0_{\mathcal{F}}(C) & \mathcal{E}^0_{\mathcal{F}}(D) \end{pmatrix}. \tag{102}$$

Further, we introduce the *-homomorphisms $\Phi_{J,\mathcal{S}}$ and $\Phi_{J,\mathcal{F}}$ by

$$\Phi_{J,\mathcal{S}} : \mathcal{L}(l^2(\mathbb{Z}^+))^{2\times2} \to \mathcal{L}(l^2(\mathbb{Z})), \quad \begin{pmatrix} A & B \\ C & D \end{pmatrix} \mapsto A + BJ + JC + JDJ, \tag{103}$$

$$\Phi_{J,\mathcal{F}} : \qquad \mathcal{F}^{2\times2}_+ \to \mathcal{F}, \qquad \begin{pmatrix} (A_n) & (B_n) \\ (C_n) & (D_n) \end{pmatrix} \mapsto (A_n + B_nJ + JC_n + JD_nJ). \tag{104}$$

In what follows, we will be concerned with these mappings and with the *-homomorphisms $\mathcal{E}_{\mathcal{S}}$, $\mathcal{E}_{\mathcal{F}}$ and Φ_η. Recall that they act between the following C*-algebras:

$$\begin{aligned}
S(PC) &\xrightarrow{\mathcal{E}_{\mathcal{S}}} \Sigma \xrightarrow{\Phi_\eta} \Sigma^0_2 \xrightarrow{\mathcal{E}''_{\mathcal{S}}} \mathcal{L}(l^2(\mathbb{Z}^+))^{2\times2} \xrightarrow{\Phi_{J,\mathcal{S}}} \mathcal{L}(l^2(\mathbb{Z})), \\
\mathcal{F}(PC) &\xrightarrow{\mathcal{E}_{\mathcal{F}}} \Xi \xrightarrow{\Phi_\eta} \Xi^0_2 \xrightarrow{\mathcal{E}''_{\mathcal{F}}} \mathcal{F}^{2\times2}_+ \xrightarrow{\Phi_{J,\mathcal{F}}} \mathcal{F}.
\end{aligned} \tag{105}$$

(See also Proposition 7.2 and Corollary 8.1.) Finally, let $\mathcal{E}'_{\mathcal{S}}$ and $\mathcal{E}'_{\mathcal{F}}$ be the mappings defined by $\mathcal{E}'_{\mathcal{S}} := \Phi_{J,\mathcal{S}} \circ \mathcal{E}''_{\mathcal{S}} \circ \Phi_\eta : \Sigma \to \mathcal{L}(l^2(\mathbb{Z}))$ and $\mathcal{E}'_{\mathcal{F}} := \Phi_{J,\mathcal{F}} \circ \mathcal{E}''_{\mathcal{F}} \circ \Phi_\eta : \Xi \to \mathcal{F}$. Hence, by simple computations, we have

$$\begin{aligned}
\mathcal{E}'_{\mathcal{S}}(A) &= E_{-n}AE_n \quad \text{for all } n \geq 1, \ A \in \Sigma, \\
\mathcal{E}'_{\mathcal{F}}(A) &= (E_{-n}AE_n) \quad \text{for } A \in \Xi.
\end{aligned} \tag{106}$$

Lemma 9.5 *Let* $\rho_S = \mathcal{E}'_S \circ \mathcal{E}_S : \mathcal{S}(PC) \to \mathcal{L}(l^2(\mathbb{Z}))$. *Then*

(a) $\rho_S(AB) - \rho_S(A)\rho_S(B) \in \mathcal{I}_1$ *for all* $A, B \in \mathcal{S}(PC)$.

(b) $\rho_S(A) - A \in \mathcal{I}_1$ *for all* $A \in \mathcal{S}(PC)$.

Proof. (a): We have $\rho_S = \Phi_{J,S} \circ \mathcal{E}''_S \circ \Phi_\eta \circ \mathcal{E}_S$. By Proposition 7.2 and Corollary 8.1, $\Phi_\eta \circ \mathcal{E}_S : \mathcal{S}(PC) \to \Xi_2^0$ is a *-homomorphism "onto". Hence it suffices to show that the mapping $\Phi_{J,S} \circ \mathcal{E}''_S : \Xi_2^0 \to \mathcal{L}(l^2(\mathbb{Z}))$ is multiplicative modulo $\mathcal{I}_1$. Indeed, by Lemma 9.2, we have for all $A, B \in \Xi_2^0$ that

$$\mathcal{E}''_S(A)\,\mathcal{E}''_S(B) - \mathcal{E}''_S(AB) \;=\; \begin{pmatrix} A' & B' \\ C' & D' \end{pmatrix}$$

with $A', D' \in \mathcal{I}_1$ and $B', C' \in \mathcal{K}$. The assertion follows now from Lemma 9.4(b).

(b): We first consider the generating elements of $\mathcal{S}(PC)$. If $A = P$, then obviously $\rho_S(A) = P$. Again, if $A = L(1 \pm \sigma)$, then $\rho_S(A) = L(1 \pm \sigma)$, cf. (55). Finally, if $A = L(d)$ where d is a function on $\mathbb{T}$ which is continuous at 1 and vanishes there, then $\rho_S(A) = 0$, by (59), and $A \in \mathcal{I}_1$, by an approximation argument. In order to prove the assertion for all elements of $\mathcal{S}(PC)$, we consider the identity

$$\rho_S(AB) - AB \;=\; \big[\rho_S(AB) - \rho_S(A)\rho_S(B)\big] + \big[\rho_S(A) - A\big]\big[\rho_S(B) - B\big] +$$
$$\big[\rho_S(A) - A\big]B + A\big[\rho_S(B) - B\big].$$

From (a) it follows that, if $A \in \mathcal{S}(PC)$ and $B \in \mathcal{S}(PC)$ satisfy (b), then so does AB. Since ρ_S is moreover linear and continuous, (b) holds for all elements of $\mathcal{S}(PC)$. ∎

Lemma 9.6 *Let* $\rho_{\mathcal{F}} = \mathcal{E}'_{\mathcal{F}} \circ \mathcal{E}_{\mathcal{F}} : \mathcal{F}(PC) \to \mathcal{F}$. *Then*

(a) $\rho_{\mathcal{F}}(A_n B_n) - \rho_{\mathcal{F}}(A_n)\rho_{\mathcal{F}}(B_n) \in \mathcal{J}_1$ *for all* $(A_n), (B_n) \in \mathcal{F}(PC)$.

(b) $\rho_{\mathcal{F}}(A_n) - (A_n) \in \mathcal{J}_1$ *for all* $(A_n) \in \mathcal{F}(PC)$.

Proof. (a): We have $\rho_{\mathcal{F}} = \Phi_{J,\mathcal{F}} \circ \mathcal{E}''_{\mathcal{F}} \circ \Phi_\eta \circ \mathcal{E}_{\mathcal{F}}$. Since $\Phi_\eta \circ \mathcal{E}_{\mathcal{F}}$ is a *-homomorphism from $\mathcal{F}(PC)$ onto Ξ_2^0, it suffices to show that the mapping $\Phi_{J,\mathcal{F}} \circ \mathcal{E}''_{\mathcal{F}} : \Xi_2^0 \to \mathcal{F}$ is multiplicative modulo $\mathcal{J}_1$. Indeed, for $A, B \in \Xi_2^0$ we have (Lemma 9.3)

$$\mathcal{E}''_{\mathcal{F}}(A)\,\mathcal{E}''_{\mathcal{F}}(B) - \mathcal{E}''_{\mathcal{F}}(AB) \;=\; \begin{pmatrix} (A_n) & (B_n) \\ (C_n) & (D_n) \end{pmatrix}$$

with $(A_n), (D_n) \in \mathcal{J}_1$ and $(B_n), (C_n) \in \mathcal{J}$. Now we employ Lemma 9.4(c).

(b): Consider first the generation elements of $\mathcal{F}(PC)$. The case where $(A_n) \in \mathcal{J}$ is evident. For $(A_n) = (P_n A P_n)$ with $A \in \mathcal{S}(PC)$ we obtain from (106) that $\rho_{\mathcal{F}}(P_n A P_n) =$

$(P_n \, \rho_S(A) \, P_n)$, and we employ Lemma 9.4(a) and Lemma 9.5(b). To prove the assertion for all elements, we proceed in the same manner as in the proof of Lemma 9.5(b). ∎

In order to resume the results we have so far obtained, recall the following notion: A *sequence* $0 \to \mathfrak{A} \xrightarrow{\alpha} \mathfrak{B} \xrightarrow{\beta} \mathfrak{C} \to 0$ of C*-algebras $\mathfrak{A}$, $\mathfrak{B}$, $\mathfrak{C}$ (not necessarily with unit element) is called *short exact* if α and β are *-homomorphisms satisfying $\ker \alpha = \{0\}$, $\ker \beta = \operatorname{Im} \alpha$ and $\mathfrak{C} = \operatorname{Im} \beta$. Further, ρ is called a *continuous cross-section* of β if ρ is a linear and continuous mapping $\mathfrak{C} \to \mathfrak{B}$ with $\beta \circ \rho = \operatorname{id}$.

Proposition 9.7

(a) *The sequence* $0 \to \mathcal{I}_1 \xrightarrow{\operatorname{id}} S(PC) \xrightarrow{\mathcal{E}_S} \Sigma \to 0$ *is short exact, and* $\mathcal{E}'_S : \Sigma \to S(PC)$ *is a continuous cross-section of* $\mathcal{E}_S$.

(b) *The sequence* $0 \to \mathcal{J}_1 \xrightarrow{\operatorname{id}} \mathcal{F}(PC) \xrightarrow{\mathcal{E}_{\mathcal{F}}} \Xi \to 0$ *is short exact, and* $\mathcal{E}'_{\mathcal{F}} : \Xi \to \mathcal{F}(PC)$ *is a continuous cross-section of* $\mathcal{E}_{\mathcal{F}}$.

Proof. Since (a) and (b) are similar, we prove only (b). With regard to Proposition 7.2, we have to show that $\ker \mathcal{E}_{\mathcal{F}} \subseteq \mathcal{J}_1$, $\mathcal{E}_{\mathcal{F}} \circ \mathcal{E}'_{\mathcal{F}} = \operatorname{id}$ and $\mathcal{E}'_{\mathcal{F}} : \Xi \to \mathcal{F}(PC)$. The latter assertion, i. e., that $\mathcal{E}'_{\mathcal{F}}$ maps *into* $\mathcal{F}(PC)$ is a consequence of Lemma 9.6(b) and the fact that $\operatorname{Im} \mathcal{E}'_{\mathcal{F}} = \operatorname{Im}(\mathcal{E}'_{\mathcal{F}} \circ \mathcal{E}_{\mathcal{F}})$. That $\mathcal{E}_{\mathcal{F}} \circ \mathcal{E}'_{\mathcal{F}} = \operatorname{id}$ follows from (106) and from the definition of $\mathcal{E}_{\mathcal{F}}$. We are going to show that $\ker \mathcal{E}_{\mathcal{F}} \subseteq \mathcal{J}_1$. Let $(A_n) \in \ker \mathcal{E}_{\mathcal{F}}$. Again by Lemma 9.6(b), we have $0 = (\mathcal{E}'_{\mathcal{F}} \circ \mathcal{E}_{\mathcal{F}})(A_n) = (A_n) + (B_n)$ with $(B_n) \in \mathcal{J}_1$. Hence $(A_n) \in \mathcal{J}_1$. ∎

An implication of the preceding proposition is that $\mathcal{F}_1(PC)$ is *-isomorphic to Ξ, which means that the *-homomorphism $\tilde{\mathcal{E}}_{\mathcal{F}}$ is actually a *-isomorphism. However, we are interested in $\mathcal{L}(H)$–valued case, see (48). For this purpose, we cite the following result on tensor products from [5], Proposition 9.1. A proof is in [6], Proposition 2.

Proposition 9.8 *If* $0 \to \mathfrak{A} \xrightarrow{\alpha} \mathfrak{B} \xrightarrow{\beta} \mathfrak{C} \to 0$ *is a short exact sequence of C*-algebras such that* β *has a continuous cross-section* ρ *and* $\mathfrak{D}$ *is a C*-algebra, then the sequence*

$$0 \longrightarrow \mathfrak{A} \otimes \mathfrak{D} \xrightarrow{\alpha \otimes \operatorname{id}} \mathfrak{B} \otimes \mathfrak{D} \xrightarrow{\beta \otimes \operatorname{id}} \mathfrak{C} \otimes \mathfrak{D} \longrightarrow 0$$

is short exact.

Corollary 9.9 *The sequence*

$$0 \longrightarrow \mathcal{J}_{1,\mathcal{L}(H)} \xrightarrow{\operatorname{id}} \mathcal{F}(PC_{\mathcal{L}(H)}) \xrightarrow{\mathcal{E}_{\mathcal{F},\mathcal{L}(H)}} \Xi_{\mathcal{L}(H)} \longrightarrow 0$$

is short exact. Furthermore, the mapping $\mathcal{E}'_{\mathcal{F},\mathcal{L}(H)} : \Xi_{\mathcal{L}(H)} \to \mathcal{F}(PC_{\mathcal{L}(H)})$ *defined by* $\mathcal{E}'_{\mathcal{F},\mathcal{L}(H)}(A) := (E_{-n} A E_n)$ *is a continuous cross-section of* $\mathcal{E}_{\mathcal{F},\mathcal{L}(H)}$.

Proof.   Recall the identifications of tensor products we have made in the proof of Proposition 7.2(a). Then we have $\mathcal{E}'_{\mathcal{F},\mathcal{L}(H)} \cong \mathcal{E}'_{\mathcal{F}} \otimes \mathrm{id}$, which implies by definition that $\mathcal{E}'_{\mathcal{F},\mathcal{L}(H)}$ maps into $\mathcal{F}(PC_{\mathcal{L}(H)}) \cong \mathcal{F}(PC) \otimes \mathcal{L}(H)$ and that $\mathcal{E}'_{\mathcal{F},\mathcal{L}(H)}$ is a continuous cross-section of $\mathcal{E}_{\mathcal{F},\mathcal{L}(H)} \cong \mathcal{E}_{\mathcal{F}} \otimes \mathrm{id}$ (see Proposition 9.7(b)). Again from Proposition 9.7(b) and from Proposition 9.8, the sequence

$$0 \longrightarrow \mathcal{J}_1 \otimes \mathcal{L}(H) \xrightarrow{\mathrm{id}\otimes\mathrm{id}} \mathcal{F}(PC) \otimes \mathcal{L}(H) \xrightarrow{\mathcal{E}_{\mathcal{F}}\otimes\mathrm{id}} \Xi \otimes \mathcal{L}(H) \longrightarrow 0$$

is short exact. This proves the assertion.   ∎

Corollary 9.10 $\tilde{\mathcal{E}}_{\mathcal{F},\mathcal{L}(H)}$ *is a* *-isomorphism from* $\mathcal{F}_1(PC_{\mathcal{L}(H)})$ *onto* $\Xi_{\mathcal{L}(H)}$, *and*

$$\tilde{\mathcal{E}}^{-1}_{\mathcal{F},\mathcal{L}(H)} \;:\; A \mapsto \mathcal{E}'_{\mathcal{F},\mathcal{L}(H)}(A) + \mathcal{J}_{1,\mathcal{L}(H)}, \tag{107}$$

A similar results can be obtained also for $\tilde{\mathcal{E}}_{\mathcal{S},\mathcal{L}(H)}$. However, we will make no use of it. Now, in the second part of this section, we are going to show that $\Phi_{\tau,\mathcal{S}}$ and $\Phi_{\tau,\mathcal{F}}$ are actually *-isomorphisms (see (48)). We need the following auxiliary result. For a function $f \in L^\infty$, let $\sigma_n f$ denote the *Fejér-Cesaro mean*

$$(\sigma_n f)(e^{i\phi}) \;:=\; \sum_{k=-n}^{n} \left(1 - \tfrac{|k|}{n+1}\right) f_k\, e^{ik\phi}, \qquad \phi \in [0, 2\pi). \tag{108}$$

Lemma 9.11 *Let* $\tau \in \mathbb{T}$ *and* $\xi \in M^0_\tau(QC)$. *Then for all* $q \in QC$ *there is a sequence* $\{k_n\}^\infty_{n=1} \subseteq \mathbb{Z}^+$ *with* $k_n \to \infty$ *and* $(\sigma_{2k_n-1}q)(\tau) \to \xi(q)$ *às* $n \to \infty$.

Proof.   For $\lambda \in (1, \infty)$ and $f \in L^\infty$, let $m_\lambda f$ denote the moving average :

$$(m_\lambda f)(e^{i\phi}) \;:=\; \frac{\lambda}{2\pi} \int_{\phi-\pi/\lambda}^{\phi+\pi/\lambda} f(e^{i\vartheta})\, d\vartheta, \qquad \phi \in [0, 2\pi). \tag{109}$$

Then $(m_\lambda\,\cdot)(\tau)$ is a linear bounded functional on QC, and we know that ([1], 3.30 and 3.34)

$$M^0_\tau(QC) \;=\; M(QC) \cap \mathrm{clos}_{QC^*}\big\{(m_\lambda\,\cdot)(\tau) \,:\, \lambda \in (1, \infty)\big\}.$$

Let $\xi \in M^0_\tau(QC)$ and $q \in QC$. For $\varepsilon \to 0$, consider neighbourhoods U_ε of ξ in QC^*,

$$U_\varepsilon \;=\; \big\{\rho \in QC^* \,:\, |\rho(q) - \xi(q)| + |\rho(\chi_1) - \xi(\chi_1)| < \varepsilon\big\},$$

where χ_1 is the function $\chi_1(t) = t$, $t \in \mathbb{T}$. In this way we find a sequence $\{\lambda_n\}^\infty_{n=1} \subseteq (1, \infty)$ with $(m_{\lambda_n}q)(\tau) \to \xi(q)$ and $(m_{\lambda_n}\chi_1)(\tau) \to \xi(\chi_1) = \tau$ as $n \to \infty$. The latter relation implies that $\lambda_n \to \infty$. Now set $k_n = [(\lambda_n+1)/2] \in \mathbb{Z}^+$ and $\lambda'_n = 2k_n - 1$. Since $\lambda_n/\lambda'_n \to 1$, we obtain by simple computations that $|(m_{\lambda_n}q)(\tau) - (m_{\lambda'_n}q)(\tau)| \to 0$. Hence $(m_{2k_n-1}q)(\tau) \to \xi(q)$, and the assertion follows now from [1], 3.13, 3.14 and 3.27.   ∎

Proposition 9.12 *The kernels of $\Phi_{\tau,\mathcal{F}}$ and $\Phi_{\tau,\mathcal{S}}$ are trivial.*

Proof. We first consider $\Phi_{\tau,\mathcal{F}}$. Assume that $\ker \Phi_{\tau,\mathcal{F}} \neq \{0\}$. Since $\mathcal{F}_1(PC_{\mathcal{L}(H)})$ and $\Xi_{\mathcal{L}(H)}$ are *-isomorphic via $\tilde{\mathcal{E}}_{\mathcal{F},\mathcal{L}(H)}$, the ideal $\ker \Phi_{\tau,\mathcal{F}}$ corresponds to the non-trivial ideal $\mathcal{J} := \tilde{\mathcal{E}}_{\mathcal{F},\mathcal{L}(H)}(\ker \Phi_{\tau,\mathcal{F}})$ of $\Xi_{\mathcal{L}(H)}$. Take $A \in \mathcal{J}\backslash\{0\}$. For $x, y \in L_H^2([-1,1])$, define the operator $K_{xy} \in \mathcal{K}(L_H^2([-1,1]))$ by

$$K_{xy} z := x \cdot \langle y, z\rangle_{L_H^2([-1,1])} \qquad (z \in L_H^2([-1,1])).$$

Obviously, $K_{xy} \in \mathcal{K}(L^2([-1,1])) \otimes \mathcal{K}(H) \subseteq \Xi \otimes \mathcal{L}(H) \cong \Xi_{\mathcal{L}(H)}$, by Corollary 8.8. Since $A \neq 0$, there are $x_1, x_2 \in L_H^2([-1,1])$ for which $\langle x_2, Ax_1\rangle \neq 0$. Now choose an $h \in H$ with $\|h\| = 1$, and put $x = \chi_{[-1,1]} \in L^2([-1,1])$. Straightforward computations yield that

$$K_{xh,x_2} A K_{x_1,xh} = \langle x_2, Ax_1\rangle K_{xh,xh}.$$

Since $\mathcal{J}$ is an ideal of $\Xi_{\mathcal{L}(H)}$, we have $K_{xh,xh} \in \mathcal{J}$. From Corollary 9.9 and 9.10 we obtain

$$\tilde{\mathcal{E}}_{\mathcal{F},\mathcal{L}(H)}^{-1}(K_{xh,xh}) = (E_{-n} K_{xh,xh} E_n) + \mathcal{J}_{1,\mathcal{L}(H)} \in \ker \Phi_{\tau,\mathcal{F}},$$

and thus (see Proposition 7.1)

$$(K_n) := (Y_\tau E_{-n} K_{xh,xh} E_n Y_\tau^{-1}) \in \mathcal{J}_{\xi,\mathcal{L}(H)}. \tag{110}$$

Given $\varepsilon > 0$, by Lemma 6.3 and the definition of $\mathcal{J}_{\xi,\mathcal{L}(H)}$, there is a sequence (A_n) for which

$$\|(A_n) - (K_n)\|_{\mathcal{F}} \leq \varepsilon, \tag{111}$$

$$(A_n) = \sum_{i=1}^{k} (A_n^{(i)})(P_n f_i P_n) + (B_n'),$$

where $(A_n^{(i)}) \in \mathcal{F}(PQC_{\mathcal{L}(H)})$, $f_i \in QC$, $\xi(f_i) = 0$ and $(B_n') \in \mathcal{J}$. Consider the open neighbourhood U of ξ in $M(QC)$,

$$U := \left\{ \xi' \in M(QC) : |\xi'(f_i)| < \varepsilon k^{-1} \|(A_n^{(i)})\|_{\mathcal{F}}^{-1} \; \forall i \right\}.$$

By the Gelfand-Naimark Theorem, there is an $f \in QC$ for which $\xi(f) = 1$, $\xi'(f) \in [0,1]$ if $\xi' \in U$, and $\xi'(f) = 0$ if $\xi' \notin U$. Hence $\|f\| = 1$ and $\sum_i \|(A_n^{(i)})\| \cdot \|f f_i\| \leq \varepsilon$. Therefore,

$$(A_n)(P_n f P_n) = \sum_{i=1}^{k} (A_n^{(i)})(P_n f_i f P_n) + (B_n),$$

$$\|(A_n)(P_n f P_n) - (B_n)\|_{\mathcal{F}} \leq \varepsilon, \tag{112}$$

where $(B_n) = (P_n K P_n + W_n L W_n + C_n') \in \mathcal{J}$. Now choose a $\varkappa$ such that $\|Q_\varkappa K\|_A \leq \varepsilon$ and $\|Q_\varkappa L\|_A \leq \varepsilon$, and set

$$(R_n) := (P_n Q_\varkappa P_n)(W_n Q_\varkappa W_n) = (W_n Q_\varkappa W_n)(P_n Q_\varkappa P_n).$$

Therefore,

$$\begin{aligned} R_n B_n &= R_n P_n K P_n + R_n W_n L W_n + C_n \\ &= W_n Q_{\varkappa} W_n Q_{\varkappa} K P_n + P_n Q_{\varkappa} W_n Q_{\varkappa} L W_n + C_n, \end{aligned}$$

where $(C_n) \in \mathcal{N}$, and we have

$$\| (R_n)(B_n) - (C_n) \|_{\mathcal{F}} \leq 2\varepsilon. \tag{113}$$

Observing $\|R_n\| = 1$ and $\|f\| = 1$, we readily obtain from (111)–(113) that for all $n \geq 0$

$$\| R_n K_n P_n f P_n \|_{\mathcal{L}(l_H^2(\mathbb{Z}_n))} \leq 4\varepsilon + \| C_n \|_{\mathcal{L}(l_H^2(\mathbb{Z}_n))}.$$

Hence

$$\limsup_{n \to \infty} \| R_n K_n P_n f P_n \|_{\mathcal{L}(l_H^2(\mathbb{Z}_n))} \leq 4\varepsilon. \tag{114}$$

Let z_n and z_n^* be the linear bounded operators

$$z_n : \ \mathbb{C} \to l_H^2(\mathbb{Z}_n) \qquad \lambda \mapsto \left(\lambda h / \sqrt{2n} \right)_{i=-n}^{n-1},$$

$$z_n^* : \ l_H^2(\mathbb{Z}_n) \to \mathbb{C} \qquad (x_i)_{i=-n}^{n-1} \mapsto \sum_{i=-n}^{n-1} \langle h, x_i \rangle_H / \sqrt{2n}.$$

Then $\|z_n\| = \|z_n^*\| = 1$. Straightforward computations yield that

$$(K_n) = (Y_\tau z_n z_n^* Y_\tau^{-1}).$$

Therefore, we have

$$\begin{aligned} \| R_n K_n P_n f P_n \|_{\mathcal{L}(l_H^2(\mathbb{Z}_n))} &\geq |z_n^* Y_\tau^{-1} R_n K_n P_n f P_n Y_\tau z_n| \\ &= |z_n^* R_n z_n| \cdot |z_n^* Y_\tau^{-1} f Y_\tau z_n| \\ &= \frac{2n-4\varkappa}{2n} \cdot \sum_{i=-2n+1}^{2n-1} f_i \tau^i \frac{2n-|i|}{2n} \qquad \text{(if } n \geq 2\varkappa) \\ &= \frac{2n-4\varkappa}{2n} \cdot |(\sigma_{2n-1} f)(\tau)|, \end{aligned}$$

see (108). Since $\xi \in M_\tau^0(QC)$ and $\xi(f) = 1$, we obtain from Lemma 9.11, that

$$\limsup_{n \to \infty} \| R_n K_n P_n f P_n \|_{\mathcal{L}(l_H^2(\mathbb{Z}_n))} \geq \limsup_{n \to \infty} |(\sigma_{2n-1} f)(\tau)| \geq |\xi(f)| = 1,$$

which contradicts (114) for sufficiently small ε. Hence $\ker \Phi_{\tau, \mathcal{F}}$ is trivial.

Now we consider $\Phi_{\tau, \mathcal{S}}$. Let $A + \mathcal{I}_{1, \mathcal{L}(H)} \in \ker \Phi_{\tau, \mathcal{S}}$, where $A \in \mathcal{S}(PC_{\mathcal{L}(H)})$. Then $Y_\tau A Y_\tau^{-1} \in \mathcal{I}_{\xi, \mathcal{L}(H)}$. From Lemma 6.5 we obtain $(Y_\tau P_n A P_n Y_\tau^{-1}) \in \mathcal{J}_{\xi, \mathcal{L}(H)}$, and hence, since $(P_n A P_n) \in \mathcal{F}(PC_{\mathcal{L}(H)})$, we have $(P_n A P_n) + \mathcal{J}_{1, \mathcal{L}(H)} \in \ker \Phi_{\tau, \mathcal{F}}$. By what we have just proved, $(P_n A P_n) \in \mathcal{J}_{1, \mathcal{L}(H)}$. Now we apply the homomorphism $\mathcal{P}$ to this sequence (see

Corollary 5.4(b)). Observing that $\mathcal{P}$ maps $\mathcal{J}_{1,\mathcal{L}(H)}$ into $\mathcal{I}_{1,\mathcal{L}(H)}$, we obtain that $A \in \mathcal{I}_{1,\mathcal{L}(H)}$. Hence the kernel of $\Phi_{\tau,S}$ is trivial. ∎

10. Final Result

Let us recall scheme (48) and extend it by the canonical homomorphisms from $\mathcal{F}(PQC_{\mathcal{L}(H)})$ and $\mathcal{S}(PQC_{\mathcal{L}(H)})$ onto the local algebras :

$$
\begin{array}{ccccccc}
\mathcal{F}(PQC_{\mathcal{L}(H)}) & \longrightarrow & \mathcal{F}_\xi(PQC_{\mathcal{L}(H)}) & \overset{\Phi_{\tau,\mathcal{F}}}{\underset{}{\rightleftarrows}} & \mathcal{F}_1(PC_{\mathcal{L}(H)}) & \overset{\tilde{\mathcal{E}}_{\mathcal{F},\mathcal{L}(H)}}{\underset{}{\rightleftarrows}} & \Xi_{\mathcal{L}(H)} \\
\downarrow & & \downarrow & & \downarrow & & \downarrow \\
\mathcal{S}(PQC_{\mathcal{L}(H)}) & \longrightarrow & \mathcal{S}_\xi(PQC_{\mathcal{L}(H)}) & \overset{\Phi_{\tau,S}}{\underset{}{\rightleftarrows}} & \mathcal{S}_1(PC_{\mathcal{L}(H)}) & \overset{\tilde{\mathcal{E}}_{S,\mathcal{L}(H)}}{\underset{}{\rightleftarrows}} & \Sigma_{\mathcal{L}(H)}
\end{array}
$$

Combining Proposition 7.1 and 9.12, we see that $\Phi_{\tau,\mathcal{F}}$ and $\Phi_{\tau,S}$ are *-isomorphisms. Further, by Proposition 7.2 and Corollary 9.10, also $\tilde{\mathcal{E}}_{\mathcal{F},\mathcal{L}(H)}$ (and $\tilde{\mathcal{E}}_{S,\mathcal{L}(H)}$) are *-isomorphisms. Now we are able to define *-homomorphisms from $\mathcal{S}(PQC_{\mathcal{L}(H)})$ and $\mathcal{F}(PQC_{\mathcal{L}(H)})$ onto $\Sigma_{\mathcal{L}(H)}$ and $\Xi_{\mathcal{L}(H)}$, respectively. Its properties are described in the following theorem and can be proved straightforwardly. (Recall that Φ_ξ has been defined in [7]; see also the last paragraph in Section 2.)

Theorem 10.1 *Let* $\tau \in \mathbb{T}$ *and* $\xi \in M_\tau^0(QC)$. *Then the mappings*

$$
\Psi_{S,\xi} \,:\, \mathcal{S}(PQC_{\mathcal{L}(H)}) \to \Sigma_{\mathcal{L}(H)}, \quad A \mapsto (\tilde{\mathcal{E}}_{S,\mathcal{L}(H)} \circ \Phi_{\tau,S}^{-1})(A + \mathcal{I}_{\xi,\mathcal{L}(H)}) \tag{115}
$$
$$
\Psi_{\mathcal{F},\xi} \,:\, \mathcal{F}(PQC_{\mathcal{L}(H)}) \to \Xi_{\mathcal{L}(H)}, \quad (A_n) \mapsto (\tilde{\mathcal{E}}_{\mathcal{F},\mathcal{L}(H)} \circ \Phi_{\tau,\mathcal{F}}^{-1})((A_n) + \mathcal{J}_{\zeta,\mathcal{L}(H)}) \tag{116}
$$

*are *-homomorphisms. Moreover, we have* $\mathcal{J} \subseteq \ker \Psi_{\mathcal{F},\xi}$ *and*

$$
\begin{aligned}
\Psi_{S,\xi}(P) &= \chi_{[0,\infty)} & &\tag{117}\\
\Psi_{S,\xi}(a) &= a(\tau+0)\frac{(I - S_\mathbb{R})}{2} + a(\tau-0)\frac{(I + S_\mathbb{R})}{2} & (a \in PC_{\mathcal{L}(H)}) &\tag{118}\\
\Psi_{S,\xi}(a) &= \Phi_\xi(a) & (a \in QC_{\mathcal{L}(H)}^s) &\tag{119}\\
\Psi_{\mathcal{F},\xi}(P_n A P_n) &= \chi_{[-1,1]}\, \Psi_{S,\xi}(A)\, \chi_{[-1,1]} & (A \in \mathcal{S}(PQC_{\mathcal{L}(H)}). &\tag{120}
\end{aligned}
$$

Now we formulate the main result. Again, we use the *-isomorphisms $\mathcal{P}$ and $\mathcal{W}$; see Corollary 5.4(b).

Theorem 10.2 *Let* $(A_n) \in \mathcal{F}(PQC_{\mathcal{L}(H)})$ *Then the sequence* (A_n) *is stable if and only if*

(a) $\mathcal{P}(A_n)$ *is invertible in* $\mathcal{S}(PQC_{\mathcal{L}(H)})$,

(b) $\mathcal{W}(A_n)$ *is invertible in* $\mathcal{S}(PQC_{\mathcal{L}(H)})$, *and*

(c) $\Psi_{\mathcal{F},\xi}(A_n)$ *is invertible in* $\Xi_{\mathcal{L}(H)}$ *for all* $\xi \in M_\tau^0(QC)$ *with* $\tau \in \mathbb{T}$.

Proof. Combine Corollary 5.6 and 6.4, Theorem 6.6, and use the above mentioned fact that $\Phi_{\tau,\mathcal{F}}$ and $\tilde{\mathcal{E}}_{\mathcal{F},\mathcal{L}(H)}$ are *-isomorphisms for $\xi \in M^0_\tau(QC)$. The latter means that $(A_n) + \mathcal{J}_{\xi,\mathcal{L}(H)}$ is invertible in $\mathcal{F}_\xi(PQC_{\mathcal{L}(H)})$ if and only if $\Psi_{\mathcal{F},\xi}(A_n)$ is invertible in $\Xi_{\mathcal{L}(H)}$. ∎

Remark 10.3 We want to discuss $\mathcal{L}(H)$–valued quasicontinuous functions once more. The definitions of $\mathcal{F}(PQC_{\mathcal{L}(H)}), \mathcal{S}(PQC_{\mathcal{L}(H)})$ and $PQC_{\mathcal{L}(H)}$ are based on the C*-algebra $QC^s_{\mathcal{L}(H)}$. It has already been pointed out (see Section 2 or [7]) that $QC^s_{\mathcal{L}(H)}$ is strictly contained in $QC_{\mathcal{L}(H)}$ if dim $H = \infty$. Therefore, one may ask whether or not Theorem 10.1 and Theorem 10.2 remain valid if a C*-algebra $\mathfrak{B}$ with $QC^s_{\mathcal{L}(H)} \subseteq \mathfrak{B} \subseteq QC_{\mathcal{L}(H)}$ is taken as the basis of the definitions of $\mathcal{F}(PQC_{\mathcal{L}(H)}), \mathcal{S}(PQC_{\mathcal{L}(H)})$ and $PQC_{\mathcal{L}(H)}$. In fact, it is possible to consider a C*-algebra $\mathfrak{B}$ which is larger than $QC^s_{\mathcal{L}(H)}$, although the most desirable aim – to handle $QC_{\mathcal{L}(H)}$ – we do not achieve. However, we note that the assertions concerning the homomorphisms $\mathcal{P}$ and $\mathcal{W}$ remain true also if $QC_{\mathcal{L}(H)}$ is considered.

The significance of the C*-algebra $QC_{\mathcal{L}(H)}$ is certainly justified by the relationship stated in Proposition 3.5. Further, the C*-algebra $QC^s_{\mathcal{L}(H)}$, which can be identified with $QC \otimes \mathcal{L}(H)$, is characterized by the property that it is the largest closed algebra which is locally trivial at all points $\xi \in M(QC)$. We have made use of the local triviality of $QC^s_{\mathcal{L}(H)}$ only in Proposition 7.1, in order to show that the *-homomorphisms $\Phi_{\tau,\mathcal{S}}$ and $\Phi_{\tau,\mathcal{F}}$ are "onto". Actually, we did only use that $QC^s_{\mathcal{L}(H)}$ is locally trivial at all points $\xi \in M^0_\tau(QC)$, $\tau \in \mathbb{T}$. One can show that the homomorphisms fail to be "onto" if one considers a closed algebra $\mathfrak{B}$ with $QC^s_{\mathcal{L}(H)} \subseteq \mathfrak{B} \subseteq QC_{\mathcal{L}(H)}$ which is not locally trivial at these points.

In the proof Theorem 6.6, which settles the situation for $\xi \in M_\tau(QC)\backslash M^0_\tau(QC)$, it is shown that for each $p \in PC_{\mathcal{L}(H)}$ there is an $a \in \mathcal{L}(H)$ such that $p - a \in \mathcal{I}_{\xi,\mathcal{L}(H)}$. (This characterization is similar to that of the local triviality of $QC^s_{\mathcal{L}(H)}$. Therefore, one could say that $PC_{\mathcal{L}(H)}$ is "locally trivial" at $\xi \in M_\tau(QC)\backslash M^0_\tau(QC)$.) For each function $p \in PC_{\mathcal{L}(H)}$ which is furthermore continuous at τ, there exists obviously an $a \in \mathcal{L}(H)$ such that $p - a \in \mathcal{I}_{\xi,\mathcal{L}(H)}$ for all $\xi \in M_\tau(QC)$. These observations lead us to the following modification of the above results:

For a subset $M \subseteq \mathbb{T}$, let $PC_{\mathcal{L}(H),M}$ be the C*-algebra of all functions $p \in PC_{\mathcal{L}(H)}$ which are continuous at all points of $\mathbb{T}\backslash M$. Further, let $\mathfrak{B}_M$ be a C*-algebra with $QC^s_{\mathcal{L}(H)} \subseteq \mathfrak{B}_M \subseteq QC_{\mathcal{L}(H)}$ which is locally trivial at all points ξ of

$$\bigcup_{\tau \in M} M^0_\tau(QC). \tag{121}$$

Then Theorem 10.1 remains valid for all $\xi \in M^0_\tau(QC)$, $\tau \in M$ (instead of $\tau \in \mathbb{T}$), and Theorem 10.2 remains valid with the modification that one considers in (c) only the *-homomorphisms $\Psi_{\mathcal{F},\xi}$ with $\xi \in M^0_\tau(QC)$, $\tau \in M$. (However, the assertion stated in Remark 6.7 can no longer be guaranteed.)

We present some properties of the above considered algebras $\mathfrak{B}_M$. First of all, among all closed algebras $\mathfrak{B}_M$ with $QC^s_{\mathcal{L}(H)} \subseteq \mathfrak{B}_M \subseteq QC_{\mathcal{L}(H)}$ which are locally trivial at all points ξ of (121), there exists a largest closed algebra, denoted by $\mathfrak{B}^*_M$, which is moreover a C*-algebra. Further, we have

$$\mathfrak{B}^*_\emptyset = QC_{\mathcal{L}(H)}, \tag{122}$$

$$\mathfrak{B}^*_{\{\tau\}} = \left\{ q \in QC_{\mathcal{L}(H)} : \Phi(q) \text{ is continuous at all points } \xi \in M^0_\tau(QC) \right\}, \tag{123}$$

$$\mathfrak{B}^*_M = \bigcap_{\tau \in M} \mathfrak{B}^*_{\{\tau\}} \qquad \text{if } M \subseteq \mathbb{T}, \, M \neq \emptyset. \tag{124}$$

Here Φ was introduced in [7], Section 2. Finally, the following non-trivial results, for which we omit a proof, are known to the authors:

(i) Let $q \in QC_{\mathcal{L}(H)}$ and $\tau \in \mathbb{T}$. Then $q \in \mathfrak{B}^*_{\{\tau\}}$ if and only if the set
$$\left\{ (m_\lambda q)(\tau) : \lambda \in (1,\infty) \right\} \text{ is precompact in } \mathcal{L}(H).[3]$$

(ii) We have $QC^s_{\mathcal{L}(H)} \subseteq \mathfrak{B}^*_\mathbb{T} \subseteq QC_{\mathcal{L}(H)}$. Both inclusions are proper if $\dim H = \infty$. For $\dim H < \infty$, equality holds.

References

[1] A. BÖTTCHER, B. SILBERMANN, Analysis of Toeplitz operators. – Akademie Verlag, Berlin, 1989, and Springer Verlag, Berlin, 1990.

[2] A. BÖTTCHER, B. SILBERMANN, Operator-valued Szegö-Widom limit theorems. – In: Toeplitz operators and related topics: The Harold Widom anniversary Volume. – Operator Theory 71, Birkhäuser Verlag, Basel, Boston, Berlin, 1994, 33 – 53.

[3] A. BÖTTCHER, B. SILBERMANN, Infinite Toeplitz and Hankel matrices with operator-valued entries. – SIAM Journal on Mathematical Analysis (to appear).

[4] M. COSTABEL, An inverse for the Gohberg-Krupnik symbol map. – Proc. Royal Soc. of Edinburgh 87A(1980), 153 – 165.

[5] R. G. DOUGLAS, Banach algebra techniques in the theory of Toeplitz operators. – CBMS Lecture Notes 15, Amer. Math. Soc., Providence, R.I., 1973.

[6] R. G. DOUGLAS, R. HOWE, On the C*-algebra of Toeplitz operators on the quarter-plane. – Trans. Amer. Math. Soc. 158, No. 1 (1971), 203 – 217.

[7] T. EHRHARDT, S. ROCH, B. SILBERMANN, Symbol calculus for singular integrals with operator-valued PQC-coefficients. – this volume.

[8] I. GOHBERG, M. A. KAASHOEK, Projection method for Block Toeplitz operators with operator-valued symbols. – In: Toeplitz operators and related topics: The Harold Widom anniversary Volume. – Operator Theory 71, Birkhäuser Verlag, Basel, Boston, Berlin, 1994, 79 – 104.

---

[3] The moving average $(m_\lambda q)(\tau)$ has been introduced for the scalar case in (109). The definition for the operator-valued case is similar. Instead of the moving average, one can also consider e. g. the Fejer means.

[9] R. HAGEN, S. ROCH, B. SILBERMANN, Spectral Theory of Approximation Methods for Convolution Equations. – Operator Theory 74, Birkhäuser Verlag, Basel, Boston, Berlin, 1995.

[10] R. V. KADISON, J. R. RINGROSE, Fundamentals of the theory of operator algebras. Vol. I (1983) and Vol. II (1986). – Pure and Applied Mathematics 100. Orlando, New York, London etc.: Academic Press (Harcourt Brace Jovanovich, Publishers).

[11] L. B. PAGE, Bounded and compact vectorial Hankel operators. – Trans. Amer. Math. Soc. **150** (1970), 529-539.

[12] S. ROCH, B. SILBERMANN, Toeplitz-like operators, quasicommutator ideals, numerical analysis. – Part I: Math. Nachr. **120** (1985), 141 – 173; Part II: Math. Nachr. **134** (1987), 381 – 391.

[13] S. ROCH, B. SILBERMANN, A symbol calculas for finite sections of singular integral operators with flip and piecewise quasicontinuous coefficients. – J. Funct. Anal.**78** (1988), 2, 365 – 389.

[14] S. ROCH, B. SILBERMANN, Limiting sets of eigenvalues and singular values of Toeplitz matrices. – Asymptotic Analysis **8** (1994), 293 – 309.

[15] D. SARASON, Toeplitz operators with piecewise quasicontinuous symbols. – Indiana Univ. Math. J. **26:5** (1977), 817-838.

[16] D. SARASON, Function theory on the unit circle. – Virginia Polytechnic Institute and State Univ., Blacksburg, 1978.

[17] B. SILBERMANN, Asymptotics for Toeplitz operators with piecewise quasicontinuous symbols and related questions. – Math. Nachr. **125** (1986), 179-190.

[18] B. SILBERMANN, Toeplitz-like operators and their finite sections (to appear).

[19] I. B. SIMONENKO, CHIN NGOK MINH, A local method in the theory of one-dimensional singular integral equations with piecewise continuous coefficients. Fredholmness. – Izd. Rostov-na-Donu, 1986 (Russian).

[20] E. M. STEIN, Singular integrals and differentiability properties of functions. – Princeton Univ. Press, Princeton, New Jersey, 1970.

[21] H. WIDOM, Asymptodic behaviour of block Toeplitz determinants. II. – Adv. Math. **21** (1976), 1 – 29.

Torsten Ehrhardt and Bernd Silbermann
Technische Universität Chemnitz-Zwickau
Fakultät für Mathematik
D 09107 Chemnitz
Germany
ehrhardt@mathematik.tu-chemnitz.de
silbermn@mathematik.tu-chemnitz.de

Steffen Roch
Universität Leipzig
Mathematisches Institut
Augustusplatz 10 – 11
D 04109 Leipzig
Germany
roch@miaix550.mathematik.uni-leipzig.de

MSC 1991:   Primary   47B35
            Secondary   65R20, 46E40, 45E10, 45F15, 46L05

Operator Theory
Advances and Applications, Vol. 90
© 1996 Birkhäuser Verlag Basel/Switzerland

DISTRIBUTION OF ZEROS OF ORTHOGONAL FUNCTIONS RELATED TO THE NEHARI PROBLEM

Robert L. Ellis and Israel Gohberg

In this paper, we prove continuous analogues of results concerning the distribution of zeros of orthogonal matrix functions related to the Nehari problem.

0   INTRODUCTION

In [K66], M.G. Krein proved a now well-known theorem concerning the distribution of zeros of orthogonal polynomials associated with Toeplitz matrices. (For a different presentation, see [EGL88].) A continuous analogue of this theorem was formulated and proved in [KL85]. Later, Krein's original theorem was extended in [GL88] and [AG88] to orthogonal matrix polynomials. A continuous analogue of the extended theorem was obtained in [EGL92] and an alternative proof was given in [D94]. The orthogonal matrix polynomials and their continuous analogues that appeared in these works are related to the Carathéodory extension problem. The same type of relation was found to exist between a new class of orthogonal functions and the Nehari extension problem [EG92]. For the discrete scalar case, an analogue of Krein's Theorem for these orthogonal functions was proved in [EG92]. The latter results were generalized to the discrete matrix case in [EGL95].

In the present paper, we take the next step and obtain a theorem about the distribution of zeros in the appropriate continuous matrix case. Before stating the main theorem, we rewrite the equations that determine the orthogonal functions in a form that obviates the necessity of using distributions.

The formal continuous analogue of equation (4.1) in [EGL95], which there defines the orthogonal functions, is the system of equations

$$\text{(0.1)} \qquad \phi_a(t) + \int_a^\infty k(t+s-a)\psi_a(-s)\,ds = \delta(t-a)I \qquad (t \geq a)$$

and

$$\text{(0.2)} \qquad \int_a^\infty k(t+s-a)^*\phi_a(s)\,ds + \psi_a(-t) = 0 \qquad (t \geq a).$$

Here k is a given $m \times m$ matrix-valued function in $L_1^{m \times m}(a, \infty)$, δ is the Dirac delta function, and I is the $m \times m$ identity matrix. The solutions ϕ_a and ψ_a are to belong to $L_1^{m \times m}(a, \infty)$ and $L_1^{m \times m}(-\infty, -a)$, respectively, which are the spaces of integrable $m \times m$ matrix-valued functions on the indicated intervals. In order to eliminate the delta function, we make the substitutions

(0.3)
$$g_a(t) = \phi_a(t) - \delta(t-a)I \qquad (t \geq a)$$

and

(0.4)
$$h_a(-t) = \psi_a(-t) \qquad (t \geq a).$$

Then (0.1) and (0.2) may be rewritten as

$$g_a(t) + \int_a^\infty k(t+s-a)h_a(-s)\,ds = 0 \qquad (t \geq a)$$

and

$$\int_a^\infty k(t+s-a)^* g_a(s)\,ds + h_a(-t) = -k(t)^* \qquad (t \geq a).$$

With the substitutions in (0.3) and 0.4, the formal analogues of the functions α and β that are defined in (4.3) of [EGL95] and that there determine the orthogonal functions, are given by

$$\Phi_a(\lambda) = e^{i\lambda a}I + \int_a^\infty e^{i\lambda t} g_a(t)\,dt \qquad (\mathrm{Im}\,\lambda \geq 0)$$

and

$$\Psi_a(\lambda) = \int_a^\infty e^{-i\lambda t} h_a(-t)\,dt \qquad (\mathrm{Im}\,\lambda \leq 0).$$

Here $\mathrm{Im}\,\lambda$ denotes the imaginary part of a complex number λ. In a similar way, we obtain from equation (4.2) in [EGL95] the system of equations

$$\gamma_a(t) + \int_a^\infty k(t+s-a)\chi_a(-s)\,ds = -k(t) \qquad (t \geq a)$$

and

$$\int_a^\infty k(t+s-a)^* \gamma_a(s)\,ds + \chi_a(-t) = 0 \qquad (t \geq a).$$

The corresponding functions are

$$\Omega_a(\lambda) = \int_a^\infty e^{i\lambda t} \gamma_a(t)\,dt \qquad (\mathrm{Im}\,\lambda \geq 0)$$

and

$$\Theta_a(\lambda) = e^{-i\lambda a}I + \int_a^\infty e^{-i\lambda t} \chi_a(-t)\,dt \qquad (\mathrm{Im}\,\lambda \leq 0).$$

The version of Krein's theorem that we will prove in this paper concerns the distribution of zeros of the determinants of the matrix functions Φ_a and Θ_a. We state it next as the main theorem.

MAIN THEOREM *Let* $k \in L_1^{m \times m}(a, \infty)$. *Assume that there exist solutions* g_a *in* $L_1^{m \times m}(a, \infty)$ *and* h_a *in* $L_1^{m \times m}(-\infty, -a)$ *of the equations*

$$(0.5) \qquad g_a(t) + \int_a^\infty k(t + s - a) h_a(-s)\, ds = 0 \qquad (t \geq a)$$

and

$$(0.6) \qquad \int_a^\infty k(t + s - a)^* g_a(s)\, ds + h_a(-t) = -k(t)^* \qquad (t \geq a).$$

Assume also that there exist solutions γ_a *in* $L_1^{m \times m}(a, \infty)$ *and* χ_a *in* $L_1^{m \times m}(-\infty, -a)$ *of the equations*

$$(0.7) \qquad \gamma_a(t) + \int_a^\infty k(t + s - a) \chi_a(-s)\, ds = -k(t) \qquad (t \geq a)$$

and

$$(0.8) \qquad \int_a^\infty k(t + s - a)^* \gamma_a(s)\, ds + \chi_a(-t) = 0 \qquad (t \geq a).$$

Define

$$(0.9) \qquad \Phi_a(\lambda) = e^{i\lambda a} I + \int_a^\infty e^{i\lambda t} g_a(t)\, dt \qquad (\mathrm{Im}\, \lambda \geq 0)$$

and

$$(0.10) \qquad \Theta_a(\lambda) = e^{-i\lambda a} I + \int_a^\infty e^{-i\lambda t} \chi_a(-t)\, dt \qquad (\mathrm{Im}\, \lambda \leq 0).$$

Then $\Phi_a(\lambda)$ *and* $\Theta_a(\lambda)$ *are invertible for all real* λ *and, counting multiplicities, the number of zeros of* $\det \Phi_a$ *(respectively,* $\det \Theta_a$*) in the upper (respectively, lower) half plane is finite and equals the number of negative eigenvalues of the operator*

$$(0.11) \qquad T = \begin{pmatrix} I & K \\ K^* & I \end{pmatrix}$$

on $L_1^{m \times m}(a, \infty) \times L_1^{m \times m}(-\infty, -a)$. *Here*

$$(K\psi)(t) = \int_a^\infty k(t + s - a)\psi(-s)\, ds \qquad (t \geq a)$$

and

$$(K^*\phi)(-t) = \int_a^\infty k(t + s - a)^* \phi(s)\, ds \qquad (t \geq a)$$

for $\phi \in L_1^{m \times m}(a, \infty)$ *and* $\psi \in L_1^{m \times m}(-\infty, -a)$.

In addition to the Introduction, this paper contains two sections. The first contains the necessary auxiliary results, and the second is devoted to proving the main theorem. The proof is based on methods and results from [EGL92], [EGL96] and [GH75]

1 AUXILLIARY THEOREMS

In this section, we prove several auxilliary theorems that will be needed in Section 2 for the proof of the main theorem.

THEOREM 1.1.   *Let $k \in L_1^{m \times m}(a, \infty)$. Assume that $g_a \in L_1^{m \times m}(a, \infty)$ and $h_a \in L_1^{m \times m}(-\infty, -a)$ satisfy the equations*

$$(1.1) \qquad\qquad g_a(t) + \int_a^\infty k(t + s - a) h_a(-s)\, ds = 0 \qquad\qquad (t \geq a)$$

and

$$(1.2) \qquad\qquad \int_a^\infty k(t + s - a)^* g_a(s)\, ds + h_a(-t) = -k(t)^* \qquad\qquad (t \geq a).$$

Define

$$(1.3) \qquad\qquad \Phi_a(\lambda) = e^{i\lambda a} I + \int_a^\infty e^{i\lambda t} g_a(t)\, dt \qquad\qquad (\operatorname{Im} \lambda \geq 0)$$

and

$$(1.4) \qquad\qquad \Psi_a(\lambda) = \int_a^\infty e^{-i\lambda t} h_a(-t)\, dt \qquad\qquad (\operatorname{Im} \lambda \leq 0).$$

Then the following equality holds:

$$(1.5) \qquad\qquad \Phi_a(\lambda)^* \Phi_a(\lambda) - \Psi_a(\lambda)^* \Psi_a(\lambda) = I \qquad\qquad (-\infty < \lambda < \infty).$$

PROOF   Let λ be any real number. From (1.3) and (1.4) it follows that

$$
\begin{aligned}
\Phi_a(\lambda)^* \Phi_a(\lambda) - \Psi_a(\lambda)^* \Psi_a(\lambda) &= \left(e^{-i\lambda a} I + \int_a^\infty e^{-i\lambda t} g_a(t)^*\, dt \right) \left(e^{i\lambda a} I + \int_a^\infty e^{i\lambda t} g_a(t)\, dt \right) \\
&\quad - \int_a^\infty e^{i\lambda t} h_a(-t)^*\, dt \int_a^\infty e^{-i\lambda t} h_a(-t)\, dt \\
&= I + \int_a^\infty e^{i\lambda(a-t)} g_a(t)^*\, dt + \int_a^\infty e^{i\lambda(t-a)} g_a(t)\, dt \\
&\quad + \int_a^\infty \int_a^\infty e^{i\lambda(s-t)} g_a(t)^* g_a(t)\, dt\, ds \\
&\quad - \int_a^\infty \int_a^\infty e^{i\lambda(t-s)} h_a(-t)^* h_a(-s)\, dt\, ds.
\end{aligned}
$$

(1.6)

Substituting $u = t - s$ and then reversing the order of integration, we have

$$
\begin{aligned}
\int_a^\infty \int_a^\infty e^{i\lambda(s-t)} g_a(t)^* g_a(s)\, dt\, ds &= \int_a^\infty \int_{a-s}^\infty e^{-i\lambda u} g_a(u + s)^* g_a(s)\, du\, ds \\
&= \int_{-\infty}^0 \int_{a-u}^\infty e^{-i\lambda u} g_a(u + s)^* g_a(s)\, ds\, du \\
&\quad + \int_0^\infty \int_a^\infty e^{-i\lambda u} g_a(u + s)^* g_a(s)\, ds\, du.
\end{aligned}
$$

In the next to last integral, we replace u by $-u$ and then substitute v for $s - u$ to obtain

$$\int_a^\infty \int_a^\infty e^{i\lambda(s-t)} g_a(t)^* g_a(s)\, dt\, ds = \int_0^\infty \int_{a+u}^\infty e^{i\lambda u} g_a(-u+s)^* g_a(s)\, ds\, du$$

$$+ \int_0^\infty \int_a^\infty e^{-i\lambda u} g_a(u+s)^* g_a(s)\, ds\, du$$

(1.7)

$$= \int_0^\infty \int_a^\infty e^{i\lambda u} g_a(v)^* g_a(u+v)\, dv\, du$$

$$+ \int_0^\infty \int_a^\infty e^{-i\lambda u} g_a(u+s)^* g_a(s)\, ds\, du.$$

Similarly, reversing the order of integration and substituting $u = s - t$, we have

$$\int_a^\infty \int_a^\infty e^{i\lambda(t-s)} h_a(-t)^* h_a(-s)\, dt\, ds = \int_a^\infty \int_a^\infty e^{i\lambda(t-s)} h_a(-t)^* h_a(-s)\, ds\, dt$$

$$= \int_a^\infty \int_{a-t}^\infty e^{-i\lambda u} h_a(-t)^* h_a(-u-t)\, du\, dt.$$

Reversing the order of integration again, we obtain

$$\int_a^\infty \int_a^\infty e^{i\lambda(t-s)} h_a(-t)^* h_a(-s)\, dt\, ds = \int_{-\infty}^0 \int_{a-u}^\infty e^{-i\lambda u} h_a(-t)^* h_a(-u-t)\, dt\, du$$

$$+ \int_0^\infty \int_a^\infty e^{-i\lambda u} h_a(-t)^* h_a(-u-t)\, dt\, du$$

$$= \int_0^\infty \int_{a+u}^\infty e^{i\lambda u} h_a(-t)^* h_a(u-t)\, dt\, du$$

$$+ \int_0^\infty \int_a^\infty e^{-i\lambda u} h_a(-t)^* h_a(-u-t)\, dt\, du.$$

Letting $v - t - u$ in the next to last integral, we find that

(1.8)
$$\int_a^\infty \int_a^\infty e^{i\lambda(t-s)} h_a(-t)^* h_a(-s)\, dt\, ds = \int_0^\infty \int_a^\infty e^{i\lambda u} h_a(-v-u)^* h_a(-v)\, dv\, du$$

$$+ \int_0^\infty \int_a^\infty e^{-i\lambda u} h_a(-t)^* h_a(-t-u)\, dt\, du.$$

From (1.6)–(1.8), we obtain

$$\Phi_a(\lambda)^* \Phi_a(\lambda) - \Psi_a(\lambda)^* \Psi_a(\lambda) - I$$

$$= \int_0^\infty e^{i\lambda t} \left[g_a(a+t) + \int_a^\infty g_a(s)^* g_a(s+t)\, ds - \int_a^\infty h_a(-s-t)^* h_a(-s)\, ds \right] dt$$

$$+ \int_0^\infty e^{-i\lambda t} \left[g_a(a+t)^* + \int_a^\infty g_a(s+t)^* g_a(s)\, ds - \int_a^\infty h_a(-s)^* h_a(-s-t)\, ds \right] dt.$$

Therefore, it suffices to prove that

(1.9)
$$g_a(a+t) + \int_a^\infty g_a(s)^* g_a(s+t)\, ds - \int_a^\infty h_a(-s-t)^* h_a(-s)\, ds = 0.$$

By (1.1) we have

$$g_a(a+t) = - \int_a^\infty k(t+s) h_a(-s)\, ds$$

and

$$\int_a^\infty g_a(s)^* g_a(s+t)\, ds = -\int_a^\infty \int_a^\infty g_a(s)^* k(s+t+u-a) h_a(-u)\, du\, ds$$
$$= -\int_a^\infty \int_a^\infty g_a(u)^* k(s+t+u-a) h_a(-s)\, du\, ds.$$

Therefore,

$$g_a(a+t) + \int_a^\infty g_a(s)^* g_a(s+t)\, ds$$
$$= -\int_a^\infty \left[k(t+s) + \int_a^\infty g_a(u)^* k(s+t+u-a)\, du \right] h_a(-s)\, ds.$$

The equation in (1.2) implies that

$$k(t+s) + \int_a^\infty g_a(u)^* k(s+t+u-a)\, du = -h_a(-t-s)^*.$$

Therefore

$$g_a(a+t) + \int_a^\infty g_a(s)^* g_a(s+t)\, ds = \int_a^\infty h_a(-t-s)^* h_a(-s)\, ds,$$

which coincides with (1.9).

THEOREM 1.2. *Let $k \in L_1^{m \times m}(a, \infty)$. Assume that $\gamma_a \in L_1^{m \times m}(a, \infty)$ and $\chi_a \in L_1^{m \times m}(-\infty, -a)$ satisfy the equations*

$$\gamma_a(t) + \int_a^\infty k(t+s-a) \chi_a(-s)\, ds = -k(t) \qquad (t \geq a) \tag{1.10}$$

and

$$\int_a^\infty k(t+s-a)^* \gamma_a(s)\, ds + \chi_a(-t) = 0 \qquad (t \geq a). \tag{1.11}$$

Define

$$\Omega_a(\lambda) = \int_a^\infty e^{i\lambda t} \gamma_a(t)\, dt \qquad (\mathrm{Im}\,\lambda \geq 0)$$

and

$$\Theta_a(\lambda) = e^{-i\lambda a} I + \int_a^\infty e^{-i\lambda t} \chi_a(-t)\, dt \qquad (\mathrm{Im}\,\lambda \leq 0).$$

Then the following equality holds:

$$\Theta_a(\lambda)^* \Theta_a(\lambda) - \Omega_a(\lambda)^* \Omega_a(\lambda) = I \qquad (-\infty < \lambda < \infty). \tag{1.12}$$

PROOF   This follows from Theorem 1.1. Indeed, let

$$g_a(t) = \chi_a(-t) \qquad (t \geq a)$$

and

$$h_a(-t) = \gamma_a(t) \qquad (t \geq a).$$

Then (1.10) and (1.11) imply that

$$g_a(t) + \int_a^\infty k(t + s - a)^* h_a(-s)\, ds = 0 \qquad\qquad (t \geq a)$$

and

$$\int_a^\infty k(t + s - a) g_a(s)\, ds + h_a(-t) = -k(t) \qquad\qquad (t \geq a).$$

Applying Theorem 1.1 with k replaced by k^*, we obtain

and $\qquad\qquad \Theta_a(-\lambda)^* \Theta_a(-\lambda) - \Omega_a(-\lambda)^* \Omega_a(-\lambda) = I \qquad\qquad (-\infty < \lambda < \infty)$

which implies (1.12).

COROLLARY 1.3.   *Let* Φ_a *and* Θ_a *be as in Theorems 1.1 and 1.2. Then*

$$\det \Phi_a(\lambda) \neq 0 \qquad\qquad (-\infty < \lambda < \infty)$$

and

$$\det \Theta_a(\lambda) \neq 0 \qquad\qquad (-\infty < \lambda < \infty).$$

PROOF   Suppose there exists a real number λ such that $\Phi_a(\lambda)$ is not invertible. Then there exists a nonzero vector x in $\mathbf{C}^m$ such that

$$\Phi_a(\lambda)x = 0.$$

Multiplying (1.5) from the left by x^* and from the right by x, we obtain

$$-x^* \Psi_a(\lambda)^* \Psi_a(\lambda)x = x^* x > 0$$

which is a contradiction. Therefore, $\det \Phi_a(\lambda) \neq 0$ for all real λ. The proof that $\det \Theta_a(\lambda) \neq 0$ for λ real is similar.

In the next theorem, we show that the existence of solutions of (0.5)–(0.8) implies that the operator

$$\begin{pmatrix} I & K \\ K^* & I \end{pmatrix}$$

is invertible on $L_1^{m \times m}(a, \infty) \times L_1^{m \times m}(-\infty, -a)$.

THEOREM 1.4   *Let* $k \in L_1^{m \times m}(a, \infty)$. *Assume that there are solutions* $g_a \in L_1^{m \times m}(a, \infty)$ *and* $h_a \in L_1^{m \times m}(-\infty, -a)$ *of the equations*

(1.13) $\qquad\qquad g_a(t) + \int_a^\infty k(t + s - a) h_a(-s)\, ds = 0 \qquad\qquad (t \geq a)$

and

(1.14) $$\int_a^\infty k(t+s-a)^* g_a(s)\,ds + h_a(-t) = -k(t)^* \qquad (t \geq a).$$

Assume also that there exist solutions $\gamma_a \in L_1^{m \times m}(a,\infty)$ *and* $\chi_a \in L_1^{m \times m}(-\infty,-a)$ *of the equations*

(1.15) $$\gamma_a(t) + \int_a^\infty k(t+s-a)\chi_a(-s)\,ds = -k(t) \qquad (t \geq a)$$

and

(1.16) $$\int_a^\infty k(t+s-a)^* \gamma_a(s)\,ds + \chi_a(-t) = 0 \qquad (t \geq a).$$

Then the operator

$$T = \begin{pmatrix} I & K \\ K^* & I \end{pmatrix}$$

is invertible on $L_1^{m \times m}(a,\infty) \times L_1^{m \times m}(-\infty,-a)$, *where*

$$(K\Psi)(t) = \int_a^\infty k(t+s-a)\Psi(-s)\,ds \qquad (t \geq a)$$

and

$$(K^*\Phi)(-t) = \int_a^\infty k(t+s-a)^*\phi(s)\,ds \qquad (t \geq a).$$

PROOF   Since T is a Fredholm operator with index 0, it suffices to prove that $\operatorname{Ker} T = \{0\}$. As a first step in this direction ,we will show that for every $(\phi,\psi)^T$ in $\operatorname{Ker} T$, ϕ and ψ are absolutely continuous with $\phi' \in L_1^{m \times m}(a,\infty)$ and $\psi' \in L_1^{m \times m}(-\infty,-a)$. Let W_1 be the dense subspace of $L_1^{m \times m}(a,\infty)$ consisting of the absolutely continuous functions in $L_1^{m \times m}(a,\infty)$ with derivative in $L_1^{m \times m}(a,\infty)$. On W_1, we define a second norm by

(1.17) $$\|\phi\|_{W_1} = \|\phi\|_{L_1} + \|\phi'\|_{L_1}.$$

Here and in the remainder of the paper, $\| \cdot \|_{L_1}$ denotes the norm in $L_1^{m \times m}(a,\infty)$ or $L_1^{m \times m}(-\infty,-a)$. Let W_2 and $\| \cdot \|_{W_2}$ be the analogous subspace of $L_1^{m \times m}(-\infty,-a)$ and the corresponding norm. Then W_1 and W_2 are Banach spaces for $\| \cdot \|_{W_1}$ and $\| \cdot \|_{W_2}$.
For any $\psi \in W_2$,

$$(K\psi)(t) = \int_a^\infty k(t+s-a)\psi(-s)\,ds$$
$$= \int_{a+t}^\infty k(s-a)\psi(-s+t)\,ds \qquad (t \geq a).$$

Since ψ is differentiable almost everywhere, $K\psi$ is also differentiable almost everywhere, and

(1.18)
$$\frac{d}{dt}(K\psi)(t) = \int_{a+t}^\infty k(s-a)\psi'(-s+t)\,dt - k(t)\psi(-a)$$
$$= \int_a^\infty k(t+s-a)\psi'(-s)\,ds - k(t)\psi(-a) \qquad (t \geq a).$$
$$= (K\psi')(t) - k(t)\psi(-a)$$

It follows that $(K\psi)'$ is in $L_1^{m\times m}(a,\infty)$. Thus, K maps W_2 into W_1. Also observe that

$$\psi(-a) = \int_a^\infty \psi'(-s)\, ds$$

so that

(1.19) $$\left\|\psi(-a)\right\| \leq \|\psi'\|_{L_1}.$$

Here the norm on the left is the norm of the matrix $\psi(-a)$. From (1.18) and (1.19) we have

$$\left\|(K\psi)'\right\|_{L_1} \leq \|K\psi'\|_{L_1} + \|k\|_{L_1}\|\psi'\|_{L_1}$$
$$\leq (\|K\| + \|k\|_{L_1})\,\|\psi'\|_{L_1}$$

so that

$$\|K\psi\|_{W_1} = \|K\psi\|_{L_1} + \|(K\psi)'\|_{L_1}$$
$$\leq \|K\|\,\|\psi\|_{L_1} + (\|K\| + \|k\|_{L_1})\,\|\psi'\|_{L_1}$$
$$\leq (\|K\| + \|k\|_{L_1})\,\|\psi\|_{W_2}.$$

Therefore, the restriction K_2 of K to W_2 is a bounded linear operator from W_2 into W_1 with the norms $\|\cdot\|_{W_2}$ and $\|\cdot\|_{W_1}$, respectively, and

$$\|K_2\| \leq \|K\| + \|k\|_{L_1}.$$

Next, we prove that K_2 is a compact operator. For this, let $\{\psi_n\}_0^\infty$ be a bounded sequence in W_2. Since

(1.20) $$\|\psi_n\|_{W_2} = \|\psi_n\|_{L_1} + \|\psi_n'\|_{L_1}$$

both $\{\psi_n\}_0^\infty$ and $\{\psi_n'\}_0^\infty$ are bounded in $L_1^{m\times m}(-\infty,-a)$. Therefore, since K is a compact operator from $L_1^{m\times m}(-\infty,-a)$ into $L_1^{m\times m}(a,\infty)$, there is a subsequence $\{\psi_{n_k}\}_0^\infty$ such that $\{K\psi_{n_k}\}_0^\infty$ and $\{K\psi_{n_k}'\}_0^\infty$ converge in $L_1^{m\times m}(a,\infty)$. For convenience, we replace $\{\psi_n\}_0^\infty$ by the subsequence $\{\psi_{n_k}\}_0^\infty$. Thus, we may assume that $\{K\psi_n\}_0^\infty$ and $\{K\psi_n'\}_0^\infty$ converge in $L_1^{m\times m}(a,\infty)$. From (1.19) it follows that $\{\psi_n(-a)\}_0^\infty$ is a bounded sequence of $m\times m$ matrices. Therefore, there exists a subsequence $\{\psi_{n_k}\}_0^\infty$ of $\{\psi_n\}_0^\infty$ for which $\left\{\psi_{n_k}(-a)\right\}_0^\infty$ converges. From (1.18) it now follows that $\{(K\psi_{n_k})'\}_0^\infty$ converges in $L_1^{m\times m}(a,\infty)$. Since both $\{K\psi_{n_k}\}_0^\infty$ and $\{(K\psi_{n_k})'\}_0^\infty$ converge in $L_1^{m\times m}(a,\infty)$, it follows that $\{K\psi_{n_k}\}_0^\infty$ is a Cauchy sequence in W_1 and hence converges in W_1. We conclude that K_2 is a compact operator from W_2 to W_1. Analogous results can be proved for K^*. Namely,

(1.21) $$\frac{d}{dt}(K^*\phi)(-t) = -\int_a^\infty k(t+s-a)^*\phi'(s)\, ds - k(t)^*\phi(a) \qquad (t \geq a)$$

and the restriction K_1^* of K^* to W_1 is a compact operator from W_1 into W_2. Therefore the operator

$$\begin{pmatrix} 0 & K_2 \\ K_1^* & 0 \end{pmatrix}$$

is a compact operator on $W_1 \times W_2$ with the product norm determined by $\| \cdot \|_1$ and $\| \cdot \|_2$. We conclude that the operator

$$\tilde{T} = \begin{pmatrix} I & K_2 \\ K_1^* & I \end{pmatrix}$$

is a Fredholm operator on $W_1 \times W_2$ with index 0. Therefore

(1.22) $\dim \operatorname{Ker} \tilde{T} = \operatorname{codim} \operatorname{Im}(\tilde{T})$

where $\operatorname{Im}(\cdot)$ denotes the range of an operator. Since T is also a Fredholm operator on $L_1^{m \times m}(a, \infty) \times L_1^{m \times m}(-\infty, -a)$, we have

(1.23) $\dim \operatorname{Ker} T = \operatorname{codim} \operatorname{Im}(T)$.

But $\operatorname{Ker} \tilde{T} \subseteq \operatorname{Ker} T$, so that

(1.24) $\dim \operatorname{Ker} \tilde{T} \leq \dim \operatorname{Ker} T$.

Let us show that

(1.25) $\operatorname{codim} \operatorname{Im}(\tilde{T}) \geq \operatorname{codim} \operatorname{Im}(T)$

where the codimension on the left is computed in $W_1 \times W_2$ and the codimension on the right is computed in $L_1^{m \times m}(a, \infty) \times L_1^{m \times m}(-\infty, -a)$. This is immediate if $\operatorname{codim} \operatorname{Im}(T) = 0$. Otherwise, let $y_1, \ldots, y_n$ be a basis for a subspace M such that

$$L_1^{m \times m}(a, \infty) \times L_1^{m \times m}(-\infty, -a) = \operatorname{Im}(T) \oplus M.$$

If $y_1, \ldots, y_n$ happen to belong to $W_1 \times W_2$, then since $\operatorname{Im}(\tilde{T}) \subseteq \operatorname{Im}(T)$, it would follow that

(1.26) $\operatorname{codim} \operatorname{Im}(\tilde{T}) \geq n = \operatorname{codim} \operatorname{Im}(T)$.

If $y_1, \ldots, y_n$ do not all belong to $W_1 \times W_2$, then since $W_1 \times W_2$ is dense in $L_1^{m \times m}(a, \infty) \times L_1^{m \times m}(-\infty, -a)$, we may replace $y_1, \ldots, y_n$ by vectors $x_1, \ldots, x_n$ in $W_1 \times W_2$ for which

$$L_1^{m \times m}(a, \infty) \times L_1^{m \times m}(-\infty, -a) = \operatorname{Im}(T) \oplus M'$$

where

$$M' = \operatorname{span} \{x_1, \ldots, x_n\}.$$

Then (1.26) follows as before. Thus, (1.25) holds. From (1.22)–(1.25), we deduce that

$$\dim \operatorname{Ker} T = \dim \operatorname{Ker} \tilde{T}$$

so that

$$\operatorname{Ker} T = \operatorname{Ker} \tilde{T}.$$

Therefore, for all $(\phi, \psi)^T \in \operatorname{Ker} T$, ϕ and ψ are absolutely continuous with $\phi' \in L_1^{m \times m}(a, \infty)$ and $\psi' \in L_1^{m \times m}(-\infty, -a)$.

Now let us show that $\operatorname{Ker} T = \{0\}$. Let $(\phi, \psi)^T \in \operatorname{Ker} T$. Then

(1.27) $$\phi(t) + \int_a^\infty k(t + s - a)\psi(-s)\,ds = 0 \qquad (t \geq a)$$

and

(1.28) $$\int_a^\infty k(t + s - a)^*\phi(s)\,ds + \psi(-t) = 0 \qquad (t \geq a).$$

Mutiplying (1.27) by $g_a(t)^*$ and integrating, we obtain

$$\int_a^\infty g_a(t)^*\phi(t)\,dt + \int_a^\infty \int_a^\infty g_a(t)^*k(t + s - a)\psi(-s)\,ds\,dt = 0.$$

Reversing the order of integration and using the adjoint of (1.14), we find that

(1.29) $$\int_a^\infty g_a(t)^*\phi(t)\,dt + \int_a^\infty \Big(-h_a(-s)^* - k(s)\Big)\psi(-s)\,ds = 0.$$

Multiplying (1.28) by $h_a(-t)^*$ and integrating, we have

$$\int_a^\infty \int_a^\infty h_a(-t)^*k(t + s - a)^*\phi(s)\,ds\,dt + \int_a^\infty h_a(-t)^*\psi(-t)\,dt = 0.$$

Reversing the order of integration and using the adjoint of (1.13), we obtain

(1.30) $$\int_a^\infty -g_a(s)^*\phi(s)\,ds + \int_a^\infty h_a(-t)^*\psi(-t)\,dt = 0.$$

Adding (1.29) and (1.30) yields

$$-\int_a^\infty k(s)\psi(-s)\,ds = 0.$$

This and (1.27) with $t = a$ imply that

$$\phi(a) = 0.$$

Multiplying (1.28) by $\chi_a(-t)^*$ and integrating, we obtain

$$\int_a^\infty \int_a^\infty \chi_a(-t)^*k(t + s - a)^*\phi(s)\,ds\,dt + \int_a^\infty \chi_a(-t)^*\psi(-t)\,dt = 0.$$

Taking the adjoint in the last equation and reversing the order of integration in the double integral yeilds

$$\int_a^\infty \int_a^\infty \phi(s)^*k(t + s - a)\chi_a(-t)\,dt\,ds + \int_a^\infty \psi(-t)^*\chi_a(-t)\,dt = 0.$$

Using (1.15), we obtain

(1.31) $$\int_a^\infty \phi(s)^*\Big(-\gamma_a(s) - k(s)\Big)\,ds + \int_a^\infty \psi(-t)^*\chi_a(-t)\,dt = 0.$$

Multiplying (1.27) by $\gamma_a(t)^*$ and integrating, we find that

$$\int_a^\infty \gamma_a(t)^*\phi(t)\,dt + \int_a^\infty \int_a^\infty \gamma_a(t)^*k(t + s - a)\psi(-s)\,ds\,dt = 0.$$

Taking the adjoint and reversing the order of integration yields

$$\int_a^\infty \phi(t)^* \gamma_a(t)\, dt + \int_a^\infty \int_a^\infty -\psi(-s)^* k(t+s-a)^* \gamma_a(t)\, dt\, ds = 0.$$

From this and (1.16), we obtain

(1.32) $$\int_a^\infty \phi(t)^* \gamma_a(t)\, dt + \int_a^\infty \psi(-s)^* \big(-\chi_a(-s)\big)\, ds = 0.$$

Adding (1.31) and (1.32), we have

$$-\int_a^\infty \phi(s)^* k(s)\, ds = 0.$$

This and (1.28) with $t = a$ imply that

$$\psi(-a) = 0.$$

Differentiating in (1.27) and (1.28) and using (1.18) and (1.21), we obtain

(1.33) $$\phi'(t) + \int_a^\infty k(t+s-a)\psi'(-s)\, ds - k(t)\psi(-a) = 0 \qquad (t \ge a)$$

and

(1.34) $$-\int_a^\infty k(t+s-a)^* \phi'(s)\, ds - k(t)^* \phi(a) - \psi'(-t) = 0 \qquad (t \ge a).$$

If we multiply (1.34) by -1 and use the fact that $\psi(-a) = \phi(a) = 0$, it follows from (1.33) and (1.34) that

$$\begin{pmatrix} I & K \\ K^* & I \end{pmatrix} \begin{pmatrix} \phi' \\ \psi' \end{pmatrix} = \begin{pmatrix} 0 \\ 0 \end{pmatrix}$$

so that $(\phi', \psi')^T \in \operatorname{Ker} T$. By induction, $(\phi^{(n)}, \psi^{(n)})^T \in \operatorname{Ker} T$, $\phi^{(n)}(a) = 0$ and $\psi^{(n)}(-a) = 0$ for any $n \ge 0$, where the superscript (n) refers to the n-th derivative. Since $\operatorname{Ker} T$ is finite-dimensional, it follows that $(\phi, \psi)^T$ satisfies an equation of the form

and $$\sum_{j=0}^n c_j \begin{pmatrix} \phi^{(j)}(t) \\ \psi^{(j)}(-t) \end{pmatrix} = \begin{pmatrix} 0 \\ 0 \end{pmatrix} \qquad (t \ge a).$$

Since $\phi^{(k)}(a) = \psi^{(k)}(-a) = 0$ for $0 \le k \le n$, we conclude that $\phi = 0$ and $\psi = 0$. This implies that $\operatorname{Ker} T = \{0\}$ and hence that T is invertible.

2 PROOF OF THE MAIN THEOREM

The first step in the proof is to reduce to the case in which k is continuous and has compact support.

Let $\{k_n\}_0^\infty$ be a sequence of continuous functions on (a, ∞) having compact support such that

$$\lim_{n \to \infty} \|k_n - k\|_{L_1} = 0.$$

Define corresponding operators on $L_1^{m \times m}(-\infty, -a)$ and $L_1^{m \times m}(a, \infty)$, respectively, by

$$(K_n \psi)(t) = \int_a^\infty k_n(t + s - a)\psi(-s)\, ds \qquad (t \geq a)$$

and

$$(K_n^* \phi)(-t) = k_n(t + s - a)^* \phi(s)\, ds \qquad (t \geq a).$$

Then for any $\psi \in L_1^{m \times m}(-\infty, -a)$,

$$\left\|(K - K_n)\psi\right\|_{L_1} = \int_a^\infty \left\| \int_a^\infty \left[k(t + s - a) - k_n(t + s - a)\right]\psi(-s)\, ds \right\|\, dt$$

$$\leq \int_a^\infty \int_a^\infty \left\|k(t + s - a) - k_n(t + s - a)\right\|\, dt \left\|\psi(-s)\right\|\, ds$$

$$\leq \|k - k_n\|_{L_1} \|\psi\|_{L_1}.$$

Therefore, K_n converges to K in $\mathcal{L}\left(L_1^{m \times m}(-\infty, -a), L_1^{m \times m}(a, \infty)\right)$. Similarly, K_n^* converges to K^* in $\mathcal{L}\left(L_1^{m \times m}(a, \infty), L_1^{m \times m}(-\infty, -a)\right)$. Let

$$T^{(n)} = \begin{pmatrix} I & K_n \\ K_n^* & I \end{pmatrix}.$$

Then $T^{(n)} \to T$ in $\mathcal{L}\left(L_1^{m \times m}(a, \infty) \times L_1^{m \times m}(-\infty, -a)\right)$. From (0.5)–(0.8) and Theorem 1.4, T is invertible. Therefore, if n is sufficiently large, then $T^{(n)}$ is invertible and has the same number of negative eigenvalues as T. This number is finite since T is the identity plus a compact operator. We assume from now on that n is so large that $T^{(n)}$ is invertible with the same number of negative eigenvalues as T.

In the remainder of the proof, we will, for convenience, let

$$\ell(-t) = -k(t)^* \qquad (t \geq a)$$

and

$$\ell_n(-t) = -k_n(t)^* \qquad (t \geq a).$$

From (0.5) and (0.6) we have

(2.1)
$$\begin{pmatrix} g_a \\ h_a \end{pmatrix} = T^{-1} \begin{pmatrix} 0 \\ \ell \end{pmatrix}$$

and we define

(2.2)
$$\begin{pmatrix} g_n \\ h_n \end{pmatrix} = \left(T^{(n)}\right)^{-1} \begin{pmatrix} 0 \\ \ell_n \end{pmatrix}.$$

Then $\ell_n \to \ell$ in $L_1^{m \times m}(-\infty, -a)$ and

$$\left(T^{(n)}\right)^{-1}\begin{pmatrix} 0 \\ \ell_n \end{pmatrix} - T^{-1}\begin{pmatrix} 0 \\ \ell \end{pmatrix} = \left(\left(T^{(n)}\right)^{-1} - T^{-1}\right)\begin{pmatrix} 0 \\ \ell \end{pmatrix}$$

$$+ \left(\left(T^{(n)}\right)^{-1} - T^{-1}\right)\begin{pmatrix} 0 \\ \ell_n - \ell \end{pmatrix}$$

$$+ T^{-1}\begin{pmatrix} 0 \\ \ell_n - \ell \end{pmatrix}.$$

From this, (2.1) and (2.2), it follows that

$$\lim_{n \to \infty} \begin{pmatrix} g_n \\ h_n \end{pmatrix} = \lim_{n \to \infty} \left(T^{(n)}\right)^{-1}\begin{pmatrix} 0 \\ \ell_n \end{pmatrix}$$

$$= T^{-1}\begin{pmatrix} 0 \\ \ell \end{pmatrix} = \begin{pmatrix} g_a \\ h_a \end{pmatrix}$$

so that

$$\lim_{n \to \infty} g_n = g_a$$

in $L_1^{m \times m}(a, \infty)$. Define

$$\Phi_a^{(n)}(\lambda) = e^{i\lambda a}I + \int_a^\infty e^{i\lambda t}g_n(t)\,dt \qquad (\text{Im}\,(\lambda) \geq 0).$$

Then for λ real,

$$\left\|\Phi_a^{(n)}(\lambda) - \Phi_a(\lambda)\right\| = \left\|\int_a^\infty e^{i\lambda t}\left(g_n(t) - g(t)\right)dt\right\|$$

$$\leq \int_a^\infty \left\|g_n(t) - g(t)\right\|dt$$

$$= \left\|g_n - g\right\|_{L_1}.$$

Therefore

$$\lim_{n \to \infty} \Phi_a^{(n)} = \Phi_a$$

uniformly on $\mathbf{R}$. Thus, for n sufficiently large,

$$\Phi_a^{(n)} = \Phi_a + (\Phi_a^{(n)} - \Phi_a)$$

and

$$\left\|\Phi_a(\lambda)^{-1}(\Phi_a^{(n)} - \Phi_a)(\lambda)\right\| < 1.$$

It now follows from Rouché's Theorem that $\det \Phi_a^{(n)}$ has the same number of zeros in the upper half plane as $\det \Phi_a$ does, provided n is sufficiently large.

From the results we have proved so far, it suffices to prove the theorem when k is replaced by k_n for n sufficiently large. Thus, from now on we will assume that k is continuous on (a, ∞) with compact support $[a, b]$. It follows from (0.5) that g_a also has its support in $[a, b]$, and from (0.6), that h_a has its support in $[-b, -a]$.

The next step in the proof is to convert the operator T to an operator of the form $I - B$, where B is an integral operator on the interval $[0, 2(b - a)]$ with a kernel depending on the difference of arguments.

Let $c = b - a$ and define a mapping $M : L_1^{m \times m}(a, b) \times L_1^{m \times m}(-b, -a) \to L_1^{m \times m}(0, 2)$ by

(2.3)
$$M \left(\begin{array}{c} g_+ \\ g_- \end{array} \right) = g$$

where

(2.4)
$$g(t) = \begin{cases} g_+(t + a), & \text{for } 0 \le t < c; \\ g_-(t - 2b + a), & \text{for } c \le t \le 2c. \end{cases}$$

Then

$$\|g\| = \|g_+\| + \|g_-\|$$

and M^{-1} is given by

$$M^{-1}(g) = \left(\begin{array}{c} g_+ \\ g_- \end{array} \right),$$

where

$$g_+(x) = g(x - a) \quad \text{for } a \le x \le b$$

and

$$g_-(x) = g(x + 2b - a) \quad \text{for } -b \le x \le -a.$$

Thus, M is a Banach space isomorphism.

Define a function $\kappa \in L_1^{m \times m}(-2c, 2c)$ by

(2.5)
$$\kappa(x) = \begin{cases} -k(x + 2b - a), & \text{for } -2c \le x \le -c; \\ 0, & \text{for } -c < x < c; \\ -k(-x + 2b - a)^*, & \text{for } c \le x \le 2c. \end{cases}$$

Let $B : L_1^{m \times m}(0, 2c) \to L_1^{m \times m}(0, 2c)$ be the corresponding integral operator defined by

$$(B\phi)(t) = \int_0^{2c} \kappa(t - s)\phi(s) \, ds.$$

We will next prove that

(2.6)
$$T = M^{-1}(I - B)M.$$

For any $(g_+, g_-) \in L_1^{m \times m}(a, b) \times L_1^{m \times m}(-b, -a)$, let $g = M(g_+, g_-)$. Then

$$(Bg)(t) = \int_0^{2c} \kappa(t - s)g(s) \, ds$$
$$= \int_0^c \kappa(t - s)g_+(s + a) \, ds + \int_c^{2c} \kappa(t - s)g_-(s - 2b + a) \, ds$$

$(0 \le t \le 2c).$

For $0 \le t < c$ and $0 \le s < c$, we have $-c < t - s < c$, so that $\kappa(t - s) = 0$. For $0 \le t \le c$ and $c \le s \le 2c$, we have $-2c \le t - s \le 0$. But for such s and t, $\kappa(t - s) \ne 0$ only for $t - s \le -c$, that is, for $s \ge c + t$. Therefore,

$$(Bg)(t) = \int_{c+t}^{2c} \kappa(t-s)g_-(s-2b+a)\,ds \qquad (0 \le t < c).$$

For $c < t \le 2c$ and $c < s \le 2c$, we have $-c < t - s < c$, so that $\kappa(t - s) = 0$. For $c < t \le 2c$ and $0 \le s \le c$, we have $0 < t - s \le 2c$. For such s and t, $\kappa(t - s) \ne 0$ only for $t - s \ge c$, that is, for $s \le t - c$. Therefore,

$$(Bg)(t) = \int_0^{t-c} \kappa(t-s)g_+(s+a)\,ds \qquad (c < t \le 2c).$$

Now let

$$h = M(h_+, h_-)$$

where

$$\begin{pmatrix} h_+ \\ h_- \end{pmatrix} = \begin{pmatrix} 0 & -K \\ -K^* & 0 \end{pmatrix} \begin{pmatrix} g_+ \\ g_- \end{pmatrix}.$$

Then

$$h_+(t) = -\int_a^b k(t+s-a)g_-(-s)\,ds$$

and

$$h_-(-t) = -\int_a^b k(t+s-a)^* g_+(s)\,ds.$$

For $0 \le t < c$,

$$h(t) = h_+(t+a) = -\int_a^b k(t+s)g_-(-s)\,ds.$$

Since $k(t+s) = 0$ for $b - t < s \le b$, we have

$$\begin{aligned}
h(t) &= -\int_a^{b-t} k(t+s)g_-(-s)\,ds \\
&= -\int_{c+t}^{2c} k(t-s+2b-a)g_-(s-2b+a)\,ds \\
&= \int_{c+t}^{2c} \kappa(t-s)g_-(s-2b+a)\,ds \\
&= (Bg)(t)
\end{aligned} \qquad (0 \le t < c).$$

For $c < t \le 2c$,

$$h(t) = h_-(t-2b+a) = -\int_a^b k(-t+2c+s)^* g_+(s)\,ds.$$

Since $k(-t+2c+s)^* = 0$ for $-t+2c+s > b$, that is, for $s > t - c + a$, we have

$$\begin{aligned}
h(t) &= -\int_a^{t-c+a} k(-t+2c+s)^* g_+(s)\,ds \\
&= -\int_0^{t-c} k(-t+s+2b-a)^* g_+(s+a)\,ds \\
&= \int_0^{t-c} \kappa(t-s)g_+(s+a)\,ds \\
&= (Bg)(t)
\end{aligned} \qquad (c < t \le 2c).$$

Therefore

$$h(t) = (Bg)(t) \qquad\qquad (0 \le t \le 2c).$$

This implies that

$$M \begin{pmatrix} 0 & -K \\ -K^* & 0 \end{pmatrix} = BM$$

which is equivalent to (2.6). In particular, T and $I - B$ have the same number of negative eigenvalues. By (0.5) and (0.6), $(g_a, h_a)^T$ is a solution of

$$(2.7) \qquad\qquad T \begin{pmatrix} g_a \\ h_a \end{pmatrix} = \begin{pmatrix} 0 \\ \ell \end{pmatrix}$$

where

$$\ell(-t) = -k(t)^*.$$

Next we will show that $(g_a, h_a)^T$ is transformed by M into a solution g of

$$(2.8) \qquad\qquad (I - B)g = \kappa.$$

For this, we first apply (2.6) and (2.7) to obtain

$$M^{-1}(I - B)M \begin{pmatrix} g_a \\ h_a \end{pmatrix} = \begin{pmatrix} 0 \\ \ell \end{pmatrix}.$$

Thus

$$(2.9) \qquad\qquad (I - B)g = h,$$

where

$$(2.10) \qquad\qquad g = M \begin{pmatrix} g_a \\ h_a \end{pmatrix} \quad \text{and} \quad h = M \begin{pmatrix} 0 \\ \ell \end{pmatrix}.$$

By (2.4) and (2.5),

$$h(t) = 0 = \kappa(t) \qquad\qquad (0 \le t < c)$$

and

$$h(t) = \ell(t - 2b + a) = -k(-t + 2b - a)^* = \kappa(t) \qquad\qquad (c < t \le 2c).$$

Therefore, $h = \kappa$, so that (2.9) implies (2.8). By (2.4) and (2.10), the solution g of (2.8) is given by

$$(2.11) \qquad\qquad g(t) = \begin{cases} g_a(t + a), & \text{for } 0 \le t < c; \\ h_a(t - 2b + a), & \text{for } c \le t \le 2c. \end{cases}$$

The next step in the proof is to use the results of [EGL92] to deduce the statement in the theorem concerning the number of zeros of $\det \Phi_a$ in the upper half plane. Define $g(t) = 0$ for t outside $[0, 2c]$ and let

$$\hat{g}(\lambda) = \int_0^\infty g(t)e^{i\lambda t}\, dt = \int_0^{2c} g(t)e^{i\lambda t}\, dt \qquad\qquad (\text{Im }\lambda \geq 0).$$

For λ real, it follows from (2.11) that

$$
\begin{aligned}
I + \hat{g}(\lambda) &= I + \int_0^{2c} g(t)e^{i\lambda t}\, dt \\
&= I + \int_0^c g_a(t+a)e^{i\lambda t}\, dt + \int_c^{2c} h_a(t - 2b + a)e^{i\lambda t}\, dt \\
\text{(2.12)}\qquad &= I + \int_a^b g_a(t)e^{i\lambda(t-a)}\, dt + \int_{-b}^{-a} h_a(t)e^{i\lambda(t+2b-a)}\, dt \\
&= e^{-i\lambda a}\left(e^{i\lambda a} I + \int_a^\infty e^{i\lambda t} g_a(t)\, dt \right) + \int_{-\infty}^{-a} h_a(t)e^{i\lambda(t+2b-a)}\, dt \\
&= e^{-i\lambda a}\Phi_a(\lambda) + e^{i\lambda(2b-a)}\Psi_a(\lambda)
\end{aligned}
$$

where Ψ_a is as in (1.4). By Corollary 1.3, $\Phi_a(\lambda)$ is invertible for all real λ. Suppose $I + \hat{g}(\lambda)$ is not invertible for some real λ, and let x be a nonzero vector in $\mathbf{C}^m$ such that $\left[I + \hat{g}(\lambda)\right]x = 0$. Then (2.12) implies that

$$e^{-i\lambda a}\Phi_a(\lambda)x = -e^{i\lambda(2b-a)}\Psi_a(\lambda)x$$

so that

$$\Phi_a(\lambda)x = -e^{2i\lambda b}\Psi_a(\lambda)x.$$

But since λ and b are real, this implies that

$$x^*\Phi_a(\lambda)^*\Phi_a(\lambda)x = x^*\Psi_a(\lambda)^*\Psi_a(\lambda)x.$$

Therefore, it follows from (1.5) that

$$
\begin{aligned}
x^* I x &= x^*\Phi_a(\lambda)^*\Phi_a(\lambda)x - x^*\Psi_a(\lambda)^*\Psi_a(\lambda)x \\
&= x^*\Psi_a(\lambda)^*\Psi_a(\lambda)x - x^*\Psi_a(\lambda)^*\Psi(\lambda)x \\
&= 0.
\end{aligned}
$$

This contradiction to $x \neq 0$ proves that $I + \hat{g}(\lambda)$ is invertible for λ real. Now write

$$\text{(2.13)}\qquad\qquad I + \hat{g}(\lambda) = W(\lambda) + S(\lambda)$$

where

$$\text{(2.14)}\qquad\qquad W(\lambda) = e^{-i\lambda a}\Phi_a(\lambda) \quad\text{and}\quad S(\lambda) = e^{i\lambda(2b-a)}\Psi_a(\lambda).$$

For λ real,

$$W(\lambda)^*W(\lambda) - S(\lambda)^*S(\lambda) = \Phi_a(\lambda)^*\Phi_a(\lambda) - \Psi_a(\lambda)^*\Psi_a(\lambda) = I.$$

Since $\Phi_a(\lambda)$ and hence $W(\lambda)$ are invertible for λ real by Corollary 1.3, we may rewrite the last equality as

$$I - \left[S(\lambda)W(\lambda)^{-1}\right]^*\left[S(\lambda)W(\lambda)^{-1}\right] = \left[W(\lambda)^{-1}\right]^*W(\lambda)^{-1}.$$

Sine the right side is positive definite, it follows that

$$\left\|S(\lambda)W(\lambda)^{-1}\right\| < 1 \qquad\qquad (-\infty < \lambda < \infty).$$

It follows from this, (2.13), and Rouché's Theorem that the functions $\det(I+\hat{g})$ and $\det W$ have the same number of zeros in the upper-half plane. From (2.14) we deduce that $\det(I + \hat{g})$ and $\det \Phi_a$ have the same number of zeros in the upper half plane.

Now we apply Theorem 3.3 in [EGL92], which implies that $\det(I + \hat{g})$ has as many zeros in the upper half plane as $I - B$ has negative eigenvalues. Therefore, $\det \Phi_a$ has as many zeros in the upper half plane as T has negative eigenvalues.

Finally, we prove the statement in the theorem about $\det \Theta_a$. Observe that if we let

(2.15) $\qquad\qquad g_a(t) = \chi_a(-t) \quad\text{and}\quad h_a(-t) = \gamma_a(t) \qquad\qquad (t \geq a),$

then (0.7) and (0.8) become (0.6) and (0.5) with k replaced by k^*. Since

$$\begin{pmatrix} I & K^* \\ K & I \end{pmatrix} = \begin{pmatrix} 0 & I \\ I & 0 \end{pmatrix}\begin{pmatrix} I & K \\ K^* & I \end{pmatrix}\begin{pmatrix} 0 & I \\ I & 0 \end{pmatrix},$$

replacing k by k^* does not change the number of negative eigenvalues of the operator in (0.11). Also (2.15), (0.9) and (0.10) imply that

$$\Theta_a(\lambda) = \Phi_a(-\lambda).$$

Therefore, the number of zeros of $\det \Theta_a$ in the lower half plane equals the number of zeros of $\det \Phi_a$ in the upper half plane. Thus, the statement about $\det \Theta_a$ follows from the statement about $\det \Phi_a$.

References

[AG88]   Alpay, D. and I. Gohberg. On orthogonal matrix polynomials. In *Operator Theory: Advances and Applications*, Volume 34, pages 25–46, Basel, 1988. Birkhäuser Verlag.

[D94]   Dym, H. *On the zeros of some continuous analogues of matrix orthogonal polynomials and a related extension problem with negative squares*. Comm. Pure and Applied Math. **47** (1994), 207–256.

[EG92]   Ellis, R.L. and I. Gohberg. *Orthogonal systems related to infinite Hankel matrices*. J. Functional Analysis **109** (1992), 155–198.

[EGL88] Ellis, R.L., I. Gohberg, and D.C. Lay. *On two theorems of M.G. Krein concerning polynomials orthogonal on the unit circle*. Integral Equations and Operator Theory **11** (1988), 87–104.

[EGL92] Ellis, R.L., I. Gohberg, and D.C. Lay. Distribution of zeros of matrix-valued continuous analogues of orthogonal polynomials. In *Operator Theory: Advances and Applications*, Volume 58, pages 26–70, Basel, 1992. Birkhäuser Verlag.

[EGL95] Ellis, R.L., I. Gohberg, and D.C. Lay. *Infinite analogues of block Toeplitz matrices and related orthogonal functions.* Integral Equations and Operator Theory **22** (1995), 375–419.

[EGL96] Ellis, R.L., I. Gohberg, and D.C. Lay. *On a class of block Toeplitz matrices.* To appear in *Linear Algebra Appl.*

[GH75] Gohberg, I. and G. Heinig. *On matrix-valued integral operators on a finite interval with kernels depending on the difference of the arguments.* Rev. Roumaine Math. Pures et Appl. **20(1)** (1975), 55-73. [Russian].

[GL88] Gohberg, I. and L. Lerer. Matrix generalizations of M.G. Krein theorems on orthogonal polynomials. In *Operator Theory: Advances and Applications*, Volume 34, pages 137–202, Basel, 1988. Birkhäuser Verlag.

[K66] Krein, M.G. *On the distribution of the roots of polynomials which are orthogonal on the unit circle with respect to an alternating weight.* Teor. Funkciĭ Funkcional Anal. i Priloẑen **2** (1966), 131–137. [Russian].

[KL85] Krein, M.G. and H. Langer. *On some continuation problems which are closely related to the theory of operators in spaces Π_x. IV: Continuous analogues of orthogonal functions on the unit circle with respect to an indefinite weight and related continuation problems for some classes of functions.* J. Operator Theory **13** (1985), 299–417.

Robert L. Ellis
Department of Mathematics
University of Maryland
College Park, MD 20942

Israel Gohberg
School of Mathematical Sciences
Raymond and Beverly Sackler
Faculty of Exact Sciences
Tel Aviv University
Tel Aviv, Ramat Aviv 69978   ISRAEL

MSC 1991
Primary: 47N20, 30C15
Secondary: 47G10, 45P05, 47B38

Operator Theory
Advances and Applications, Vol. 90
© 1996 Birkhäuser Verlag Basel/Switzerland

SEMI-FREDHOLM PROPERTIES OF CERTAIN SINGULAR INTEGRAL OPERATORS

Yuri I. Karlovich* and Ilya M. Spitkovsky*†

A necessary and sufficient condition is established for n- and d- normality of singular integral operators $P_+ + GP_-$ with G being an almost periodic matrix function with absolutely convergent Fourier series. This result is used to develop the semi-Fredholm theory in case G is semi almost periodic, but its asymptotics at $\pm\infty$ is described by almost periodic matrix functions of the above mentioned type.

0. INTRODUCTION.

A bounded linear operator A acting on a Banach space X is called *semi-Fredholm* if its range Im A is closed and at least one of the *defect numbers* $\alpha(A) = \dim \operatorname{Ker} A$ and $\beta(A) = \operatorname{codim} \operatorname{Im} A$ is finite. Semi-Fredholm operators are further distinguished according to the value of their *index* ind $A = \alpha(A) - \beta(A)$. Namely, A is (strictly) n-normal if ind $A < +\infty$ $(= -\infty)$, (strictly) d-normal if ind $A > -\infty$ $(= +\infty)$, and *Fredholm* if ind A is finite.

This paper is devoted to singular integral operators of the form

$$(0.1) \qquad\qquad R_G = P_+ + GP_-,$$

where G is an $n \times n$ matrix function in $L^\infty(\mathbb{R}) = L^\infty$ (further restrictions on G will be imposed later), $P_\pm = \frac{1}{2}(I \pm S)$, and S is the Cauchy singular integral operator

$$(0.2) \qquad\qquad (S\phi)(t) = \frac{1}{\pi i} \int_{\mathbb{R}} \phi(\tau) \frac{d\tau}{\tau - t}, \qquad t \in \mathbb{R}.$$

Mainly, we will deal with operators (0.1) acting on $L^2(\mathbb{R})$, but weighted spaces $L^p(\mathbb{R}, \rho)$ with general $p \in (1, \infty)$ will pop up occasionally. Recall therefore that a weight ρ is a measurable

* Research supported by the NATO grant CRG 950332
† Partially supported by the NSF grant DMS 9401848

a.e. positive function on $\mathbb{R}$, $L^p(\mathbb{R}, \rho) = \{f : \rho f \in L^p(\mathbb{R})\}$, and the norm on $L^p(\mathbb{R}, \rho)$ is defined as

$$\|f\|_{L^p(\mathbb{R},\rho)} = \|\rho f\|_{L^p(\mathbb{R})}.$$

According to [7], the operators (0.1) are bounded on $L^p(\mathbb{R}, \rho)$ if and only if $\rho \in W_p$, that is,

$$\sup_{a,b \in \mathbb{R}} \frac{1}{b-a} \left(\int_a^b \rho(t)^p \, dt \right)^{1/p} \left(\int_a^b \rho(t)^{-q} \, dt \right)^{1/q} < \infty,$$

where, as usual, $q = p/(p-1)$. Only W_p-weights are allowed in what follows. The most important of them for our purposes is a *power* weight $\rho(t) = 1 + |t|^\beta$. The latter belongs to W_p if and only if $-1/p < \beta < 1/q$, and the weighted space $L^p(\mathbb{R}, \rho)$ with such a weight ρ will be denoted later by $L^{p,\beta}$.

It is a well known algebraic fact that the semi-Fredholm properties (and even the defect numbers $\alpha(A)$, $\beta(A)$) are the same for R_G and the *Toeplitz operator* $T_G = P_-G|\operatorname{Im} P_-$. There is an extensive literature on the operators R_G, T_G, see [2, 3, 17] and references there. In particular, the following result holds.

THEOREM 0.1. (Hartman-Wintner) *For any* $G \in L^\infty$ *the semi-Fredholmness of* R_G *implies* $G^{-1} \in L^\infty$.

To formulate further results on semi-Fredholmness, we need to introduce the Hardy classes $H_p^\pm$ of functions analytic in the upper/lower half-plane $\Pi^\pm$. In particular, $H_\infty^\pm$ stands for the class of functions analytic and bounded in $\Pi^\pm$. A simple sufficient condition of semi-Fredholmness is given by the following

LEMMA 0.2. *Let*

$$(0.3) \qquad\qquad G = G_+ G_-,$$

where $G_+^{\pm 1}, G_-^{\pm 1} \in L^\infty$. *If, in addition,* $G_+ \in H_\infty^+$, $G_-^{-1} \in H_\infty^-$ $(G_+^{-1} \in H_\infty^+$, $G_- \in H_\infty^-)$, *then the operator* R_G *is right (respectively, left) invertible in all the spaces* $L^p(\mathbb{R}, \rho)$, $p \in (1, \infty)$, $\rho \in W_p$.

Proof. Consider the case of right invertibility. The conditions $G_+ \in H_\infty^+$ and $G_-^{-1} \in H_\infty^-$ imply that $P_-G_+P_- = P_-G_+$ and $P_-G_-^{-1}P_- = G_-^{-1}P_-$ in all the spaces $L^p(\mathbb{R}, \rho)$. Therefore,

$$P_-GP_- \cdot P_-G_-^{-1}P_-G_+^{-1}P_- = P_-G_+G_-(P_-G_-^{-1}P_-)G_+^{-1}P_- =$$

$$P_-G_+G_-G_-^{-1}P_-G_+^{-1}P_- = P_-G_+G_+^{-1}P_- = P_-.$$

In other words, $Y = P_-G_-^{-1}P_-G_+^{-1}P_-$ is a right inverse to the operator T_G. Then, of course, $P_+ + (I - P_+G)Y$ is a right inverse to R_G. ∎

The sufficient condition given by Lemma 0.2 is not necessary, even for scalar symbols G. In the latter case, a necessary and sufficient condition was obtained by A. Devinatz [5] and H. Widom [23], and several variations of it can be found in [2, Sections 2.20–2.23]. Yet another variation reads as follows.

THEOREM 0.3. (Widom-Devinatz) *Let G be a scalar-valued function in L^∞. Then the operator R_G is right (left) invertible in $L^2(\mathbb{R})$ if and only if G admits a factorization*

$$(0.4) \qquad\qquad G(x) = G_+(x)\Lambda(x)G_-(x),$$

in which

$$\frac{1}{z+i}G_+^{\pm 1} \in H_2^+, \quad \frac{1}{z-i}G_-^{\pm 1} \in H_2^-,$$

the operator $G_-^{-1}P_-G_+^{-1}$ is bounded on $L^2(\mathbb{R})$, and Λ is an inner function in Π^+ (Π^-).

Remind that an *inner* function in $\Pi^\pm$ is, by definition, a function in $H_\infty^\pm$ with absolute values on $\mathbb{R}$ a.e. equal 1.

The authors are not aware of analogous criteria for the operators R_G with **matrix** symbols $G \in L^\infty$. However, we were able to establish it for $G \in AP_W$ — almost periodic matrix symbols with absolutely convergent Fourier series. This result is expounded in Section 1. It implies, in particular, that for such matrices the sufficient condition of Lemma 0.2 is also necessary. Additional properties of $G_\pm$ and Λ, established along the way, lead to a necessary and sufficient condition for the operators R_G with semi almost periodic symbols to be semi-Fredholm. The latter criterion, as well as the history of this subject, is contained in Section 3. Section 2 is devoted to the behavior of semi-Fredholm properties of the operators R_G with almost periodic symbols G when this symbol is multiplied by $e^{i\alpha x}$.

1. OPERATORS WITH ALMOST PERIODIC SYMBOLS.

1.1. Known results. Recall that the set AP of Bohr almost periodic functions is defined as the Banach subalgebra of L^∞ generated by the singletons $e_\lambda(x) = e^{i\lambda x}$, $\lambda \in \mathbb{R}$. In its turn, AP_W is the set of absolutely convergent series

$$(1.1) \qquad\qquad f = \sum_j a_j e_{\lambda_j}, \qquad \lambda_j \in \mathbb{R},\ a_j \in \mathbb{C}.$$

Of course, AP_W is a dense (non-closed) subalgebra of AP. Properties of AP functions are discussed, for example, in [16]. For our purposes, the following facts are needed:

For every $f \in AP$, there exists the *Bohr mean value*

$$(1.2) \qquad\qquad \mathbf{M}(f) = \lim_{T\to\infty} \frac{1}{2T}\int_{-T}^{T} f(x)\,dx.$$

In particular, there exist $\hat{f}(\lambda) = \mathbf{M}(fe_{-\lambda})$ — the *Fourier coefficients* of f. For a given $f \in AP$, the set $\Omega(f) = \{\lambda \in \mathbb{R}: \hat{f}(\lambda) \neq 0\}$, called the *Fourier spectrum* of f, is not more than countable. It allows one to associate a formal (not necessarily convergent) *Fourier series*

(1.3)
$$\sum_{\lambda \in \Omega(f)} \hat{f}(\lambda)e^{i\lambda x}$$

with every $f \in AP$. Naturally, (1.1) is a particular case of (1.3).

Later we will need the sets $AP^{\pm} = \{f \in AP: \Omega(f) \subseteq \mathbb{R}_{\pm}\}$ and $AP_W^{\pm} = AP^{\pm} \cap AP_W$. Here, as usual, $\mathbb{R}_{\pm} = \{t \in \mathbb{R}: \pm t \geq 0\}$.

If $f \in AP$ and $f^{-1} \in L^{\infty}$, then automatically $f^{-1} \in AP$. According to Bohr's theorem, for all such functions there exists an (obviously, unique) real number $\kappa(f)$ called the *mean motion* of f such that $\log fe_{-\kappa(f)} \in AP$. In other words, every invertible $f \in AP$ can be represented as

(1.4)
$$f(x) = e^{i\kappa(f)x}e^{b(x)},$$

where $b \in AP$.

The semi-Fredholm properties of the operators R_G with scalar AP symbols G are determined completely by the mean motion of G. Namely, the following theorem holds [4, 6].

THEOREM 1.1. (Coburn-Douglas, Gohberg-Feldman) *Let G be a scalar almost periodic function. Then the operator R_G is semi-Fredholm in $L^p(\mathbb{R}, \rho)$ for $\rho \in W_p$ if and only if $G^{-1} \in L^{\infty}$ (or, equivalently, $G^{-1} \in AP$). If the latter condition is satisfied, then R_G is strictly d-normal (and even right invertible) when $\kappa(G) > 0$, strictly n-normal (and even left invertible) when $\kappa(G) < 0$, and Fredholm (and even invertible) when $\kappa(G) = 0$.*

Note that in the representation (0.4) for $G \in AP$ it is always possible to choose Λ in the form of an AP singleton. Moreover, this singleton has to be equal to $e_{\kappa(G)}$. The factors $G_{\pm}$ are not necessarily almost periodic. If, however, $G \in AP_W$, then, for the above-mentioned choice of Λ, automatically $G_+^{\pm 1} \in AP_W^+$ and $G_-^{\pm 1} \in AP_W^-$. In other words, along with Theorem 1.1, the following (simpler) result holds:

THEOREM 1.2. *Let G be a scalar function in AP_W. Then the operator R_G is, respectively, 1) d-normal (right invertible), 2) n-normal (left invertible), or 3) Fredholm (invertible) in $L^p(\mathbb{R}, \rho)$, $\rho \in W_p$, if and only if G admits a factorization*

(1.5)
$$G(x) = G_+(x)e^{i\alpha x}G_-(x),$$

in which $G_+^{\pm 1} \in AP_W^+$, $G_-^{\pm 1} \in AP_W^-$, and 1) $\alpha \geq 0$, 2) $\alpha \leq 0$, or 3) $\alpha = 0$.

The Fredholm theory of operators R_G with **matrix** AP_W symbols has been also (to some extent) completed. It reads:

THEOREM 1.3. *Let G be an $n \times n$ matrix function in AP_W. Then the following statements are equivalent:*

    1) *The operator R_G is Fredholm in $L^2(\mathbb{R})$.*

    2) *The operator R_G is invertible in all the spaces $L^p(\mathbb{R}, \rho)$, $p \in (1, \infty)$, $\rho \in W_p$.*

    3) *G admits a representation (0.3) with*

(1.6)
$$G_+^{\pm 1} \in AP^+, \quad G_-^{\pm 1} \in AP^-.$$

    4) *G admits a representation (0.3) with*

(1.7)
$$G_+^{\pm 1} \in AP_W^+, \quad G_-^{\pm 1} \in AP_W^-.$$

    Implications 2)→1) and 4)→3) are trivial; 3)→2) follows from Lemma 0.2 since $AP^\pm \subset H_\infty^\pm$, and the implication 1)→4) was established by one of the authors in [8].

    We will call the representation (0.3) a *canonical AP (AP_W) factorization* of G, provided that conditions (1.6) (respectively, (1.7)) are satisfied.

    Finally, recall an auxiliary result from the abstract Fredholm theory.

    LEMMA 1.4. *A bounded linear operator T acting on a Hilbert space $\mathfrak{H}$ is d-normal (right-invertible) if and only if the operator*

$$\hat{T} = \begin{bmatrix} I & T^* \\ T & 0 \end{bmatrix},$$

acting on the direct sum $\mathfrak{H} \oplus \mathfrak{H}$, is Fredholm (respectively, invertible).

    This lemma is formulated, for example, in [21], and probably in many other places. In [21], it is proved by direct computation of $\operatorname{Im} \hat{T}$ and $\operatorname{Ker} \hat{T}$, but it can be also deduced from 1) the simple observation that T is d-normal (right invertible) if and only if TT^* is Fredholm (invertible), and 2) the matrix equality

$$\begin{bmatrix} I & 0 \\ T_1 & I \end{bmatrix} \begin{bmatrix} I & 0 \\ 0 & T_1 T_2 \end{bmatrix} \begin{bmatrix} I & T_2 \\ 0 & -I \end{bmatrix} = \begin{bmatrix} I & T_2 \\ T_1 & 0 \end{bmatrix},$$

showing that $T_1 T_2$ is Fredholm (invertible) simultaneously with the block operator $\begin{bmatrix} I & T_2 \\ T_1 & 0 \end{bmatrix}$.

    When applied to $\mathfrak{H} = H_2^-$ and $T = T_G$, Lemma 1.4 implies that the operator T is d-normal if and only if $\hat{T} = T_{\hat{G}}$ is Fredholm. Here $\hat{G}$ is the double-size matrix function given by the formula

(1.8)
$$\hat{G} = \begin{bmatrix} I & G^* \\ G & 0 \end{bmatrix}.$$

1.2. d-normality. In this subsection, a result analogous to Theorem 1.3 is obtained for d-normal operators.

THEOREM 1.5.   Let $G \in AP_W$ be an $n \times n$ matrix function.   The following statements are equivalent:

1) G admits a representation (0.3), where

(1.9) $$G_+ \in AP^+, \ G_-^{-1} \in AP^-; \ G_+^{-1}, G_- \in AP.$$

2) G admits a representation (0.3) with

(1.10) $$G_+ \in AP_W^+, \ G_-^{-1} \in AP_W^-; \ G_+^{-1}, G_- \in AP_W.$$

3) G admits a representation

(1.11) $$G = \Phi_+ U \Phi_-,$$

where

(1.12) $$\Phi_+^{\pm 1} \in AP_W^+, \quad \Phi_-^{\pm 1} \in AP_W^-, \ \text{and} \ U \in AP_W^+ \ \text{is unitary.}$$

4) G admits a representation (1.11) where, in addition to (1.12), U is of a block diagonal form

(1.13) $$U = \begin{bmatrix} V & 0 \\ 0 & I \end{bmatrix}$$

with $\mathbf{M}(V) = 0$.

5) The operator R_G is d-normal in $L^2(\mathbb{R})$.

6) The operator R_G is right invertible in $L^p(\mathbb{R}, \rho)$ for all $p \in (1, \infty)$ and $\rho \in W_p$.

Proof. Implications 6)→5), 4)→3), and 2)→1) are obvious. Representation (0.3) can be obtained from (1.11) by choosing $G_+ = \Phi_+ U$, $G_- = \Phi_-$. Therefore, 3)→2). Implication 1)→6) follows from Lemma 0.2.

If 5) holds, then the operator T_G is also d-normal. According to Lemma 1.4, the operator $T_{\hat{G}}$ is then Fredholm. Here the matrix $\hat{G}$ is given by (1.8), and belongs to AP_W together with G.

Theorem 1.3 implies then that $\hat{G}$ admits a canonical AP_W-factorization

(1.14) $$\hat{G} = \hat{G}_+ \hat{G}_-, \quad \left(\hat{G}_+\right)^{\pm 1} \in AP_W^+, \left(\hat{G}_-\right)^{\pm 1} \in AP_W^-.$$

Further, $\hat{G}$ is invertible due to Theorem 0.1 and the d-normality of R_G (it is also clear from (1.14)). Because of its algebraic structure (1.8), $\hat{G}(x)$ is Hermitian, with zero signature for all $x \in \mathbb{R}$. According to [22], the factorization (1.14) may be re-written in the more specific form

(1.15) $$\hat{G} = XJX^*,$$

where $J = \begin{bmatrix} I & 0 \\ 0 & -I \end{bmatrix}$ and $X^{\pm 1} \in AP_W^+$. In terms of the partition $X = (X_{ij})_{i,j=1}^2$ of X into $n \times n$ blocks, (1.15) is equivalent to the system

(1.16) $$X_{11}X_{11}^* - X_{12}X_{12}^* = I,$$
(1.17) $$X_{21}X_{11}^* - X_{22}X_{12}^* = G,$$
(1.18) $$X_{21}X_{21}^* - X_{22}X_{22}^* = 0.$$

From (1.18) and the polar representations of X_{21}, X_{22} it follows that

(1.19) $$X_{22} = X_{21}Y, \text{ where } Y \text{ is unitary.}$$

Then

$$X = \begin{bmatrix} I & 0 \\ 0 & X_{21} \end{bmatrix} \begin{bmatrix} X_{11} & X_{12} \\ I & Y \end{bmatrix} = \begin{bmatrix} I & 0 \\ 0 & X_{21} \end{bmatrix} \begin{bmatrix} I & X_{11} \\ 0 & I \end{bmatrix} \begin{bmatrix} 0 & X_{12} - X_{11}Y \\ I & Y \end{bmatrix},$$

and the invertibility of X implies, in particular, that

(1.20) $$X_{21}^{-1} \in AP_W,$$

and $(X_{12} - X_{11}Y)^{-1} \in AP_W$. Moreover,

$$X^{-1} = \begin{bmatrix} -Y(X_{12} - X_{11}Y)^{-1} & X_{21}^{-1} + Y(X_{12} - X_{11}Y)^{-1}X_{11}X_{21}^{-1} \\ (X_{12} - X_{11}Y)^{-1} & -(X_{12} - X_{11}Y)^{-1}X_{11}X_{21}^{-1} \end{bmatrix},$$

and therefore in fact

(1.21) $$(X_{12} - X_{11}Y)^{-1} \in AP_W^+.$$

Note that from (1.20) and (1.19) follows the invertibility of X_{22} in AP_W. Put now $G_+ = X_{22}$, $G_- = (X_{11}Y - X_{12})^*$. Then $G_+ \in AP_W^+$ since $X \in AP_W^+$, $G_+^{-1} \in AP_W$ as mentioned above, $G_- \in AP_W^-$ because $X, Y \in AP_W$ and, finally, $G_-^{-1} \in AP_W^-$ due to (1.21).

On the other hand, (1.17) and (1.19) yield:

$$G = X_{22}Y^*X_{11}^* - X_{22}X_{12}^* = X_{22}(X_{11}Y - X_{12})^* = G_+G_-.$$

Therefore, 5)→2).

Let now 2) hold. The matrix function $G_+ G_+^*$, being positive definite and invertible in APW, admits a factorization

(1.22) $$G_+ G_+^* = \Phi_+ \Phi_+^*,$$

where $\Phi_+^{\pm 1} \in AP_W^+$ [22, Corollary 1]. From (1.22) it follows that $U_1 = \Phi_+^{-1} G_+$ is a unitary matrix function in AP_W^+.

Analogously,

$$G_-^* G_- = \Phi_-^* \Phi_-,$$

where $\Phi_-^{\pm 1} \in AP_W^-$, and $U_2 = \Phi_- G_-^{-1}$ is unitary and belongs to AP_W^-.

From (0.3) and our definition of U_1, U_2 it follows that

$$G = \Phi_+ U_1 U_2^{-1} \Phi_- = \Phi_+ U \Phi_-,$$

where $U = U_1 U_2^{-1}$ is unitary and belongs to AP_W^+ together with U_1 and $U_2^{-1} = U_2^*$. So, 2)$\rightarrow$3).

It is left to show that 3)$\rightarrow$4). To this end, suppose that (1.11), (1.12) hold. From the definition (1.2) of the Bohr mean value it follows that for a unitary U, $\mathbf{M}(U)$ ($\in \mathbb{C}^{n \times n}$) is a contraction. Hence, in its singular value decomposition $\mathbf{M}(U) = W_2 Z W_1$ with W_1, W_2 unitary and $Z = \text{diag}[z_1, \dots, z_n]$, z_j are located in $[0, 1]$. Without loss of generality, $0 \leq z_1 \leq \cdots \leq z_n \leq 1$. By changing Φ_+ to $\Phi_+ W_2$ and Φ_- to $W_1 \Phi_-$, we also may (and will) suppose that $\mathbf{M}(U)$ itself equals Z.

Denote by k the number of z_j which equal 1. Then $Z = \begin{bmatrix} Z_0 & 0 \\ 0 & I_k \end{bmatrix}$ where $0 \leq Z_0 < I_{n-k}$, and in the corresponding partition $U = \begin{bmatrix} U_{00} & U_{01} \\ U_{10} & U_{11} \end{bmatrix}$ of U, $\mathbf{M}(U_{11}) = I$. Hence, its Fourier series (1.3) takes the form

(1.23) $$U_{11}(x) = I + \sum A_k e^{i \lambda_k x}, \qquad \lambda_k > 0,$$

and $\mathbf{M}(U_{11} U_{11}^*) = I + \sum A_k A_k^*$. On the other hand, the unitarity of U implies that $U_{11} U_{11}^* = I - U_{10} U_{10}^*$ is a contraction, so that $\mathbf{M}(U_{11} U_{11}^*) \leq I$. From this it follows that in fact all the coefficients A_k in (1.23) equal zero, that is, $U_{11} = I$ and $U_{10} = 0$. Using the unitarity of U again, we may conclude that, in addition, $U_{01} = 0$ and U_{00} is a unitary matrix (with $\mathbf{M}(U_{00}) = Z_0$).

If $Z_0 = 0$ (which happens if and only if $\mathbf{M}(U)^* \mathbf{M}(U)$ is a projection), we can just set $V = U_{00}$. For the general case of nonzero Z_0, note that $U_{00} - Z_0 = U_{00}(I - U_{00}^* Z_0)$, $U_{00}^* Z_0$ is a strict contraction, and therefore $U_{00} - Z_0$ is invertible. Moreover,

(1.24) $$(U_{00} - Z_0)^{-1} = (I - U_{00}^* Z_0)^{-1} U_{00}^* = \sum_{k=0}^{\infty} (U_{00}^* Z_0)^k U_{00}^* \in AP_W^-.$$

Together with $(U_{00} - Z_0)^{-1}$, the matrix function $X_- = I + (U_{00} - Z_0)^{-1} Z_0$ belongs to AP_W^-. On the other hand, $X_- = (U_{00} - Z_0)^{-1} U_{00}$, so that also

$$X_-^{-1} = U_{00}^*(U_{00} - Z_0) = I - U_{00}^* Z_0 \in AP_W^-.$$

Hence,

$$U_{00} = U_{00} - Z_0 + Z_0 = (U_{00} - Z_0) X_-,$$

where $X_-^{\pm 1} \in AP_W^-$. Consider now the positive Hermitian matrix function $R = (U_{00} - Z_0)(U_{00} - Z_0)^*$. Being invertible in AP_W together with $U_{00} - Z_0$, this matrix admits an AP-factorization of the form $R = X_+ X_+^*$, $X_+^{\pm 1} \in AP_W^+$ [22, Corollary 1]. Put $V = X_+^{-1}(U_{00} - Z_0)$. Then $V \in AP_W^+$, $V^* V = (U_{00} - Z_0)^*(X_+ X_+^*)^{-1}(U_{00} - Z_0) = I$, so that V is unitary, and $M(V) = M(X_+)^{-1} M(U_{00} - Z_0) = 0$.

The factorization (1.11) may be re-written in the form

$$G = \Phi_+ \begin{bmatrix} X_+ & 0 \\ 0 & I \end{bmatrix} \begin{bmatrix} V & 0 \\ 0 & I \end{bmatrix} \begin{bmatrix} X_- & 0 \\ 0 & I \end{bmatrix} \Phi_-,$$

where $\Phi_+ \begin{bmatrix} X_+ & 0 \\ 0 & I \end{bmatrix}$ is invertible in AP_W^+ and $\begin{bmatrix} X_- & 0 \\ 0 & I \end{bmatrix} \Phi_-$ is invertible in AP_W^-. Therefore, 3)→4). ∎

COROLLARY 1.6. *If $G \in AP$ and the operator R_G is d-normal in $L^2(\mathbb{R})$, then $R_{\det G}$ is right invertible in (all the spaces) $L^p(\mathbb{R}, \rho)$. If, in addition, $R_{\det G}$ is invertible, then so is R_G.*

Proof. Consider first the case $G \in AP_W$. According to Theorem 1.5, the d-normality of R_G implies the existence of the factorization (0.3) which, in turn, leads to

(1.25) $\det G = (\det G_+)(\det G_-).$

From (1.9) it follows that

$$\det G_+ \in AP^+, \ (\det G_-)^{-1} \in AP^-; \ (\det G_+)^{-1}, \det G_- \in AP.$$

In other words, condition 1) of Theorem 1.5, if satisfied for G, is also satisfied for $\det G$. But then condition 6) of this theorem is also satisfied for $\det G$, that is, $R_{\det G}$ is right invertible.

If $R_{\det G}$ is invertible, then, since $\det G \in AP_W$ along with G, Theorem 1.2 implies that

$$\det G = \phi_+ \phi_-, \qquad \phi_+^{\pm 1} \in AP^+, \ \phi_-^{\pm 1} \in AP^-.$$

Comparing this with (1.25):

$$(1.26) \qquad \phi_+^{-1} \det G_+ = (\det G_-)^{-1}\phi_-.$$

Since the left hand side of (1.26) lies in AP^+ and the right hand side in AP^-, both of them are in fact constant:

$$\phi_+^{-1} \det G_+ = (\det G_-)^{-1}\phi_- = c.$$

The constant c is non-zero due to the invertibility of ϕ_-, $\det G_-$, so that

$$(\det G_+)^{-1} = 1/c\phi_+ \in AP^+, \quad \det G_- = \phi_-/c \in AP^-.$$

Hence, in (0.3) instead of (1.9) the stronger condition (1.6) holds. According to Theorem 1.3, this implies the invertibility of R_G.

For a general $G \in AP$, consider its approximation $G_1 \in AP_W$ so close to G in L^∞ norm that R_{G_1} is d-normal,

$$(1.27) \qquad \text{ind } R_{G_1} = \text{ind } R_G,$$

and $\kappa(\det G) = \kappa(\det G_1)$. According to the already proved case of Corollary 1.6, the operator $R_{\det G_1}$ is right invertible. Hence (Theorem 1.1), $\kappa(\det G_1) \geq 0$. Our choice of G_1 implies that $\kappa(\det G) \geq 0$, so that (Theorem 1.1 again) $R_{\det G}$ is right invertible.

If $R_{\det G}$ is invertible, then by the same token we may establish the invertibility of R_{G_1}. From here and (1.27) follows the Fredholmness of R_G, which, in turn, is equivalent to its invertibility ([8, Theorem 4]). ∎

1.3. n-normality. By a standard use of duality arguments, the corresponding results for n-normal operators may be established. They read as follows:

THEOREM 1.7. *For any $n \times n$ matrix function $G \in AP_W$, the following statements are equivalent:*

1) *G admits a representation (0.3) with*

$$G_+^{-1} \in AP^+, \; G_- \in AP^-; \; G_+, G_-^{-1} \in AP.$$

2) *G admits a representation (0.3) with*

$$G_+^{-1} \in AP_W^+, \; G_- \in AP_W^-, \; G_+, G_-^{-1} \in AP_W.$$

3) *G admits a representation (1.11) with*

$$(1.28) \qquad \Phi_+^{\pm 1} \in AP_W^+, \quad \Phi_-^{\pm 1} \in AP_W^-, \text{ and } U \in AP_W \text{ unitary.}$$

4) G admits a representation (1.11) where, in addition to (1.28), U is of a block diagonal form (1.13) with $\mathbf{M}(V) = 0$.

5) The operator $R_G = P_+ + GP_-$ is n-normal in $L^2(\mathbb{R})$.

6) The operator R_G is left invertible in $L^p(\mathbb{R}, \rho)$ for all $p \in (1, \infty)$ and $\rho \in W_p$.

COROLLARY 1.8. Let G be an $n \times n$ matrix function in AP such that the operator R_G is n-normal in $L^2(\mathbb{R})$. Then the operator $R_{\det G}$ is left invertible (in all the spaces $L^p(\mathbb{R}, \rho)$). If, in fact, $R_{\det G}$ is invertible, then so is R_G.

Earlier [9, 11], we introduced an AP-factorization of a matrix functions G as its representation in the form

$$(1.29) \qquad\qquad G(x) = G_+(x)\Lambda(x)G_-(x),$$

where $G_+^{\pm 1} \in AP^+$, $G_-^{\pm 1} \in AP^-$, and $\Lambda(x) = \text{diag}[e^{i\lambda_1 x}, \dots, e^{i\lambda_n x}]$, $\lambda_1, \dots, \lambda_n \in \mathbb{R}$.

Among other things, it was established in [11] that the partial AP-indices $\lambda_1, \dots,$ λ_n are defined by G uniquely, and several classes of AP-factorizable matrix functions were described (see [13, 14] for later developments on the subject of AP-factorizability). If G is AP-factorizable, then, clearly, R_G is d- (n-) normal if and only if all its partial AP-indices are non-negative (respectively, non-positive). Therefore, in the case of an AP-factorizable G Theorems 1.5 and 1.7 become trivial.

The importance of Theorems 1.5, 1.7 lies in the fact that they are applicable to matrix functions $G \in AP_W$ which do not admit an AP-factorization (the existence of such matrices, even in the 2×2 case, was observed as early as in [10], see also [15, 11, 14]). The corresponding example will be discussed in Section 2, where it is needed for some other purposes. Note that the unitary matrix function V (and therefore U) in Theorems 1.5, 1.7 can be chosen diagonal if and only if the given matrix G is AP-factorizable.

Because of Theorem 1.1, the semi-Fredholm properties of the operators R_G with scalar $G \in AP$ are the same in all the spaces $L^p(\mathbb{R}, \rho)$ ($p \in (1, \infty)$, $\rho \in W_p$). We do not know if this is the case for matrix symbols G. More precisely, Theorems 1.5, 1.7 guarantee that the semi-Fredholmness of R_G with $G \in AP_W$ in $L^2(\mathbb{R})$ implies its semi-Fredholmness in other spaces $L^p(\mathbb{R}, \rho)$, but it is not clear if the converse implication holds. From Theorem 3.4 below it follows that it is true at least for $p = 2$ and power weights $\rho(t) = 1 + |t|^\beta$, $|\beta| < 1/2$.

2. $\mathcal{N}_G$ AND $\mathcal{D}_G$.

For an arbitrary square matrix function $G \in AP$, denote by $\mathcal{N}_G$ ($\mathcal{D}_G$) the set of all $\alpha \in \mathbb{R}$ such that an operator $R_{\alpha, G} = P_+ + e_\alpha GP_-$ is n- (respectively, d-) normal in $L^2(\mathbb{R})$.

2.1. General properties. Of course, if $G^{-1} \notin L^\infty$, then $(e_\alpha G)^{-1} \notin L^\infty$ as well, and the operator $R_{\alpha, G}$ cannot be semi-Fredholm. In other words, for G with $G^{-1} \notin L^\infty$ the sets $\mathcal{N}_G$, $\mathcal{D}_G$ are void. The next lemma shows that this is the only case when this happens.

LEMMA 2.1. *Let $G \in AP$ be an invertible matrix function. Then $\mathcal{N}(G)$, $\mathcal{D}(G)$ are non-empty.*

Proof. Note first of all that for any G

$$(2.1) \qquad \mathcal{D}_G = -\mathcal{N}_{G^*}.$$

Therefore, it suffices to consider the case of n-normality only.

Introduce an almost periodic polynomial matrix function G_0 such that

$$\|G - G_0\| < \frac{1}{2}\|G^{-1}\|^{-1}.$$

Then $G_0 G^{-1} = I + (G_0 - G)G^{-1}$ is invertible in AP because $\|(G_0 - G)G^{-1}\| < 1$. From this it follows that $G = FG_0$, where $F = (I + (G_0 - G)G^{-1})^{-1}$. Due to the inequality

$$\|F - I\| \leq \frac{\|G_0 - G\| \|G^{-1}\|}{1 - \|G_0 - G\| \|G^{-1}\|} < \frac{1/2}{1 - 1/2} = 1,$$

the operator R_F is invertible in $L^2(\mathbb{R})$. Since G_0 is an almost periodic polynomial, its Fourier spectrum $\Omega(G_0)$ is bounded. Denote (one of) its upper bound(s) by $-\gamma$. Then, of course, $e_\gamma G_0 \in AP^-$, $P_+ e_\gamma G_0 P_- = 0$, and $R_{\gamma,G} = R_F R_{\gamma,G_0}$. The operator $R_{\gamma,G}$ is therefore n-normal simultaneously with R_{γ,G_0}. In other words, $\gamma \in \mathcal{N}(G)$, so that $\mathcal{N}_G \neq \emptyset$. ∎

In the rest of this section, we always suppose that G is invertible (in L^∞, and therefore in AP).

Observe now that for any $r > 0$ $R_{\alpha - r,G} = R_{\alpha,G} R_{e_{-r}}$, and the factor $R_{e_{-r}}$ is strictly n-normal. Therefore, the n-normality of $R_{\alpha,G}$ implies the strict n-normality of all operators $R_{\alpha - r,G}$, $r > 0$. In other words, the set $\mathcal{N}_G$ is either of the form $(-\infty, \alpha_0)$, or $(-\infty, \alpha_0]$, $\alpha_0 > -\infty$, and for all $\alpha < \alpha_0$ operators $R_{\alpha,G}$ are strictly n-normal.

By the same token, $\mathcal{D}_G$ has the form $(\beta_0, +\infty)$ or $[\beta_0, +\infty)$, $\beta_0 < +\infty$, and the operators $R_{\beta,G}$ are strictly d-normal for all $\beta > \beta_0$. Since strict d-normality and strict n-normality are mutually exclusive,

$$(-\infty, \alpha_0) \cap (\beta_0, +\infty) = \emptyset.$$

The latter relation means that $\alpha_0 \leq \beta_0$, in particular, both α_0 and β_0 are finite numbers. Yet another restriction is imposed on the structure of $\{\mathcal{N}_G, \mathcal{D}_G\}$ by the following:

LEMMA 2.2. *Let G be an AP matrix function for which the sets $\mathcal{N}_G$ and $\mathcal{D}_G$ are adjacent (that is, $\alpha_0 = \beta_0$). Then, if one of these sets is closed, the other is closed as well.*

Proof. Considering $e_{-\alpha_0} G$ instead of G, we may restrict ourselves to the case when $\mathcal{N}_G \supset (-\infty, 0)$, $\mathcal{D}_G \supset (0, +\infty)$. Also, due to (2.1), we may without loss of generality suppose that the set given to be closed is $\mathcal{D}_G$, that is, the operator R_G is d-normal.

From Corollaries 1.6, 1.8, it follows that the operator $R_{\det G}$ is right invertible, but the operators $R_{-\gamma,\det G}$ are left invertible for all $\gamma > 0$. Theorem 1.1 implies now that $\kappa(\det G) \geq 0$, but $\kappa(e_{-n\gamma} \det G) = -n\gamma + \kappa(\det G) \leq 0$ for all $\gamma > 0$. The latter is possible only if $\kappa(\det G) = 0$.

Using Theorem 1.1 again, we find that the operator $R_{\det G}$ is invertible. According to Corollary 1.6, so is R_G. This means that $0 \in \mathcal{N}_G$, and the set $\mathcal{N}_G$ is therefore closed. ∎

2.2. Existence results. The properties discovered in the previous subsection leave us with six possible choices for the structure of the pair $\{\mathcal{N}_G, \mathcal{D}_G\}$, namely:

(i) $\{(-\infty, \alpha_0], [\alpha_0, +\infty)\}$,   (ii) $\{(-\infty, \alpha_0], [\beta_0, +\infty)\}$,

(iii) $\{(-\infty, \alpha_0), (\alpha_0, +\infty)\}$,   (iv) $\{(-\infty, \alpha_0], (\beta_0, +\infty)\}$,

(v) $\{(-\infty, \alpha_0), [\beta_0, +\infty)\}$,   (vi) $\{(-\infty, \alpha_0), (\beta_0, +\infty)\}$,

where in (ii) and (iv)–(vi) $\alpha_0 < \beta_0$.

We will show now that all these possibilities actually occur and discuss the circumstances under which they happen.

Observe first of all that for scalar AP functions the complete description of $\{\mathcal{N}_G, \mathcal{D}_G\}$ follows from Theorem 1.1.

THEOREM 2.3. *Let G be an invertible scalar AP function. Then $\{\mathcal{N}_G, \mathcal{D}_G\}$ has the configuration* (i), *with α_0 being equal to $-\kappa(G)$.*

For AP-factorizable matrix functions, the answer is also easy.

THEOREM 2.4. *Let G admit an AP-factorization* (1.29). *Then*

$$\{\mathcal{N}_G, \mathcal{D}_G\} = \{(-\infty, -\lambda_{\max}], [-\lambda_{\min}, +\infty)\},$$

where $\lambda_{\min}$ $(\lambda_{\max})$ is the minimal (respectively, maximal) partial AP index of G.

Hence, for AP-factorizable matrix functions only configurations (i) and (ii) occur, the first of them only when the partial AP-indices of G coincide. In particular, G admits a canonical AP-factorization if and only if $\{\mathcal{N}_G, \mathcal{D}_G\} = \{(-\infty, 0], [0, +\infty)\}$.

The next theorem provides a description of matrix functions G for which a pair $\{\mathcal{N}_G, \mathcal{D}_G\}$ is of the type (iii).

THEOREM 2.5. 1) *Suppose that the operator $R_{\alpha_0,G}$ is not Fredholm in $L^2(\mathbb{R})$, but for every $\epsilon > 0$ there exists a representation*

(2.2) $$G = G_+^{(\epsilon)} U^{(\epsilon)} G_-^{(\epsilon)},$$

in which $\left(U^{(\epsilon)}\right)^{\pm 1} \in AP$, $\Omega\left(U^{(\epsilon)}\right) \subset [-\alpha_0 - \epsilon, -\alpha_0 + \epsilon]$, $\left(G_+^{(\epsilon)}\right)^{\pm 1} \in AP^+$, and $\left(G_-^{(\epsilon)}\right)^{\pm 1} \in AP^-$. Then (iii) *holds.*

2) *For $G \in AP_W$ (iii) implies that G is not AP-factorizable though representation (2.2) exists for all $\epsilon > 0$. Moreover, $G_\pm^{(\epsilon)}$ can be chosen in $AP_W^\pm$.*

Proof. As in Lemma 2.2, it suffices to consider the case $\alpha_0 = 0$.

1) If G admits the representation (2.2) with $\Omega\left(U^{(\epsilon)}\right) \subset (-\epsilon, \epsilon)$, then $e_\epsilon G$ is a product of two multiples: $e_\epsilon G_+^{(\epsilon)} U^{(\epsilon)} \in AP^+$, invertible in AP but not necessarily in AP^+, and $G_-^{(\epsilon)}$, invertible in AP^-. According to Theorem 1.5, the operator $R_{\epsilon,G}$ is d-normal, in other words, $\epsilon \in \mathcal{D}_G$. This reasoning holds for all $\epsilon > 0$, so that in fact $(0, +\infty) \in \mathcal{D}_G$.

Analogously, $(-\infty, 0) \in \mathcal{N}_G$, and it is left only to show that $\mathcal{N}_G$, $\mathcal{D}_G$ do not contain 0.

Since the operator R_G is not Fredholm, 0 cannot lie in $\mathcal{N}_G \cap \mathcal{D}_G$. On the other hand, 0 cannot lie in exactly one of them either, due to Lemma 2.2. This completes the proof of 1).

2) Let $G \in AP_W$ and (iii) hold with $\alpha_0 = 0$. Then, in particular, the operator R_G is not Fredholm, so that (Theorem 1.3) G is not AP-factorizable.

On the other hand, for every $\epsilon > 0$, the operator $R_{\epsilon,G}$ is d-normal, and Theorem 1.5 implies the existence of the factorization

$$(2.3) \qquad e_\epsilon G = G_+^{(1)} U_+ G_-^{(1)},$$

where $\left(G_+^{(1)}\right)^{\pm 1} \in AP_W^+$, $\left(G_-^{(1)}\right)^{\pm 1} \in AP_W^-$, and $U_+ \in AP_W^+$.

Analogously, by the n-normality of $R_{-\epsilon,G}$ and Theorem 1.7,

$$(2.4) \qquad e_{-\epsilon} G = G_+^{(2)} U_- G_-^{(2)},$$

where $\left(G_+^{(2)}\right)^{\pm 1} \in AP_W^+$, $\left(G_-^{(2)}\right)^{\pm 1} \in AP_W^-$, and $U_- \in AP_W^-$. From (2.3) and (2.4) follows:

$$e_{-\epsilon} \left(G_+^{(2)}\right)^{-1} G_+^{(1)} U_+ = e_\epsilon U_- G_-^{(2)} \left(G_-^{(1)}\right)^{-1}.$$

The left side of the latter equation is an AP_W matrix function having its Fourier spectrum in $[-\epsilon, +\infty)$, the right side has its Fourier spectrum in $(-\infty, \epsilon]$, and so in fact they equal a matrix X with $\Omega(X) \subset [-\epsilon, \epsilon]$.

To obtain a desired representation (2.2) of G we may put now $G_+^{(\epsilon)} = G_+^{(2)}$, $U^{(\epsilon)} = X$, and $G_-^{(\epsilon)} = G_-^{(1)}$. ∎

Of·course, Theorem 2.5 by itself does not prove the existence of matrix functions G with $\{\mathcal{N}_G, \mathcal{D}_G\}$ of the type (iii). However, in [14, Theorem 3.1] we established that matrix functions

$$(2.5) \qquad G(x) = \begin{bmatrix} e^{i\lambda x} & 0 \\ c_{-1} e^{-i\nu x} - c_0 + c_1 e^{i\alpha x} & e^{-i\lambda x} \end{bmatrix}$$

with $\alpha, \nu > 0$, $\lambda = \alpha + \nu$, $\beta = \nu/\alpha$ being irrational, and $|c_1^\beta c_{-1}| = |c_0^{\beta+1}| \neq 0$, are not AP-factorizable. In the course of the proof, factorizations (2.2) of matrices (2.5) were constructed for all $\epsilon > 0$. Hence, matrix functions of this type exist, even for $n = 2$.

Note now that if a matrix G is block diagonal:

$$G = \begin{bmatrix} G_1 & 0 \\ 0 & G_2 \end{bmatrix},$$

then $\mathcal{N}_G = \mathcal{N}_{G_1} \cap \mathcal{N}_{G_2}$, $\mathcal{D}_G = \mathcal{D}_{G_1} \cap \mathcal{D}_{G_2}$. This simple observation allows one to construct matrix functions G of the remaining types (iv)–(vi) as direct sums of matrices of the types (i) and (iii). By doing that, it is possible to produce 3×3 matrices of the types (iv), (v), and 4×4 matrices of the type (vi). We do not know whether or not there exist 2×2 or 3×3 matrices of the type (vi). However, as we will show in the next subsection, there are no 2×2 matrices $G \in AP_W$ of the types (iv), (v).

2.3. Reducibility of AP-matrices. Let us say that an $n \times n$ matrix function $G \in AP$ is *reducible* if there exist matrix functions F_+ and F_-, invertible in AP_W^+ and AP_W^- respectively, such that $F_+ G F_-$ is a block diagonal matrix $\mathrm{diag}[F_1, F_2]$, with the sizes of both diagonal blocks strictly less than n.

THEOREM 2.6. *Let G be an $n \times n$ matrix function in AP_W for which at least one of the sets $\mathcal{N}_G$, $\mathcal{D}_G$ is closed. Then G is reducible.*

Proof. The reducibility property is not changed if G is multiplied by e_α, or when G^* is considered instead of G. Therefore, without loss of generality we may suppose that $\mathcal{N}_G = (-\infty, 0]$. Since $G \in AP_W$ and the operator R_G is n-normal, Theorem 1.7 guarantees that by splitting off the invertible factor in AP_W^+ from the left, and the invertible factor in AP_W^- from the right, G can be reduced to the form $\mathrm{diag}[U_-, I]$, where $U_- \in AP_W^-$ and $\mathbf{M}(U_-) = 0$. The only case left is therefore when the diagonal block I is trivial, in other words, when G itself is a matrix in AP_W^- with $\mathbf{M}(G) = 0$.

For such a matrix G, let us consider its approximation by an almost periodic polynomial G_0, as we did in the proof of Lemma 2.1. It is well known (see, for example, [16]) that this can always be done in such a way that $\Omega(G_0) \subset \Omega(G)$. Since $\Omega(G_0)$ is a finite set, the latter condition implies that an upper bound $-\gamma$ for it can be chosen to be negative. But then (see the proof of Lemma 2.1), $\mathcal{N}_G \ni \gamma > 0$. The contradiction obtained shows that the case of a trivial block I does not occur, that is, G is always reducible. ∎

Note that for $G \in AP_W$ the "reducing multiples" F_+, F_- can always be chosen in AP_W. With such a choice, a block diagonalized matrix $F_+ G F_-$ also lies in AP_W.

COROLLARY 2.7. *Let G be a 2×2 matrix in AP_W. Then $\{\mathcal{N}_G, \mathcal{D}_G\}$ cannot be of the form (iv) or (v).*

Proof. According to Theorem 2.6, if one of the sets $\mathcal{N}_G$, $\mathcal{D}_G$ is closed, then G is reducible. Being a 2×2 matrix, it is then of the form $F_+ \operatorname{diag}[g_1, g_2] F_-$, where g_1, g_2 are *scalar* AP_W-functions, and is therefore AP-factorizable. Hence, it is of the type (i) or (ii), but not (iv) or (v). $\blacksquare$

3. OPERATORS WITH SEMI ALMOST PERIODIC SYMBOLS.

3.1. Known results. The set SAP of semi almost periodic functions was introduced in [20] as the Banach subalgebra of L^∞ generated by AP and $C(\dot{\mathbb{R}})$ — the set of functions continuous on the two-point compactification of $\mathbb{R}$. Clearly, all functions of the form

(3.1) $$f = (1 - u)f^{(-)} + uf^{(+)} + f^{(0)},$$

where $f^{(\pm)} \in AP$, $u, f^{(0)} \in C(\dot{\mathbb{R}})$, $f^{(0)}(\pm\infty) = u(-\infty) = 0$, $u(+\infty) = 1$, belong to SAP. As was shown in [20], the converse is also true, that is, functions of the form (3.1) exhaust the algebra SAP. Moreover, the functions $f^{(\pm)}$ are defined by f uniquely, and the mappings $f \mapsto f^{(\pm)}$ are homomorphisms of SAP onto AP. We will call $f^{(\pm)}$ the *almost periodic representatives* of f at $\pm\infty$.

To formulate a semi-Fredholmness criterion for the operators R_G with **scalar** symbols $G \in SAP$, note first of all that, due to Theorem 0.1, G has to be invertible in L^∞. For $G \in SAP$ this implies automatically that $G^{-1} \in SAP$ and its almost periodic representatives $G^{(\pm)}$ are invertible in AP. Denote the mean motions $\kappa(G^{(\pm)})$ by $\kappa_\pm(G)$. Also, put $\mathbf{d}_\pm(G) = \mathbf{d}(G^{(\pm)})$, where for an invertible scalar function $f \in AP$ the number $\mathbf{d}(f)$ is defined with reference to the representation (1.4) as

(3.2) $$\mathbf{d}(f) = e^{\mathbf{M}(b)}.$$

THEOREM 3.1. (Sarason-Saginashvili) *Let G be an invertible scalar function in SAP. Then the operator R_G is* 1) *strictly d-normal,* 2) *strictly n-normal, or* 3) *Fredholm in $L^{p,\beta}$ if and only if, respectively,*

1) $\kappa_\pm(G) \geq 0$, $\kappa_+(G) + \kappa_-(G) > 0$,
2) $\kappa_\pm(G) \leq 0$, $\kappa_+(G) + \kappa_-(G) < 0$,
3) $\kappa_+(G) = \kappa_-(G) = 0$, $\dfrac{1}{2\pi} \arg \dfrac{\mathbf{d}_+(G)}{\mathbf{d}_-(G)} \neq \dfrac{1}{p} + \beta$.

This theorem was proved in [20] for L^2, and in [19] for general $L^{p,\beta}$ (see also [1] for Fredholm criteria in case of general $\rho \in W_p$).

Our goal (partially achieved in this section) is to establish a corresponding result in the case of **matrix** functions $G \in SAP$. As a first step towards this goal, we need the following relation between R_G and the operators $R_{G^{(\pm)}}$ generated by the almost periodic representatives $G^{(\pm)}$ of G (the latter, of course, are defined entry-wise).

THEOREM 3.2. *If G is a square matrix function in SAP, and the operator R_G is d- (n-) normal in the space $L^{2,\beta}$, then the operators $R_{G^{(\pm)}}$ are right- (respectively, left-) invertible.*

For scalar $G \in SAP$, Theorem 3.2 is a simple corollary of Theorems 1.1 and 3.1 and, moreover, is valid in all the spaces $L^{p,\beta}$. The essence of Theorem 3.2 is, however, that it can be proved without apriori knowledge of the semi-Fredholm and Fredholm criteria for the operators R_G with G in SAP (and even AP). Theorem 3.2 for $\beta = 0$ is formulated in [8], where also an outline of its proof (based on the theory of limit operators) is given. A detailed proof of the analogous fact for Wiener-Hopf operators, unitarily equivalent to T_G, can be found in [12]. Transition to the case of general β is based on the following trick, going back to Paatashvili [18].

LEMMA 3.3. *For any $G \in L^\infty$, the operator R_G is d-normal (n-normal, Fredholm) on the weighted space $L^{p,\beta}$ if and only if the operator R_{G_β}, where*

$$G_\beta(x) = \left(\frac{x+i}{x-i} \right)^\beta G(x),$$

is d-normal (respectively, n-normal, Fredholm) on the unweighted space $L^p(\mathbb{R})$.

In the particular case $G \in SAP$, $G_\beta \in SAP$ as well, and, moreover, $G_\beta^{(-)} = G^{(-)}$, $G_\beta^{(+)} = e^{-2\pi i \beta} G^{(+)}$. Therefore, the operators $R_{G_\beta^{(\pm)}}$ and $R_{G^{(\pm)}}$ are right-/left- invertible simultaneously, and Theorem 3.2 in its general form follows from the particular case $\beta = 0$.

3.2. Criteria for d- and n-normality. Since semi-Fredholm criteria in Section 1 were formulated for matrices in AP_W only, we will have to impose the corresponding condition on the almost periodic representatives $G^{(\pm)}$ of the SAP matrix functions G considered here. The set of such semi almost periodic matrices is usually denoted by SAP_W.

THEOREM 3.4. *Let $G \in SAP_W$ be an $n \times n$ matrix function. Then the operator R_G is d-normal in the weighted space $L^{2,\beta}$ if and only if*
  i) $\det G(x) \neq 0$ *for all $x \in \mathbb{R}$,*
  ii) *the almost periodic representatives $G^{(\pm)}$ of G admit factorizations*

(3.3) $$G^{(+)} = G_+^{(+)} U^{(+)} G_-^{(+)}, \quad G^{(-)} = G_+^{(-)} U^{(-)} G_-^{(-)}$$

with

(3.4) $$G_+^{(\pm)}, \left(G_+^{(\pm)} \right)^{-1} \in AP_W^+, \quad G_-^{(\pm)}, \left(G_-^{(\pm)} \right)^{-1} \in AP_W^-,$$

(3.5) $$U^{(\pm)} = \begin{bmatrix} V^{(\pm)} & 0 \\ 0 & I_{n_\pm} \end{bmatrix},$$

$V^{(\pm)} \in AP_W^+$ *being unitary and* $\mathbf{M}(V^{(\pm)}) = 0$,

      iii) *if* $\min\{n_+, n_-\} > 0$, *the matrix* $[\Phi_{00}^+\Phi_{00}^- - \chi I, \Phi_{01}^+, \Phi_{00}^+\Phi_{01}^-]$ *has full row rank for all numbers* χ *on the ray* $\Sigma_\beta = \{\sigma \exp \pi i(1 + 2\beta): \sigma \geq 0\}$.

      Here $\Phi_{ij}^{\pm}$ denote blocks of the partitions $\Phi^{\pm} = \begin{bmatrix} \Phi_{11}^{\pm} & \Phi_{10}^{\pm} \\ \Phi_{01}^{\pm} & \Phi_{00}^{\pm} \end{bmatrix}$ of the matrices $\Phi^+ =$

$\mathbf{M}(G_+^{(-)})^{-1}\mathbf{M}(G_+^{(+)})$, $\Phi^- = \mathbf{M}(G_-^{(+)})\mathbf{M}(G_-^{(-)})^{-1}$ with $\Phi_{00}^{\pm}$ being of the size $n_{\mp} \times n_{\pm}$.

      Proof. If R_G is d-normal, then Condition i) holds due to Theorem 0.1 and the continuity of G on $\mathbb{R}$. Further, Theorem 3.2 implies the right invertibility of the operators $R_{G(\pm)}$. From here and Theorem 1.5 follows ii).

      It is left to show that, when i) and ii) are satisfied, the condition iii) is necessary and sufficient for R_G to be d-normal. For $V^{(\pm)}$ of the form $\mathrm{diag}[e^{i\lambda_1^{\pm}x}, \ldots, e^{i\lambda_{n-\pm}^{\pm}x}]$, $\lambda_j^{\pm} > 0$, this fact was established in [11, pp. 303–305], where d-normality of the operators R_G with $G \in SAP$ was considered in the case of AP-factorizable $G^{(\pm)}$. The reasoning there was based on the properties $\mathbf{M}(V^{(\pm)}) = \mathbf{M}((V^{(\pm)})^{-1}) = 0$ (which hold in our situation as well), and not on the specific algebraic structure of $V^{(\pm)}$. Therefore, it remains valid in our setting. ∎

      The analogous result for n-normality reads:

      THEOREM 3.5. *Let* $G \in SAP_W$ *be an* $n \times n$ *matrix function. Then the operator* R_G *is n-normal in the weighted space* $L^{2,\beta}$ *if and only if*

      i) $\det G(x) \neq 0$ *for all* $x \in \mathbb{R}$,

      ii) *the almost periodic representatives* $G^{(\pm)}$ *of* G *admit factorizations* (3.3) *satisfying conditions* (3.4), (3.5), *with* $V^{(\pm)} \in AP_W^-$ *unitary and* $\mathbf{M}(V^{(\pm)}) = 0$,

      iii) *if* $\min\{n_+, n_-\} > 0$, *the matrix*

$$\begin{bmatrix} \Phi_{00}^+\Phi_{00}^- - \chi I \\ \Phi_{10}^- \\ \Phi_{10}^+\Phi_{00}^- \end{bmatrix}$$

has full column rank for all $\chi \in \Sigma_\beta$.

      Note that the rank condition iii) of Theorems 3.4, 3.5 should be checked only for (finitely many) χ lying in the intersection of Σ_β with the spectrum of $\Phi_{00}^+\Phi_{00}^-$; for all other values of $\chi \in \Sigma_\beta$ it holds automatically.

      Observe also that the mean motion of $\det G^{(\pm)}$ is strictly positive (negative) if $n_{\pm} < n$ and in (3.3) the multiples $U^{(\pm)}$ are of the form (3.5) with $V^{(\pm)} \in AP^+$ (respectively, AP^-). From here and Theorem 3.1 it follows that in the case where $\min\{n_+, n_-\} < n$, the operator $R_{\det G}$ is either strictly d-normal (if $V^{(\pm)} \in AP^+$) or strictly n-normal (if $V^{(\pm)} \in AP^-$). Therefore, the simultaneous existence of the factorizations (3.3) with $V^{(\pm)} \in AP_W^+$ and $V^{(\pm)} \in AP_W^-$ is possible only when $n_+ = n_- = n$. In this case all the matrices Φ_{01}^+, $\Phi_{00}^+\Phi_{01}^-$, Φ_{10}^-, $\Phi_{10}^+\Phi_{00}^-$ vanish, $\Phi_{00}^{\pm} = \Phi^{\pm}$, and condition iii) of both Theorems 3.4,

3.5 takes the form: $\sigma\left(\mathbf{M}(G_+^{(-)})^{-1}\mathbf{M}(G_+^{(+)})\mathbf{M}(G_-^{(+)})\mathbf{M}(G_-^{(-)})^{-1}\right) \cap \Sigma_\beta = \emptyset$, or, equivalently,

$\sigma\left(\mathbf{M}(G_-^{(-)})^{-1}\mathbf{M}(G_+^{(-)})^{-1}\mathbf{M}(G_+^{(+)})\mathbf{M}(G_-^{(+)})\right) \cap \Sigma_\beta = \emptyset$. The latter condition can be rewritten

as $\sigma\left(\mathbf{d}(G^{(-)})^{-1}\mathbf{d}(G^{(+)})\right) \cap \Sigma_\beta = \emptyset$, if we use the notation $\mathbf{d}(F) = \mathbf{M}(F_+)\mathbf{M}(F_-)$ for a ma-

trix function F having a canonical AP-factorization $F = F_+ F_-$. Note that for scalar AP_W

functions $F = f$ this notation agrees with the definition (3.2) of $\mathbf{d}(f)$.

From here follows:

THEOREM 3.6. *Let $G \in SAP_W$ be an $n \times n$ matrix function. Then the operator*

R_G is Fredholm in the weighted space $L^{2,\beta}$ if and only if

i) $\det G(x) \neq 0$ *for all $x \in \mathbb{R}$,*

ii) *the almost periodic representatives $G^{(\pm)}$ of G admit canonical AP-factorizations*

$G^{(+)} = G_+^{(+)}G_-^{(+)}$, $G^{(-)} = G_+^{(-)}G_-^{(-)}$,

iii) *for all the eigenvalues ξ_j of the matrix $\mathbf{d}(G^{(-)})^{-1}\mathbf{d}(G^{(+)})$,*

$$
(3.6) \qquad\qquad \frac{1}{2\pi}\arg\xi_j \neq \frac{1}{2} + \beta.
$$

For matrix functions G with apriori AP-factorizable $G^{(\pm)}$, the corresponding result

in all the spaces $L^{p,\beta}$, $p \in (1,\infty)$, was established in [11]; the right hand side of (3.6) in this

setting should be changed to $\frac{1}{p} + \beta$. A generalization of Theorem 3.6 to the case of arbitrary

$G \in SAP$ is given in [8, Theorem 16]; the statement there remains exactly the same, but, of

course, it requires an appropriate extension of the definition of the canonical AP-factorization

and the mapping $\mathbf{d}\colon F \mapsto \mathbf{d}(F)$.

Comparing now the statements of Theorems 3.4, 3.5, on one hand, and Theorem

3.6, on the other, we arrive at the criterion of strict semi-Fredholmness.

COROLLARY 3.7. *The operator R_G with $G \in SAP_W$ is strictly d- (n-) normal in*

$L^{2,\beta}$ if and only if conditions i)–iii) of Theorem 3.4 (respectively, 3.5) are satisfied, and in

(3.5) $\min\{n_+, n_-\} < n$.

3.3. Sufficient conditions for semi-Fredholmness. Consider now a matrix

function $G \in SAP_W$ for which factorizations (3.3) with $G_\pm^{(\pm)}$ satisfying (3.4) are known,

$U^{(\pm)} \in AP_W^\pm$ are unitary, but not necessarily of the block diagonal form (3.5). As was shown

in the proof of Theorem 1.5, the additional condition (3.5) always can be arranged for, but

the corresponding changes of the multiples $G_\pm^{(\pm)}$ are not easy to trace. Therefore, it would

be interesting to have a necessary and sufficient condition for d-normality of R_G formulated

in terms of the above mentioned factorizations. The next theorem provides such a result

under some additional conditions on $\mathbf{M}(G_\pm^{(\pm)})$. On the other hand, the absolute convergence

of the Fourier series of the almost periodic matrix functions involved $(G_\pm^{(\pm)}, U^{(\pm)})$ is not used

in the proof of this theorem, and is therefore dropped from its conditions.

THEOREM 3.8. *Let the almost periodic representatives $G^{(\pm)}$ of the matrix function $G \in SAP$ admit factorizations*

$$(3.7) \qquad G^{(+)} = G_+^{(+)} U^{(+)} G_-^{(+)}, \quad G^{(-)} = G_+^{(-)} U^{(-)} G_-^{(-)}$$

in which $U^{(\pm)} \in AP^+$ are unitary,

$$G_+^{(\pm)}, \left(G_+^{(\pm)}\right)^{-1} \in AP^+, \quad G_-^{(\pm)}, \left(G_-^{(\pm)}\right)^{-1} \in AP^-,$$

and, in addition,

$$(3.8) \qquad \mathbf{M}(G_-^{(-)})^* \mathbf{M}(G_-^{(-)}) \text{ is a scalar multiple of } \mathbf{M}(G_-^{(+)})^* \mathbf{M}(G_-^{(+)}),$$
$$\mathbf{M}(G_+^{(-)}) \mathbf{M}(G_+^{(-)})^* \text{ is a scalar multiple of } \mathbf{M}(G_+^{(+)}) \mathbf{M}(G_+^{(+)})^*.$$

Then the operator R_G is d-normal in $L^{2,\beta}$ if and only if
  i) $\det G(x) \neq 0$ for all $x \in \mathbb{R}$ and
  ii) $\|X_+ - e^{2\pi i \beta} X_-\| < 2$,

where

$$X_+ = \left(\mathbf{M}(G_+^{(+)})\mathbf{M}(G_+^{(+)})^*\right)^{-1/2} \mathbf{M}(G_+^{(+)})\mathbf{M}(U^{(+)})\mathbf{M}(G_-^{(+)}) \left(\mathbf{M}(G_-^{(+)})^*\mathbf{M}(G_-^{(+)})\right)^{-1/2},$$

$$X_- = \left(\mathbf{M}(G_+^{(-)})\mathbf{M}(G_+^{(-)})^*\right)^{-1/2} \mathbf{M}(G_+^{(-)})\mathbf{M}(U^{(-)})\mathbf{M}(G_-^{(-)}) \left(\mathbf{M}(G_-^{(-)})^*\mathbf{M}(G_-^{(-)})\right)^{-1/2}.$$

Note that a *scalar multiple* of a given matrix A is, by definition, its product αA with any $\alpha \in \mathbb{C}$. By $\|.\|$ in ii) is denoted the *operator norm* of a matrix, that is, $\|A\|$ is the maximum eigenvalue of $(A^*A)^{1/2}$.

Proof. Condition i) is, of course, necessary, and is presumed satisfied in what follows. According to [11, Corollary 3.1], the d-normality of R_G is not affected by the substitutions $G^{(+)} \mapsto N_+^{(+)} G^{(+)} N_-^{(+)}$, $G^{(-)} \mapsto N_+^{(-)} G^{(-)} N_-^{(-)}$, as long as $N_+^{(\pm)}$ ($N_-^{(\pm)}$) are invertible elements of AP^+ (respectively, AP^-), and $\mathbf{M}(N_+^{(+)}) = \mathbf{M}(N_+^{(-)})$, $\mathbf{M}(N_-^{(+)}) = \mathbf{M}(N_-^{(-)})$. Choosing $N_+^{(+)} = \mathbf{M}(G_+^{(+)})^{-1} G_+^{(+)}$, $N_+^{(-)} = \mathbf{M}(G_+^{(-)})^{-1} G_+^{(-)}$, $N_-^{(+)} = G_-^{(+)} \mathbf{M}(G_-^{(+)})^{-1}$, $N_-^{(-)} = G_-^{(-)} \mathbf{M}(G_-^{(-)})^{-1}$, we may therefore suppose that in (3.7), the multiples $G_\pm^{(+)}$, $G_\pm^{(-)}$ are constant:

$$G_\pm^{(+)} = \mathbf{M}(G_\pm^{(+)}), \quad G_\pm^{(-)} = \mathbf{M}(G_\pm^{(-)}).$$

Using polar representations

$$\mathbf{M}(G_+^{(\pm)}) = R_+^{(\pm)} V_+^{(\pm)}, \quad \mathbf{M}(G_-^{(\pm)}) = V_-^{(\pm)} R_-^{(\pm)},$$

where

$$R_+^{(\pm)} = \left(\mathbf{M}(G_+^{(\pm)})\mathbf{M}(G_+^{(\pm)})^*\right)^{1/2}, \quad R_-^{(\pm)} = \left(\mathbf{M}(G_-^{(\pm)})^*\mathbf{M}(G_-^{(\pm)})\right)^{1/2},$$

we may then rewrite (3.7) in a form

$$G^{(\pm)} = R_+^{(\pm)}(V_+^{(\pm)}U^{(\pm)}V_-^{(\pm)})R_-^{(\pm)}.$$

Multiplication of G by the constant matrices $R_+^{(-)-1}$ on the left and $R_-^{(-)-1}$ on the right does not change the d-normality of R_G. This reduces the general case to the situation where

$$(3.9) \qquad G^{(+)} = R_+^{(-)-1}R_+^{(+)}(V_+^{(+)}U^{(+)}V_-^{(+)})R_-^{(+)}R_-^{(-)-1}, \quad G^{(-)} = V_+^{(-)}U^{(-)}V_-^{(-)}.$$

Because of (3.8), the matrices $R_+^{(-)-1}R_+^{(+)}$ and $R_-^{(+)}R_-^{(-)-1}$ are (positive) scalar multiples of I. Moreover, $V_+^{(+)}U^{(+)}V_-^{(+)}$ and $V_+^{(-)}U^{(-)}V_-^{(-)}$ are unitary matrix functions in AP^+, and

$$\mathbf{M}(V_+^{(\pm)}U^{(\pm)}V_-^{(\pm)}) = V_+^{(\pm)}\mathbf{M}(U^{(\pm)})V_-^{(\pm)} = X_\pm.$$

Therefore, the case (3.9) finally simplifies to

$$G^{(+)} = \mu U^{(+)}, \ G^{(-)} = U^{(-)} \qquad (\mu > 0),$$

with $X_\pm$ in the condition ii) being $\mathbf{M}(U^{(\pm)})$. Moreover, due to Lemma 3.3, we may restrict ourselves to the case $\beta = 0$; the general result follows automatically.

In turn, Lemma 1.4 guarantees that the operator R_G is d-normal on the unweighted space $L^2(\mathbb{R})$ if and only if the operator $R_{\hat{G}}$ is Fredholm. The matrix $\hat{G}$ given by (1.8) lies in SAP together with G, and its representatives at $\pm\infty$ are given by the formula

$$\hat{G}^{(\pm)} = \begin{bmatrix} I & G^{(\pm)*} \\ G^{(\pm)} & 0 \end{bmatrix}.$$

Since $U^{(+)}$ is unitary,

$$(3.10) \qquad \hat{G}^{(+)} = \begin{bmatrix} I & \mu U^{(+)*} \\ \mu U^{(+)} & 0 \end{bmatrix} = \begin{bmatrix} I & 0 \\ \mu U^{(+)} & -\mu^2 I \end{bmatrix}\begin{bmatrix} I & \mu U^{(+)*} \\ 0 & I \end{bmatrix}.$$

Formula (3.10) provides a canonical AP-factorization of $\hat{G}^{(+)}$. Hence,

$$\mathbf{d}(\hat{G}^{(+)}) = \begin{bmatrix} I & 0 \\ \mu X_+ & -\mu^2 I \end{bmatrix}\begin{bmatrix} I & \mu X_+^* \\ 0 & I \end{bmatrix} = \begin{bmatrix} I & \mu X_+^* \\ \mu X_+ & \mu^2(X_+X_+^* - I) \end{bmatrix}.$$

Analogously,

$$\mathbf{d}(\hat{G}^{(-)}) = \begin{bmatrix} I & X_-^* \\ X_- & (X_-X_-^* - I) \end{bmatrix}.$$

According to Theorem 3.6, the operator $R_{\hat{G}}$ is Fredholm if and only if, in addition to the already satisfied condition i), the matrix

$$Z = \mathbf{d}\left(\hat{G}^{(-)}\right)^{-1}\mathbf{d}\left(\hat{G}^{(+)}\right)$$

has no negative eigenvalues. Since

$$\mathbf{d}\left(\hat{G}^{(-)}\right)^{-1} = \begin{bmatrix} (I - X_-^* X_-) & X_-^* \\ X_- & -I \end{bmatrix} = \begin{bmatrix} X_-^* & -I \\ -I & 0 \end{bmatrix}\begin{bmatrix} -X_- & I \\ -I & 0 \end{bmatrix},$$

the matrix Z is similar to

$$\begin{bmatrix} -X_- & I \\ -I & 0 \end{bmatrix}\mathbf{d}\left(\hat{G}^{(+)}\right)\begin{bmatrix} X_-^* & -I \\ -I & 0 \end{bmatrix} = \begin{bmatrix} -X_- & I \\ -I & 0 \end{bmatrix}\begin{bmatrix} I & \mu X_+^* \\ \mu X_+ & \mu^2(X_+ X_+^* - I) \end{bmatrix}\begin{bmatrix} X_-^* & -I \\ -I & 0 \end{bmatrix} =$$

$$\begin{bmatrix} -X_- + \mu X_+ & -\mu X_- X_+^* + \mu^2(X_+ X_+^* - I) \\ -I & -\mu X_+^* \end{bmatrix}\begin{bmatrix} X_-^* & -I \\ -I & 0 \end{bmatrix} =$$

$$\begin{bmatrix} -X_- X_-^* + \mu(X_+ X_-^* + X_- X_+^*) + \mu^2(I - X_+ X_+^*) & X_- - \mu X_+ \\ -(X_- - \mu X_+)^* & I \end{bmatrix} = \begin{bmatrix} -XX^* + \mu^2 I & X \\ -X^* & I \end{bmatrix},$$

where

(3.11) $$X = X_- - \mu X_+.$$

Therefore, Z has no negative eigenvalues if and only if the same property holds for the matrix $\begin{bmatrix} \mu^2 I - XX^* & X \\ -X^* & I \end{bmatrix}$, in other words, when $\begin{bmatrix} (\lambda + \mu^2)I - XX^* & X \\ -X^* & (1+\lambda)I \end{bmatrix}$ is invertible for all $\lambda > 0$. The latter matrix is invertible simultaneously with

(3.12) $$(\lambda + \mu^2)I - XX^*\left(1 - \frac{1}{1+\lambda}\right) = \left(\frac{(1+\lambda)(\lambda + \mu^2)}{\lambda}I - XX^*\right)\frac{\lambda}{1+\lambda}.$$

The function $\lambda \mapsto \frac{1+\lambda}{\lambda}(\lambda + \mu^2)$ maps $(0, +\infty)$ onto $[(1+\mu)^2, +\infty)$. Hence, the matrix (3.12) is invertible for all $\lambda > 0$ if and only if

(3.13) $$\sigma(XX^*) \cap \left[(1+\mu)^2, +\infty\right) = \emptyset.$$

Since $||X||^2 \in \sigma(XX^*) \subset [0, ||X||^2]$, condition (3.13) means exactly that

(3.14) $$||X|| < 1 + \mu.$$

On the other hand, the matrices $X_\pm = \mathbf{M}(U^{(\pm)})$ are contractions, and therefore (3.11) implies

(3.15) $$||X|| \le ||X_-|| + \mu||X_+|| \le 1 + \mu.$$

Hence, the only case when (3.14) is not satisfied is when all the inequalities in (3.15) degenerate to equalities, in other words, when there exists a vector $\xi(\neq 0)$ such that $||X_\pm\xi|| = ||\xi||$ and $X_+\xi = -X_-\xi$.

Hence, property (3.14) is satisfied either for all $\mu > 0$ or for none of them. So, it is equivalent to its particular case for $\mu = 1$, in which it takes the form $||X_+ - X_-|| < 2$. ∎

Of course, the analogous result for n-normality also holds. It is formulated exactly the same way as Theorem 3.8, with the only difference being that $U^{(\pm)}$ belong to AP^-, not AP^+.

REFERENCES

[1] A. Böttcher, Yu. Karlovich, and I. Spitkovsky, *Toeplitz operators with semi-almost periodic symbols on spaces with Muckenhoupt weight*, Integr. Equat. and Oper. Theory **18** (1994), 261–276.

[2] A. Böttcher and B. Silbermann, *Analysis of Toeplitz operators*, Springer-Verlag, Berlin, Heidelberg, New York, 1990.

[3] K. F. Clancey and I. Gohberg, *Factorization of matrix functions and singular integral operators*, Birkhäuser, Basel and Boston, 1981.

[4] L. Coburn and R. G. Douglas, *Translation operators on the half-line*, Proc. Nat. Acad. Sci. USA **62** (1969), 1010–1013.

[5] A. Devinatz, *Toeplitz operators on H^2 spaces*, Trans. Amer. Math. Soc. **12** (1964), 304–317.

[6] I. Gohberg and I. Feldman, *On Wiener-Hopf integro-difference equations*, Soviet Math. Dokl. **9** (1968), 1312–1316.

[7] R. Hunt, B. Muckenhoupt, and R. Wheeden, *Weighted norm inequalities for the conjugate function and Hilbert transform*, Trans. Amer. Math. Soc. **176** (1973), 227–251.

[8] Yu. I. Karlovich, *On the Haseman problem*, Demonstratio Math. **26** (1993), 581–595.

[9] Yu. I. Karlovich and I. M. Spitkovsky, *On the Noether property for certain singular integral operators with matrix coefficients of class SAP and the systems of convolution equations on a finite interval connected with them*, Soviet Math. Dokl. **27** (1983), 358–363.

[10] ———, *Factorization of almost periodic matrix functions and Fredholm theory of Toeplitz operators with semi almost periodic matrix symbols*, Linear and Complex Analysis Problem Book: 199 research problems, Lecture Notes in Mathematics **1043** (1984), 279–282.

[11] ———, *Factorization of almost periodic matrix-valued functions and the Noether theory for certain classes of equations of convolution type*, Izv. Akad. Nauk SSSR, Ser. Mat **53** (1989), no. 2, 276–308 (in Russian), English translation in Mathematics of the USSR, Izvestiya **34** (1990), 281–316.

[12] ———, *(Semi)-Fredholmness of convolution operators on the spaces of Bessel potentials*, Operator Theory: Advances and Applications, vol. 71, Birkhäuser-Verlag, 1994, pp. 122–152.

[13] ———, *Almost periodic factorization: An analogue of Chebotarev's algorithm*, Contemporary Math. **189** (1995), 327–352.

[14] ———, *Factorization of almost periodic matrix functions*, J. Math. Anal. Appl. **193** (1995), 209–232.

[15] N. Krupnik and I. Feldman, *Relations between factorization and invertibility of finite Toeplitz matrices*, Izvestiya Akademii Nauk Moldavskoi SSR. Serya fiziko-tehnicheskih i matematicheskih nauk (1985), no. 3, 20–26 (in Russian).

[16] B. M. Levitan, *Almost periodic functions*, GITTL, Moscow, 1953 (in Russian).

[17] G. S. Litvinchuk and I. M. Spitkovsky, *Factorization of measurable matrix functions*, Birkhäuser Verlag,

Basel and Boston, 1987.

[18] V. A. Paatashvili, *On the discontinuous problem of linear conjugation*, Soobshch. Akad. Nauk Gruzin. SSR **36** (1964), no. 3, 539–540 (in Russian).

[19] A. I. Saginashvili, *Singular integral equations with coefficients having discontinuities of semi-almost periodic type*, Trudy Tbiliss. Mat. Inst. Razmadze **66** (1980), 84–95 (in Russian), English translation: Amer. Math. Soc. Transl. **127**, no. 2 (1986).

[20] D. Sarason, *Toeplitz operators with semi-almost periodic symbols*, Duke Math. J. **44** (1977), no. 2, 357–364.

[21] I. Spitkovsky, *Noether criteria for block triangular operators and related problems of factorization theory*, No. 2543-81 dep., VINITI, Moscow, 1981.

[22] ———, *On the factorization of almost periodic matrix functions*, Math. Notes **45** (1989), no. 5–6, 482–488.

[23] H. Widom, *Inversion of Toeplitz matrices. III*, Notices Amer. Math. Soc. **7** (1960), 63.

Yuri Karlovich

Hydroacoustic Department
Marine Hydrophysical Institute
Ukrainian Academy of Sciences
Preobrazhenskaya Str. 3
270100 Odessa, Ukraine

Current address:

TU Chemnitz-Zwickau
Fakultät für Mathematik
09107 Chemnitz
Germany

Ilya Spitkovsky

Department of Mathematics
The College of William and Mary
Williamsburg, Virginia 23187–8795
USA

MSC 1991: Primary 45E10
         Secondary 42A75, 47A53, 47A68

Operator Theory
Advances and Applications, Vol. 90
© 1996 Birkhäuser Verlag Basel/Switzerland

ON CANONICAL FACTORIZATION OF DISSIPATIVE AND POSITIVE MATRIX FUNCTIONS RELATIVE TO NON-SIMPLE CONTOURS

Ilya Krupnik[1], Naum Krupnik, Vladimir Matsaev

Let G_+ be an open set on the complex plane bounded by a non-simple curve Γ and $z_0 \in G_+$. It is proved that any dissipative continuous matrix function of the form $A(t) = (t - z_0)^{-1} A_0 + B_+(t)(t \in \Gamma)$, where A_0 is a constant matrix and $B_+(z)$ is analytic in G_+, admits a canonical factorization. Also it is shown that for any non-simple contour Γ there exist 2×2 rational dissipative matrix functions and 2×2 Hölder continuous positive matrix functions which admit non-canonical factorization.

Let Γ be a rectifiable counter-clockwise oriented closed contour which separates the extended complex plane into two open sets: a bounded set G_+ and unbounded set G_-. For simplicity, we suppose that $0 \in G_+$.

Denote by $C(\Gamma)$ the set of all continuous functions on Γ, by $C^+(\Gamma)$ $(C^-(\Gamma))$ the subset of $C(\Gamma)$ consisting of the functions which admit holomorphic extension to $G_+(G_-)$ and by $X_{n \times n}$ $(X \subset C(\Gamma))$ the set of all $n \times n$ matrices, with entries from X.

A matrix $A(t)$ $(\in C(\Gamma)_{n \times n})$ is said to admit a (right) factorization relative to Γ if there exist matrix functions $A_\pm(t) \subset C^\pm(\Gamma)_{n \times n}$ and a diagonal matrix

$$D = \mathrm{diag}\,(t^{\kappa_1}, t^{\kappa_2}, \ldots, t^{\kappa_n})$$

such that

$$\det A_\pm(z) \neq 0 \ (z \in G_\pm \cup \Gamma, \quad \kappa_j \in \mathbb{Z} \ (j = 1, \ldots, n))$$

and

$$A(t) = A_-(t)\, D(t)\, A_+(t) \qquad (t \in \Gamma).$$

If $\kappa_1 = \cdots = \kappa_n = 0$, then the *factorization* is said to be *canonical*.

In this paper we prove the following statements which are known for the case when Γ is a simple contour.

---

[1]The first author acknowledges support from Professor P. Lancaster on a grant from the Natural Sciences and Engineering Research Council of Canada

Throughout this paper we assume that Γ consists of a finite number of disjoint simple closed smooth curves.

Theorem 1.   *Suppose that $A(t)$ is a continuous $n \times n$ matrix function on Γ and has a form*

$$(1) \qquad\qquad A(t) = \frac{1}{t - z_0} A_0 + B_+(t) \qquad (t \in \Gamma)$$

where $z_0 \in G_+$, $B_+(t) \in C^+(\Gamma)_{n \times n}$ and A_0 is a constant matrix. If $A(t)$ is a dissipative matrix, then A admits a canonical factorization relative to contour Γ.

For the case of simple contour Γ, this theorem was proved by I. Gohberg and J. Leiterer [GL], (see also [CG, page 88]).

Theorem 2.   *Let Γ be a non-simple contour. Then there exists a rational dissipative 2×2 matrix-function which admits a non-canonical factorization relatively Γ.*

For the case of a simple contour which is not a circle, this theorem was proved by A. Markus and V. Matsaev [MM1], (see also [M, Th. 27.5] and [CG, Th. 3.1]).

Theorem 3.   *Let Γ be a non-simple contour. Then there exists a 2×2 matrix-function that is positive, Hölder continuous on Γ and admits a non-canonical factorization relatively Γ.*

For a simple contour which is not a circle, this theorem is proved by A. Markus and V. Matsaev [MM2].

Using the same arguments as in [KMM, section 4], we can deduce from Theorem 2 the following statement:

Theorem 4.   *Let Γ be the same as in theorem 2. Then there exists a matrix polynomial $L(\lambda)$ with $(L(\lambda) f, f) \neq 0$ on Γ $(f \neq 0)$ such that spectral factorization of $L(\lambda)$ with respect to G_+ does not exist.*

The proofs of Theorems 1,2 and 3 are given in sections 1,2, and 3 respectively.

The authors are grateful to Israel Gohberg and Alexander Markus for useful discussions.

§1. Proof of Theorem 1.

Denote by P the analytical projection from $L_2^n(\Gamma)$ onto $L_2^+(\Gamma)^n := P(L_2^n(\Gamma))$ defined by $P = \frac{1}{2}(I + S)$ where

$$(S\varphi)(t) = \frac{1}{\pi i} \int_\Gamma \frac{\varphi(\tau)d\tau}{\tau - t} \qquad (t \in \Gamma).$$

First we prove the following statement: If the matrix A admits a factorization, then the factorization is canonical. Suppose that matrix A admits a noncanonical factorization. Since ind det $A(t) = 0$, matrix A has positive as well as negative partial indices. It follows from here that the equation $PAf_+ = 0$ has in the space $L_2^+(\Gamma)^n$ a nontrivial solution f_+, hence, $Af_+ = g_-$, where $g_- \in \operatorname{Im} Q$ $(Q = I - P)$. Without loss of generality we can suppose that $z_0 = 0$. Then,

$$(1.1) \qquad\qquad b_+ f_+ + \frac{1}{z} A_0 f_+ = g_-$$

and hence,

$$(1.2) \qquad g_-(t) = (Qg_-)(t) = \frac{1}{t} A_0 f_+(0) = \frac{1}{t} C \qquad (t \in \Gamma)$$

where C is a constant vector. Since A is a continuous matrix function invertible on Γ, the vector $f_+(t) = A^{-1} g_-(t)$ is also continuous on Γ.

Let $\langle u, v \rangle$ denote the standard scalar product in $\mathbb{C}^n$.

The matrix A is dissipative, hence, the winding number of the continuous function $h(t) = \langle f_+, A f_+ \rangle$ is zero:

$$\operatorname{ind} h(t) = 0.$$

But, $h(t) = \langle f_+, g_- \rangle = \dfrac{h_+(t)}{t} = \dfrac{1}{|t|^2} h_+(t)\, t$, and it follows from here that ind $h(t) > 0$.

The contradiction shows that the factorization is canonical.

Now we prove the existence of factorization. Let $\tilde{G}_+$ be an open subset of G_+ with the boundary $\tilde{\Gamma}$ such that $z_0 \in \tilde{G}_+, 0 \in \tilde{G}_+, \tilde{\Gamma} \subset G_+$ and $\tilde{\Gamma}$ consists of a finite number of simple closed smooth curves. One can also assume that $A(t)$ is dissipative for all $t \in G_+ \backslash \tilde{G}_+$. Since the matrix function is analytic on $\tilde{\Gamma}$ and det $A(t) \neq 0$ $(t \in \tilde{\Gamma})$, it admits a factorization. As was shown above, the factorization is canonical

$$(1.3) \qquad\qquad A(t) = \tilde{A}_-(t)\, \tilde{A}_+(t) \qquad (t \in \tilde{\Gamma}).$$

Set

$$A_+(z) = \begin{cases} \tilde{A}_-^{-1}(z)\, A(z), & z \in G_+ \backslash \tilde{G}_+ \\ \tilde{A}_+(z), & z \in \tilde{G}_+ \cup \tilde{\Gamma}. \end{cases}$$

It follows from (1.3) that the matrix function $A_+(z)$ is well defined on G_+, it is analytic on G_+ and det $A_+(z) \neq 0$ $(z \in G_+)$. Thus, the equality $A(t) = \tilde{A}_-(t)\, A_+(t)$ gives a canonical factorization of the matrix $A(t)$ relative to contour Γ. □

In the proof of Theorem 1 for simple contours (see [CG, p. 88]), the following intermediate result was obtained.

Let Γ be a simple contour, then there exits a measurable function $\rho(t)$ $(t \in \Gamma)$, such that $0 < m \leq \rho(t) \leq M$ and in the space $L_2(\Gamma, \rho)$ with the weight ρ the subspaces $\operatorname{Im} P$ and $\{\frac{\lambda}{t}\}$ $(\lambda \in \mathbb{C})$ are orthogonal.

It is interesting to mention that this statement is a characteristic property of simple contours. Namely, the following statement is true:

Let Γ be a union of a finite number of disjoint simple contours. If there exists a weight $\rho(t) \geq 0$ such that the one-dimentional subspace $span\left\{\frac{1}{t}\right\}$ is orthogonal to $\operatorname{Im} P$ in the space $L_2(\Gamma, \rho)$, then Γ is a simple contour.

The proof of this statement uses the same arguments as in [K, Th. 1], where the following result was obtained:

If in a space $L_2(\Gamma, \rho)$ the subspaces $\operatorname{Im} P$ and $\operatorname{Im} Q$ are orthogonal, then Γ is a circle and ρ is a constant.

§2. Proof of Theorem 2.

We start with the following:

Lemma 2.1. *Let Γ be a non-simple contour defined in Theorem 2. Then there exist two Hölder continuous vectors $f_+ = (f_1, f_2)$ and $g_- = (g_1, g_2)$ on Γ, such that $f_+ \in \operatorname{Im} P$, $g_- \in \operatorname{Im} Q$ $(Q = I - P)$, and the scalar product $\langle f_+, g_- \rangle$ is positive.*

Proof. Consider two cases.

1. The set G_- is not connected. Take some point a in some bounded component of G_- and set

$$(2.1) \qquad f_+(z) = \left(\frac{1}{z-a}, 1\right) \qquad (z \in \Gamma)$$

$$(2.2) \qquad g_-(z) = \begin{cases} \left(\dfrac{1}{z-a}, 0\right) & z \in \Gamma_0 \\ (0, \ 1) & z \in \Gamma \setminus \Gamma_0, \end{cases}$$

where Γ_0 is the boundary of the unbounded component of G_-.

2. The set G_- is connected. Since Γ is a non-simple contour, the set G_+ is not connected. Take two points a and b from two different components of G_+, and set

$$(2.3) \qquad g_-(z) = \left(\frac{1}{z-a}, \frac{1}{z-b}\right), \qquad z \in \Gamma$$

$$(2.4) \qquad f_+(z) = \begin{cases} \left(0, \dfrac{1}{z-b}\right), & z \in \Gamma_a \\ \left(\dfrac{1}{z-b}, 0\right), & z \in \Gamma \setminus \Gamma_a, \end{cases}$$

where Γ_a is the boundary of the component $G_a^+ \subset G_+$ which contains the point a.

The vectors f_+ and g_- defined by (2.1) - (2.4) satisfy all the needed conditions.   □

Proof of Theorem 2. It is known (see, for example, [SL], Theorem 2 on page 116, and Note 1 on page 115) that the set of rational functions is dense in $C(\Gamma)$ and the set of rational functions without poles in $G_+ \cup \Gamma$ is dense in $C_+(\Gamma)$. It follows from here and Lemma 2.1 that there exist rational vector functions $r_+(t) = (p(t), q(t))$ and $r_-(t) = (a(t), b(t))$ such that $r_+ \in \operatorname{Im} P$, $r_- \in \operatorname{Im} Q$ $(P + Q = I)$ and

$$(2.6) \qquad\qquad \operatorname{Re}\langle r_+, r_-\rangle = \operatorname{Re}(p\bar{a} + q\bar{b}) > 0 \qquad \text{on } \Gamma.$$

Moreover, it follows from the proof of Lemma 2.1 that $|f_1(t)\, g_2(t) - f_2(t)\, g_1(t)| \geq \delta > 0$ on Γ, hence, we can suppose that $|p(t)\, b(t) - q(t)\, a(t)| > 0$ on Γ and, hence, the vectors $r_+(t)$ and $r_-(t)$ are linear independent for each $t \in \Gamma$.

Set $\beta = \sqrt{|p|^2 + |q|^2}$ and

$$(2.7) \qquad\qquad X = \begin{bmatrix} \dfrac{a\bar{p} + b\bar{q}}{\beta^2} & \dfrac{\bar{a}q - \bar{b}p}{\beta^2} \\[2ex] \dfrac{bp - aq}{\beta^2} & \alpha \end{bmatrix}$$

It follows from (2.6) that X is a dissipative matrix on Γ.

Let B be an operator acting in $\mathbb{C}^2$ (for each fixed $t \in \Gamma$) which in orthogonal basis

$$u = \frac{1}{\beta}\,(p, q) \ , \quad v = \frac{1}{\beta}\,(-\bar{q}, \bar{p})$$

is defined by the matrix X. Since X is a dissipative matrix B is a dissipative operator.

It is readily checked that the matrix Y of operator B in basis r_+, r_- has the following form:

$$(2.8) \qquad\qquad Y = \begin{bmatrix} 0 & \varphi \\ 1 & \psi \end{bmatrix}$$

where φ and ψ are some continuous functions on Γ.

Let C be an operator which in basis r_+, r_- has the form (2.8) with rational φ, ψ and $\|B - C\| < \varepsilon$. For $\varepsilon > 0$ small enough, C is also a dissipative operator (for each $t \in \Gamma$). Now we pass to standard basis $e_1 = (1, 0); e_2 = (0, 1)$ and let A be the matrix of operator C in the basis e_1, e_2. Since r_+, r_-, φ, ψ are rational functions, the matrix A is rational dissipative matrix and $Ar_+ = r_-$.

Show that A does not admit a right canonical factorization relative to Γ. Suppose that $A = A_- A_+$, then $A_- A_+ r_+ = r_-$, $A_+ r_+ = A_-^{-1} r_- = 0$. It follows from here that $r_+ = r_- = 0$ which is impossible because of (2.6).   □

§3. Proof of Theorem 3.

Let Γ be a non-simple contour and let $f_+ = (f_1, f_2), g_- = (g_1, g_2)$ be the vectors defined by (2.1) - (2.4). Set

(3.1)
$$\gamma = \sqrt{|f_1|^2 + |f_2|^2},$$

$$U = \frac{1}{\gamma} \begin{bmatrix} \overline{f_1} & \overline{f_2} \\ -f_2 & f_1 \end{bmatrix}, \quad V = \frac{1}{\gamma} \begin{bmatrix} f_1 & -\overline{f_2} \\ f_2 & \overline{f_1} \end{bmatrix}$$

and

$$B = \frac{1}{\gamma^2} \begin{bmatrix} \overline{f_1}\, g_1 + \overline{f_2}\, g_2 & \overline{f_1}\, \overline{g_2} - \overline{f_2}\, \overline{g_1} \\ f_1\, g_2 - f_2\, g_1 & M \end{bmatrix}$$

where M is a constant which provides $\det B > 0$. Finally, we denote by A the matrix $A = VBU$.

The matrices U and V are unitary and $UV = I$. Hence A is positive. It is readily checked that $Af_+ = g_-$. It follows from here (see the last step in the proof of Theorem 2) that A does not admit a (right) canonical factorization. □

Note that the vectors f_+ and g_- in Lemma 2.1 are explicitly constructed and, hence, the matrix A is also presented explicitely. Here is an illustrative example:

Let $\Gamma = \{|t + 2| = 1\} \cup \{|t + 2| = 3\}$. for this contour we can take

$$A(t) = \begin{cases} \begin{bmatrix} 6 & -\dfrac{3}{t+2} \\ -\dfrac{t+2}{3} & \dfrac{1}{3} \end{bmatrix} & (|t + 2| = 3) \\[20pt] \begin{bmatrix} 1 & -\dfrac{1}{t+2} \\ -(t+2) & 2 \end{bmatrix} & (|t + 2| = 1). \end{cases}$$

This matrix admits a non-canonical factorization: $A = A_- D A_+$, where

$$A_+(t) = \begin{bmatrix} 1 & -\dfrac{1}{t+2} \\ 0 & 1 \end{bmatrix}, \quad D = \begin{bmatrix} t & 0 \\ 0 & t^{-1} \end{bmatrix}$$

and

$$A_-(t) = \begin{cases} \begin{bmatrix} \dfrac{6}{t} & \dfrac{3t}{t+2} \\ -\dfrac{t+2}{3t} & 0 \end{bmatrix} & (|t + 2| = 3) \\[20pt] \begin{bmatrix} \dfrac{1}{t} & 0 \\ -\dfrac{t+2}{t} & t \end{bmatrix} & (|t + 2| = 1). \end{cases}$$

We conclude this section with the following remark and its application.

Remark 3.1. *Let A be a Hölder continuous dissipative $n \times n$ matrix function on a simple contour Γ, then there exists a Hölder continuous on Γ matrix function $B_+ \in C^+(\Gamma)_{n \times n}$ such that $\det B_+(z) \neq 0$ ($z \in G_+ \cup \Gamma$) and the matrix AB_+ is positive.*

Indeed, without loss of generality we can suppose that Γ is a circle. In this case, A admits a canonicl factorization $A = A_- A_+$, and we can take $B_+ = A_+^{-1} A_-^*$. □

It would be interesting to find out whether the statement given in Remark 3.1 can be extended to non-simple contours.

Let us return to simple contour Γ. Remark 3.1 shows that Theorem 3.1 for simple contours is a corollary from Theorem 2.1. We mention that the proof of Theorem 3.1 (proposed in [MM2]) is easier than the proof of Theorem 2.1, and hence, its independent proof also makes sense.

References

[CG]   Clancey, K., Gohberg, I., *Factorization of Matrix Functions and Singular Integral Operators*, OT, vol. 3, Birkhäuser-Verlag, Basel, 1981.

[GL]   Gohberg, I., Leiterer, J., *General theorems on canonical factorization of operator functions relative to a contour*, Mat. Issled **3 (25)** (1972), 87-134.

[K]   Krupnik, N., *The conditions of selfadjointness of the operator of singular integration*, Integr. Equat. and Oper. Th. **14** (1991), 760-763.

[KMM] Krupnik, I., Markus A., Matsaev, V., *Factorization of matrix functions and characteristic properties of the circle*, Integr. Equat. and Oper. Th. **17** (1993), 554-566.

[M]   Markus, A., *Introduction to the spectral theory of polynomial operator pencils*, Amer. Math. Soc. (1988), Providence.

[MM1] Markus, A., Matsaev, V., *Two remarks about factorization of matrix-valued functions*, Mat. Issled **N42** (1976), 216-223, (Russian).

[MM2] ———, *The failure of factorication of positive matrix functions on noncircular contours*, Linear Algebra and its Applications **208/209** (1994), 231-237.

[SL]   Smirnov, V., Lebedev, N., *Functions of a Complex Variable Constructive Theory*, London, 1968.

Ilya Krupnik   Naum Krupnik

Dept. of Mathematics & Statistics   Dept. of Mathematics & Computer Science

University of Calgary   Bar-Ilan University

2500 University Drive N.W.   52900 Ramat-Gan, Israel

Alberta, Canada

Vladimir Matsaev

School of Mathematical Sciences

Raymond & Beverly Sackler Faculty of Exact Sciences

69978 Tel-Aviv University

Tel-Aviv, Israel

MSC 47A79

Operator Theory
Advances and Applications, Vol. 90
© 1996 Birkhäuser Verlag Basel/Switzerland

ASYMPTOTIC INVERTIBILITY OF TOEPLITZ OPERATORS

Bernd Silbermann

The purpose of this lecture is to exemplify with Toeplitz operators some ideas behind recent methods for studying stability of operator sequences and related problems. We demonstrate that several notions, strategies, and techniques commonly employed in operator theory have useful analogues in numerical analysis. In particular, we discuss a "symbol calculus" for sequences of finite sections of Toeplitz operators and embark on its consequences for the asymptotic behavior of the pseudospetra and the Moore-Penrose inverses of large truncated Toeplitz matrices.

1. Introduction

A variety of problems in analysis and applications leads to the following question: Suppose we are given a sequence $\{A_n\}$ of linear operators acting on some Hilbert space H which is related to some mathematical object. This object can be an operator, a function, a measure, and so on. Then one asks for the relations between $\{A_n\}$ and the related object.

Example 1. One can have in mind an operator equation

$$Ax = y \quad (x, y \in H)$$

which one tries solving by some "reasonable" approximation method

$$A_n x_n = y_n \quad (x_n, y_n \in H)$$

with y_n tending to y for $n \to \infty$. Of course, the sequence $\{A_n\}$ has to approximate A. The case of strong convergence of $\{A_n\}$ to A is of special interest. If it is known that the operator A is invertible and that the sequence $\{A_n\}$ is stable, then $A_n^{-1} y_n$ tends to the uniquely defined solution x of the equation $Ax = y$. Recall that stability means the following:

1. There is an n_0 such that the operators A_n are invertible for $n \geq n_0$;

2. $\sup\limits_{n \geq n_0} \|A_n^{-1}\| < \infty$.

Example 2. There are models in statistical physics which lead to Toeplitz matrices of very high size which can easily exceed 10^{22}, the number of involved particles. The following strategy is often very useful. Replace large matrices by their infinite limit operators and hope that they can tell us something about their finite approximations.

In the last two-three decades considerable progress has been made in understanding the interplay between convolution-like operators and their finite approximations. I will demonstrate this by means of Toeplitz operators and their finite sections, because in this setting the underlying theory occurs in a quite clear and imaginable manner. In the course of the presentation, some new results will be cited.

2. Operators

For a function $a \in L^\infty$ on the complex unit circle **T**, denote by $\{a_n\}_{n \in \mathbf{Z}}$ the sequence of its Fourier coefficients. Then the infinite matrices

$$(a_{i-j})_{i,j=0}^\infty \quad \text{and} \quad (a_{i+j+1})_{i,j=0}^\infty$$

induce bounded operators on $l^2 := l^2(\mathbf{Z}_+)$, which are called the Toeplitz and the Hankel operators with the generating function a. The properties of $T(a)$ and $H(a)$ are quite different. For instance, $\|T(a)\| = \|a\|_\infty$ and $T(a)$ is compact if and only if $a \equiv 0$, whereas $H(a)$ is compact if only $a \in C$, or more generally, $a \in QC$. By C and QC we denote the algebra of all continuous and all quasicontinuous functions defined on the unit circle **T**, respectively. The identity

$$T(ab) = T(a)T(b) + H(a)H(\tilde{b}) \quad (a, b \in L^\infty)$$

with $\tilde{b}(t) := b(1/t)$ is crucial in the theory of Toeplitz operators. It shows for instance that, if $b \in C$, the Toeplitz operators $T(a)$ and $T(b)$ commute up to a compact operator.

We are however mainly interested in the finite sections of Toeplitz operators. They are defined to be the matrices $T_n(a) := (a_{i-j})_{i,j=0}^n$ or, equivalently, the operators $P_n T(a) P_n$, where P_n is the orthoprojection defined by the rule

$$\{x_0, x_1, ..., x_n, x_{n+1}, ...\} \mapsto \{x_0, ..., x_n, 0, 0, ...\}.$$

Example 3. (Example 2 specified). The free energy of an one-dimensional Gaussian model (on the half axis) in statistical physics is given by

$$F_{n+1} = \frac{1}{2\beta}\left(\log\frac{\beta}{\pi} + \frac{1}{n+1}D_n\right),$$

where $D_n := \det(a_{|i-j|})_{i,j=1}^n$, β is the temperature and a_j are the Fourier coefficients of some non-negative even measure μ on $[-\pi, \pi]$. F_{n+1} ist not known since n is nearly equal to 10^{22} in real systems. The idea is to replace F_{n+1} by $F := \lim F_{n+1}$ in case the limit exists. The result is the more precise the better we understand the asymptotics of the term $F - F_{n+1}$. In some cases this question is directly connected with the so-called stability problem for the finite sections $T_n(a)$ of some Toeplitz operators $T(a)$ $(a \in L^\infty)$.

This problem differs formally a little bit from the stability problem proposed in the beginning. We call $\{T_n(a)\}$ $(a \in L^\infty)$ stable if, for n large enough, the matrices $T_n(a)$ are invertible and $\sup_{n \geq n_0} ||T_n^{-1}(a)|| < \infty$. This notion of stability is equivalent to the former one; this is easily seen by introducing $A_n := T_n(a) + I - P_n$.

Now let us mention two important consequences of the stability of $\{T_n(a)\}$:

- $T(a)$ is invertible on l^2,

- the strong limit of $T_n^{-1}(a)$ equals $T^{-1}(a)$, i. e. s-lim $T_n^{-1}(a) = T^{-1}(a)$.

The last property is often called the asymptotic invertibility of $T(a)$ by means of $\{T_n(a)\}$. So we have: The stability of $\{T_n(a)\}$ is equivalent to the asymptotic invertibility of $T(a)$. Only in 1987 S. Treil gave an example of an invertible Toeplitz operator which is not asymptotically invertible. This example underlines the importance of the question which invertible Toeplitz operators are asymptotically invertible. The first positive results go back to G. Baxter (1963) and to I. Gohberg/ I. Feldman (1965). In the beginning, they considered merely Toeplitz operators $T(a)$ with continuous generating functions a. For $a \in C + H^\infty$ the asymptotic invertiblity of $T(a)$ was studied for instance by A. Devinatz/M. Shinbrot (1969), G. Ambartsumyan (1973) and H. Widom (1976). For a being piecewise continuous I. Gohberg obtained the asymptotic invertibility of invertible Toeplitz operators already in 1967. These historical remarks are far from being complete. Up to now questions closely related to the asymptotic invertibility of Toeplitz operators have been investigated. About 15 years ago a new method entered the scene whose development led to a variety of remarkable results. I have in mind Banach algebra techniques, especially C^*-algebra techniques.

Literature: [Ba], [G/F], [Vl/Vo], [S1], [B/S1], [B/S2], [W]

3. Finite sections and algebras

We denote by F the collection of all sequences $\{A_n\}$ with $A_n : \operatorname{im} P_n \to \operatorname{im} P_n$ such that $||\{A_n\}|| := \sup_n ||A_n|| < \infty$. On defining $\{A_n\} + \{B_n\} := \{A_n + B_n\}, \{A_n\}\{B_n\} := \{A_n B_n\}$ and $\{A_n\}^* := \{A_n^*\}$ we make F into a C^*-algebra. Let G refer to the set of all sequences $\{A_n\} \in F$ with $||A_n|| \to 0$ as $n \to \infty$. It is easy to see that G is a closed two-sided ideal in F. The following proposition is easy to prove.

PROPOSITION 3.1. $\{A_n\} \in F$ *is stable if and only if the coset* $\{A_n\} + G$ *is invertible in* F/G.

This proposition tells us that stability is equivalent to the invertibility of some elements in a suitably constructed algebra. On the other hand F or F/G are very complicated algebras. They contain (in some sense) all bounded linear operators acting on l^2. Therefore there is no hope to get invertibility criteria in F/G. So the question we are left with is what is special for the finite sections of Toeplitz operators from the algebraic point of view. The answer is contained in an observation due to H. Widom:

$$T_n(ab) - T_n(a)T_n(b) = P_n H(a)H(\tilde{b})P_n + W_n H(\tilde{a})H(b)W_n \qquad (1)$$

with $W_n : l^2 \to l^2, \{x_k\} \mapsto \{x_n, x_{n-1}, ..., x_0, 0, 0, ...\}$. Looking for a connection between G and (1) we observe that the ideal G is, in a sense, too small: sequences of the form $\{P_n K P_n + W_n L W_n\}$ with K, L compact do, in general, not belong to G. Thus, in order to develop a theory of stability of finite sections in analogy to the Fredholm theory, it would be desirable to have an ideal J that contains all sequences of the form $\{P_n K P_n + W_n L W_n\}$ with K, L being compact. But there is no such ideal in F. This algebra is, again in some sense, too large. We therefore shall construct a smaller algebra possessing, on the one hand, such an ideal and containing, on the other hand, sufficiently many interesting elements, in particular all elements of the form $\{T_n(a)\}$.

For this aim, we introduce the collection $\mathcal{A}$ of all sequences $\{A_n\} \in F$ for which there exist A and $\tilde{A}$ belonging to $\mathcal{L}(l^2)$ such that

$$A_n P_n \to A, \quad A_n^* P_n \to A^*, \quad \tilde{A}_n P_n \to \tilde{A}, \quad \tilde{A}_n^* P_n \to \tilde{A}^*$$

strongly, where $\tilde{A}_n := W_n A_n W_n$. It is easy to see that $\mathcal{A}$ is closed and actually forms a C^*-subalgebra of F. If K is compact, then $\{P_n K P_n\}$ and $\{W_n K W_n\}$ are in $\mathcal{A}$ (notice that W_n converges weakly to 0). The identity $W_n T_n(a) W_n = T_n(\tilde{a})$ implies that $\{T_n(a)\}$ belongs to $\mathcal{A}$.

Let J be the collection of all elements $\{A_n\} \in F$ of the form

$$\{A_n\} = \{P_n K P_n + W_n L W_n + C_n\},$$

where K and L are compact operators, and $\{C_n\} \in G$. Clearly, J is a subset of $\mathcal{A}$. Even much more is true.

PROPOSITION 3.2. *J is a closed two-sided ideal of $\mathcal{A}$.*

The importance of $\mathcal{A}$ and J is revealed by the following theorem.

THEOREM 3.1. *Let $A \in \mathcal{L}(l^2)$ and let $\{A_n\} \in \mathcal{A}$ be any sequence such that $A_n P_n \to A$ strongly. Abbreviate the strong limit of $\{W_n A_n W_n\}$ to $\tilde{A}$. Then $\{A_n\}$ is stable if and only if the operators A and $\tilde{A}$ are invertible and the coset $\{A_n\} + J$ is invertible in $\mathcal{A}/J$.*

Here is the proof of the sufficiency potion. Suppose A and $\tilde{A}$ are invertible on l^2 and $\{A_n\} + J$ is invertible in $\mathcal{A}/J$. Then there is a sequence $\{B_n\} \in \mathcal{A}$ such that

$$A_n B_n = P_n + P_n K P_n + W_n L W_n + C_n, \tag{2}$$

where K and L are compact and $\{C_n\} \in G$. Passage to the limit $n \to \infty$ gives $AB = I + K$, and if we multiply (2) by W_n from the left and the right and then pass to the limit $n \to \infty$, we arrive at the equality $\tilde{A}\tilde{B} = I + L$, where $\tilde{B} = s - \lim W_n B_n W_n$. Hence $R := A^{-1} - B$ and $T := \tilde{A}^{-1} - \tilde{B}$ are compact. Put $B_n' := B_n + P_n R P_n + W_n T W_n$. Then $\{B_n'\}$ is in $\mathcal{A}$ and $A_n B_n' = P_n + C_n$, where $\{C_n'\} \in G$. It can be shown analogously that $\{B_n' A_n\} - \{P_n\} \in G$. So the sufficiency is proved. This proof immediately also gives the following.

PROPOSITION 3.3. *If $\{A_n\} + J$ is invertible then A and $\tilde{A}$ are Fredholm.*

I already mentioned that $\{T_n(a)\}$ is in $\mathcal{A}$. It is of interest to have an imagination on what sequences generally belong to $\mathcal{A}$. There is no answer. However, if $\{A_n\}$ is of the form $\{P_n A P_n\}$ with $A = \sum_{j=1}^{l} \prod_{i=1}^{k} T(a_{ij})$ $(a_{ij} \in L^\infty(\mathbf{T}))$, then $\{A_n\}$ belongs to $\mathcal{A}$. This result is not trivial.

Applying Theorem 3.1 to $\{T_n(a)\}$ we have to ensure the invertibility of $T(a)$, $T(\tilde{a})$ (notice that in the scalar case the invertibility of $T(a)$ implies the invertibility of $T(\tilde{a})$) and the invertibility of the coset $\{T_n(a)\} + J$. The last question is crucial. How can we study the invertibility of $\{T_n(a)\} + J$? The construction proposed up to now is mainly directed to the application of so-called local principles! I cannot give the main ideas here, but I like to mention that a local principle is something like the Gelfand theory, however mainly for noncommutative algebras. So it was shown that there is a variety of function classes with the property that solely the Fredholmness of $T(a)$ implies the invertibility of $\{T_n(a)\} + J$. Such classes are $C, C + H^\infty, PC, PQC$, and functions locally sectorial over C or QC (here PC and PQC stand for the algebras of piecewise continuous and piecewise quasicontinuous functions, respectively).

THEOREM 3.2. *If a belongs to one of the function classes mentioned, then the invertibility of $T(a)$ is equivalent to its asymptotic invertibility.*

This theorem can be extended to the block case. Because the invertibility of $T(a)$ does in general not imply the invertibility of $T(\tilde{a})$, the asymptotic invertibility of $T(a)$ is equivalent to the invertibility of both $T(a)$ and $T(\tilde{a})$.

I want to finish this section with noticing that the above approach indicates that ideas widely used in operator theory (especially local principles) are also applicable to problems of numerical analysis.

Literature: [B/S1], [B/S2], [S 1], [S 2], [R/S]

4. Symbols

Theorem 3.1 possesses a remarkable reformulation which is the key to some further development. First of all we introduce C^*−algebra homomorphisms W_i $(i = 1, 2, 3)$ by the rules $(\{A_n\} \in \mathcal{A})$

$$W_1 : \{A_n\} \mapsto s - \lim A_n P_n, \quad W_2 : \{A_n\} \mapsto s - \lim W_n A_n W_n, \quad W_3 : \{A_n\} \mapsto \{A_n\} + J.$$

Consider the direct sum $\mathcal{L}(l^2) \dotplus \mathcal{L}(l^2) \dotplus \mathcal{A}/J$ and provide it with the norm $\|(A, B, C)\| := \max\{\|A\|, \|B\|, \|C\|\}$. This direct sum actually forms a C^*−algebra, and

$$\{A_n\} \mapsto (W_1\{A_n\}, W_2\{A_n\}, W_3\{A_n\})$$

is a C^*−algebra homomorphism from $\mathcal{A}$ into that direct sum. We denote it by smb. Because the image of the C^*−algebra homomorphism is closed, it forms a C^*−algebra which we denote by smb$\mathcal{A}$.

THEOREM 4.1. *The C^*−algebra $\mathcal{A}/G$ is isometrically isomorphic to smb $\mathcal{A}$.*

This theorem has an unexpected consequence I am now going to explain.

THEOREM 4.2. *Let $\{A_n\} \in \mathcal{A}$ be an arbitrarily given sequence such that*

$$\max\{\|W_1\{A_n\}\|, \|W_2\{A_n\}\|\} \geq \|W_3\{A_n\}\|. \tag{3}$$

Then

$$\lim_{n \to \infty} \|A_n\| = \|\text{smb}\{A_n\}\|.$$

Proof outline. Using the Banach-Steinhaus Theorem we obtain

$$\max\{||W_1\{A_n\}||, ||W_2\{A_n\}||\} \leq \lim_{n\to\infty} \inf ||A_n||,$$

whence

$$||\mathrm{smb}\{A\}|| \leq \lim_{n\to\infty} \inf ||A_n||$$

by (3). On the other hand, $||\mathrm{smb}\{A_n\}|| = \inf_{\{C_n\}\in G} ||\{A_n\} + \{C_n\}||$. Consequently, given some $\varepsilon > 0$ there is a sequence $\{C_n\} \in G$ such that $||A_n|| - ||C_n|| \leq ||A_n + C_n|| \leq ||\mathrm{smb}\{A_n\}|| + \varepsilon$. Taking the upper limit we obtain

$$\lim_{n\to\infty} \sup ||A_n|| \leq ||\mathrm{smb}\{A_n\}|| + \varepsilon,$$

which completes the proof.

If $T(a)$ is invertible and a belongs to $C, C + H^\infty, PC, QC$, or is locally sectorial over C, then (3) is satisfied for sequence $\{T_n^{-1}(a)\}$, and so we have

$$\lim_{n\to\infty} ||T_n^{-1}(a)|| = ||T^{-1}(a)||.$$

That the inequality (3) is in force in these cases is by no means trivial. For the algebra $C + H^\infty$, for instance, it rests on the decomposition

$$\mathcal{B} := \mathrm{alg}\{\{T_n(a)\} : a \in C + H^\infty\} = \{\{T_n(a)\} : a \in C + H^\infty\} \dot{+} J,$$

where $\mathrm{alg}\{\cdot\}$ means the smallest closed subalgebra of $\mathcal{A}$ containing all sequences $\{T_n(a)\}$ with $a \in C + H^\infty$. Indeed, the quotient algebra $\mathcal{B}/J$ is isometrically isomorphic to $C + H^\infty$, and $||T^{-1}(a)|| \geq ||a^{-1}|| = ||(\{T_n(a)\} + J)^{-1}||$.

The ideas we have outlined contain a general element: The main tool in proving such results, for example the inequality (3), is to consider suitable subalgebras $\mathcal{C}$ of $\mathcal{A}$ and to describe $\mathcal{C}/J$.

Literature: [S 2], [S 3], [B], [R/S].

5. Pseudospectra

The relationship between the spectra of $T_n(a)$ and the spectrum of $T(a)$ is an interesting but very complicated question. Only recently it has become clear that the so-called ε-pseudospectra of $T_n(a)$ behave very nicely for the function classes listed in Section 3 (with the exception of the last one). The ε-pseudospectrum of a linear operator A acting continuously on some Hilbert space $\mathcal{H}$ is defined by

$$\mathrm{sp}_\varepsilon(A) := \{\lambda \in \mathbf{C} : ||(A - \lambda I)^{-1}|| \geq 1/\varepsilon\}$$

Here the convention is used that $||(A - \lambda I)^{-1}|| := \infty$ for λ belonging to the spectrum of A. Now we ask how $\mathrm{sp}_\varepsilon(T_n(a))$ (more generally: $\mathrm{sp}_\varepsilon(A_n)$ for $\{A_n\} \in \mathcal{A}$) is related to $\mathrm{sp}_\varepsilon(T(a))$ (more generally: to $\mathrm{sp}_\varepsilon(W_1\{A_n\})$ or something else). A first piece of information can be obtained in terms of the limiting set (= partial limiting set) of the sequence $\{\mathrm{sp}_\varepsilon(A_n)\}$. By definition, the limiting set $\lim_{n\to\infty} M_n$ of a sequence of subsets M_n of the complex plane is the collection of all complex numbers which are a partial limit of a sequence $\{t_n\}$ of numbers $t_n \in M_n$. So the question we will deal with is the following: Describe the limiting set of $\{\mathrm{sp}_\varepsilon(A_n)\}$, $\{A_n\} \in \mathcal{A}$. This question was taken up by L. Reichel and L. N. Trefethen (1992) for $\{\mathrm{sp}_\varepsilon(T_n(a))\}$, a belonging to the Wiener algebra. In 1994, A. Böttcher observed that the problem can successfully be tackled if only the following two claims would be true:

- Let $\mathcal{B}$ be a C^*−algebra with identity e, let $a \in \mathcal{B}$, suppose that $a - \lambda e$ is invertible for all λ in some open subset U of the complex plane, and assume $||(a - \lambda e)^{-1}|| \leq C$ for all $\lambda \in U$. Then $||(a - \lambda e)^{-1}|| < C$ for all $\lambda \in U$.

- If $\lambda \notin \mathrm{sp}(\mathrm{smb}\{A_n\})$, then $||(A_n - \lambda P_n)^{-1}|| \to ||(\mathrm{smb}\{A_n - \lambda P_n\})^{-1}||$.

That the first claim is true was communicated to us by A. Daniluk. The second claim is in force for a variety of sequences $\{A_n\} \in \mathcal{A}$ due to Theorem 4.2.

THEOREM 5.1. *Suppose* $\{A_n\} \in \mathcal{A}$ *and*

$$\max\{||W_1\{A_n - \lambda P_n\}^{-1}||, ||W_2\{A_n - \lambda P_n\}^{-1}||\} \geq ||W_3\{A_n - \lambda P_n\}^{-1}||$$

for all $\lambda \notin \mathrm{sp}(\mathrm{smb}\{A_n\})$. *Then*

$$\lim_{n \to \infty} \mathrm{sp}_\varepsilon(A_n) = \mathrm{sp}_\varepsilon(\mathrm{smb}\{A_n\}).$$

This theorem applies, for instance, to Toeplitz operators $T(a)$ with a belonging to C, $C + H^\infty$, PC, PQC, ... The consequence is remarkable: $\lim_{n \to \infty} \mathrm{sp}_\varepsilon(T_n(a)) = \mathrm{sp}_\varepsilon T(a)$.

Literature: [R/T], [B], [R/S]

6. Asymptotic Moore-Penrose invertibility

Recall that a bounded linear operator $A : \mathcal{H} \to \mathcal{H}$ is called Moore-Penrose invertible if there is a bounded linear operator $B : \mathcal{H} \to \mathcal{H}$ such that $BAB = B$, $ABA = A$, $AB = (AB)^*$, $BA = (BA)^*$. If the Moore-Penrose inverse of A exists then it is uniquely determined, and its standard notation is A^+. The Moore-Penrose inverse of an operator A exists if and only if this operator is normally solvable, and in this case one has

$$A^+ = (A^*A + P_{\ker A})^{-1} A^*$$

where $P_{\ker A}$ is the orthogonal projection onto the kernel of A. Let A be a Toeplitz operator $T(a)$ which is Fredholm and consider the finite sections $\{T_n(a)\}$. Is it true that $\{T_n^+(a)\}$ converges strongly to $T^+(a)$? There are examples which show that even for rational generating functions the answer is *no* in general: Take $a(t) = t - \alpha$, $0 < |\alpha| < 1$, for instance. So it is surprising that the following problem has solutions for a variety of sequences $\{A_n\} \in \mathcal{A}$: Find all sequences $\{B_n\} \in \mathcal{A}$ such that

$$||A_n B_n A_n - A_n|| \to 0, \quad ||B_n A_n B_n - B_n|| \to 0,$$
$$||(A_n B_n)^* - A_n B_n|| \to 0, \quad ||(B_n A_n)^* - B_n A_n|| \to 0. \qquad (4)$$

Clearly, if $\{B_n\} \in \mathcal{A}$ is a solution of this problem, then $\{B_n\}$ converges strongly to the Moore-Penrose inverse of $W_1\{A_n\}$.

THEOREM 6.1. *If $\{A_n\} \in \mathcal{A}$ has the property that $\{A_n\} + J$ is invertible in $\mathcal{A}/J$, then the problem (4) has a solution $\{B_n\}$. Moreover, the sequence $\{B_n\}$ is unique up to sequences in the ideal G.*

We only remark that the sequence $\{B_n\}$ can be chosen as

$$B_n = (A_n^* A_n + P_n P_{\ker A} P_n + W_n P_{\ker \tilde{A}} W_n)^{-1} A_n^*$$

for n large enough, where $A := W_1\{A_n\}, \tilde{A} := W_2\{A_n\}$. The proof of Theorem 6.1 is based on the fact that the invertibility of $\{A_n\} + J$ implies the Fredholmness of A and $\tilde{A}$, and it makes also use of Theorem 4.1.

As already mentioned in Section 3, $\{T_n(a)\} + J$ is invertible if only $T(a)$ is Fredholm and a belongs to one of the following function classes: C, $C + H^\infty$, PC, PQC, functions being locally sectorial over C or QC. Therefore, for the related sequences $\{T_n(a)\}$ problem (4) has solutions! This result is accomplished by the following one.

THEOREM 6.2. *Let $T(a)$ be Fredholm an suppose that a belongs to one of the function classes quoted above. If there is an n_0 such that*

$$\ker T(a) \subseteq \operatorname{im} P_{n_0}, \quad \ker T(\tilde{a}) \subseteq \operatorname{im} P_{n_0}$$

then

$$P_{\ker T_n(a)} = P_n P_{\ker T(a)} P_n + W_n P_{\ker T(\tilde{a})} W_n \text{ for all } n \text{ large enough,}$$

and the Moore-Penrose inverses of $T_n(a)$ converge strongly to the Moore-Penrose inverse of $T(a)$.

Literature: [Hei/He], [S 4], [R/S]

References

[Ba]   G. Baxter, *A norm inequality for a finite-section Wiener-Hopf equation.* Illinois J.Math. 7, 17-103 (1963).

[B]   A. Böttcher, *Pseudospectra and singular values of large convolution operators*, J. Integral Equations Appl. 6, 267-301 (1994).

[B/S 1]   A. Böttcher, B. Silbermann, *Analysis of Toeplitz operators.* Akademie-Verlag Berlin, 1989, and Springer-Verlag, Berlin, 1990.

[B/S 2]   A. Böttcher, B. Silbermann, *The finite section method for Toeplitz operators on the quarter-plane with piecewise continuous symbols.* Math. Nachr. 110 (1983), 279-291.

[G/F]   I. Gohberg, I. Feldman, *Convolution equations and projection methods for their solution.* Nauka, Moscow, 1971 (Russian); Engl. transl.: Amer. Math. Soc. Transl. of Math. Monographs 41, Providence, R. I., 1974.

[Hei/He] G. Heinig, F. Hellinger, *The finite section method for Moore-Penrose inversion of Toeplitz operators.* Integr. Equations and Operator Theory 19 (1994), 419-446.

[R/S]   S. Roch, B. Silbermann, C^*-algebra techniques in numerical analysis. J. Operator Theory (submitted).

[R/T]   L. Reichel, L. N. Trefethen, Eigenvalues and pseudo-eigenvalues of Toeplitz matrices. Linear Algebra Appl. 162 (1992), 153-185.

[S 1]   B. Silbermann, Lokale Theorie des Reduktionsverfahrens für Toeplitz-operatoren. Math. Nachrichten 104 (1981), 137-146.

[S 2]   B. Silbermann, Local objects in the theory of Toeplitz operators. Integr. Equations and Operator Theory 9 (1986), 706-738.

[S 3]   B. Silbermann, On the limiting set of singular values of Toeplitz matrices. Linear Algebra Appl. 182 (1993), 35-43.

[S 4]   B. Silbermann, Asymptotic Moore-Penrose inversion of Toeplitz operators. Linear Algebra Appl. (submitted).

[Vl/Vo] V. S. Vladimirov, and I. V. Volovich, On a model of statistical physics. Teor. Matem. Fiz. 54:1, 8-22 (1983) (Russian).

[W]   H. Widom, Asymptotic behavior of block Toeplitz matrices and determinants. II. Adv. Math. 21, (1976), 1-29.

Technical University Chemnitz-Zwickau
Department of Mathematics
D - 09107 Chemnitz
Germany

AMS subject classification: 47B35

Operator Theory
Advances and Applications, Vol. 90
© 1996 Birkhäuser Verlag Basel/Switzerland

CALCULATION OF THE NORM OF POLYNOMIALS
OF TWO ADJOINT PROJECTIONS

Y. SPIGEL

Let $\psi(X,Y)$ be a polynomial , P a projection acting in a Hilbert space and φ the function associated with $\psi(X,Y)$ in [2].

In the present paper a subset of polynomials is described so that for each of them the problem of calculating the norm $\| \psi(P,P^*) \|$ can be reduced to the same problem for more simple polynomials.

0. INTRODUCTION

Let Γ be a closed contour on the plane C and ρ a weight so that the linear one-dimensional singular integral operator S:

$$(0.1) \qquad (S\varphi)(t) = \frac{1}{\pi i} \int \frac{\varphi(t)}{\tau - t} d\tau \quad (t \in \Gamma)$$

is bounded in the Hilbert space $H = L_2(\Gamma, \rho)$ (for details see [1]).

Denote by P the analytical projection $\frac{1}{2}(I + S)$. Let $\psi(X,Y)$ be a polynomial of two non-commuting variables. I. Feldman, N. Krupnik and A. Markus [2] have associated with each polynomial $\psi(X,Y)$ (and respectively with an operator $\psi(P,P^*)$)a function $\varphi(x)$ and proved the following statement.

If the function $\varphi(x)$ $(x \geq 0)$ is non-decreasing, then the equality

$$(0.2) \qquad \|\psi(P,P^*)\| = \varphi(\|P\|^2 - 1)$$

holds in any Hilbert space H for any nontrivial (i.e. $P \neq 0, I$) projection (not only for analytical projection).

But it is not always easy to construct the explicit form of $\varphi(x)$ and to verify that $\varphi(x)$ is non-decreasing (see examples in section 3). For the case when the function $\varphi(x)$ is non-monotonic one can turn to [3].

In this paper we describe a subset of polynomials so that for each of them the problem of calculating the norm $\|\psi(P, P^*)\|$ can be reduced to the same problem for more simple polynomials.

The paper consists of three sections. In the first one we recall some necessary results from [2].

The main results of this paper are presented in section 2. Some examples and additional remarks are given in section 3. I am deeply grateful to Professor N. Krupnik for his interest in my work and for useful discussions.

1. SOME NECESSARY RESULTS

Let H be a Hilbert space and $L(H)$ the set of all bounded linear operators acting in H. By $\Phi(H)$ we denote the set of all nontrivial projections $P \in \Phi(H)$. Let $\psi(X, Y)$ be an arbitrary polynomial of two non-commuting variables.

Any operator $\psi(P, P^*)$ $(P \in \Phi(H))$ can be presented in the following form:

$$(1.1) \quad \psi(P, P^*) = \sum_{k=0}^{n}(PP^*)^k(a_k I + b_k P) + \sum_{k=0}^{n} P^*(PP^*)^k(c_k I + d_k P)(a_k, b_k, c_k, d_k \in C).$$

Introduce four functions $A_{ij}(x)$ $(i, j = 1, 2; \ x \geq 0)$

$$A_{11} = \sum_{k=0}^{n}(a_k + b_k + c_k + d_k)(x + 1)^k, A_{12} = \sum_{k=0}^{n}(b_k + d_k)(x + 1)^k ,$$

$$(1.2) \qquad A_{21} = \sum_{k=0}^{n}(c_k + d_k)(x + 1)^k, A_{22} = x \sum_{k=0}^{n} d_k(x + 1)^k + a_0.$$

Let $r(x)$ and $s(x)$ be the functions

$$r(x) = | A_{11}(x) |^2 + | A_{22}(x) |^2 + x(| A_{12} |^2 + | A_{21}(x) |^2)$$

$$s(x) = 2 | A_{11}(x) \cdot A_{22}(x) - x \cdot A_{12}(x) \cdot A_{21}(x) |$$

Now denote by $\varphi(x)$ the function:

$$(1.3) \qquad \varphi(x) = \frac{1}{2} \cdot [(r(x) + s(x))^{\frac{1}{2}} + (r(x) - s(x))^{\frac{1}{2}}] .$$

In the paper [2] the following statements have been proved

1) If $\dim H = 2$, then for an arbitrary projection P $(P \in \Phi(H); \ P \neq 0; I)$ and for an arbitrary polynomial $\psi(X, Y)$

$$(1.4) \qquad \|\psi(P, P^*)\| = \varphi(\|P\|^2 - 1)$$

2) If the function $\varphi(x)$ is non-decreasing, then equality (0.2) holds for any P $(P \neq 0; I)$

3) The function $\varphi(x)$ acossiated with $\psi(X,Y)$ is non-decreasing not for all polynomials $\psi(X,Y)$. For example, if $\psi(X,Y) = 1+(X-Y)^2$, then the function $\varphi(x)$ associated with the polynomial is $\varphi(x) = |x-1|$.

4)If $\psi(X,Y) = a + bX + cY$ and $\varphi(x)$ is the function associated with $\psi(X,Y)$, then $\varphi(x)$ is always non-decreasing.

2. THE MAIN RESULTS

THEOREM 1. *Let $P \in \Phi(H)$ and let $\psi(X,Y)$ be an arbitrary polynomial with associated function $\varphi(x)$. Let $\psi_L(X,Y)$ be a linear function with the following property*

(2.1) $$\psi_L(P, P^*) \cdot [\psi_L(P, P^*)]^* = k \cdot I (k > 0)$$

and with the associated function $\varphi_L(x)$. Then

1) *for function $\varphi_1(x)$ associated with polynomial $\psi_1(X,Y) = \psi_L(X,Y) \cdot \psi(X,Y)$ the equality*

$$\varphi_1(x) = \varphi_L(x) \cdot \varphi(x)$$

holds;

2) *if $\dim H = 2$ and $P \in \Phi(H)$ $(P \neq 0; I)$ the equality*

(2.2) $$\|\psi_L(P, P^*) \cdot \psi(P, P^*)\| = \|\psi_L(P, P^*)\| \cdot \|\psi(P, P^*)\|$$

holds.

3) *if the function $\varphi(x)$ is non-decreasing, the equality (2.2) holds for any Hilbert space H and any $P \in \Phi(H)$ $(P \neq 0, I)$.*

PROOF. If $\dim H = 2$ it follows from [2] that

$$[\psi_L(P, P^*)]^* \cdot \psi_L(P, P^*) = \varphi_L(x) \cdot I (x = \| P \|^2 - 1).$$

Now we have

$$[\psi_1(P, P^*)]^* \cdot \psi_1(P, P^*) = [\psi_L(P, P^*)]^* \cdot \psi_L(P, P^*) \cdot [\psi(P, P^*)]^* \cdot \psi(P, P^*)$$

(2.3) $$[\psi_L(P, P^*)]^* \cdot \psi_L(P, P^*) \cdot [\psi(P, P^*)]^* \cdot \psi(P, P^*) = \varphi_L^2(x) \cdot [\psi(P, P^*)]^* \cdot \psi(P, P^*)$$

In the case $\dim H = 2$ $\|\psi(P, P^*)\| = \varphi(x)$ for any polynomial $\psi(X, Y)$ and associated function $\varphi(x)$ [2] so from (2.3) it follows that

$$\varphi_1(x) = \|\psi_1(P, P^*)\| = \varphi_L(x) \cdot \|\psi(P, P^*)\| = \varphi_L(x) \cdot \varphi(x).$$

If the function $\varphi(x)$ is non-decreasing (it is clear that the function $\varphi_L(x)$ is always non-decreasing), then the function $\varphi_1(x)$ is non-decreasing too. Hence, the equality (2.2) holds for any space H and any $P \in \Phi(H)$. The theorem is proved.

THEOREM 2. Let $\psi_L(X, Y) = a \cdot I + b \cdot X + c \cdot Y (a, b, c \in C, |b|^2 + |c|^2 \neq 0)$ and $\dim H = 2$. Let $P \in \Phi(H)(P \neq 0, I)$, then

$$(2.4) \qquad \psi_L(P, P^*) \cdot [\psi_L(P, P^*)]^* = k \cdot I (k \in R)$$

if and only if

$$(2.5) \qquad a = t, b = -\frac{t+1}{\bar{t}+1}, \ c = 1 \text{ (up to any complex factor) } (t \in C, \ t \neq -1)$$

or

$$(2.6) \qquad a = -1, b = t, c = 1 \text{ (up to any complex factor) } (t \in C, \ |t| = 1)$$

If L is one of the following operators

$$(2.7) \qquad t \cdot I - \frac{t+1}{\bar{t}+1} \cdot P + P^* \ (t \in C, \ t \neq -1)$$

or

$$(2.8) \qquad -I + t \cdot P + P^* \ (t \in C, \ |t| = 1)$$

and $\varphi_L(x)$ is the function associated with L ,then

$$(2.9) \qquad \varphi_L(x) = \sqrt{|t|^2 + x}.$$

PROOF. If $P \neq 0, I$, then we can take such an orthogonal basis in H that the projections P and P^* have the following representations:

$$P = \begin{bmatrix} 1 & z \\ 0 & 0 \end{bmatrix}, P^* = \begin{bmatrix} 1 & 0 \\ \bar{z} & 0 \end{bmatrix} \quad (z \in C)$$

in this basis and

$$(2.10) \ \psi_L(P, P^*) \cdot [\psi_L(P, P^*)]^* = \begin{bmatrix} |a + b + c|^2 + |b|^2 \cdot |z|^2 & ((a + b + c) \cdot \bar{c} + b \cdot \bar{a}) \cdot z \\ (c \cdot (\overline{a + b + c}) + a \cdot \bar{b}) \cdot \bar{z} & |c|^2 \cdot |z|^2 + |a|^2 \end{bmatrix}$$

From here we have the following two conditions

$$\begin{cases} |a + b + c|^2 + |b|^2 \cdot |z|^2 = |c|^2 \cdot |z|^2 + |a|^2 \\ (a + b + c) \cdot \bar{c} + b \cdot \bar{a} = 0 \end{cases}$$

and this system is equivalent to the system

$$(2.11) \qquad \begin{cases} (a + b + c) \cdot \bar{c} + b \cdot \bar{a} = 0 \\ |b| = |c| \end{cases}$$

It is not difficult to see that from (2.11) only two cases (2.5) and (2.6) follow. From formulas (1.1)-(1.4), the result is the equality (2.8). The Theorem is proved.

COROLLARY 1. *The operators*

$$(2.12) \qquad\qquad T = -I + P + P^*$$

and

$$(2.13) \qquad\qquad K = P - P^*$$

satisfy the condition (2.4) and, hence,

$$(2.14) \qquad\qquad \varphi_T(x) = \sqrt{x + 1}$$

$$(2.15) \qquad\qquad \varphi_K(x) = \sqrt{x}.$$

COROLLARY 2. *If* $\psi_L(X, Y) = \psi_T(X, Y) = -I + X + Y$, *then under conditions of Theorem 2 we get for operators* $T = -I + P + P^*$ *and* $\psi(P, P^*)$ *that*

$$(2.16) \qquad\qquad \|T \cdot \psi(P, P^*)\| = \|P\| \cdot \|\psi(P, P^*)\|$$

If $\psi_L(X, Y) = \psi_K(X, Y) = X - Y$, *then under conditions of Theorem 2 we get for operators* $K = P - P^*$ *and* $\psi(P, P^*)$ *that*

$$(2.17) \qquad\qquad \|K \cdot \psi(P, P^*)\| = (\|P\|^2 - 1)^{\frac{1}{2}} \cdot \|\psi(P, P^*)\|$$

PROOF.   It follows from the equalities (2.2),(2.14) and (2.15) ($x = \|P\|^2 - 1$).

COROLLARY 3.   Let $P \in \Phi(H)$ and $\beta, \gamma \in C, n \in N$, then the following equality

$$(2.18) \qquad \|(P - P^*)^n \cdot (\beta \cdot P + \gamma \cdot P^*)\| = (\|P\|^2 - 1)^{\frac{n}{2}} \cdot \|\beta \cdot P + \gamma \cdot P^*\|$$

is valid.

PROOF.   Recall [2, pp.76-77] that the function associated with polynomial $\beta \cdot X + \gamma \cdot Y$ is non-decreasing. Now the equality (2.18) can be proved by induction and by the equality (2.17).

In order to formulate the next corollary we need some definition. Let $A, B, C \in L(H)$ and $R \in \{P, P^*\}$. Suppose that the operator B is of the form $B = R \cdot R^* \cdot A$, then the operator $R^* \cdot A$ is named left reduced operator for B. Similarly we say that $A \cdot R^*$ is the right reduced operator for operator $C = A \cdot R^* \cdot R$.

COROLLARY 4.   Let $P \in \Phi(H)$ and $\psi(X, Y)$ a polynomial without linear part. Suppose that $\psi_l(P, P^*)$ ($\psi_r(P, P^*)$) is the left (right) reduced operator for $\psi(P, P^*)$ and let the corresponding associate function $\varphi_l(x)$ ($\varphi_r(x)$) be non-decreasing. Then for any $P \in \Phi(H)$ the following equalities hold

$$(2.19) \qquad \begin{aligned} \|\psi(P, P^*)\| &= \|P\| \cdot \|\psi_l(P, P^*)\| \\ \|\psi(P, P^*)\| &= \|P\| \cdot \|\psi_r(P, P^*)\| \end{aligned}$$

In particular if $\dim H = 2$, then the equalities (2.19) hold for any (not only non-decreasing) functions $\varphi_l(x)$ ($\varphi_r(x)$).

PROOF.   If the polynomial $\psi(X, Y)$ without linear part it is possible to represent the operator $\psi(P, P^*)$ in one of the following forms (it follows from (1.1)):

$$\begin{aligned} \psi(P, P^*) &= P \cdot P^* \cdot \psi_1(P, P^*) + P^* \cdot P \cdot \psi_2(P, P^*) \\ \psi(P, P^*) &= \psi_3(P, P^*) \cdot P \cdot P^* + \psi_4(P, P^*) \cdot P^* \cdot P \end{aligned}$$

It is easy to see that for operator $T = -I + P + P^*$ the following equalities:

$$(2.20) \qquad P \cdot P^* = P \cdot T = T \cdot P^*, P^* \cdot P = T \cdot P = P^* \cdot T$$

hold.

Hence we can represent $\psi(P, P^*)$ in one of the following forms

$$\begin{aligned} \psi(P, P^*) &= T \cdot (P^* \cdot \psi_1(P, P^*) + P \cdot \psi_2(P, P^*)) = T \cdot \psi_l(P, P^*) \\ \psi(P, P^*) &= (\psi_3(P, P^*) \cdot P + \psi_4(P, P^*) \cdot P^*) = \psi_r(P, P^*) \cdot T. \end{aligned}$$

Now from Theorem 1 and Corollary 2 we get formulas (2.19).

COROLLARY 5. *Let $P \in \Phi(H)$ and $\beta, \gamma \in C, n \in N$, then the following* equalities

(2.21)
$$\|\beta \cdot (P \cdot P^*)^n + \gamma \cdot (P^* \cdot P)^n\| = \|P\|^{2 \cdot n - 1} \cdot \|\beta \cdot P^* + \gamma \cdot P\|$$
$$\|\beta \cdot (P \cdot P^*)^n \cdot P + \gamma \cdot (P^* \cdot P)^n \cdot P^*\| = \|P\|^{2 \cdot n} \cdot \|\beta \cdot P + \gamma \cdot P^*\|$$

are valid.

PROOF. We start with the polynomial $\psi_1(X, Y) = \beta \cdot Y \cdot X + \gamma \cdot X \cdot Y$. As the polynomial $\psi_1(X, Y)$ is without a linear part and the function associated with polynomial $\beta \cdot X + \gamma \cdot Y$ is non-decreasing, then it follows from (2.16) that

$$\|\beta \cdot P \cdot P^* + \gamma \cdot P^* \cdot P\| = \|P\| \cdot \|\beta \cdot P + \gamma \cdot P^*\|$$

and the function associated with polynomial $\beta \cdot Y \cdot X + \gamma \cdot X \cdot Y$ is also non-decreasing. Now for polynomial $\beta \cdot Y \cdot X \cdot Y + \gamma \cdot X \cdot Y \cdot X$ in a similar way we get $\|\beta \cdot P^* \cdot P \cdot P^* + \gamma \cdot P \cdot P^* \cdot P\| = \|P\|^2 \cdot \|\beta \cdot P + \gamma \cdot P^*\|$ and the function associated with polynomial $\beta \cdot Y \cdot X \cdot Y + \gamma \cdot X \cdot Y \cdot X$ is non-decreasing. Extending in this way one can prove the first equality (2.21). Analogously we can also prove the second equality (2.21).

3. ADDITIONAL REMARKS AND EXAMPLES

For which operator $\psi(P, P^*)$ can we use the convenient formulas (2.19)? We can see this from the following two remarks.

REMARK 3.1. *Let $\psi(X, Y)$ be an arbitrary polynomial. If*

(3.1)
$$\psi(1, 0) = \psi(0, 1) = 0,$$

then it is possible to represent the operator $\psi(P, P^*)$ in one of the following forms:

(3.2)
$$\psi(P, P^*) = T \cdot (-\psi(0, 0) \cdot I + \psi_l(P, P^*))$$
$$\psi(P, P^*) = (-\psi(0, 0) \cdot I + \psi_r(P, P^*)) \cdot T,$$

where $\psi_l(P, P^*)$ ($\psi_r(P, P^*)$) is the left (right) reduced operator for operator $\psi(P, P^*) + \psi(0, 0) \cdot T$.

PROOF. From conditions (3.1) we see that the linear part of polynomial $\psi(X, Y)$ is $\psi(0, 0) \cdot (1 - X - Y)$. Now from here and equalities (2.20) representations (3.2) and (3.3) follow.

REMARK 3.2. *Let $\psi(X, Y)$ be an arbitrary polynomial. If*

(3.3)
$$\psi(0, 0) = \psi(1, 1) = 0,$$

then it is possible to represent the operator $\psi(P, P^*)$ in the following form

(3.4) $$\psi(P, P^*) = [\psi_1(P, P^*)] \cdot (P - P^*),$$

where the polynomial $\psi_1(X, Y)$ is more simple than the polynomial $\psi(X, Y)$.

  PROOF. The operator $\psi(P, P^*)$ can be represented in form (1.1). From conditions (3.4) it follows that

$$a_0 = 0, b_0 = -\sum_{k=1}^{n}(a_k + b_k) - \sum_{k=0}^{n}(c_k + d_k).$$

Now by direct calculation one can verify the following equalities :

(3.5)
$$
\begin{aligned}
(P \cdot P^*)^k - P &= \left(\sum_{j=0}^{k-1}(P \cdot P^*)^j\right) \cdot (T - P) \cdot (P - P^*), \\
(P \cdot P^*)^k \cdot P - P &= \left(\sum_{j=0}^{k-1}(P \cdot P^*)^j\right) \cdot (P \cdot T - P) \cdot (P - P^*), \\
(P^* \cdot P)^k - P &= \left(\sum_{j=0}^{k-1}(P^* \cdot P)^j\right) \cdot (T - P) \cdot (P - P^*), \\
(P^* \cdot P)^k \cdot P^* - P &= \left(\sum_{j=0}^{k-1}(P \cdot P^*)^j \cdot (P \cdot T - P) + I\right) \cdot (P - P^*).
\end{aligned}
$$

  This proves representation (3.4) and moreover it is possible to find the explicit form of $\psi_1(P, P^*)$.

  Let us consider some examples . We assume that $P \neq 0, I$.

  $1^0.\psi(X, Y) = Y \cdot X \cdot Y + X \cdot Y \cdot X - (Y \cdot X)^2 - (X \cdot Y)^2$.

  We start with polynomial $\psi_1(X, Y) = Y + X - Y \cdot X - X \cdot Y$. According to formulas (1.1)-(1.3) we can find that the function associated with this polynomial is $\varphi_1(x) = x$. Now it is clear that $\varphi(x) = (\sqrt{x+1})^2 \cdot x = x \cdot (x + 1)$ is the function associated with polynomial $\psi(X, Y)$. As $\varphi(x)$ is non-decreasing function so for all spaces H and all $P \in \Phi(H)$ we have $\|\psi(P, P^*)\| = \|P\|^2 \cdot \left(\|P\|^2 - 1\right)$.

  $2^0.\psi(X, Y) = -1 + X + Y + \beta \cdot Y \cdot X + \gamma \cdot X \cdot Y$.

  It is not difficult to find the explicit form (1.3) of the corresponding function $\varphi(x)$, but it is not very easy to prove that this function is non-decreasing.

  If we will use Remark 3.1 we can see that $\psi(P, P^*) = -I + P + P^* + \beta \cdot P^* \cdot P + \gamma \cdot P \cdot P^* + T \cdot (I + \beta \cdot P + \gamma \cdot P^*)$, where $T = -I + P + P^*$.

  If $\varphi(x)$ is the function associated with linear polynomial $\psi(X, Y)$, then $\varphi(x) = \sqrt{x+1} \cdot \varphi_1(x)$ and $\varphi_1(x)$ is the function associated with linear polynomial $\psi_1(X, Y) = 1 + \beta \cdot X + \gamma \cdot Y$. In paper [2] it is proved that $\varphi_1(x)$ is non-decreasing function, then $\varphi(x) = \sqrt{x+1} \cdot \varphi_1(x)$ is also a non-decreasing function. Hence,

$$\| -I + P + P^* + \beta \cdot P^* \cdot P + \gamma \cdot P \cdot P^* \| = \|P\| \cdot \|I + \beta \cdot P + \gamma \cdot P^*\|$$

for all spaces H and all $P \in \Phi(H)$.

$3^0. \psi(X, Y) = -2X + X \cdot Y + Y \cdot X.$

As $\psi(0,0) = \psi(1,1) = 0$, we can use Remark 3.2 and represent the operator $\psi(P, P^*)$ in the following form :

$$-2 \cdot P + P \cdot P^* + P^* \cdot P = (-I - P + P^*)(P - P^*).$$

The operator $-I - P + P^*$ is the operator (2.8) with $t = -1$. Now from formula (2.17) it follows that

$$\| -2 \cdot P + P \cdot P^* + P^* \cdot P \| = \| P - P^* \| \cdot \| -I - P + P^* \| = \left(\|P\|^2 - 1 \right)^{\frac{1}{2}} \cdot \|P\|$$

4^0. Let S be the singular integral operator defined by (0.1) acting in the space $L_2(\Gamma, \rho)$ with weight $\rho = |t - t_0|^\beta$ $(|\beta| < 1)$. It is well known [1] that

$$\|S\| = \|S\|_{L_2(\Gamma, |t - t_0|^\beta)} = \cot \frac{\pi \cdot (1 - \beta)}{4}.$$

From the equality $S = 2 \cdot P - I$ one can find that $\|P\|^2 - 1 = \tan^2 \frac{\pi \cdot \beta}{2}$.

Now if the function $\varphi(x)$ associated with the polynomial $\psi(X, Y)$ is non-decreasing, then,

$$\| \psi(P, P^*) \|_{L_2(\Gamma, \rho)} = \varphi \left(\tan^2 \frac{\pi \cdot \beta}{2} \right).$$

In particular $\|(S + S^*)^n\| = 2^n \cdot \|P\|^n$ and, hence,

(3.6) $$\|(S + S^*)^n\| = \left(\frac{2}{\cos \frac{\pi \cdot \beta}{2}} \right)^n$$

Similarly $\|(S - S^*)^n\| = \|[2 \cdot (P - P^*)]^n\| = 2^n \cdot \left(\|P\|^2 - 1 \right)^{\frac{n}{2}}$ and ,hence,

(3.7) $$\| (S - S^*)^n \| = \left(2 \cdot \tan \frac{\pi \cdot \beta}{2} \right)^n.$$

REFERENCES

1. I. Gohberg and N. Krupnik, One-dimensional Linear Singular Integral Equations, vol. II, General Theory and Applications OT54, Birkhäuser Verlag, Basel, 1992.
2. I. Feldman, A. Markus and N. Krupnik, On the Norm of Polynomials of two Adjoint Projections , Integral Equations and Operator Theory,14, 69-90 (1991).

3. E. Spigel, On the Norm of Polynomials of two Adjoint Projections in Hilbert Spaces, Integral Equations and Operator Theory, 22, 232-241 (1995).

Department of Mathematics
Holon Center for Technological Education
affiliated with Tel-Aviv University
52 Golomb St., P.O.Box 305
Holon 58102
Israel
e-mail spigel@milk.cteh.ac.il

AMS: 47A30

Toeplitz Lectures 1995

Joseph A. Ball
Zero-pole interpolation problems for rational matrix functions and applications
Zero-pole interpolation problems for meromorphic matrix functions on a closed Riemann surface

Bernd Silbermann
Asymptotic invertibility of Toeplitz operators
Toeplitz determinants with one large Fisher-Hartwig singularity

Mark Vishik
Evolution equations that arise in mathematical physics and their attractors
Non-autonomous evolutionary equations with translation-compact symbols and their attractors

Lectures of the Joint German-Israeli Workshop

Daniel Alpay and I. Gohberg
Inverse spectral problems for differential operators with rational scattering matrix functions

Asher Ben-Artzi and I. Gohberg
Discrete Bohl exponents and spectrum of block weighted shifts

Luise Blank
Collocation method for integral equations

A. Böttcher and S.M. Grudsky
Toeplitz operators with discontinuous symbols: problems, methods, and results beyond piecewise continuity

H. Dym
Trace formulas for some classes of Wiener-Hopf like operators

T. Ehrhardt, S. Roch, and B. Silbermann
Finite sections of Toeplitz operators with piecewise quasicontinuous generating functions

S.D. Eidelman
Qualitative properties of the positive solutions of linear evolution equations

J. Elschner
Optimal order approximation methods for boundary integral equations on polygons

Israel Feldman
On explicit factorization and applications

Israel Gohberg
About the past and the future

P. Junghanns
Recent developments in the numerical analysis of Cauchy singular integral equations on an interval

M.A. Kaashoek
The state space method for solving singular integral equations and applications

Yu.I. Karlovich
Singular integral operators with shifts

Yu.I. Karlovich and I. Spitkovsky
Semi-Fredholmness of singular integral operators with (semi) almost periodic matrix symbols

B.A. Kon
Operators of convolution type as resultant operators

Naum Krupnik
Banach algebras generated by singular integral operators

A. Kulesko
On the spectrum of a matrix pencil and two-side infinite periodic Jacobi matrices

L. Lerer
On Toeplitz and Wiener-Hopf operators with contourwise rational symbols

V.A. Marchenko
Factorization of Weyl function, inverse problems, and nonlinear equations

A. Markus and V. Matsaev
Factorization of matrix functions and characteristic properties of the circle

E. Meister
On systems of Wiener-Hopf equations arising in the theory of diffraction by parallel half-planes

Boris Rubin
Wavelet type representations of singular integrals and reproducing formulas

Eliahu Shamir and B. Rubin
Extending Calderon's reproducing formula to singular integral transforms

E. Spigel
Some formulas for calculating the norm of polynomials of two projections

A. Volpert
Elliptic boundary value problems and one-dimensional singular integral equations

L. von Wolfersdorf
Nonlinear singular integral equations by monotonicity methods

E. Wegert
A geometric approach to singular integral equations

H. Widom
Determinants of a class of integral operators